Renewable Energy - Technologies and Applications

Edited by Tolga Taner, Archana Tiwari and Taha Selim Ustun

Published in London, United Kingdom

IntechOpen

Supporting open minds since 2005

Renewable Energy - Technologies and Applications
http://dx.doi.org/10.5772/intechopen.84929
Edited by Tolga Taner, Archana Tiwari and Taha Selim Ustun

Contributors
İzzet Yüksek, İlker Karadağ, Mostefa Ghassoul, Kudiyarasan Swamynathan, Chawalit Ngamcharussrivichai, Veeramuthu Ashokkumar, Zhengang Liu, Jetsada Posom, Kanvisit Maraphum, Arthit Phuphaphud, Ľubomír Šooš, Omer K. Ahmed, Bhupendra S. Rawat, Poonam Negi, Bharti Ramola, Pradeep Chandra Pant, Sammy N. Aso, Swapan Suman, Shalini Gautam, Awani Bhushan, Nomendra Tomar, Antonio Jr. Gonzales, Farid Touati, Kamel Benhmed, Amith M. A. Khandakar, Christian Kim Sorino, Muhammad E.H. Chowdhury, Md. Obaidullah, Jacques De Ruyck, Nicolás Muñoz-Galeano, Oswaldo A. Arráez-Cancelliere, Jesús M Lopez-Lezama, Dan Serbanescu, Ibrahim Yuksel, Hasan Arman, Omer Yuksek, Selçuk Sarıkoç, Nestor Gonzalez, Jose Miguel Garcia-Guzman, L. Alberto Contreras-Aguilar, J. Merced-Lozano-García, A. Pizano-Martine, Ivan Shorstkii, Angel Molina-Garcia, Ana Fernandez-Guillamon, Emilio Gomez-Lazaro, Eduard Muljadi, Mustafa Mosbah, Ridha Djamel Mohammedi, Salem Arif, Cihan Dogruoz, Kaan Yamanturk, Jushan Chin, Jin Dang, P. Sivakumar, K. Karthikeyan, Evgeny Koshevoi, Wentao Jiao, Nana Peng

First published in London, United Kingdom, 2021 by IntechOpen
IntechOpen is the global imprint of INTECHOPEN LIMITED, registered in England and Wales, registration number: 11086078, 5 Princes Gate Court, London, SW7 2QJ, United Kingdom
Printed in Croatia

British Library Cataloguing-in-Publication Data
A catalogue record for this book is available from the British Library

Additional hard and PDF copies can be obtained from orders@intechopen.com

Renewable Energy - Technologies and Applications
Edited by Tolga Taner, Archana Tiwari and Taha Selim Ustun
p. cm.
Print ISBN 978-1-83881-000-9
Online ISBN 978-1-83881-001-6
eBook (PDF) ISBN 978-1-83881-002-3

For EU product safety concerns:
IN TECH d.o.o., Prolaz Marije Krucifikse Kozulić 3, 51000 Rijeka, Croatia, info@intechopen.com or visit our website at intechopen.com.

Meet the editors

Dr. Tolga Taner is the head of the Department of Motor Vehicles and Transportation Technology at Aksaray University, Turkey. He received a BS in Mechanical Engineering in 1998; an MS in Mechanical Engineering from Pamukkale University, Denizli, Turkey, in 2002; and a Ph.D. in Mechanical Engineering from the Gazi University of Engineering Faculty, Ankara, Turkey, in 2013. In 2018, he received the title of Associate Professor from the Inter-University Council (UAK). He also worked as a part-time lecturer in the Department of Technical Programs, Middle East Technical University, Turkey, from 2003 to 2006. His current research interests include exergy, renewable energy, and PEM fuel cells. He has published many scientific and conference papers and books. In addition, Dr. Taner has been a manager and researcher for many research projects.

Dr. Archana Tiwari is Associate Professor at Amity University, India. Her research interests include renewable sources of energy from microalgae and further utilizing the residual biomass for the generation of value-added products, bioremediation through microalgae and microbial consortium, antioxidative enzymes and stress, and nutraceuticals from microalgae. She has been working on algal biotechnology for the last two decades. She has published her research in many international journals and has authored many books and chapters with renowned publishing houses. She has also delivered talks as an invited speaker at many national and international conferences. Dr. Tiwari is the recipient of several awards including Researcher of the Year and Distinguished Scientist.

Dr. Taha Selim Ustun received a Ph.D. in Electrical Engineering from Victoria University, Melbourne, Australia. He is a researcher with the Fukushima Renewable Energy Institute, AIST (FREA), where he leads the Smart Grid Cybersecurity Laboratory. Prior to that, he was a faculty member with the School of Electrical and Computer Engineering, Carnegie Mellon University, Pittsburgh, PA, USA. His current research interests include power systems protection, communication in power networks, distributed generation, microgrids, electric vehicle integration, and cybersecurity in smart grids. He serves on the editorial boards of IEEE Access, IEEE Transactions on Industrial Informatics, Energies, Electronics, Electricity, World Electric Vehicle and Information journals. Dr. Ustun is a member of the IEEE 2004 and 2800, IEC Renewable Energy Management WG 8, and IEC TC 57 WG17. He has been invited to run specialist courses in Africa, India, and China. He has delivered talks for the Qatar Foundation, the World Energy Council, the Waterloo Global Science Initiative, and the European Union Energy Initiative (EUEI). His research has attracted funding from prestigious programs in Japan, Australia, the European Union, and North America.

Contents

Preface

This book discusses applications of renewable energy with new technologies. It encourages both academic research and renewable energy policies worldwide, and is designed for students, researchers, scientists, and governments.

The volume contains twenty-three chapters in six sections:

Section 1 "Introduction to Renewable Energy"

Section 2 "Solar and Photovoltaic Energy"

Section 3 "Biomass Energy"

Section 4 "Hydroelectric Energy"

Section 5 "Geothermal Energy"

Section 6 "Other Applications of Renewable Energy System"

We wish to thank IntechOpen and Author Service Manager, Ms. Sara Debeuc, for her support, help, and guidance. We also thank the authors for their contributions.

Dr. Tolga Taner
Associate Professor,
Head of the Department of Motor Vehicles and Transportation Technologies,
Aksaray University,
Aksaray, Turkey

Dr. Archana Tiwari
Associate Professor,
Amity Institute of Biotechnology,
Amity University,
Noida, Uttar Pradesh, India

Dr. Taha Selim Ustun
Professor,
Fukushima Renewable Energy Institute, AIST (FREA),
Japan

Section 1

Introduction to Renewable Energy

Chapter 1

A Review of Virtual Inertia Techniques for Renewable Energy-Based Generators

Ana Fernández-Guillamón, Emilio Gómez-Lázaro, Eduard Muljadi and Ángel Molina-Garcia

Abstract

Over recent decades, the penetration of renewable energy sources (RES), especially photovoltaic and wind power plants, has been promoted in most countries. However, as these both alternative sources have power electronics at the grid interface (inverters), they are electrically decoupled from the grid. Subsequently, stability and reliability of power systems are compromised. Inertia in power systems has been traditionally determined by considering all the rotating masses directly connected to the grid. Thus, as the penetration of renewable units increases, the inertia of the power system decreases due to the reduction of directly connected rotating machines. As a consequence, power systems require a new set of strategies to include these renewable sources. In fact, 'hidden inertia,' 'synthetic inertia' and 'virtual inertia' are terms currently used to represent an artificial inertia created by inverter control strategies of such renewable sources. This chapter reviews the inertia concept and proposes a method to estimate the rotational inertia in different parts of the world. In addition, an extensive discussion on wind and photovoltaic power plants and their contribution to inertia and power system stability is presented.

Keywords: frequency control, grid stability, inertia, power systems, inverter-interfaced renewable energy sources

1. Introduction

Imbalances between generation and consumption cause frequency variations in a power system [1]. To maintain frequency in its nominal value, power systems rely on synchronous machines connected to the grid, which store kinetic energy automatically extracted in response to a sudden power imbalance [2]. However, due to the new environmental policies and the limited fossil fuel reserves, conventional generators are being replaced by renewable energy sources (RES)-based generators [3]. Among the different RES available, the most promising for electrical power generation are PV and wind power installations, which are inverter-interfaced RES (II-RES) [4]. However, the massive penetration of II-RES into the grid can involve several issues that should be taken into account [5]. First, as they depend on weather conditions, these sources are intermittent and uncertain, placing stress on power system operation [6]. Moreover, as they are connected to the grid through inverters which electrically decouple them from the grid [7], the effective inertia of the power system can be reduced [8]. This inertia reduction affects the system reliability, compromising the frequency stability [9]. The rotational inertia is related to both nadir (minimum frequency) and rate of change of frequency (ROCOF) [10]. In fact, larger nadirs and faster ROCOFs are obtained in low rotational inertia power systems, subsequently making them more sensitive to frequency deviations [11, 12]. As a result, over the last decade, several frequency control techniques have been proposed to facilitate the massive penetration of wind and PV resources into the grid [13]. In addition, recent contributions investigated the use of smart inverters with voltage and frequency support to enhance grid stability [14]. Such solutions are commonly referred to as hidden, synthetic or virtual inertia [15].

This chapter focuses on the current and future inertia concept for power systems. A methodology to estimate the current rotational inertia of power systems based on their electricity generation mix is proposed. In addition, the possibilities of wind and PV power plants to contribute to inertia and participate in frequency control are also presented. The rest of the chapter is organized as follows. The inertia analysis and swing equation of generators and current and future power systems are presented in Section 2. In Section 3, the inertia constant estimation methodology is explained, comparing the results to a previous report published by the European Network of Transmission System Operators for Electricity (ENTSO-E). Section 4 reviews different frequency control techniques for PV and wind power plants. Finally, Section 5 gives the conclusion.

2. Inertia analysis in power systems

2.1 Inertial response of a synchronous generator: inertia constant

Rotating masses of a synchronous generator store kinetic energy E_{kin} following Eq. (1), where J is the moment of inertia and ω_r is the rated rotational frequency of the machine [16]:

$$E_{kin} = \frac{1}{2} J \omega_r^2. \tag{1}$$

Moment of inertia J *is a measure of the resistance of an object to changes in its rotational motion* [17]. However, in power systems, it is common to express inertia constant H instead of moment of inertia J. Actually, the inertia constant of a generator determines the time interval during which an electrical generator can supply its rated power only by using the kinetic energy stored in its rotating masses. H is defined following Eq. (2), being S_r the rated power [18]:

$$H = \frac{E_{kin}}{S_r} = \frac{1}{2} \frac{J \ \omega_r^2}{S_r}. \tag{2}$$

Work in [10] reviews the inertia constants H of conventional power plants proposed in recent decades, which range between 2 and 10 s.

In power systems, the motion of each turbine-generator group is expressed as Eq. (3), where T_m and T_e are the mechanical torque of the turbine and the electromagnetic torque of the generator, respectively:

$$2\ H\ \frac{d\omega_r}{dt} = T_m - T_e, \tag{3}$$

However, as $P = T \cdot \omega$ and considering the initial status as 0:

$$P = P_0 + \Delta P = (T_0 + \Delta T) \cdot (\omega_{r0} + \Delta\omega_r), \tag{4}$$

where $\Delta P = \Delta P_m - \Delta P_e$ and $\Delta T = \Delta T_m - \Delta T_e$. Moreover, for small variations:

$$\Delta P \simeq T_0 \cdot \Delta\omega_r + \Delta T \cdot \omega_{r0}, \tag{5}$$

and in steady state:

$$\begin{aligned} T_{m0} &= T_{e0}, \\ \omega_{r0} &= 1\ \text{pu}. \end{aligned} \tag{6}$$

In consequence, considering small variations around the steady state, Eq. (3) can be rewritten as in Eq. (7) [19]:

$$2\ H\ \frac{d\Delta\omega_r}{dt} = \Delta P_m - \Delta P_e. \tag{7}$$

Furthermore, some electrical loads connected to the grid are also frequency-dependent, working as a load resource under frequency deviations (i.e., synchronous machines). In this way, the electrical power of those loads can be expressed as:

$$\Delta P_e = \Delta P_L + D \cdot \Delta\omega_r, \tag{8}$$

where ΔP_L is the power change of those loads independent from frequency deviations and D is the damping factor (load-frequency response). Subsequently, by including the damping factor in Eq. (7), it is modified to Eq. (9), which is usually referred to as *swing equation* and represents the motion of a synchronous generator:

$$2\ H\ \frac{d\Delta\omega_r}{dt} = \Delta P_m - (\Delta P_L + D \cdot \Delta\omega_r). \tag{9}$$

2.2 Aggregated swing equation: application to power systems

To apply the swing Eq. (9) to a power system, all synchronous generators are grouped in an equivalent rotating mass. This is carried out by determining the equivalent inertia constant H_{eq} of such generators:

$$H_{eq} = \frac{\sum_{i=1}^{SG} H_i \cdot S_{B,i}}{S_B}, \tag{10}$$

where H_i and $S_{B,i}$ are the inertia constant and rated power of synchronous generator i, SG is the total number of synchronous generators connected to the grid and S_B is the rated power of the power system.

In the same way, loads are reduced to an equivalent one with damping factor D_{eq}. If the power system under analysis is stable, an inaccurate value of D_{eq} will not have a significant impact on the study. However, under disturbance situations, the value of D_{eq} can be a major contribution [20]. As variable frequency drives become more common, the equivalent damping factor is expected to decrease [21].

2.3 Hidden and virtual inertia emulation from RES: modified equivalent inertia constant

In recent decades, several policies have promoted the penetration of RES-based generation units, which have replaced synchronous generators directly connected to the grid [22]. However, as some of them are II-RES (i.e., wind and PV), power systems with a high penetration of those RES require new frequency control strategies that emulate the behavior of conventional power plants under power imbalance conditions [23]. Such techniques are commonly referred to as hidden, synthetic, emulated or virtual inertia [15]. By including this emulation of inertia into power systems, equivalent inertia H_{eq} would be modified. Thus, it would have two different components: (i) synchronous rotating inertia coming from synchronous (conventional) generators H_S and (ii) emulated/virtual inertia coming from II-RES H_V [24, 25]. Thus, Eq. (10) would become:

$$H_{eq} = \frac{\overbrace{\sum_{i=1}^{SG} H_i \cdot S_{B,i}}^{H_S} + \overbrace{\sum_{j=1}^{VG} H_{V,j} \cdot S_{B,j}}^{H_V}}{S_B}, \quad (11)$$

where VG is the number of II-RES connected to the grid through emulation/virtual control methods and H_V is the inertia constant of the emulated/virtual generation unit. This modified equivalent inertia expressed in Eq. (11) is graphically illustrated in **Figure 1**, based on [26]. As can be seen, there are three different links between the generation units and the grid frequency: (i) rotational synchronous inertia from conventional generators, (ii) hidden inertia from VSWT and (iii) virtual inertia from PV. This is because modern VSWT have rotational inertia stored in their blades, drive train and electrical generator [27]. However, due to the inverter and maximum power point tracking (MPPT) strategy, they cannot automatically provide this inertia to the grid [28–31], being thus considered as 'hidden' from the power system point of view [32]. In fact, VSWT have inertia constants comparable to those of conventional generators, as summarized in **Figure 2**. In consequence, it is considered that the inertia provided by VSWT is 'emulated' [33].

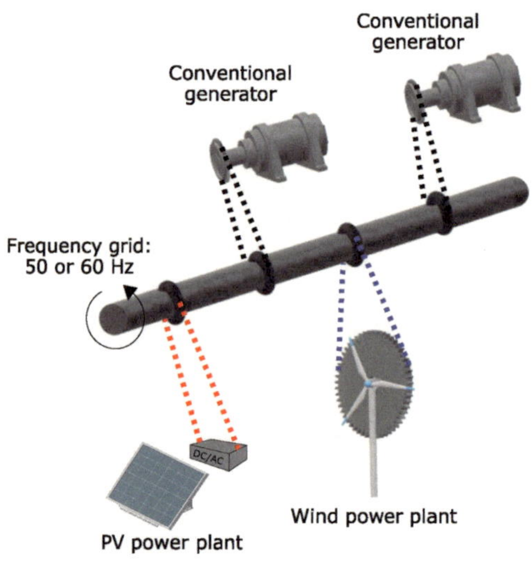

Figure 1.
Power system with synchronous, hidden and virtual inertia.

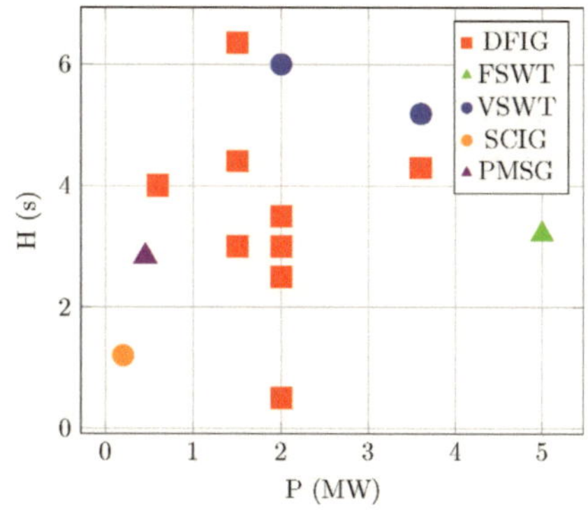

Figure 2.
Inertia constant values H *for different wind turbine technologies.*

On the other hand, PV has no rotating masses [30]. Thus, PV power plants cannot store kinetic energy and their inertia constant is $H \simeq 0$ [31]. Consequently, they cannot provide inertia unless it is synthetic/virtual, thus being usually referred to as 'emulated synthetic/virtual inertia' provided by such PV power plants [34, 35].

Due to the repercussions of II-RES with regard to the rotating inertia of power systems [36], they should start providing active power support under disturbances [37]. The specific literature includes several technologies that allow II-RES to participate in frequency control by providing additional power under disturbances [38–40].

3. Inertia estimation for power systems

Energy global statistics are provided by the International Energy Agency (IEA). Considering Eq. (10) and the electricity supply within a year presented in [41], it is possible to calculate the equivalent inertia H_{eq} in different regions of the world. According to each technology, the inertia constant H of conventional units is estimated as the mean value of those presented in [10] (i.e., $H_{coal} = 4$ s, $H_{oil} = 4$ s, $H_{gas} = 5$ s, $H_{nuclear} = 4$ s, $H_{hydro} = 3.25$ s). It is considered that II-RES are not participating in frequency control (i.e., not contributing to the system inertia).

Figure 3 depicts the generation mix change between 1996 and 2016. Over these two decades, the total electricity consumption increased by more than 80%. However, in the same time period, RES electricity generation only increased by 4%. Based on the approach previously described to estimate H_{eq}, **Figure 4** depicts the change between the inertia constant for the different continents between 1996 and

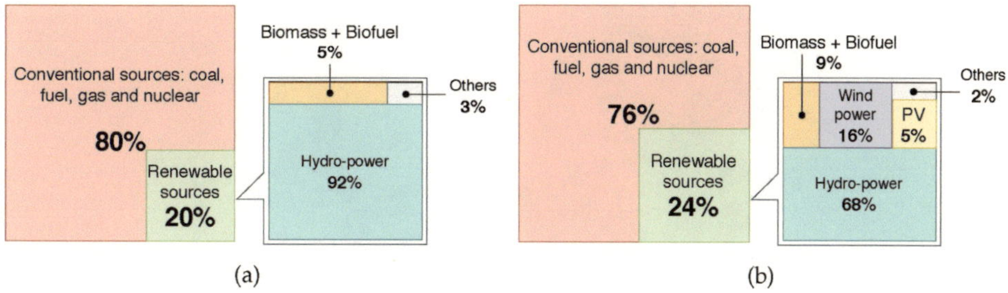

Figure 3.
Generation mix in the world: change between 1996 and 2016. (a) Generation mix in 1996. (b) Generation mix in 2016.

2016. As can be seen, the inertia reduction in Asia, the USA and South America was negligible (between 2.5 and 3%), whereas in Europe it decreased by nearly 20%.

In line with the inertia reduction suffered, RES supply in Europe increased by nearly 20% (refer to **Figure 5**). Actually, ENTSO-E has already focused on the high

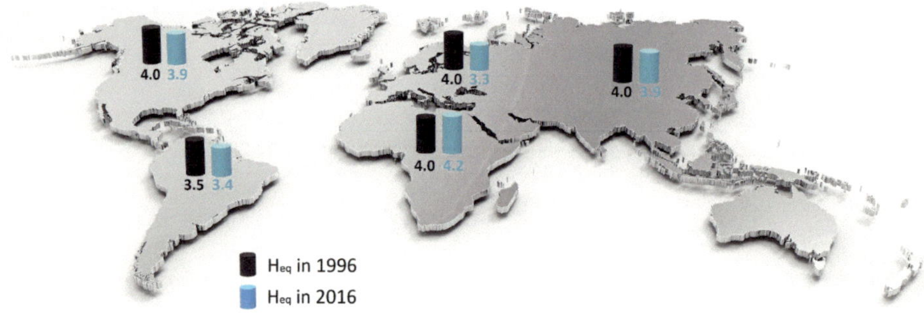

Figure 4.
Estimated equivalent inertia constants in the world by continent: change between 1996 and 2016.

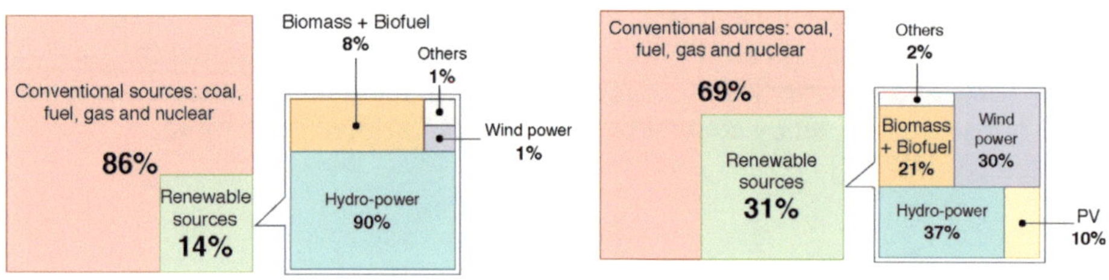

(a) (b)

Figure 5.
Generation mix in Europe: change between 1996 and 2016. (a) Generation mix in 1996. (b) Generation mix in 2016.

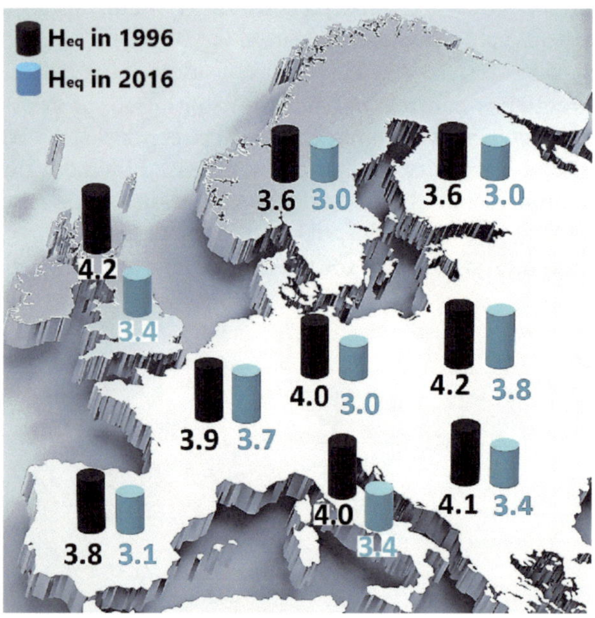

Figure 6.
Equivalent inertia constants estimated in EU-28: change between 1996 and 2016.

RES integration-low synchronous inertia problem. In one of their published reports, ENTSO-E estimated the evolution of system inertia for different TYNDP scenarios for 2030 in Europe and certain countries (i.e., the United Kingdom, France and Germany), considering that II-RES do not contribute to inertia [42]. In those estimations, H_{eq} depends on the percentage of hours in a year that II-RES are working. Thus, it is possible to compare the H_{eq} estimated in this chapter with the values obtained by ENTSO-E.

The transition of H_{eq} in a number of European countries can be seen in **Figure 6**. In [42], considering RES current generation rate: (i) H_{eq} of Europe is within range 3.8–4.5 s; (ii) H_{eq} of the United Kingdom is within range 3–4 s; (iii) H_{eq} of France is 5 s and (iv) H_{eq} of Germany is 3.5 s. Some discrepancies can be observed. The main cause of these is the values of the inertia constant of conventional plants. In fact, if the maximum value of H for all conventional plants is considered (i.e., $H_{coal} = 5$ s, $H_{oil} = 5$ s, $H_{gas} = 5$ s, $H_{nuclear} = 4$ s, $H_{hydro} = 4.75$ s), the H_{eq} results are nearly the same as those presented in [42].

4. II-RES frequency control strategies

4.1 Preliminaries

To maintain frequency within an acceptable range, generation and load in the power system must be continuously balanced [43]. In fact, frequency variations from the nominal value can cause several problems including under-/overfrequency relay operations and disconnection of some loads from the grid, among others [44]. Thus, frequency stability is an essential issue for power systems [45].

With the increase in II-RES, the equivalent inertia constant of power systems is reduced, subsequently obtaining (i) larger frequency deviations after an imbalance and (ii) higher ROCOF [7, 46]. As a consequence, II-RES should start providing active power support under disturbances [37].

4.2 PV power plant frequency control strategies

In order to provide additional active power during imbalanced situations, PV power plants can integrate different solutions, mainly based on two principal approaches: energy storage systems (ESS) or de-loading control strategies. Moreover, the technical challenge is more severe with PV power plants than with wind generation, since PV systems cannot provide any inertial response unless special countermeasures are adopted [47].

With regard to ESS, different solutions have been proposed in the literature to be applied to PV systems. Although the relevant benefits of ESS to power system's operation is widely recognized, some significant challenges can be identified: (i) the selection of a suitable technology to match the power system application requirements, (ii) an accurate evaluation of the energy storage facilities estimating both technical and economic benefits and (iii) a cost decreasing to a realistically acceptable level for deployment [48]. Among the different ESS, the battery energy storage is considered by some authors as the oldest and most mature ESS [49]. In work [50], it is concluded that the Li-Ion batteries are those that best suit frequency regulation services. Batteries are limited in power, though present a high storage ratio [51–53]; on the other hand, supercapacitors have high levels of power with low energy storage ratio. As a consequence, the battery-supercapacitor combination is proposed as an interesting ESS solution [54]. Indeed, these technologies can help to solve the problem of the 'intermittent' nature of solar PV supply [55]. Additional solutions for PV installations based on supercapacitors can be found in [56, 57]. Flywheels are another solution widely proposed as ESS, being applied from very small micro-satellites to large power systems [58]. Work in [59] points out a great benefit of flywheels backing up solar PV power plants, mainly focused on the cloud passing, which can cope with the high cycles of the flywheel technologies. Indeed, flywheels excel in short duration and high cycle applications [60]. Moreover, flywheels have a high efficiency, usually in the range between 90% and 95%, with an expected lifetime of around 15 years [61]. Different solutions propose hybrid ESS coupled to PV power plants [53], such as a battery hybridization with mechanical flywheel [62].

PV power plants usually work at the maximum power point (MPP) according to ambient temperature T and solar irradiation G [63]. However, they can work below their MPP, having thus some active power reserves (headroom) to supply in case of a frequency deviation. This approach is usually referred to as de-loading technique and is commonly proposed for PV installations [64, 65]. In this way, the PV plant is operated at P_{del}, below P_{MPP}, so that some power reserves $\Delta P = P_{MPP} - P_{del}$ are available [66, 67]. As can be seen in **Figure 7**, P_{del} can be related with two different voltages: (i) over the maximum power point voltage, $V_{del,1} > V_{MPP}$, and (ii) under the maximum power point voltage, $V_{del,2} < V_{MPP}$. However, due to stability problems, the de-loaded voltage corresponds to the higher value $V_{del,1}$ [68]. This V_{del} is then added to the MPP controller reference, in order to also de-load the inverter. This controller for de-loaded PV is modified in [69], such that the release of the reserve is directly linked to both (i) the frequency excursion and (ii) the availability of the reserve in the PV system. This controller is also proposed in [70].

4.3 Wind power plant frequency control strategies

Wind power plants can also participate in frequency control by using different solutions. Apart from the use of ESS or working with the de-loading control strategy, wind turbines can provide inertial response as conventional generators due to the rotational inertia of the blades and generator [10].

With regard to ESS, wind power plants can also include batteries [71], supercapacitors [72] and flywheels [73]. ESS are considered an alternative to compensate the lack of short-term frequency response ability of wind power plants [74]. The utility-scale battery ESS helps to reduce the ROCOF, providing frequency support and improving the system frequency response [75]. A battery ESS based on a state-machine-based coordinated control strategy is developed in [76] to support

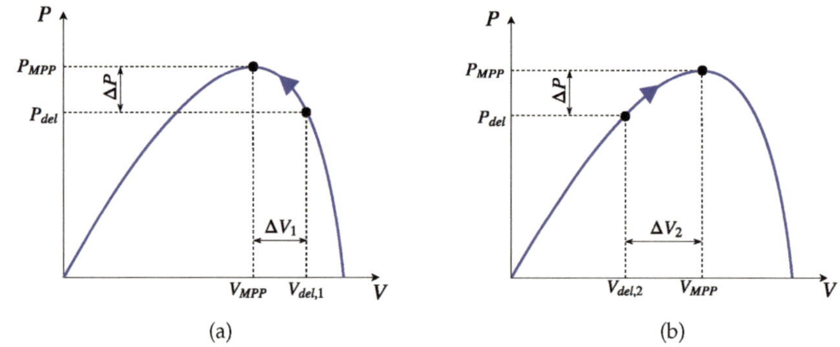

Figure 7.
De-loading techniques for PV power plants. (a) $V_{del,1} > V_{MPP}$. (b) $V_{del,2} < V_{MPP}$.

frequency response of wind power plants, including both primary and secondary frequency control. A real-time cooperation scheme by considering complementary characteristics between wind power and batteries is discussed in [77] to provide both energy and frequency regulation, considering the battery life cycle. The combination of battery and supercapacitor is considered in [78] as an effective alternative to improve the battery lifetime and enhance the system economy. In this way, an enhanced frequency response strategy is investigated in [79] to improve and regulate the wind frequency response with the integration of ultra-capacitors. With the aim of smoothing the net power injected to the grid by wind turbines (or by a wind power plant), some authors propose to use flywheels [80, 81]. Flywheels are also proposed to dynamically regulate the system equivalent inertia and damping, enhancing the frequency regulation capability of wind turbines [38, 82] and also the entire grid [83]. A coordinated regulation response of the turbine power reserves and the flywheels while participating in primary frequency control is described in [84]. Finally, other works include not only frequency response but also voltage control by using flywheels [85, 86].

In line with PV installations, wind turbines also work in the MPP according to the wind speed v_w. As a consequence, the de-loading technique is considered as a solution to provide additional active power in imbalanced situations with wind turbines, by operating them in a suboptimal point through the de-loaded control mode [87]. Wind turbines have two different possibilities to operate with the de-loading technique (refer to **Figure 8**) [32]: (i) pitch angle control and (ii) overspeed control. The pitch-angle control increases the pitch angle from β_0 to β_1 for a constant v_w; in this way, the supplied power P_{del} is below the maximum power P_{MPP}, being thus a certain amount of power ΔP that can be supplied in case of frequency contingency (**Figure 8(a)**) [88–91]. When this additional power ΔP is provided, the pitch angle has to be reduced from β_1 to β_0. The overspeed control increases the rotational speed of the rotor, shifting the supplied power P_{del} towards the right of the maximum power P_{MPP} (**Figure 8(b)**) [87, 92, 93]. As in the pitch-angle control, P_{del} is below P_{MPP} [71]. When the additional power ΔP is supplied, the rotor speed has to be reduced from Ω_{del} to Ω_{MPP}, releasing kinetic energy [39, 87, 92, 93].

In order to provide an inertial response, at least one supplementary loop control is introduced into the power controller to increase the generated power by the wind power plant. This additional loop is only activated under power imbalances (i.e., frequency deviations), supplying the kinetic energy stored in the blades and generator to the grid as an additional active power for a few seconds [94]. The droop control provides an additional active power ΔP proportional to the frequency excursion Δf (see **Figure 9**), as the primary frequency control of conventional power plants. The increase in the active power output then results in a decrease in

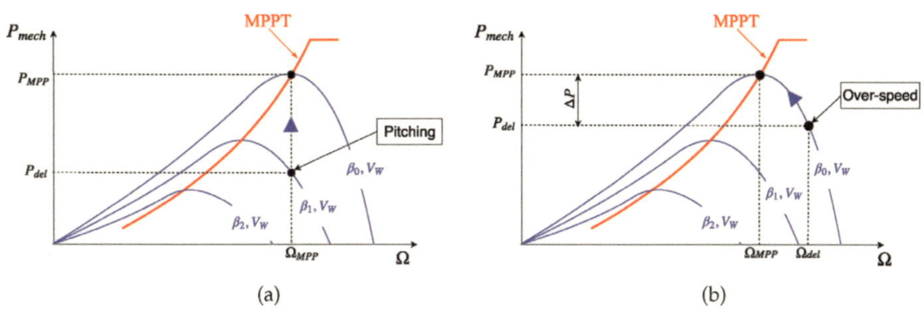

Figure 8.
De-loading techniques for wind power plants. (a) Pitch control. (b) Over-speed control.

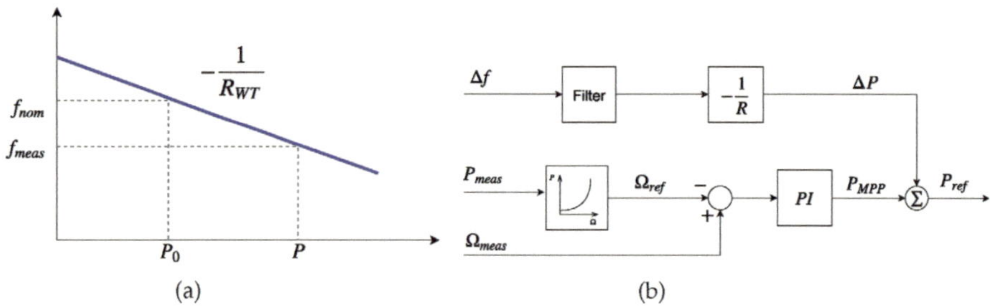

Figure 9.
Droop control for VSWTs. (a) Droop characteristic. (b) Block diagram of droop control.

the rotor speed [95–99]. ΔP can be estimated following Eq. (12), being R_{WT} the droop control setting of the wind turbine:

$$\Delta P = -\frac{\Delta f}{R_{WT}} \tag{12}$$

The hidden inertia emulation technique is based on emulating the inertial response of traditional synchronous generators. Two possibilities are found in the specific literature, as presented in **Figure 10**: (i) one loop, where the additional power is proportional to the ROCOF [100–102], and (ii) two loops, where the additional power is proportional to the ROCOF and the frequency deviation. The second strategy causes the frequency to return to its nominal value [103–105]. In both cases, the rotor and generator speeds are reduced to release the stored kinetic energy.

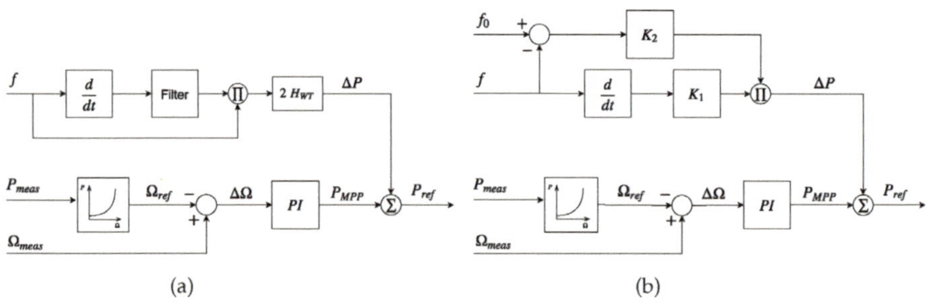

Figure 10.
Hidden inertia emulation controllers. (a) One loop. (b) Two loops.

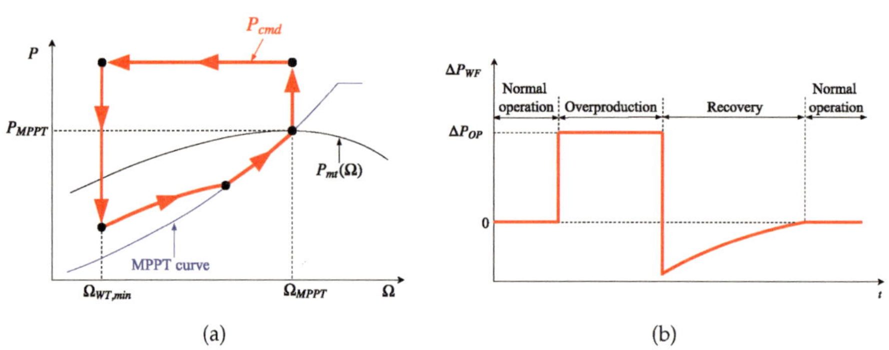

Figure 11.
Fast power reserve emulation technique [106]. (a) $P - \Omega$ curve. (b) Power variation.

The fast power reserve approach is similar to the hidden inertia emulation technique: an additional power is initially supplied, which makes the rotor speed to decrease. However, in this technique, the additional active power ΔP has been defined as a constant value independent of the system configuration and frequency deviation [106–110] or variable (depending on the frequency deviation or minimum rotor speed limits) [43, 111, 112]. The rotational speed decrease is then recovered through a recovery period, which can cause a secondary frequency dip due to the sudden decrease of the power generated by the wind power plant. As a consequence, different recovery periods have been proposed in the last decade to avoid this secondary frequency drop [43, 106, 108–111, 113, 114], even coordinating this period with ESS [115]. **Figure 11** shows the fast power reserve emulation control proposed in [106].

5. Conclusions

In this chapter, we have conducted an extensive literature review of inertia of power systems. A methodology to estimate the inertia constants of different power systems is proposed and verified with the inertia constant results of ENTSO-E. The contribution of wind and PV power plants as 'hidden inertia' and 'virtual inertia,' respectively, to participate in frequency control has also been discussed, providing significant information for their participation in frequency control.

Acknowledgements

This work was supported by the Spanish Education, Culture and Sports Ministry (FPU16/04282), Spanish Economy and Competitiveness Ministry and European Union FEDER, which supported this work under Project ENE2016-78214-C2-1-R.

Conflict of interest

The authors declare no conflict of interest.

Abbreviations

DFIG	double-fed induction generator
ESS	energy storage systems
ENTSO-E	European Network of Transmission System Operators for Electricity
FSWT	fixed-speed wind turbine
HAWT	horizontal axis wind turbine
II-RES	inverter-interfaced renewable energy sources
PMSG	permanent magnet synchronous generator
PV	photovoltaic
RES	renewable energy sources
ROCOF	rate of change of frequency
SCIG	squirrel cage induction generator
VSWT	variable speed wind turbine
WPP	wind power plant

Author details

Ana Fernández-Guillamón[1*†], Emilio Gómez-Lázaro[2†], Eduard Muljadi[3†] and Ángel Molina-Garcia[1†]

1 Department of Automatics, Electrical Engineering and Electronic Technology, Universidad Politécnica de Cartagena, Cartagena, Spain

2 Renewable Energy Research Institute and DIEEAC-EDII-AB, Universidad de Castilla-La Mancha, Albacete, Spain

3 Department of Electrical and Computer Engineering, Auburn University, Auburn, AL, USA

*Address all correspondence to: ana.fernandez@upct.es

† These authors are contributed equally.

References

[1] Babahajiani P, Shafiee Q, Bevrani H. Intelligent demand response contribution in frequency control of multi-area power systems. IEEE Transactions on Smart Grid. 2018;**9**(2): 1282-1291

[2] D'hulst R, Fernandez JM, Rikos E, Kolodziej D, Heussen K, Geibelk D, et al. Voltage and frequency control for future power systems: The ELECTRA IRP proposal. In: 2015 International Symposium on Smart Electric Distribution Systems and Technologies (EDST), IEEE. 2015. pp. 245-250

[3] Fernández-Guillamón A, Das K, Cutululis NA, Molina-García Á. Offshore wind power integration into future power systems: Overview and trends. Journal of Marine Science and Engineering. 2019;7(11):399

[4] Shah R, Mithulananthan N, Bansal R, Ramachandaramurthy V. A review of key power system stability challenges for large-scale PV integration. Renewable and Sustainable Energy Reviews. 2015;**41**(Supplement C): 1423-1436

[5] Cvetković M, Pan K, López CD, Bhandia R, Palensky P. Co-simulation aspects for energy systems with high penetration of distributed energy resources. In: AEIT International Annual Conference; 2016. IEEE. 2017. pp. 1-6

[6] Wang Y, Meng J, Zhang X, Xu L. Control of pmsg-based wind turbines for system inertial response and power oscillation damping. IEEE Transactions on Sustainable Energy. 2015;**6**(2):565-574

[7] Junyent-Ferr A, Pipelzadeh Y, Green TC. Blending hvdc-link energy storage and offshore wind turbine inertia for fast frequency response. IEEE Transactions on Sustainable Energy. 2015;**6**(3):1059-1066

[8] Yang S, Fang J, Tang Y, Qiu H, Dong C, Wang P. Synthetic-inertia-based modular multilevel converter frequency control for improved micro-grid frequency regulation. In: 2018 IEEE Energy Conversion Congress and Exposition (ECCE); IEEE. 2018. pp. 5177-5184

[9] Delille G, Francois B, Malarange G. Dynamic frequency control support by energy storage to reduce the impact of wind and solar generation on isolated power system's inertia. IEEE Transactions on Sustainable Energy. 2012;**3**(4):931-939

[10] Fernández-Guillamón A, Gómez-Lázaro E, Muljadi E, Molina-García Á. Power systems with high renewable energy sources: A review of inertia and frequency control strategies over time. Renewable and Sustainable Energy Reviews. 2019;**115**:109369

[11] Dehghanpour K, Afsharnia S. Electrical demand side contribution to frequency control in power systems: A review on technical aspects. Renewable and Sustainable Energy Reviews. 2015; **41**:1267-1276

[12] Nguyen HT, Yang G, Nielsen AH, Jensen PH. Combination of synchronous condenser and synthetic inertia for frequency stability enhancement in low inertia systems. IEEE Transactions on Sustainable Energy. 2018;**10**(3):997-1005

[13] Groß D, Bolognani S, Poolla BK, Dörfler F. Increasing the resilience of low-inertia power systems by virtual inertia and damping. In: Bulk Power Systems Dynamics and Control Symposium (IREP). 2017

[14] Ustun TS, Aoto Y. Analysis of smart inverter's impact on the distribution network operation. IEEE Access. 2019;7: 9790-9804

[15] Vokony I. Effect of inertia deficit on power system stability-synthetic inertia concepts analysis. In: 2017 6th International Youth Conference on Energy (IYCE); IEEE. 2017. pp. 1-6

[16] Ulbig A, Borsche TS, Andersson G. Impact of low rotational inertia on power system stability and operation. IFAC Proceedings Volumes. 2014;**47**(3): 7290-7297

[17] Serway RA, Jewett JW. Physics for Scientists and Engineers with Modern Physics. Cengage Learning: Brooks/ Cole; 2018

[18] Uriarte FM, Smith C, Van Broekhoven S, Hebner RE. Microgrid ramp rates and the inertial stability margin. IEEE Transactions on Power Systems. 2015;**30**(6):3209-3216

[19] Fernández-Guillamón A, Vigueras-Rodríguez A, Molina-García A. Análisis y simulación de estrategias agregadas de control de frecuencia entre grandes parques eólicos y aprovechamientos hidroélectricos [MS thesis]. Universidad Politécnica de Cartagena; 2017

[20] Huang H, Li F. Sensitivity analysis of load-damping characteristic in power system frequency regulation. IEEE Transactions on Power Systems. 2013; **28**(2):1324-1335

[21] Tielens P, Van Hertem D. The relevance of inertia in power systems. Renewable and Sustainable Energy Reviews. 2016;**55**:999-1009

[22] Fernández-Guillamón A, Vigueras-Rodríguez A, Molina-García Á. Analysis of power system inertia estimation in high wind power plant integration scenarios. IET Renewable Power Generation. 2019;**13**(15):2807-2816

[23] Muñoz-Benavente I, Hansen AD, Gómez-Lazaro E, García-Sánchez T, Fernández-Guillamón A, Molina-García Á. Impact of combined demand-response and wind power plant participation in frequency control for multi-area power systems. Energies. 2019;**12**(9):1687

[24] Gu H, Yan R, Saha TK. Minimum synchronous inertia requirement of renewable power systems. IEEE Transactions on Power Systems. 2017; **33**(2):1533-1543

[25] Tielens P, Van Hertem D. Receding horizon control of wind power to provide frequency regulation. IEEE Transactions on Power Systems. 2017; **32**(4):2663-2672

[26] Kroposki B, Johnson B, Zhang Y, Gevorgian V, Denholm P, Hodge B-M, et al. Achieving a 100% renewable grid: Operating electric power systems with extremely high levels of variable renewable energy. IEEE Power and Energy Magazine. 2017;**15**(2):61-73

[27] Du P, Matevosyan J. Forecast system inertia condition and its impact to integrate more renewables. IEEE Transactions on Smart Grid. 2018;**9**(2): 1531-1533

[28] Muyeen S, Takahashi R, Murata T, Tamura J. A variable speed wind turbine control strategy to meet wind farm grid code requirements. IEEE Transactions on Power Systems. 2010;**25**(1):331-340

[29] Zhao J, Lyu X, Fu Y, Hu X, Li F. Coordinated microgrid frequency regulation based on DFIG variable coefficient using virtual inertia and primary frequency control. IEEE Transactions on Energy Conversion. 2016;**31**(3):833-845

[30] Hosseinipour A, Hojabri H. Virtual inertia control of PV systems for dynamic performance and damping enhancement of dc microgrids with constant power loads. IET Renewable Power Generation. 2017;**12**(4):430-438

[31] Tielens P. Operation and control of power systems with low synchronous inertia [PhD thesis]. KU Leuven; 2017

[32] Yingcheng X, Nengling T. Review of contribution to frequency control through variable speed wind turbine. Renewable Energy. 2011;**36**(6): 1671-1677

[33] Fischer M, Engelken S, Mihov N, Mendonca A. Operational experiences with inertial response provided by type 4 wind turbines. IET Renewable Power Generation. 2016;**10**(1):17-24

[34] Tang ZX, Lim YS, Morris S, Yi JL, Lyons PF, Taylor PC. A comprehensive work package for energy storage systems as a means of frequency regulation with increased penetration of photovoltaic systems. International Journal of Electrical Power & Energy Systems. 2019;**110**:197-207

[35] Yang L, Hu Z, Xie S, Kong S, Lin W. Adjustable virtual inertia control of supercapacitors in PV-based ac microgrid cluster. Electric Power Systems Research. 2019;**173**: 71-85

[36] Li W, Du P, Lu N. Design of a new primary frequency control market for hosting frequency response reserve offers from both generators and loads. IEEE Transactions on Smart Grid. 2017;**9**(5):4883-4892

[37] You R, Barahona B, Chai J, Cutululis NA, Wu X. Improvement of grid frequency dynamic characteristic with novel wind turbine based on electromagnetic coupler. Renewable Energy. 2017;**113**:813-821

[38] Attya A, Dominguez-Garcia J, Anaya-Lara O. A review on frequency support provision by wind power plants: Current and future challenges. Renewable and Sustainable Energy Reviews. 2018;**81**:2071-2087

[39] Wang S, Tomsovic K. A novel active power control framework for wind turbine generators to improve frequency response. IEEE Transactions on Power Systems. 2018;**33**(6):6579-6589

[40] Ziping W, Wenzhong G, Tianqi G, Weihang Y, Zhang H, Shijie Y, et al. State-of-the-art review on frequency response of wind power plants in power systems. Journal of Modern Power Systems and Clean Energy. 2018;**6**(1): 1-16

[41] International Energy Agency. Total primary energy supply (TPES) by source, year and country. Available from: https://bit.ly/34YTcda. [Accessed: 17 October 2018]

[42] ENTSO-E. High Penetration of Power Electronic Interfaced Power Sources (HPoPEIPS). Available from: h ttps://bit.ly/2x5fZrh

[43] Fernández-Guillamón A, Sarasúa JI, Chazarra M, Vigueras-Rodríguez A, Fernández-Muñoz D, Molina-García Á. Frequency control analysis based on unit commitment schemes with high wind power integration: A Spanish isolated power system case study. International Journal of Electrical Power Energy Systems. 2020;**121**:106044

[44] Bevrani H, Daneshmand PR. Fuzzy logic-based load-frequency control concerning high penetration of wind turbines. IEEE Systems Journal. 2012; **6**(1):173-180

[45] Ozer B, Arikan O, Moral G, Altintas A. Extraction of primary and secondary frequency control from active power generation data of power plants. International Journal of Electrical Power & Energy Systems. 2015;**73**:16-22

[46] Nedd M, Booth C, Bell K. Potential solutions to the challenges of low inertia power systems with a case study concerning synchronous condensers. In: 2017 52nd International Universities Power Engineering Conference (UPEC); IEEE. 2017. pp. 1-6

[47] Wang X, Yue M. Design of energy storage system to improve inertial response for large scale PV generation.

In: 2016 IEEE Power and Energy Society General Meeting (PESGM). 2016. pp. 1-5

[48] Luo X, Wang J, Dooner M, Clarke J. Overview of current development in electrical energy storage technologies and the application potential in power system operation. Applied Energy. 2015; **137**:511-536

[49] Chen H, Cong TN, Yang W, Tan C, Li Y, Ding Y. Progress in electrical energy storage system: A critical review. Progress in Natural Science. 2009;**19**(3): 291-312

[50] Akram U, Nadarajah M, Shah R, Milano F. A review on rapid responsive energy storage technologies for frequency regulation in modern power systems. Renewable and Sustainable Energy Reviews. 2020;**120**:109626

[51] Marcos J, Storkël O, Marroyo L, Garcia M, Lorenzo E. Storage requirements for PV power ramp-rate control. Solar Energy. 2014;**99**:28-35

[52] Salim NB, Aboelsoud H, Tsuji T, Oyama T, Uchida K. Load frequency control of two-area network using renewable energy resources and battery energy storage system. Journal of Electrical Systems. 2017;**13**(2):348-365

[53] Zhao Z, Xiao H, Yang Y. Improved coordinated control strategy of hybrid energy storages in PV power smoothing. Energy Procedia. 2018;**145**:151-156

[54] Cabrane Z, Ouassaid M, Maaroufi M. Analysis and evaluation of battery-supercapacitor hybrid energy storage system for photovoltaic installation. International Journal of Hydrogen Energy. 2016;**41**(45): 20897-20907

[55] Chandra A. Supercapacitors: An alternate technology for energy storage. Proceedings of the National Academy of Sciences. 2012;**82**:79-90

[56] Taghizadeh M, Hoseintabar M, Faiz J. Frequency control of isolated WT/PV/SOFC/UC network with new control strategy for improving SOFC dynamic response. International Transactions on Electrical Energy Systems. 2015;**25**(9): 1748-1770

[57] You S, Liu Y, Tan J, Gonzalez MT, Zhang X, Zhang Y, et al. Comparative assessment of tactics to improve primary frequency response without curtailing solar output in high photovoltaic interconnection grids. IEEE Transactions on Sustainable Energy. 2018;**10**(2):718-728

[58] Mousavi GS, Faraji F, Majazi A, Al-Haddad K. A comprehensive review of flywheel energy storage system technology. Renewable and Sustainable Energy Reviews. 2017;**67**: 477-490

[59] Amiryar ME, Pullen KR. A review of flywheel energy storage system technologies and their applications. Applied Sciences. 2017;**7**:286(1-21)

[60] Pullen KR. The status and future of flywheel energy storage. Joule. 2019; **3**(6):1394-1399

[61] Akinyele D, Rayudu R. Review of energy storage technologies for sustainable power networks. Sustainable Energy Technologies and Assessments. 2014;**8**:74-91

[62] Barelli L, Bidini G, Bonucci F, Castellini L, Fratini A, Gallorini F, et al. Flywheel hybridization to improve battery life in energy storage systems coupled to res plants. Energy. 2019;**173**: 937-950

[63] Xin H, Liu Y, Wang Z, Gan D, Yang T. A new frequency regulation strategy for photovoltaic systems without energy storage. IEEE Transactions on Sustainable Energy. 2013;**4**(4):985-993

[64] Alatrash H, Mensah A, Mark E, Haddad G, Enslin J. Generator emulation controls for photovoltaic inverters. IEEE Transactions on Smart Grid. 2012;**3**(2):996-1011

[65] Zarina P, Mishra S, Sekhar P. Deriving inertial response from a non-inertial PV system for frequency regulation. In: 2012 IEEE International Conference on Power Electronics, Drives and Energy Systems (PEDES); IEEE. 2012. pp. 1-5

[66] Zarina P, Mishra S, Sekhar P. Photovoltaic system based transient mitigation and frequency regulation. In: 2012 Annual IEEE India Conference (INDICON); IEEE. 2012. pp. 1245-1249

[67] García-Gracia M, El Halabi N, Ajami H, Comech MP. Integrated control technique for compliance of solar photovoltaic installation grid codes. IEEE Transactions on Energy Conversion. 2012;**27**(3):792-798

[68] Moutis P, Vassilakis A, Sampani A, Hatziargyriou N. DC switch driven active power output control of photovoltaic inverters for the provision of frequency regulation. IEEE Transactions on Sustainable Energy. 2015;**6**(4):1485-1493

[69] Mishra S, Zarina P, Sekhar P. A novel controller for frequency regulation in a hybrid system with high PV penetration. In: 2013 IEEE Power and Energy Society General Meeting (PES); IEEE. 2013. pp. 1-5

[70] Zarina P, Mishra S, Sekhar P. Exploring frequency control capability of a PV system in a hybrid PV-rotating machine-without storage system. International Journal of Electrical Power & Energy Systems. 2014;**60**:258-267

[71] Ziping W, Wenzhong G, Tianqi G, Weihang Y, ZHANG H, Shijie Y, et al. State-of-the-art review on frequency response of wind power plants in power systems. Journal of Modern Power Systems and Clean Energy. 2017:1-16

[72] Xiong L, Li Y, Zhu Y, Yang P, Xu Z. Coordinated control schemes of super-capacitor and kinetic energy of DFIG for system frequency support. Energies. 2018;**11**(1):103

[73] Jauch C, Hippel S. Hydraulic–pneumatic flywheel system in a wind turbine rotor for inertia control. IET Renewable Power Generation. 2016; **10**(1):33-41

[74] Wen J, Liu J, Long Y, Yao W. Solution to short-term frequency response of wind farms by using energy storage systems. IET Renewable Power Generation. May 2016;**10**:669-678

[75] Gonzalez-Longatt FM, Alhejaj SM. Enabling inertial response in utility-scale battery energy storage system. In: 2016 IEEE Innovative Smart Grid Technologies-Asia (ISGT-Asia). 2016. pp. 605-610

[76] Tan J, Zhang Y. Coordinated control strategy of a battery energy storage system to support a wind power plant providing multi-timescale frequency ancillary services. IEEE Transactions on Sustainable Energy. 2017;**8**(3):1140-1153

[77] He G, Chen Q, Kang C, Xia Q, Poolla K. Cooperation of wind power and battery storage to provide frequency regulation in power markets. IEEE Transactions on Power Systems. 2017;**32**(5):3559-3568

[78] Bai L, Li F, Hu Q, Cui H, Fang X. Application of battery-supercapacitor energy storage system for smoothing wind power output: An optimal coordinated control strategy. In: 2016 IEEE Power and Energy Society General Meeting (PESGM). 2016. pp. 1-5

[79] Tan Y, Muttaqi KM, Ciufo P, Meegahapola L. Enhanced frequency response strategy for a pmsg-based

wind energy conversion system using ultracapacitor in remote area power supply systems. IEEE Transactions on Industry Applications. 2017;**53**(1): 549-558

[80] Gayathri NS, Kar IN. Smoothing of wind power using flywheel energy storage system. IET Renewable Power Generation. 2017;**11**:289-298

[81] Díaz-González F, Sumper A, Gomis-Bellmunt O, Bianchi FD. Energy management of flywheel-based energy storage device for wind power smoothing. Applied Energy. 2013;**110**: 207-219

[82] Yao J, Yu M, Gao W, Zeng X. Frequency regulation control strategy for PMSG wind-power generation system with flywheel energy storage unit. IET Renewable Power Generation. June 2017;**11**:1082-1093

[83] Zhao H, Wu Q, Hu S, Xu H, Rasmussen CN. Review of energy storage system for wind power integration support. Applied Energy. 2015;**137**:545-553

[84] Díaz-González F, Hau M, Sumper A, Gomis-Bellmunt O. Coordinated operation of wind turbines and flywheel storage for primary frequency control support. International Journal of Electrical Power Energy Systems. 2015;**68**:313-326

[85] Ahmadi R, Ghardashi F, Kabiri D, Sheykholeslami A, Haeri H. Voltage and frequency control in smart distribution systems in presence of der using flywheel energy storage system. IET Conference Proceedings. January 2013: 1307-1307

[86] Ghosh S, Kamalasadan S. An energy function-based optimal control strategy for output stabilization of integrated DFIG-flywheel energy storage system. IEEE Transactions on Smart Grid. 2017; **8**(4):1922-1931

[87] Zhang X, Zha X, Yue S, Chen Y. A frequency regulation strategy for wind power based on limited over-speed de-loading curve partitioning. IEEE Access. 2018;**6**:22938-22951

[88] Moutis P, Loukarakis E, Papathanasiou S, Hatziargyriou ND. Primary load-frequency control from pitch-controlled wind turbines. In: 2009 IEEE Bucharest PowerTech; IEEE. 2009. pp. 1-7

[89] Ma H, Chowdhury B. Working towards frequency regulation with wind plants: Combined control approaches. IET Renewable Power Generation. 2010;**4**(4):308-316

[90] Moutis P, Papathanassiou SA, Hatziargyriou ND. Improved load-frequency control contribution of variable speed variable pitch wind generators. Renewable Energy. 2012;**48**: 514-523

[91] Žertek A, Verbič G, Pantoš M. Optimised control approach for frequency-control contribution of variable speed wind turbines. IET Renewable Power Generation. 2012; **6**(1):17-23

[92] Castro LM, Fuerte-Esquivel CR, Tovar-Hernández JH. Solution of power flow with automatic load-frequency control devices including wind farms. IEEE Transactions on Power Systems. 2012;**27**(4):2186-2195

[93] Vidyanandan K, Senroy N. Primary frequency regulation by deloaded wind turbines using variable droop. IEEE Transactions on Power Systems. 2013; **28**(2):837-846

[94] Alsharafi AS, Besheer AH, Emara HM. Primary frequency response enhancement for future low inertia power systems using hybrid control technique. Energies. 2018; **11**(4):699

[95] Ye H, Pei W, Qi Z. Analytical modeling of inertial and droop responses from a wind farm for short-term frequency regulation in power systems. IEEE Transactions on Power Systems. 2016;**31**(5):3414-3423

[96] Fakhari Moghaddam Arani M, Mohamed YAI. Dynamic droop control for wind turbines participating in primary frequency regulation in microgrids. IEEE Transactions on Smart Grid. 2018;**9**(6):5742-5751

[97] Lertapanon P, Wangdee W. Analysis and modeling of wind turbine generators considering frequency controls. In: 2017 International Electrical Engineering Congress (iEECON); IEEE. 2017. pp. 1-4

[98] Huang L, Xin H, Zhang L, Wang Z, Wu K, Wang H. Synchronization and frequency regulation of DFIG-based wind turbine generators with synchronized control. IEEE Transactions on Energy Conversion. 2017;**32**(3):1251-1262

[99] Deepak M, Abraham RJ, Gonzalez-Longatt FM, Greenwood DM, Rajamani H-S. A novel approach to frequency support in a wind integrated power system. Renewable Energy. 2017; **108**:194-206

[100] Gonzalez-Longatt F, Chikuni E, Stemmet W, Folly K. Effects of the synthetic inertia from wind power on the total system inertia after a frequency disturbance. In: Power Engineering Society Conference and Exposition in Africa; Citeseer. 2012. pp. 9-13

[101] Bonfiglio A, Invernizzi M, Labella A, Procopio R. Design and implementation of a variable synthetic inertia controller for wind turbine generators. IEEE Transactions on Power Systems. 2019;**34**(1):754-764

[102] Liu K, Qu Y, Kim H-M, Song H. Avoiding frequency second dip in power unreserved control during wind power rotational speed recovery. IEEE Transactions on Power Systems. 2018; **33**(3):3097-3106

[103] Morren J, de Haan SWH, Kling WL, Ferreira JA. Wind turbines emulating inertia and supporting primary frequency control. IEEE Transactions on Power Systems. February 2006;**21**:433-434

[104] Díaz-González F, Hau M, Sumper A, Gomis-Bellmunt O. Participation of wind power plants in system frequency control: Review of grid code requirements and control methods. Renewable and Sustainable Energy Reviews. 2014;**34**:551-564

[105] Dreidy M, Mokhlis H, Mekhilef S. Inertia response and frequency control techniques for renewable energy sources: A review. Renewable and Sustainable Energy Reviews. 2017;**69**:144-155

[106] Tarnowski GC, Kjar PC, Sorensen PE, Ostergaard J. Variable speed wind turbines capability for temporary over-production. In: Power & Energy Society General Meeting, 2009. PES'09. IEEE. 2009. pp. 1-7

[107] Keung P-K, Li P, Banakar H, Ooi BT. Kinetic energy of wind-turbine generators for system frequency support. IEEE Transactions on Power Systems. 2009;**24**(1):279-287

[108] El Itani S, Annakkage UD, Joos G. Short-term frequency support utilizing inertial response of DFIG wind turbines. In: 2011 IEEE Power and Energy Society General Meeting; IEEE. 2011. pp. 1-8

[109] Hansen AD, Altin M, Margaris ID, Iov F, Tarnowski GC. Analysis of the short-term overproduction capability of variable speed wind turbines. Renewable Energy. 2014;**68**:326-336

[110] Hafiz F, Abdennour A. Optimal use of kinetic energy for the inertial support

from variable speed wind turbines. Renewable Energy. 2015;**80**:629-643

[111] Kang M, Kim K, Muljadi E, Park J-W, Kang YC. Frequency control support of a doubly-fed induction generator based on the torque limit. IEEE Transactions on Power Systems. 2016;**31**(6):4575-4583

[112] Fernández-Guillamón A, Villena-Lapaz J, Vigueras-Rodríguez A, García-Sánchez T, Molina-García Á. An adaptive frequency strategy for variable speed wind turbines: Application to high wind integration into power systems. Energies. 2018;**11**(6):1-21

[113] Liu K, Qu Y, Kim H-M, Song H. Avoiding frequency second dip in power unreserved control during wind power rotational speed recovery. IEEE Transactions on Power Systems. 2017; **33**(3):3097-3106

[114] Fernández-Guillamón A, Vigueras-Rodríguez A, Gómez-Lázaro E, Molina-García Á. Fast power reserve emulation strategy for VSWT supporting frequency control in multi-area power systems. Energies. 2018;**11**(10):2775(1-20)

[115] Wu Z, Gao DW, Zhang H, Yan S, Wang X. Coordinated control strategy of battery energy storage system and PMSG-WTG to enhance system frequency regulation capability. IEEE Transactions on Sustainable Energy. 2017;**8**(3):1330-1343

Chapter 2

Use of Renewable Energy in Buildings

İzzet Yüksek and İlker Karadağ

Abstract

Owing to factors such as high living standards and digitalization, energy use is growing. However, the proportion of renewable energy sources is also rising in all energy consumption. Given this use of renewable energy, global warming and environmental issues are still rising. Fossil-based energy species are more polluting and resource-stricken than others. Studies on environmental pollution show that fossil-based energies are the most important pollutants. Fossil-based energy source is still the most consumed type of energy. Besides, the renewable energy sources' consumption is very low. Therefore, it is important to increase the use of renewable energy, which creates cleaner and less emissions. Buildings should have the right clean energy use incentives. The needs such as heating, refrigeration, and lighting can be met by renewable energy. This research aims to explore and demonstrate how renewable energy can profit when fulfilling public building functions. Through using both conventional methods and creative methods together, the rate of use of renewable resources such as solar, wind, and geothermal resources in buildings can be increased. Therefore, major contributions to reducing the environmental issues caused by energy consumption can be made.

Keywords: renewable energy resources, sustainable buildings, sustainability, energy efficiency, renewable energy in buildings

1. Introduction

Since buildings are the major energy-consuming sectors in the world causing energy inefficiency, they can act as a promising target with the greatest potential to reach the common goal toward sustainable development. Nevertheless, too high building energy consumption will raise negative impacts on the environment, such as air pollution, greenhouse effect, urban heat island effect, and others, which can even do plenty of harm to human health and social economy development [1].

Fossil-based energies are the most important source of environmental pollutants. However, 84.7% of the total energy consumption in the world is due to fossil-based energy. Energies from renewable and nuclear sources have a share of 5 and 4% in primary energy consumption alone [2].

In a study conducted among International Energy Agency (IEA) countries, buildings are seen as an important energy consumer, and of half of the total consumed electricity, one third of natural gas is consumed in buildings. Again in this study, construction activities are held responsible for one third of the greenhouse gases generated in the world [3].

One of the effective methods of saving energy in buildings is to use renewable energy instead of limited resources. In this way, both preserving and protecting our resources for future generations and environmental values environmental values are not damaged.

The aim of this study is to emphasize the importance of using renewable energy in buildings and to investigate the possibilities of using renewable energy. For this reason, the usage of renewable energy types in buildings and their benefits have been discussed and appropriate solutions have been proposed.

2. Possibilities of using renewable energy resources in buildings

In the early stage of design, the importance of energy consumption gradually got into engineering consideration. In addition, as of the first decade of the twenty-first century, energy conservation is regarded as the most significant element in system design. Complementing the industrial revolution is certainly the best time for yet another revolution. This transition is about confronting the effects of the last century's exploitation of natural resources. The world must understand that the use of renewable energies due to harmful environmental effects of greenhouse gases and a finite supply of traditional energy supplies (fossil fuels) must be taken seriously. The strategies of reducing energy consumption should also be considered when replacing the nonrenewable energies with the renewable ones [4].

Energy is used for various reasons throughout the building's life cycle. 94.4% of the total energy used in these phases is consumed for heating/ventilation/air conditioning (HVAC) systems that provide comfort conditions during the usage phase [5].

In order to reduce this rate, passive methods and renewable energy sources should be used instead of mechanical systems to provide comfort conditions. In this way, more appropriate physical conditions for human health can be established within the buildings.

Renewable energy sources are the energy obtained from the existing energy flow in the continuous natural processes. In general, the renewable energy source is defined as the ability to renew itself at an equal rate to the energy received from the energy source or faster than the depletion rate of the source [6]. Water energy, wind energy, solar energy, wave and tidal energy, bio (organic) fuel, geothermal energy, hydrogen energy, and ocean energy are renewable energy sources.

The possibility of depletion of the most commonly used energies, such as coal and oil, has led humanity to new energy sources. When selecting energy sources, it is paid attention that it is safe, clean, economical, and most importantly a renewable source which does not harm the environment.

2.1 Use of solar systems in buildings

The sun is an unlimited source of light and heat energy. The basic principle in the designs for using solar energy in buildings is as follows.

The flow of thermal energy of the sun through conduction, convection, and radiation is used. These natural processes are managed through a building design that helps to warm up and cool the building.

The sunrays coming to the building surface are reflected, transmitted, or absorbed by the building material. In addition, the heat generated by the sun causes predictable air movements within the designed areas. This basic effect of solar heat leads to the selection of materials and design of building elements that provide a heating and cooling effect within the building, such as thickness, density (δ) (g/cm^3), heat conduction coefficient (λ) (W/m^0K) specific heat (c) (Wh/m^3 ^0K), surface absorption, and reflection coefficient smoothness or roughness, cavity, and fullness. It is possible to utilize solar energy as active and passive through the measures taken in architecture design.

Rezaie et al. determined that the most effective method for reducing carbon dioxide emissions is mixed systems, where both active and passive systems are used together [4].

2.1.1 Use of passive solar systems in buildings

The use of a building form and shell to accept, store, and distribute energy from renewable sources appropriate for buildings is commonly defined as the passive solar design. passive systems mainly use solar energy and fresh air by means of space heating, cooling, and lighting without mechanical or electronic equipment.

- Passive heating

Passive heating systems are most commonly used in passive solar architecture. With the design applications for passive solar systems, solar heat gains from solar energy can be increased during the winter months. The fundamental concept of using solar energy for heating is to design the elements that shape the exterior of the building for this purpose (the roof, walls, and floor are insulated to a high level) and to allow as much use as possible of solar radiation. The program employs three key components. Those are collectors, storers, and distributors. Solar energy is captured by the collectors and converted into heat. Storages allow heat to be used when there is no solar power. The role of the distributors is to transfer the collected energy to the storage elements and appropriate places via the collectors.

The energy obtained depends on the opening size (windows, skylight, greenhouse, etc.). The conservation of the energy obtained relies on the building envelope's thermal insulation and sealage. The storage of energy depends on the location of the building elements and their thermal performance. The house of philosopher Socrates, who lived between 470 and 399 BC, is the simplest example of the passive heating application (**Figure 1**). His house is opening into the sun, providing optimum productivity with a compact structure and trapezoidal plan scheme with the long side facing the sun, when the northern side has been reduced. The eaves on the south of the roof provide protection when the sun's orbit is in summer and allows the sun below to enter the building in winter. The roof slopes down in the back to avoid winter winds [7].

- Natural lighting

Seventeen percent of all energy used in the world is consumed for lighting purposes. With the right design, 70% of the lighting needs can be obtained from the sun. In ordinary buildings, this rate is 25%. The use of daylight as much as possible in the illumination of spaces in buildings according to visual comfort needs reduces the need for artificial lighting. It enables the buildings to consume less energy during the usage process [8]. The easiest natural lighting approach is to use the sufficient openings left in the building envelope. Below, window arrangements are seen in the traditional housing examples providing adequate natural light (**Figure 2**).

Natural lighting can be provided in places without facades suitable for direct sunlight, through roof windows or light tubes. The natural illumination can be provided by opening skylights on the roofs where the roof element covers the interior space. (**Figure 3**).

Light tubes are bringing daylight from outside to inside. The components and operating principle of a light tube are explained in **Figures 4** and **5**.

A dome-shaped collector sits at the top of the light pipe. The chimney uses light-reinforcing coatings. The light obtained is carried inside by panels at the side edges of the chimney.

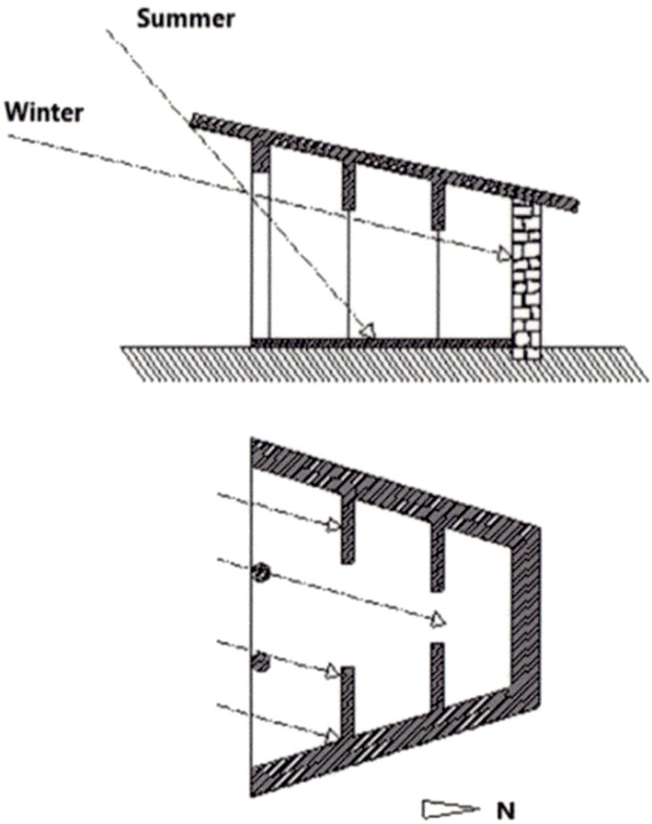

Figure 1.
Socrates's house plan and cross section [7].

Figure 2.
Window arrangements in traditional houses.

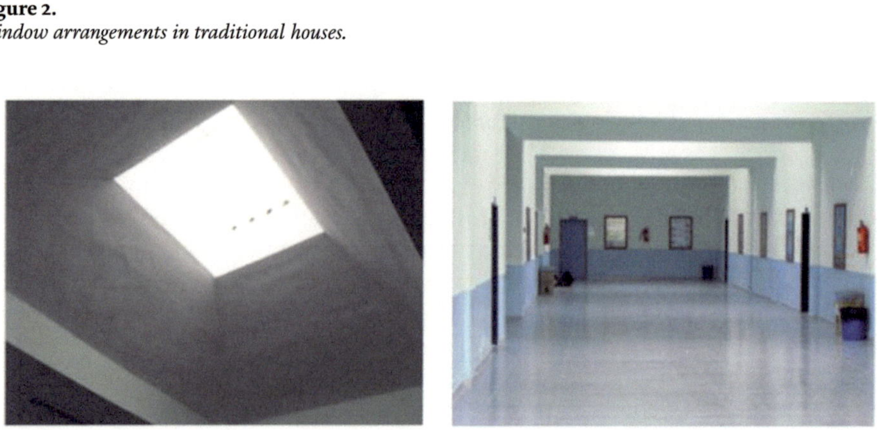

Figure 3.
Illuminated interior space with skylight [9].

Figure 4.
Operating principle of a light tube [10].

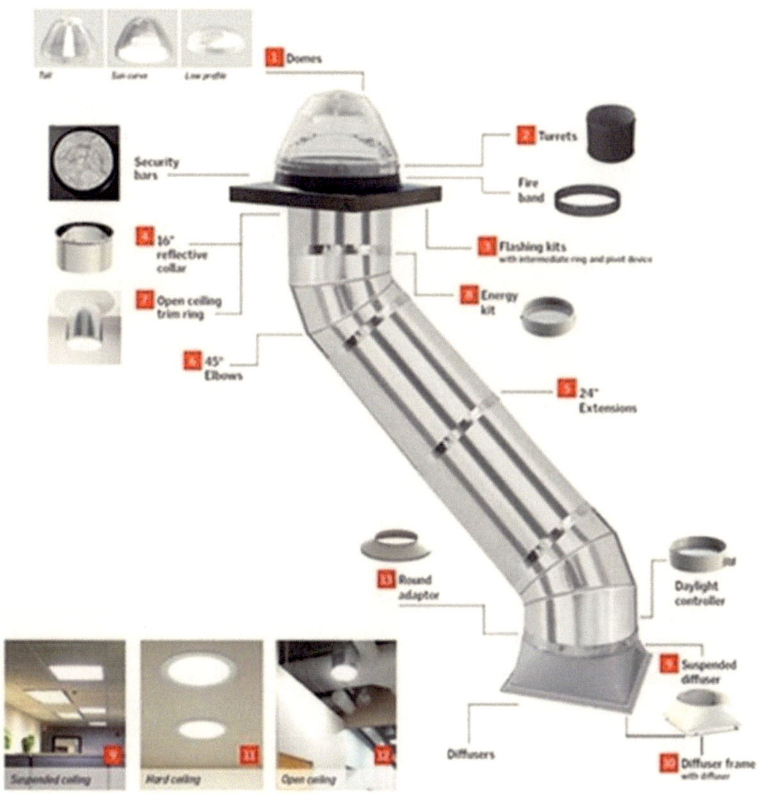

Figure 5.
Components of a light tube [11].

2.1.2 Use of active solar systems in buildings

Active solar systems are distinguished from passive systems, which use the fabric of the building, in terms of collection of solar energy, storage of collected heat, and distribution of heat to spaces. Active systems use components for collection, storage, and funs or pumps for forced distribution of collected and stored heat.

Active systems using solar energy are the systems consisting of the whole of the mechanical and/or electronic elements which convert the solar radiation absorbed by the purpose-produced collectors into the desired form of energy and allow it to be used in the structure. Through these systems, solar radiation can be transformed into heat and electrical energy [12]. These systems are converting solar radiation into energy and according to the energy they produce, they can be divided into two: solar thermal systems, which produce heat energy, and photovoltaic (PV) systems, which generate electrical energy. These systems are briefly described below.

- Solar heating systems

By converting the solar radiation into heat energy with collectors, solar heating systems can heat up water, air, etc. directly with a fluid; or, all mechanical and/or electronic systems that are used in a storage unit for evaluation and use are called "Solar Heating Systems." Solar active heating systems are used in buildings for use/heating of pool water, preheating of air-conditioning air, and space heating [12]. The general working principle of heating systems is based on the collection of heat by means of collectors, storage, and distribution of the collected heat energy to the related areas in order to use it later [13].

- Solar water heating systems

These systems convert solar radiation into heat energy. It uses some elements which store and distribute heat in the water. All solar water heating systems are based on the heating, storage and distribution of water. Depending on the characteristics of the system, the hot water produced by the transformation of solar energy can be used directly to meet the user needs such as washing or it can be used to support the traditional heating system [12].

- Photovoltaic systems

Photovoltaic (PV) systems are all components that generate electricity from solar radiation through collectors and allow the use of this energy. PV systems are used for the production of electricity in many different fields such as road lighting, lighthouses, vehicles, buildings, and power plants, with different or simple configurations. A photovoltaic system generates electrical energy, stores the generated energy when necessary, and reliably transfers it to the fields of use. Photovoltaic batteries are placed in the facades and roofs of buildings and convert the solar energy coming to these surfaces into electrical energy (**Figure 6**). The solar cells used for domestic purposes are connected to the electricity grid via an inverter, thus saving the storage of the generated electricity in the batteries.

2.2 Use of wind energy in buildings

The wind has been used as an energy source for a very long time, and it is an important source of environmental-friendly energy and has become more and more important in the recent years [15].

Figure 6.
Solar panels cover the building's south, west, and top surfaces [14].

It is possible to benefit from wind energy by using passive and active systems. These methods are described below.

2.2.1 Use of wind energy with passive systems in buildings

Passive cooling: By using passive systems, it is possible to provide comfort conditions required for human health and work efficiency in the building in certain proportions without requiring energy use. Especially, the effect of ventilation provided by natural methods is important in creating thermal comfort and indoor air quality.

The basic concept of passive cooling is to prevent the building heat gain. Planning in conjunction with this purpose should be included in the design phase of the house. For preventing building heat gain, high thermal mass and thick sectional structural elements such as a mudbrick or stone and shading elements may be provided.

Different methods of passive cooling were developed for different types of climate, such as the following.

Shading, reflection of solar heat, insulation of building element, ground cooling, wind cooling, water cooling, evaporative cooling, dehumidification, night radiant cooling, night cooling of thermal mass in buildings, exotic passive cooling methods, and seasonal cold storage [16].

Accordingly, passive cooling differs in various locations and conditions. The methods employed depend on the location and environment in question. In any application and set of conditions, not all methods will be useful. Depending on the location, environment, available materials and skills, and economic considerations, different methods of achieving passive cooling can be used separately or combined [16].

Wind catchers are the simplest example of cooling the building with natural ventilation. Thermal chimneys operate as a collector that draws in fresh air outside the building. In this method, the entrance of air from the external environment into the building is accelerated by means of a hot or warm region with an outflow. The use of wind chimneys called "badgir" is very common especially in the examples of traditional architecture of Middle Eastern countries (**Figure** 7).

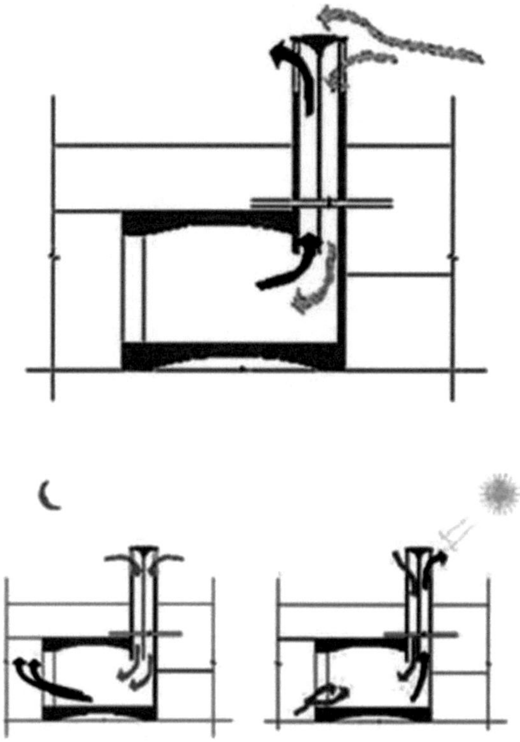

Figure 7.
Traction and suction in wind catcher [17].

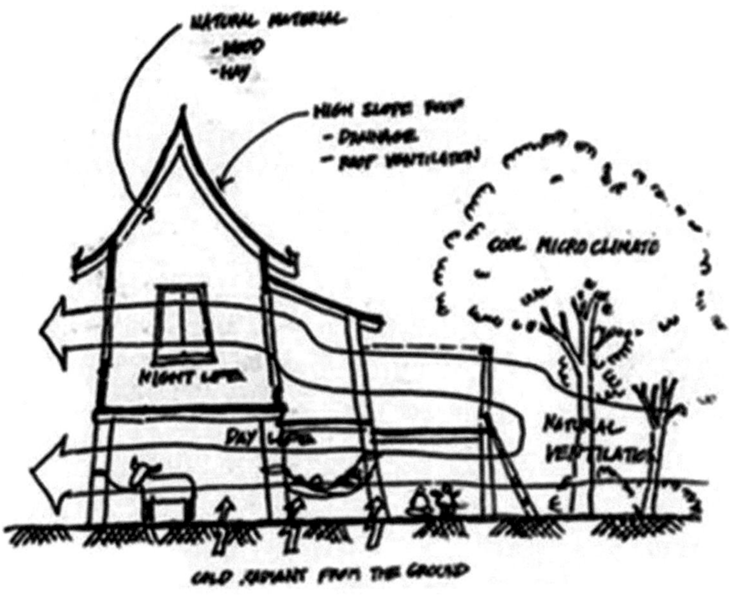

Figure 8.
A traditional tai house providing natural ventilation by taking advantage of pressure differences around the building [18].

The main strategy for passive building cooling in hot humid climates is to provide natural ventilation. The use of openable windows for natural ventilation is the most common (**Figure 8**).

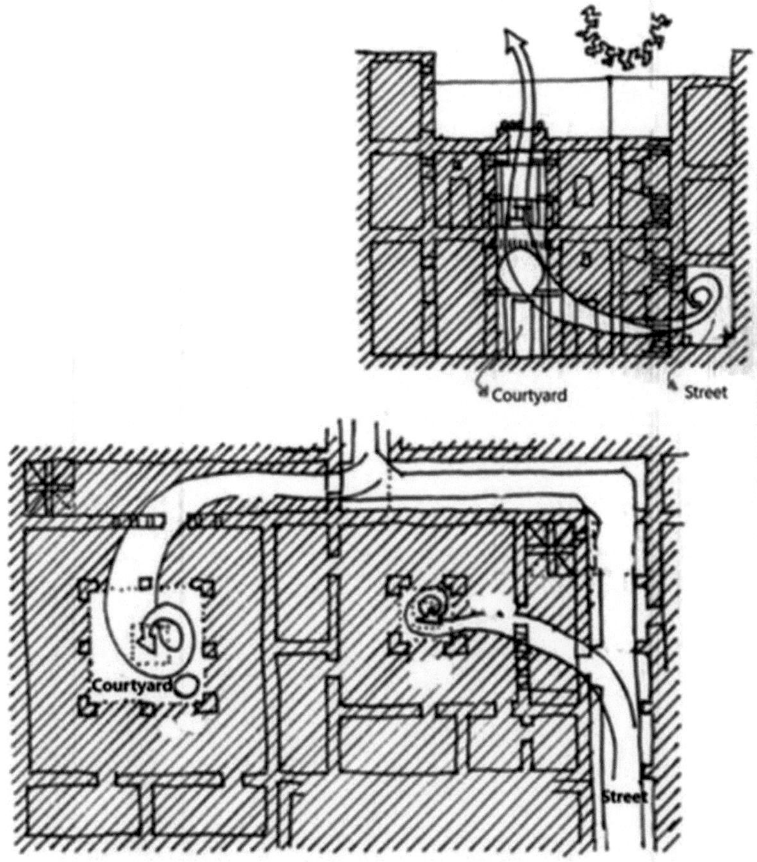

Figure 9.
Plan and section with inner courtyard showing natural ventilation through convection [19].

In addition, an atrium can be designed in the direction of the prevailing wind in the building during the design phase. While the spaces surrounding the atrium fill with cold air through the windows, the atrium collects the hot air and circulates out (**Figure 9**).

2.2.2 Use of wind energy with active systems in buildings

Wind energy is the conversion of kinetic energy of air mass into mechanical energy. Wind energy is natural, inexhaustible, does not produce waste during use, has no radioactive effect and therefore has no negative impact on nature and human health, and is a rapid energy source of technological development [18].

It is possible to obtain electricity from wind on the Earth's surface. In 2040, it is predicted that 40% of the energy of the world will be obtained from the wind [8]. Active wind energy use systems are wind turbines. Medium- and small-scale wind turbines are used in the buildings. These turbines can be placed in a suitable spot in the garden or on the roofs. In multistory high-rise buildings, there are examples of use of wind turbines integrated into the building (**Figure 10**).

2.3 Use of geothermal energy in buildings

Geothermal energy is obtained by the fact that the heat accumulated in the underground is released from the cracks to the earth. Sometimes, it can be extracted from the underground as hot water, hot water and water vapor mixture, or steam.

Geothermal energy is used in the heating and cooling of houses, greenhouses, and agriculture. According to the application methods of geothermal fluid, geothermal energy systems are applied in three different ways such as heat pumps, in-well heat exchangers, and heat pipes. Common use in buildings is in the form of heat pipes. Heat can also be extracted from the ground at "normal" temperatures using a device called a heat pump.

Another use of geothermal energy is the methods of using soil temperature. The temperature is also between 45 and 75 F (7.22–23.88 C), depending on the latitude of the earth at some level [22]. This soil temperature can be used by water or by air. The air taken through the opened chimneys is transferred to the system at different depths of the soil and the internal volume is brought to the same amount as the soil temperature. Such technology is useful in the context of winter heating and summer cooling (**Figure 11**).

Figure 10.
Wind turbines integrated into the building [20, 21].

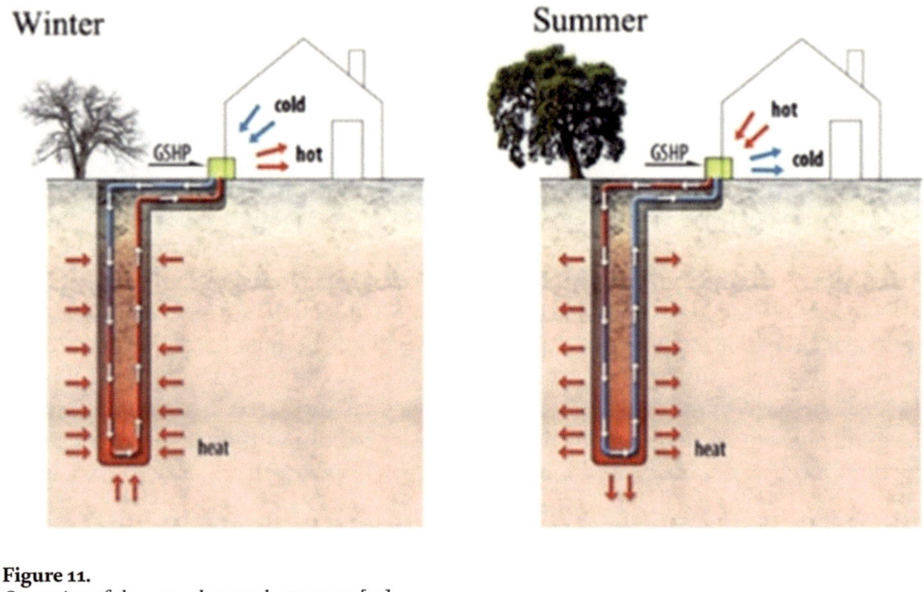

Figure 11.
Operation of the ground source heat pump [23].

2.4 Use of hydrogen energy in buildings

Hydrogen energy can be used for heating the houses, supplying hot water, cooking, and meeting electricity needs. In order to use hydrogen, it must first be produced, stored, and transported. Hydrogen can be produced from renewable energy sources such as solar, hydroelectric, wind, and geothermal.

Today, among the renewable energy sources, the solar-hydrogen hybrid system stands out as the most efficient system. In such a system, components such as photovoltaic panels, electrolyzer, fuel cell, Hydrogen (H2) storage tank, battery group, and inverter (converter) are needed. The operation of the system in the solar-hydrogen house energy system is as follows [24]:

- PV panels generate electricity from solar energy,
- H2 and O2 are produced by electrolyzer,
- gases are taken to the storage tank for ground and water heating,
- heating the air in the ventilation system by burning hydrogen flameless with catalytic hydrogen burner (1.5 kW) in winter,
- the fuel cell is activated if additional electricity is needed, and
- some of the heat released in the fuel cell is also used to heat the water.

2.5 Use of biomass energy in buildings

Bioenergy can also be called vitality energy. All living things use solar energy. For this reason, all kinds of biological materials contain energy, which is released when burned. Plants convert and store solar energy into chemical energy by photosynthesis, thus forming a source of biological mass and organic matter, called biomass [25]. Within the scope of biomass energy technology; wood (energy forests and tree residues), oilseed plants (sunflower, rapeseed, soy, etc.), carbohydrate plants (potato, wheat, corn, beet, etc.), fiber plants (flax, kenaf, hemp, sorghum, etc.), vegetable residues (branches, stalks, straw, roots, bark, etc.), animal wastes, and urban and industrial wastes are evaluated. Biomass is a renewable, strategic source of energy that can be grown everywhere, provides socioeconomic development, is environmentally friendly, can generate electricity, and provides fuel for vehicles. Biomass is evaluated in energy technology either by direct combustion or by increasing the fuel quality through various processes and by obtaining alternative biofuels (easily transportable, storable, and usable fuels) with properties equivalent to the existing fuels.

From biomass, fuel is obtained by physical processes (size reduction—crushing and grinding, drying, filtration, extraction, and briquetting) and conversion processes (biochemical and thermochemical processes) [26]. From biomass source in residences; biogas obtained by airless digestion method is used in electricity generation. Ethanol obtained by pyrolysis method is used for heating purposes, hydrogen obtained by direct burning method is used for heating (**Table 1**).

Biomass resources	Supply systems	Conversion	End use
Conventional forestry	Harvesting	Biochemical	Transportation fuels
Short rotation forestry	Collection	Combustion	Heat
Sawmill conversion products	Handling	Gasification	Electricity
Agricultural crops and residues	Delivery	Pyrolysis	Solid fuels
Oil-bearing plants	Storage	Anaerobic digestion	Renewable construction materials
Animal products		Combined heat and power	Plant-based pharmaceutical
Municipal solid waste		Heating	Renewable chemicals including polymers
Industrial waste		Deoxygenation	
		Depolymerisation	
		Hydrolysis	
		Fermantation	

Table 1.
Breakdown of the main pillars of biomass energy production [27].

3. Discussion and conclusion

Buildings have a significant share of energy consumption globally and regionally. Particularly during the usage phase of the building life cycle, a lot of energy is consumed to provide comfort conditions inside the building. The high proportion of buildings consuming energy also increases the use of fossil-based resources. Environmental issues arising from energy usage are thus also growing. Whereas, buildings appropriate to the use of renewable energy sources can be built using passive or active methods. It is clear that the use of renewable energy sources in buildings will provide environmental and economic benefits. Reducing this amount of energy as much as possible and obtaining it from renewable sources is one of the effective methods that provide buildings with energy efficiency and ecological characteristics.

Renewable energy sources can be utilized from the sun with active and passive methods for heating, cooling, ventilation, natural lighting, and obtaining hot water. Wind energy is also utilized in ventilation and cooling with active and passive systems. Geothermal energy can be used for heating and cooling purposes. It can be used in heating and hot water supply from hydrogen energy, cooking, and for supplying electricity. Biomass energy is beneficial for heating and hot water supply. These resources can be used together if necessary.

In renewable energy use, passive systems should be preferred because it is simpler and more cost-effective. In cases where passive systems are inadequate, they should be supported with active systems. By using renewable energies where these conditions are possible, the use of fossil-based energies is reduced and many environmental and economic benefits are provided. However, in order to spread the use of renewable energy sources in buildings, it is deemed necessary and important for governments to prepare the necessary laws and regulations and to have sanctions and incentives for their implementation.

Author details

İzzet Yüksek* and İlker Karadağ
Fine Arts Design and Architecture Faculty, Manisa Celal Bayar University, Manisa, Turkey

*Address all correspondence to: izzet.yuksek@cbu.edu.tr

References

[1] Pan Y, Zhang L. Data-driven estimation of building energy consumption with multi-source heterogeneous data. Applied Energy. 2020;**268**:114965. DOI: 10.1016/j.apenergy.2020.114965

[2] BP Statistical Review of World Energy. 2019. 68th edition. Available from: https://www.bp.com/content/dam/bp/business-sites/en/global/corporate/pdfs/energy-economics/statistical-review/bp-stats-review-2019-full-report.pdf

[3] International Energy Agency (IEA). Energy Efficiency: Buildings [Internet]. 2009. Available from: https://www.iea.org/topics/energyefficiency/buildings/ [Accessed: 03 March 2009]

[4] Rezaie B, Esmailzadeh E, Dincer I. Renewable energy options for buildings: Case studies. Energy and Buildings. 2010;**43**:56-65. DOI: 10.1016/j.enbuild.2010.08.013

[5] Scheuer C, Gregory A, Reppe P. Life cycle energy and environmental performance of a new university building: Modeling challenges and design implictions. Energy and Buildings. 2003;**35**(10):1049-1064. DOI: 10.1016/S0378-7788(03)00066-5

[6] General directorate of renewable energy, Turkey Minister of Energy an Natural Resources, What is Renewable Energy? [Internet]. 2009. Available from: http://www.yegm.gov.tr/genc_cocuk/Yenilenebilir_Enerji_Nedir.aspx [Accessed: 04 March 2009]

[7] Hegger M. From passive utilization to smart solar architecture. In: Schittich C, editor. Solar Architecture: Stategies, Vision, Concepts Basel, Boston, Berlin: Birkhäuser - Publishers for Architecture; 2003. pp. 12-25. DOI: 10.11129/detail.9783034615198

[8] Enerji Yolculuğunda Önemli Bir Durak, Bilgilendirme Kitabı, Diyarbakır Güneş Evi, Diyarbakır; 2008

[9] Görgülü S, Kocabey S, Yüksek İ, Dursun B. Natural lighting methods in buildings within the scope of energy efficiency: The case of Kırklareli. In: Proceedings of the International Thrace Region Development-Entrepreneurship Symposium. Kırklareli; 1-2 October 2010. pp. 97-111

[10] High-Performance Skylight System [Internet]. 2006. Available from: https://yenra.com/skylights/system.html [Accessed: 10 March 2007]

[11] Commercial Sun Tunnel Accessories [Internet]. 2009. Available from: https://www.veluxusa.com/professional/commercial/products/commercial-suntunnels [Accessed: 04 April 2009]

[12] Sakınç E. An Approach to the Evaluation of Solar Powered Factor Systems in Architecture in the Context of Sustainability [thesis]. İstanbul Yıldız Technical University; 2006

[13] Şerefhanoğlu M. Güneş Işınımlarından Yararlanma ve Korunma (Use and Protection of Sunlight). İstanbul: Yıldız Technical University Publishing; 1988

[14] Environmental Energy Innovation Building [Internet]. 2013. Available from: https://www.titech.ac.jp/english/research/stories/eei_building.html [Accessed: 06 June 2019]

[15] Eriksson S, Bernhoff H, Mats LS, Eriksson, et al. Evaluation of different turbine concepts for wind power. Renewable and Sustainable Energy Reviews. 2008;**12**:1419-1434. DOI: 10.1016/j.rser.2006.05.017

[16] Halacy D. Understanding Passive Cooling Systems. Virginia USA:

Volunteers in Technical Assistance (VITA); 1986

[17] Dehnavi M, Ghadiri MH, Mohammadi H. Study of wind catchers with square plan: Influence of physical parameters. International Journal of Modern Engineering Research. 2012;**2**(1):559-564

[18] Tantasavasdi C, Srebric J, Chen Q. Natural ventilation design for houses in Thailand. Energy and Buildings. 2001;**33-8**:815-824. DOI: 10.1016/S0378-7788(01)00073-1

[19] Hensens J. Qsours et Qasbas du Maroc. Singapore: Marka Print Pte Ltd.; 1986

[20] Wind-powered building design revealed [Internet]. 2001. Available from: https://www.newscientist.com/article/dn1292-wind-powered-building-design-revealed [Accessed: 06 June 2019]

[21] Venger Wind Unveils World's Largest Rooftop Wind Farm in Oklahoma City [Internet]. 2012. Available from: https://inhabitat.com/venger-wind-unveils-worlds-largest-rooftop-wind-farm-in-oklahoma-city/omrf-oklahoma-worlds-largest-windfarm-omni-directional-turbines-6/ [Accessed: 06 June 2019]

[22] Geoexchange Fact Sheet for Architects [Internet]. 2009. Available from: http://www.earthenergysystems.com/for_architects/ [Accessed: 04 April 2019]

[23] Johnston IW, Narsilio GA, Colls S. Emerging geothermal energy technologies. KSCE Journal of Civil Engineering. 2011;**15**(4):643-653. DOI: 10.1007/s12205-011-0005-7

[24] Tabakoğlu Ö. Hidrojen Enerjisi & Hidrojenin Binalarda Kullanımı. Diyarbakır: Energy and Ecology Panel; 2007

[25] Göksu Ç. Güneş Kent- Güneş Enerjili Yerleşim Modeli. Ankara: Güneş Kitapları; 1999

[26] Karaosmanoğlu F. Yenilenebilir Enerji Kaynakları ve Türkiye. Görüş Dergisi; 2003

[27] Korhaliller S. The UK's Biomass Energy Development Path [Internet]. 2010. Available from: https://pubs.iied.org/G02921/ [Accessed: 04 April 2019]

Section 2

Solar and Photovoltaic Energy

Chapter 3

Feasibility Analysis of Solar Power for the Safety of Fast Reactors during beyond Design Basis Events

Kudiyarasan Swamynathan, P. Sivakumar and K. Karthikeyan

Abstract

This chapter presents a new design that unites the favorable technical and ecological characteristics of the solar and nuclear power plants. The current designs of nuclear reactors promise integral configuration of the primary coolant loop, secondary coolant loop, and a number of passive safety functions and overall simplification of the reactor. The present nuclear reactor design emphasizes on the safety of the reactor core at all times, i.e., controlling the reactor, cooling the reactor core, and maintaining containment. In case of non-availability of standby emergency DGs during beyond design basis event like Fukushima incident, etc., leading to extended station blackout conditions, the passive decay heat removal system will be affected. Hence, additional DGs have been made as a mandatory requirement in nuclear power plants. In case the ADG could not be mobilized during BDBE, an additional backup power source not affected by BDBE is appreciated. Hence in addition to the diesel power sources (EDG and ADG), a new design was developed for integration of diesel power with solar power. The hybrid system was designed to improve the reliability and availability of passive heat removal system, to ensure a reliable supply without interruption, and to improve the overall system reliability (by the integration with the battery bank). This hybrid power also gives the redundant power supply to the safety critical systems. This chapter also features a detailed reliability analysis carried out for power supplies to the safety critical loads. In addition a comparison was made between PV/diesel/battery with diesel/battery. These new hybrid systems conserves diesel fuel and reduce CO2 as well as particulate emissions that are harmful to environment health. Integration of solar power to the existing battery power will increase the reliability and extended availability of the system and thereby ensures safety of the plant during crisis/calamities.

Keywords: solar power, nuclear power, diesel power, GRID power, economical, PV cell with battery, reliable power, hybrid solution

1. Introduction

Energy security is a goal that many countries are pursuing to ensure that their economies function without interruption and that their people have access to adequate, reliable and affordable supplies of modern and clean energy [1–3]. It is a pressing concern because the demand for energy is growing rapidly due to robust economic expansion, population growth, new uses of energy and income growth and yet the supplies of energy resources required to power these needs are finite and in most cases non-renewable [4–9]. Furthermore, the production, transportation and utilization of energy are a major source of greenhouse gases that cause global warming and climate change [10, 11].

BharatiyaNabhikiyaVidyut Nigam Limited (BHAVINI) is currently in the advanced stage of commissioning of the 500 MWe PFBR Prototype Fast Breeder Reactor (PFBR) at Kalpakkam. The PFBR is the forerunner of the future Fast Breeder Reactors which provides energy security to the country. The design of PFBR is indigenously developed by Indira Gandhi Center for Atomic Research (IGCAR) located at Kalpakkam.

PFBR uses sodium as a coolant to transfer heat from the reactor core to the water for steam production. PFBR is a pool type reactor having two loops of sodium viz. Primary and Secondary sodium loops. The entire bulk of the primary sodium is contained in a single large vessel called the main vessel. Two Primary Sodium Pumps (PSPs) circulates sodium in the main vessel through the reactor core. An inner vessel separates hot and cold pools of sodium. The heat transfers from primary sodium to secondary sodium through 4 Intermediate Heat Exchangers (IHXs) and then from secondary sodium to water for steam production through 8 Steam Generators (SGs). The liquid sodium being highly reactive with air, it requires additional safety measures to isolate the coolant from atmosphere. Above the free level of sodium, in main vessel argon gas is provided.

During full power operation of the reactor, sodium is drawn from the cold pool by 2 mechanical centrifugal PSPs working in parallel and is delivered to the grid plate through 4 pipes at 670 K (397°C). From there, it passes through the core and picks up the heat. Then the sodium at 820 K (547°C) flows into hot pool and enters the inlet windows of IHXs. The flow through IHX is due to sodium level difference of 1.5 m between hot and cold pool generated by main pump. It passes through the shell side of the IHX and transfers heat to the secondary sodium which passes through the tube side. The primary sodium leaves the IHX through the outlet windows and returns to the cold pool. The schematic of the reactor system is shown in **Figure 1** [12].

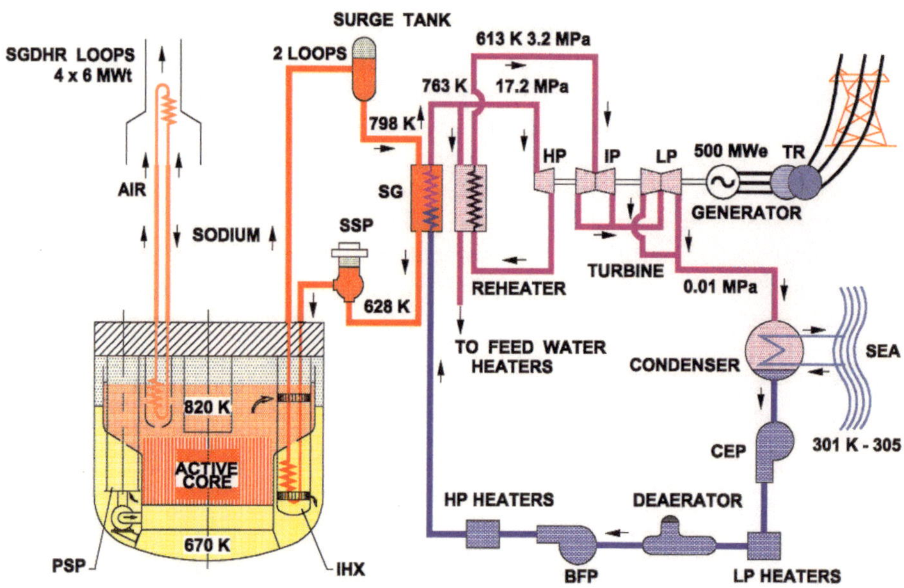

Figure 1.
Schematic of the rector system.

Variable speed AC drives are provided for the two primary and two secondary sodium pumps. The supply to these drives systems for normal operation is fed from the Class IV (Grid) normal AC power supply. When the normal AC power supply fails, the flow requirement during the initial coast down period is provided by the energy stored in the flywheels of the motor for all the four sodium pumps.

Additionally, an AC pony motor (powered by UPS supply) with over-run mechanical clutch is provided over each PSPs to provide forced core cooling during loss of off-site power supply and station black out conditions for 4 h. The over-run mechanical clutch provided disconnects the pony motor from the main drive motor when the speed of the main drive motor is greater than 17% of the rated speed. On loss of Class IV supply, PSPs coast down to 20% speed from 100% speed due to flywheel action in 40 sec and the speed reduces to 15% in 50 sec. The PSPs are provided with hydrostatic bearings and the PSPs are required to be operated at least at 15% speed to prevent damage to bearing [13]. PSPs are not envisaged to be started from rest using Emergency Diesel Generator (EDG) sets due to provisions of pony motor and the pump running under loss of offsite supply.

The power supply to the pony motor is fed from the class III 415 V bus and from a dedicated battery bank along with the associated inverter. The Pony motor is designed to drive the pump at 17% of the rated speed when the power supply to the PSP AC drive motor fails. The dedicated battery bank is designed to supply power to start the pony motor from rest and cater the load of pump lubrication oil system for 4 h. The battery is sized such that the end cell voltage is 1.85 V at the end of 4 h of operation of the pony motor. This battery is charged by battery charger fed from the Class III emergency bus power supply. Each Pony motor is provided with an inverter to convert the DC voltage into AC voltage.

No break AC and DC system power supply derived from Class III Busses. During normal operating condition, the offsite class-IV power supply is extended to battery Class I as well as UPS class-II system requirement through the Class III Emergency bus. However, during loss of offsite class-IV system, the power supply will be derived from class-III system through Emergency DGs. On failure of the offsite and onsite Emergency DG power supply, this class-I and II system will continue to feed its load from the stored energy in battery banks to meet the emergency requirement. During this condition, battery will feed the power supply requirement for bringing the reactor to safe shutdown state, ensuring decay heat removal with pony motor, passive heat removal system, etc., and also monitoring vital core parameters up to 4 h.

A detailed analysis of PFBR after Fukushima incident recommended that during a natural calamity of such higher magnitude, the existing arrangement may not be sufficient to meet the emergency requirement [14]. Hence, additional DGs/mobile DGs were provided to meet the emergency requirement. The reliability of the system will increase further with this provision. A detailed study indicated that the above system may also become insufficient when Additional Diesel Generators (ADG) could not be moved to the desired location as the access roads may not be conducive during floods or tsunami incident.

In view of the above, solar power can also be integrated with the existing ADG to improve the reliability of the system. The hybrid solar and diesel power will improve the reliability of the safety and safety critical systems on BDBE and ensures availability of these systems to mitigate the consequences of such events. These new hybrid systems conserves diesel fuel and reduce CO_2 as well as particulate emissions that are harmful to environment health. This paper also features a detailed analysis of the energy flows through the system. In addition a comparison was made between PV/diesel/battery with diesel/battery and the result shows that the capital cost of a PV/diesel hybrid solution with batteries is nearly three times higher than that of a diesel and battery combination.

2. Existing electrical configuration

Electrical system is one of the major sub-systems of PFBR which comprises normal and emergency power supply systems. Normal power supply is Generator power supply during plant operation and power supply from the grid during Startup of the plant or in case of Generator trip event. Emergency power supply is the onsite power supply to the safety related systems which supports the loads related to plant safe shutdown and to remove the decay heat. The emergency power supply system includes Class III power supply back up provided by Emergency DGs, Class II No break AC power supply and Class I No break DC power supply. In case of normal power supply failure, UPS supplies power to Class II system loads. Rectifier/ Charger with battery backup feeds the loads pertaining to Class I system until Emergency DG power supply is restored [15]. The moment Emergency DG incomer closes, these loads are fed by diesel power supply and the remaining class III safety loads are restored according to the priority. The existing normal and backup sources are shown in **Figure 2**.

Station blackout occurs when the off-site power supply fails and the all four Emergency DGs (on-site power supply) could not be deployed. Under a station blackout conditions, the Class I and Class II power supplies should be available for a minimum period of 30 min to meet the rated loads and to supply the essential loads important to the safety and controls of the reactor for the station blackout duration of 4 h. The Class I and Class II system batteries form the source of power during the station blackout condition.

Following the occurrence of the station blackout, under the reactor shut down condition, the decay heat from the core is removed by the safety-grade decay heat removal circuits. In order to help the removal of the decay heat AC Pony motors are provided for the primary sodium pumps and they are fed from their dedicated batteries and the associated inverters. Detailed analysis has been carried out for the station blackout duration. The probability of occurrence of Station Blackout (SBO) with duration of 4 h is 10^{-4} per reactor year and the probability of occurrence of an SBO with duration of 14 h is 10^{-6} per reactor year. The estimated unavailability of class III power system is 2.4×10^{-3} per reactor year for 2 out of 4 Emergency DG systems and 6.8×10^{-4} per reactor year for 1 out of 4 Emergency DG systems. The dedicated batteries supplying the AC pony motor are rated for a minimum of 4 h duration so that the clad temperature limit is not exceeded. Natural convection is adequate to limit the clad temperature during SBO beyond 4 h. The power supply

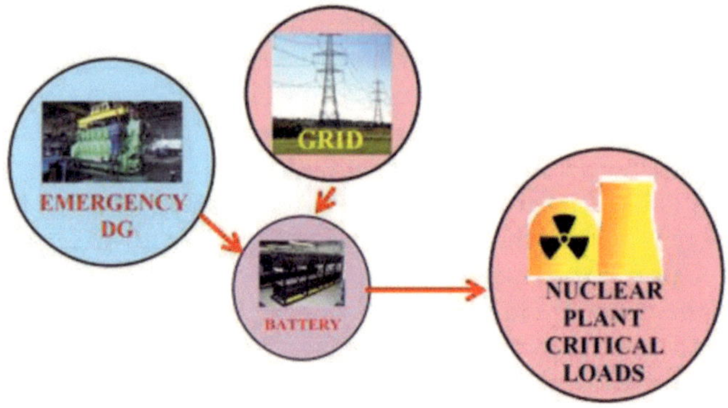

Figure 2.
Existing normal and backup sources for critical loads.

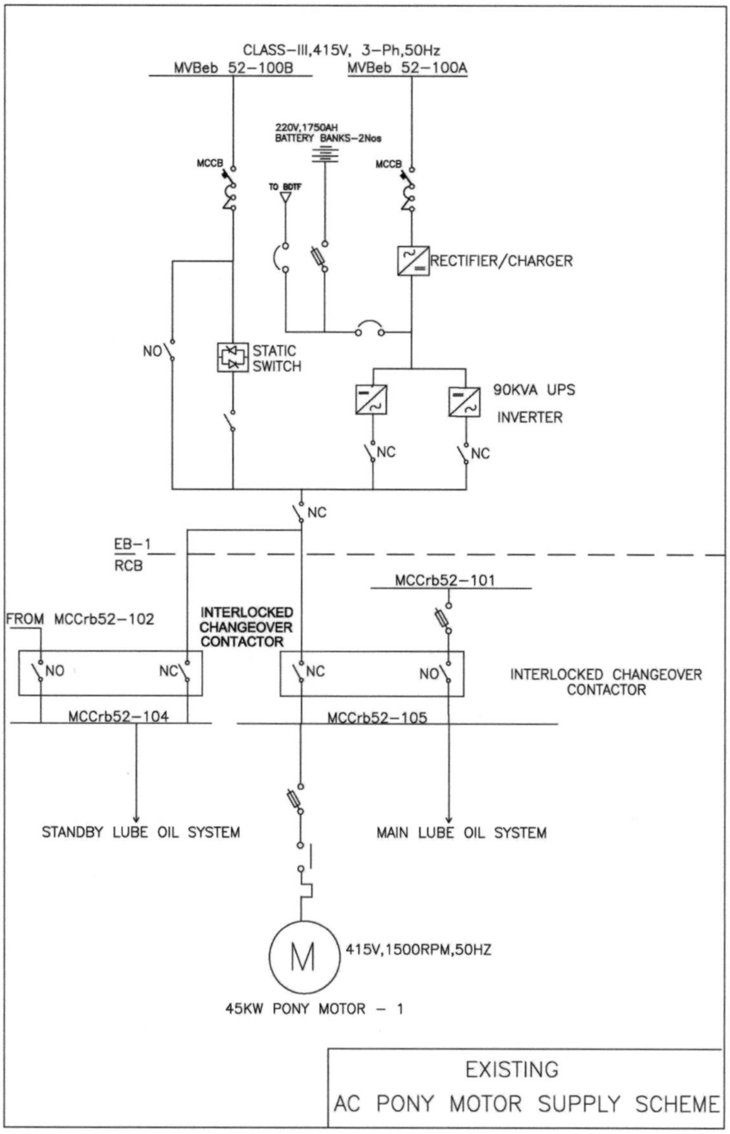

Figure 3.
Existing AC PONY motor power supply schematic.

to the pony motor is fed from two class III 415 V bus and from a dedicated battery bank along with the associated inverter. One class-III power supply is connected to the charger. Another one is connected to the bypass through static switch. In case the rectifier/charger unit is having internal problem, then the second source will feed the power supply to the pony motor [16].

Solid state converters and stationary lead acid batteries are used in the power supply circuit intended to supply power to the AC pony motor and the pump auxiliaries for the lubrication oil system. Each Pony motor is provided with an inverter to convert the DC voltage into AC voltage [17].

Further, power supply to the each main oil and standby lube oil for pony motor is received from one dedicatedclass-III Motor Control Center (MCC) panel and one standby power supply from the dedicated battery system as show in **Figure 3** [18].

3. Proposed electrical configuration

The above arrangement is generally available in existing nuclear power plants. In case of non-availability of Emergency DGs during BDBE (In view of Fukushima incident) leading to extended station blackout conditions, a safety up-gradation of emergency power supply is essential. To meet the requirement an additional two numbers of 500 kVA rating tyre mounted portable ADGs are to be moved to site from the stored location to ensure availability of power to PSP pony motor (45 kW) and monitoring, removal of decay heat from the reactor, Motors associated with Control and Safety Rod Drive Mechanism (CSRDM), Diverse Safety Rod Drive Mechanism (DSRDM) lighting in the main control room, back up control room, switchyard control room and the DG buildings.

The ADGs are planned to be located away from the main plant area, in Emergency Control Center which will be a common facility at Kalpakkam equipped to deal all types of accident/crises/emergencies. These ADGs will be mounted over the seismic pads. When requirement arises, ADGs will be brought to plant area nearby Electrical Building. The ADG is always kept on the Tyre mounted truck at an elevation which is 3.154 m above plant design basis flood level. The tyres will be raised above floor by mechanical jacks during operation. Considering the condition of access route post-accident/natural calamity this ADG will be brought into the plant area from its storage location with the help of Tractor/Hydra/Crane/JCB, etc., within the station blackout period, i.e., 4 h. ADG oil storage tanks are designed to store the fuel for 8 h of continuous operation. This ADG will feed the power to the existing 415 V busses. The power supply provision between the ADG Panel to the existing 415 V busses are permanently made available. Cables are to be laid from ADG and ADG panel when the ADGs are moved. Considering the design safety limits for driver fuel clad hotspot temperature, adequate capacity battery backup is provided to ensure effective decay heat removal [19]. The minimum coping time of 4 h is recommended considering of the combination of events. The minimum required lighting for safe movement of ADGs to locations near Electrical Building like alternate Street lights are temporary powered. The portable battery operated torches will also be pressed into service.

A solar power unit having PV cells mounted at the top of Electrical building-1, Electrical Building-2 and Control building is connected to the Pony motor Battery

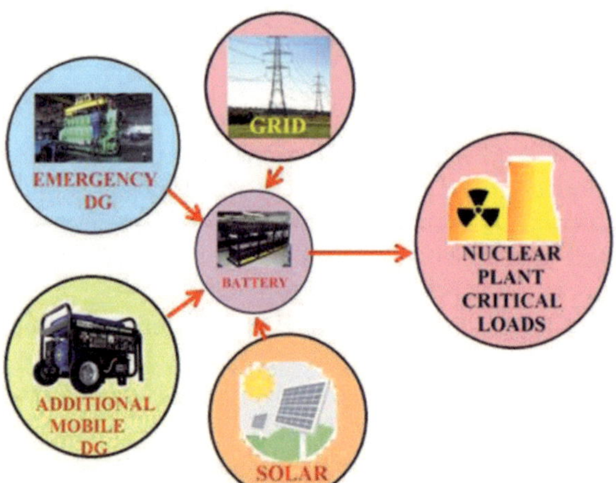

Figure 4.
Proposed backup and Normal sources for critical loads.

banks through a DC-DC Converter with surge protective devices [20]. The DC-DC converter is equipped with in an Auto synchronization facility which is provided for pumping power during day time/when solar radiation is available [21]. The synchronization is done at DC side instead of connecting at the incomer AC supply side by converting PV-DC supply to AC supply by an additional Inverter [22]. This arrangement complicates the system and an extra device reduces the reliability. This arrangement gives extra reliable power supply to the decay heat removal mission. The proposed additional mobile DGs (ADG) and Solar with the existing power sources are shown in **Figures 4** and **5**.

In addition to existing, two more redundant power supply provision are made to increase the reliability of the system. The ADGs power supply fed to the existing 415 V class-III bus through Switch Fuse Unit (SFU). A Solar power is also

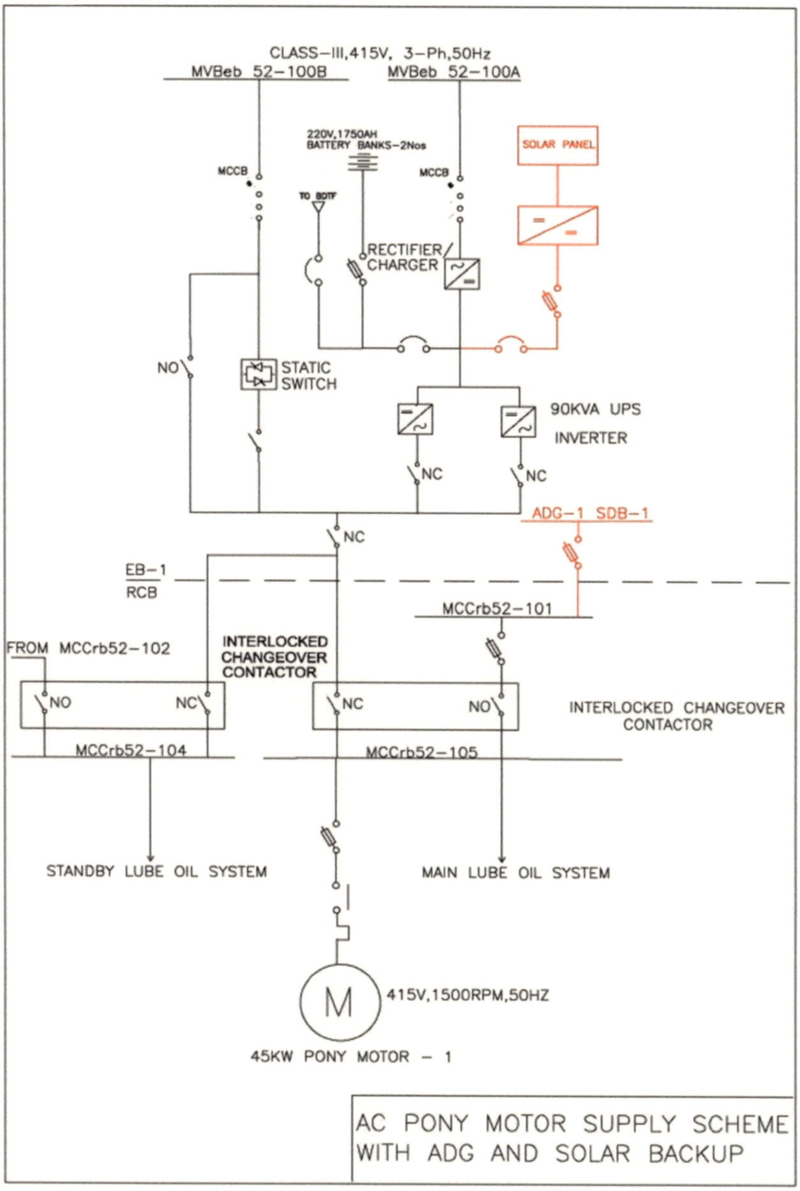

Figure 5.
Proposed AC PONY motor power supply schematic.

introduced in the proposed power supply schematic of pony motor. The solar power is directly connected to the existing battery system to charge the battery through SFU and Miniature Circuit Breaker (MCB) [23]. The solar power is will directly charge the battery as well as deliver the load through the inverter. The proposed electrical power schematic is shown in **Figure 5**.

4. Design basis external events of nuclear reactor

Nuclear reactors are prone to get affected due to various events as indicated in the **Figure 6**. The major event to affect the reactor is power failure, earthquake, flood and lightning [24]. During all the events power failure is one of the main causes to affect the reactor safety system. Hence to overcome the above issues in the modern world, integrated power supply to be arranged for reactor cooling systems [25]. In line with the above, here the solar power is integrated along with the existing additional DGs to increase the reliability of the nuclear safety [26].

In the existing system, during a BDBE, the Emergency Diesel Generator may fail and the ADGs will be lined up to feed power supply to the pony motors. If ADGs could not be shifted to the desired location within 4 h, the dedicated battery banks gets drained and there is no further backup for reactor core cooling. Hence, the proposed system will continue to feed power supply to the battery banks and the reactor safety will be ensured during beyond design basis event also.

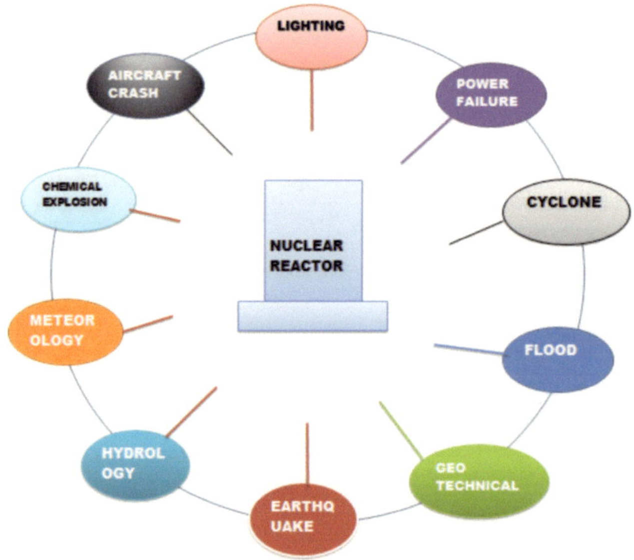

Figure 6.
Design basis external events of nuclear reactor.

5. Reliability of existing power supply scheme

Pony motor is supplied with 415 V dedicated 90 kVA UPS and also a Class III 415 V supply from emergency bus as seen earlier. Failure rate of the Pony Motor on demand has been estimated by computational methods using software called ISOGRAPH Reliability Work Bench 2008. The failure of each power supply train and its probable causes have been analyzed separately to arrive at the overall failure rate of pony motor. In the existing scheme, the UPS failure includes failure of rectifier path and battery path. With two sources of power supply failure to UPS taken for the analysis, the failure rate of Pony Motor UPS is found to be 3.85 × 10^{-5} [27]. This brings the overall failure rate of the Pony motor to 0.0149 as shown in **Figure 7** [28].

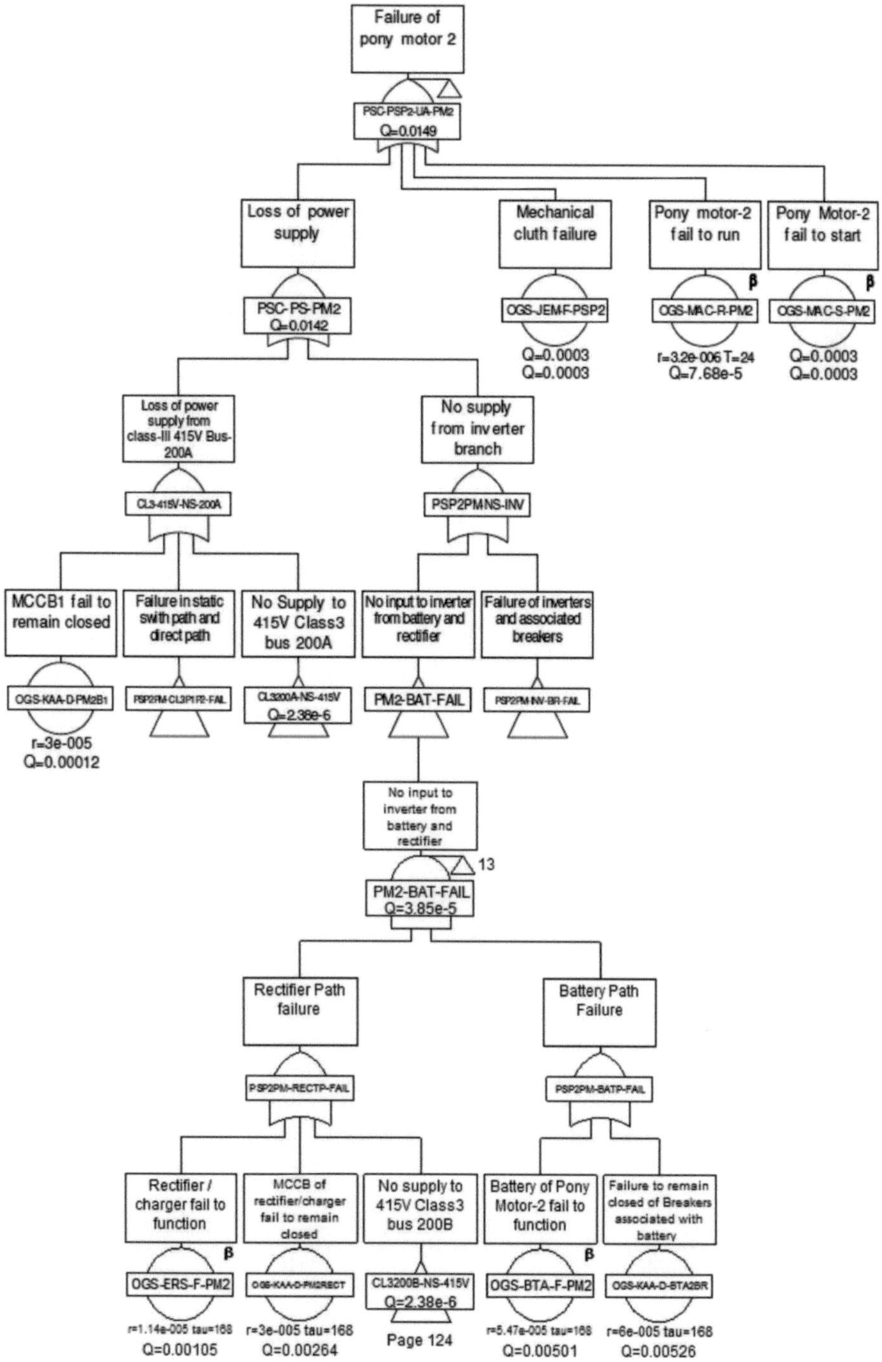

Figure 7.
Fault tree for failure of pony motor with existing power supply scheme.

6. Reliability of proposed power supply scheme

With inclusion of solar power in the existing scheme, the failure rate has been analyzed [29]. The failure rate of DC to DC converter used in solar power supply unit is 1.5×10^{-6} and for the associated components is 1×10^{-4}. The DC to DC output of solar power unit is connected before the pony motor battery inverter circuit. With this additional power source and the existing two power supply paths, the

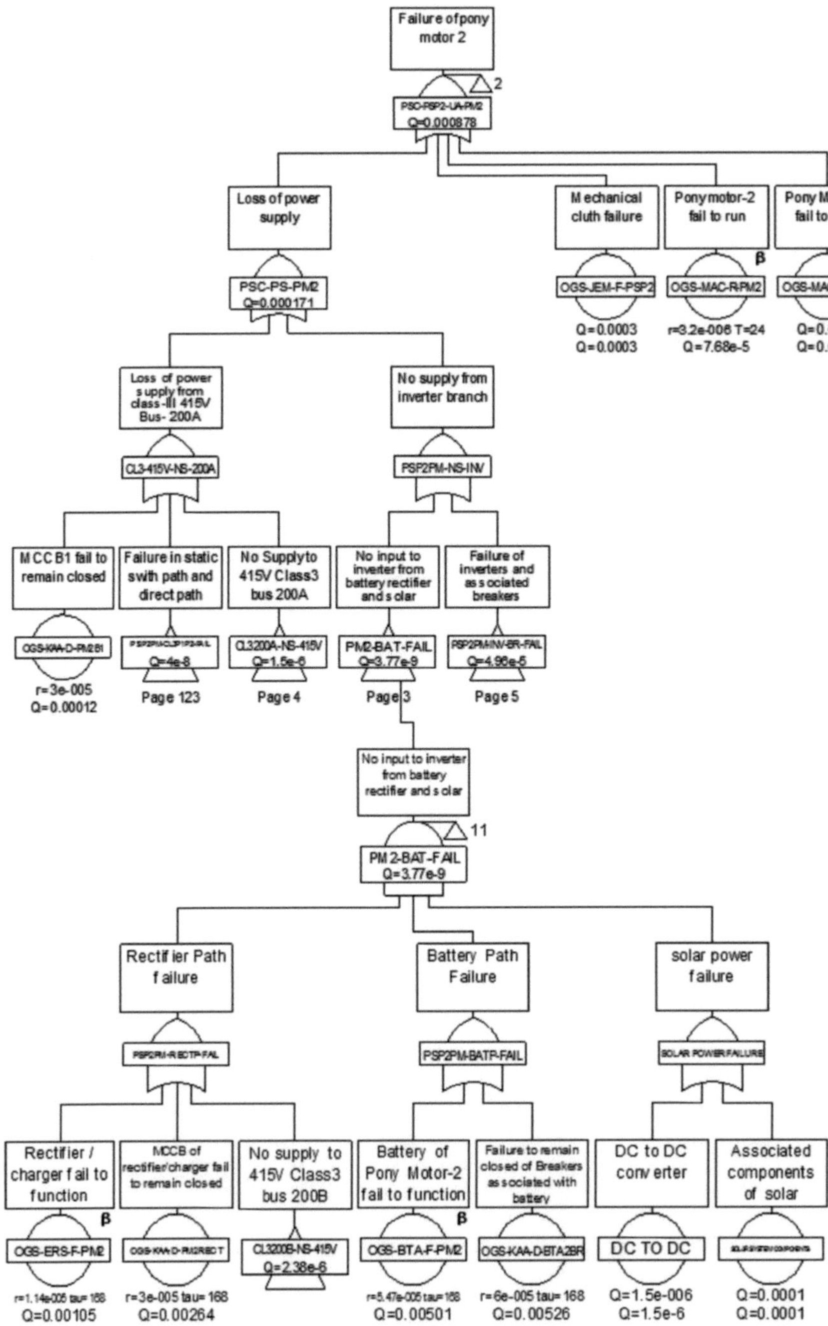

Figure 8.
Fault tree for failure of pony motor with proposed power supply scheme.

overall failure rate of the pony motor UPS is found to be 3.77×10^{-9}. This brings the overall failure rate of pony motor to 0.000878 as shown in **Figure 8** [30, 31].

7. Results and discussions

The failure rate with addition of solar power supply unit to the Pony Motor UPS system reduces the failure rate of the Pony Motor by several decades, i.e., from 0.0149 to 0.000878. With this new addition of solar power to the existing scheme, the availability of the Pony Motor is increased.

The prevailing nuclear power plants are having Grid power supply, Emergency Diesel Generators, UPS AC supply, Battery backup DC supply and ADGs for, safe shutdown and decay heat removal mission of the reactor. A complete loss of power supply for reactor cooling system was considered as BDBE for which the frequency occurrence is very remote and was neglected before Fukushima incident. However, the Fukushima incident has given a lesson to all the nuclear operators across the globe that the plant should be equipped to handle even the BDBE situations also. Hence, to overcome the above issues World Association for Nuclear Operators (WANO) has recommended that all nuclear power plant should have Emergency DGs along with an additional mobile DGs (ADG) for emergency situations. Further, to strengthen the backup power supply and to overcome the beyond design basis event solar power can be integrated with existing arrangement. This hybrid solar power will increase the reliability of the system and will reduce the non-availability of power supply failure during BDBE. The proposed arrangement reliability was also analyzed through software and found that it is increasing the reliability of the existing set up. A tropical country like India has solar radiation for 10 out of 12 months. Hence, the solar power will be used as backup power for nuclear power plant.

8. Conclusion

The prevailing nuclear power plants are having grid power supply, emergency diesel generators, UPS AC supply, battery backup DC supply, and ADGs for safe shutdown and decay heat removal mission of the reactor. A complete loss of power supply for reactor cooling system was considered as BDBE for which the frequency occurrence is very remote and was neglected before Fukushima incident. However, the Fukushima incident has given a lesson to all the nuclear operators across the globe that the plant should be equipped to handle even the BDBE situations. Hence, to overcome the above extreme events, the World Association for Nuclear Operators (WANO) has recommended that all nuclear power plant should have emergency DGs along with an additional mobile DGs (ADG) for emergency situations. Further, to strengthen the backup power supply and to overcome the beyond design basis event, solar power can be integrated with existing arrangement. This hybrid solar power will increase the reliability of the system and will reduce the nonavailability of power supply failure during BDBE.

The proposed arrangement reliability was also analyzed through software and found that it is increasing the reliability of the existing setup. A tropical country like India has solar radiation for 10 out of 12 months. Hence, the solar power will be used as a backup power for nuclear power plants. The failure rate with addition of solar power supply unit to the Pony Motor UPS system reduces the failure rate of the Pony Motor by several decades, i.e., from 0.0149 to 0.000878. With this new addition of solar power to the existing scheme, the availability of the Pony Motor is increased. The hybrid solar power supply system utilized in nuclear reactors is highly reliable to the reactor safe shutdown system during day time emergency requirement. However, during night time, the stored power supply from the batteries will cater the essential loads in discharge mode. Onset of the solar power batteries will get charged again. The batteries shall be sized to store enough power to take care of the night time requirement. In addition to this, it is proposed to integrate emergency DG, additional mobile DG (ADG), and solar power with wind power for the future nuclear reactors which may increase the reliability further thereby ensuring the plant is capable of handling any BDBE that occurred in Fukushima—Daiichi.

Author details

Kudiyarasan Swamynathan*, P. Sivakumar and K. Karthikeyan
Bharatiya Nabhikiya Vidyut Nigam Limited, Department of Atomic Energy, Kalpakkam, Tamilnadu, India

*Address all correspondence to: kudiyarasan@rediffmail.com

References

[1] Taner T. Energy and exergy analyse of PEM fuel cell: A case study of modelling and simulations. 2018;**143**:284-294. DOI: 10.1016/j.energy.2017.10.102

[2] Taner T, Sivrioglu M. A techno-economic and cost analysis of a turbine power plant: A case study of sugar plant. Renewable and Sustainable Energy Reviews. 2017;**78**:722-730. DOI: 10.1016/j.rser.2017.04.104

[3] Taner T. Optimisation processes of energy efficiency for a drying plant: A case of study for Turkey. Applied Thermal Engineering. 2015;**80**:247-260. DOI: 10.1016/j.applthermaleng.2015.01.076

[4] Taner T, Sivrioglu M. Energy-exergy analysis and optimisation of a model sugar factory in Turkey. Energy. 2015;**93**(1):641-654. DOI: 10.1016/j.energy.2015.09.007

[5] Taner T, Sivrioglu M. A techno-economic and cost analysis of a turbine plant: A case study for sugar plant. Renewable and Sustainable Energy Reviews. 2017;**78**:722-730. DOI: 10.1016/j.rser.2017.04.104

[6] Topal H, Taner T, Altıncı Y, Amirabedin E. Application of trigeneration with direct co-combustion of poultry 1 waste and coal: A case study in the poultry industry from Turkey. Thermal Science. 2017 (In Press). DOI: 10.2298/TSCI170210137T

[7] Topal H, Taner T, Naqvi SA, Altınsoy Y, Amirabedin E, Ozkaymak M. Exergy analysis of a circulating fluidized bed power plant co-firing with olive pits: A case study of power plant in Turkey. Energy. 2017;**140**:40-46. DOI: 10.1016/j.energy.2017.08.042

[8] Taner T. Economic analysis of a wind power: A case study for the Cappadocia region. Journal of Mechanical Science and Technology;**32**(3):1379-1389. DOI: 10.1007/s12206-018-0241-6

[9] Taner T, Demirci OK. Energy and economic analysis of the wind turbine Plant's drafts for the Aksaray city. Applied Ecology and Environmental Sciences. 2014;**2**(3):82-85. DOI: 10.12691/aees-2-3-2

[10] Taner T, Sivrioglu M, Topal H, Wongwises S. A model of energy management analysis, case study of sugar factory in Tukery. Sadhana-Academy Proceedings in Engineering Sciences. 2018;**43**(3):1-20. DOI: 10.1007/s12046-018-0793-2

[11] Taner T, Demirci OK. Energy and economic analysis of the wind turbine plants of the Aksaray city. Applied Ecology and Environmental Sciences. 2014;**2**(3):82-85. DOI: 10.12691/aees-2-3-2

[12] Raghupathy S, Singh OP, Govindarajan S, Chetal SC, Bhoje SB. Design of 500 Mwe prototype fast breeder reactor. Indira Gandhi Centre for Atomic Research; http://www.efn-uk.org/nuclear/nuc-lib/reactor-reports/index_files/India-PFBR-design.pdf

[13] Bhoje SB, Chetal SC, Chellapandi P, Govindarajan S, Lee SM, Kameswara R, et al. Impact of LMFBR Operating Experience on PFBR Design. Kalpakkam, India: Indira Gandhi Centre for Atomic Research; 2000. https://inis.iaea.org/collection/NCLCollectionStore/_Public/31/058/31058490.pdf

[14] Sundaram CV, Lee SM. Physics and engineering aspects of fast reactor safety. Pramana. 1985;**24**(1-2):193-210

[15] Prasad M, Theivarajan N. Normal power supply system of a nuclear power plant-modelling and simulation studies for fast bus transfer. In: IEEE, 1st International Conference on Electrical Energy Systems. Newport Beach, CA; 2011:294-299

[16] Raja K, Theivarajan N. Emergency power supply system of a nuclear power plant-modelling and simulation studies of diesel generators and load pickup on emergency transfer. In: IEEE, 1st International Conference on Electrical Energy Systems. Newport Beach, CA; 2011:302-307

[17] SubbaRaju B, VijayaKumar K, Seetha H, Jayanthi T, Satya Murty SAV. Modeling and simulation of emergency power supply system and emergency transfer scheme for PFBR operator training simulator. International Research Journal of Engineering and Technology. 2015;**02**(03):1600-1605

[18] Nawlakha R, Jasmine N, Subba Raju B, Seetha H, Jayanthi T, SatyaMurty SAV. Modeling and simulation of backup control room for PFBR operator training simulator. International Journal of Energy Engineering. 2012;**2**(4):125-130

[19] Kim J, Jae M. A study on reliability assessment of a decay heat removal system for a sodium-cooled fast reactor. Journal of Nuclear Science and Technology. 2018:534-539

[20] Yang Y, Sangwongwanich A, Blaabjerg F. Design for Reliability of power electronics for grid-connected photovoltaic systems. CPSS Transactions on Power Electronics and Applications. 2016;**1**(1):92-103

[21] Inhaber H. Is solar power more dangerous than nuclear. IAEA Bulletin;**21**(1):11-17

[22] El-Metwally M, EL-Shimy M, Elshahed M, Sayed A. Detailed Analyses of the Failure and Repair Rates of Wind and Solar-PV Systems for RAM Assessment. ICEENG-11; 2017. DOI: 10.13140/RG.2.2.18530.84160

[23] Popov D, Borissova A. Innovative configuration of a hybrid nuclear-solar tower power plant. Energy. 2017;**125**:736-746

[24] Nourbakhsh HP. Dealing with beyond-design-basis accidents in nuclear safety decisions. Presented at 12th International Probabilistic Safety Assessment and Management (PSAM 12) Conference. Honolulu, Hawaii, USA; 22-27 June 2014

[25] Shang Z, Zhu J, Zheng M. Considerations at beyond design-basis phenomenon design for new nuclear power plant seismic and Tsunami. Transactions, Smirt-22 San Francisco, California, USA, Division IV; 2013

[26] Merk B, Stanculescu A, Chellapandi P, Robert Hill R. Progress in reliability of fast reactor operation and new trends to increased inherent safety. Applied Energy. 2015;**147**:104-116

[27] Adail AS, Hassan HM, Shaat MK. Safety and reliability of the electrical system based on optimal FACTS allocation for a research reactor. Progress in Nuclear Energy. 2018;**104**:143-149

[28] Zeliang C, Singh OP, Munshi P. Uncertainty evaluation of reliability of shutdown system of a medium size fast breeder reactor. Nuclear Engineering and Design. 2016;**308**:283-296

[29] Sharma PK, Bhuvana V, Ramakrishnan M. Reliability analysis of diesel generator power supply system of prototype fast breeder reactor. Nuclear Engineering and Design. 2016;**310**:192-204

[30] Arul J, Senthil Kumar C, Athmalingam S, Singh OP, Rao KS. Reliability analysis of safety grade decay heat removal system of Indian prototype fast breeder reactor. Annals of Nuclear Energy. 2006;**33**:180-188

[31] Ramakrishnan M. Integration of functional reliability analysis and system hardware reliability through Monte Carlo simulation. Annals of Nuclear Energy. 2016;**95**:54-63

Chapter 4

Recent Advances in Photovoltaic-Trombe Wall System: A Review

Omer K. Ahmed

Abstract

Management of energy consumption for building's air conditioning is a vital issue for resource saving and environmental protection. The use of solar energy to generate electricity by solar cells is essential nowadays. However, the disadvantage of solar panels is the elevated temperature in work, especially in the hot sunny climate that leads to efficiency decline. Also, there is a problem with heating during the night and cloudy days. For the last 20 years, there has been a rapid development in the field of integrated solar technologies. A hybrid PV/Trombe wall (PV/TW) system suggested being an efficient and durable conversion system of solar energy. The design of the PV/TW system considered one of the focusing areas of the present research to make it more economically feasible. The idea of building the photovoltaic-Trombe wall has appeared as one of the green technologies. Several published works at that time are included for integrating PV/TW system. This chapter devoted to reviewing the theoretical and practical studies conducted on this system for developing and improving electrical and thermal performance.

Keywords: PV/Trombe wall, efficiency, building, review

1. Introduction

Today mankind is facing the problem of severe shortage of power supply as a result of current consumption rate and low conventional energy reserves [1, 2]. Increased human activity has led to increased energy consumption and reduced conventional fuel reserves to dangerous levels, leading to the high price of world oil [3]. Scientists and researchers have begun to look for new sources of energy to replace conventional sources and replace it with a clean and environmentally friendly option [4]. Solar energy has been a priority for these concerns because it is cheap, is nonpolluting, and does not require high technology [5].

The passive solar heating system is a technique that generates thermal energy by collecting solar radiation and then moving it into the building naturally [6]. It is one of the most economical ways for solar energy utilization, which can reduce yearly heating demand by about 25% [7]. Among the passive solar heating systems that have been developed are solar roofs, solar chimneys, and a Trombe wall [8].

The Trombe wall is a method to use solar energy for heating without using any mechanical or electrical assistance, as shown in **Figure 1** [9]. The name of the Trombe wall was taken from a French scientist named Michel Trombe, who discovered in 1880 [10]. It is considered a simple and inexpensive solution that uses solar energy in heating, which reduces heat load up to 47% [11]. A Trombe wall is always located on the southern face of buildings in the northern hemisphere to increase solar energy throughout the year. It consists of a wall of concrete or bricks for heat storage, and the wall is coated with black paint to increase the absorption of solar radiation. There are layers of glass at 5–10 cm distance from the concrete wall to increase the intensity of solar radiation. A Trombe wall is covered in the summer to prevent excessive warming [12]. The wall has two openings at the top and bottom to circulate the air. Solar heat is transferred from the Trombe wall into the room by convection heat transfer and conduction. Cold air passes from the lower opening of the wall, then the air is heated through the channel and moves into the room through the upper vent. Meanwhile, heat transfer through the thermal wall needs more time to move into the room. Although a commercial PV system is widely available, however, further research and development are necessary to improve efficiency and reduce costs. Passive solar techniques can reduce yearly heating demand by about 25%. Different architectural devices such as Trombe walls, solar roofs, solar chimneys, and others are used in buildings [13]. In the last years, modern technologies introduced that use solar cells to generate electricity. The photovoltaic cell can be integrated with buildings such as the PV-Trombe wall [14]. The increase in the solar cell temperature leads to lower efficiency as shown in **Figure 2**.

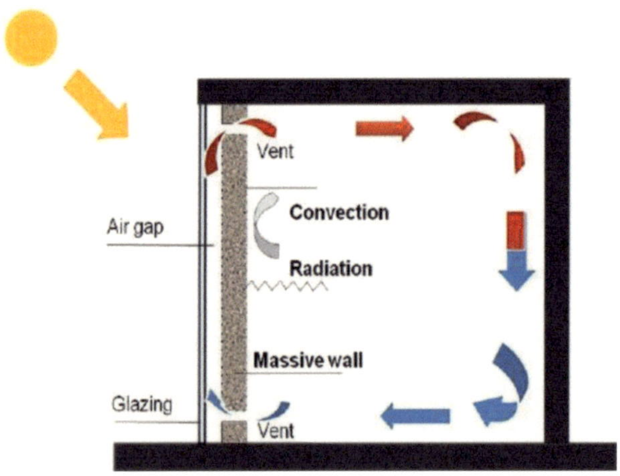

Figure 1.
Classical Trombe wall.

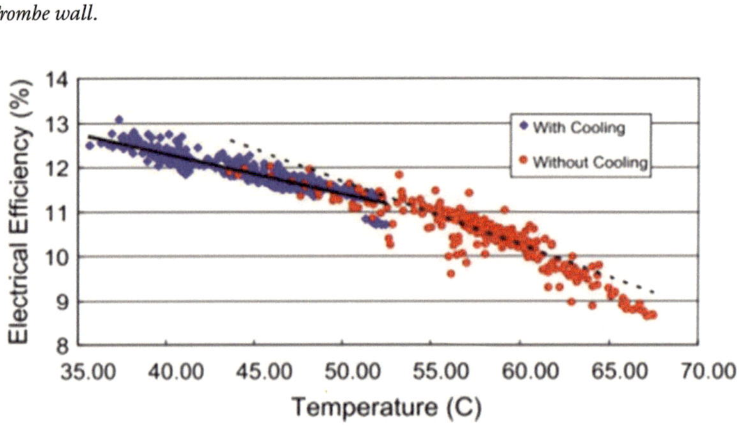

Figure 2.
Electrical efficiency as a function of solar cell temperature.

2. Building-integrated photovoltaic/thermal systems

In early 1990s, the idea of building-integrated photovoltaic/thermal (BIPV/T) systems emerged. It has attracted increasing attention since 2000 because of its potential to facilitate the design of net-zero energy buildings by improving solar energy use [15]. Generally, a BIPV/T system has the following features:

1. BIPV/T system is physically attached to buildings.

2. The system produces electricity.

3. The system provides thermal energy ready to be collected and used by the building.

Figure 3 shows the PV modules installed on the roof of a room with an air duct at an angle of 34° to the horizontal, which corresponds to the latitude of Srinagar, India.

Chow et al. [16] presented a comparative study of three different parameters for BIPV technology in China. The results showed no significant difference in electricity production as the indoor space with 24-hour temperature supplied continuous cooling of the solar cells through the outer façade. For semi-equatorial climate, PV/C and PV/T systems are better than the BIPV system because of their ability to limit the increase of space temperature. The PV/C system considered a better option because of its effectiveness in reducing the cooling power of elimination and has a simple design.

Hu et al. [17] compared three types of systems [i.e., solar cells affixed to glass (BIPVGTW), solar cell attached to the mass wall (BIPVMTW), and PV-blind-integrated Trombe wall system (BIPVBTW)] as illustrated in **Figure 4**, and the results showed that the electrical performance of BIPVGTW was better and the system adopted electrifying performance at the blindness angle. In terms of thermal performance, the BIPVBTW system increased room temperature during the winter season higher than other systems. The BIPVGTW system is better than BIPVBTW and BIPVMTW systems.

Nagano et al. [18] studied the experimental thermal/PV hybrid exterior wallboards that incorporate PV cells, as shown in **Figure 5**. The clapboard-shaped hybrid wallboards permit modular assembly that is more suitable for building applications than former PV systems. Solar heat collected in the form of heated air

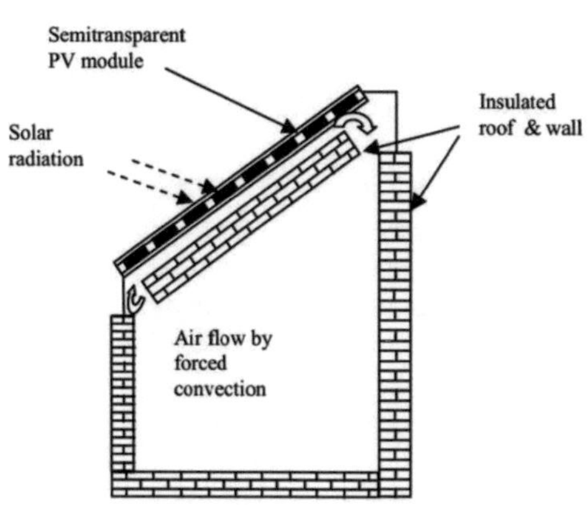

Figure 3.
Photovoltaic-thermal system integrated on the roof.

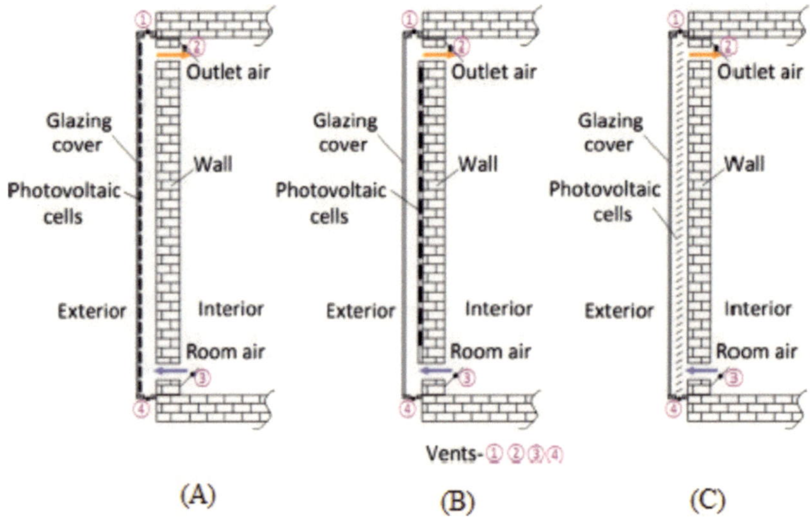

Figure 4.
Three models of BIPV Trombe wall systems. (A) BIPVGTW. (B) BIPVMTW. (C) BIPVBTW.

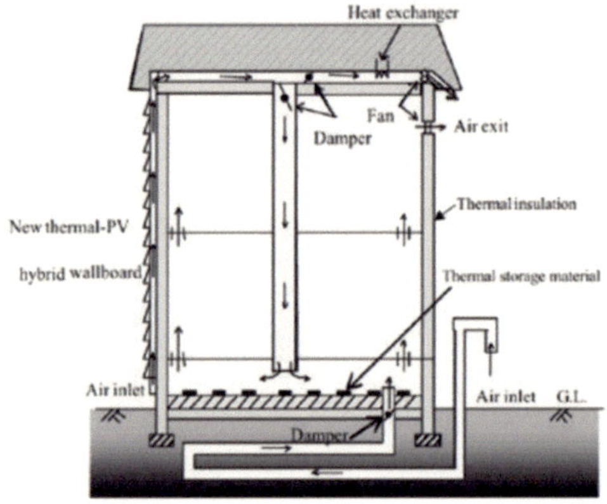

Figure 5.
The concept of a new type of hybrid wallboard.

circulating in the air duct between the thermal insulation of the exterior walls and the composite wallboard.

3. Photovoltaic-Trombe wall (PV/TW)

For the last 20 years, there has been a rapid development in the field of integrated solar technologies. The idea of building the photovoltaic-Trombe wall has appeared as one of the green technologies. The concept is to use solar cells with a Trombe wall to generate electricity as well as heating. The photovoltaic/thermal (PV/T) systems convert solar radiation into heat and electricity together, such as PV-Trombe wall (PV/TW) system.

The design of the PV/TW system is like the original design of the classical Trombe wall, just replacing the thermal wall by a solar cell, as shown in **Figure 6** [19]. They are less expensive, environmentally friendly, and more suitable for a given house or building, and it is more efficient than the separate solar thermal

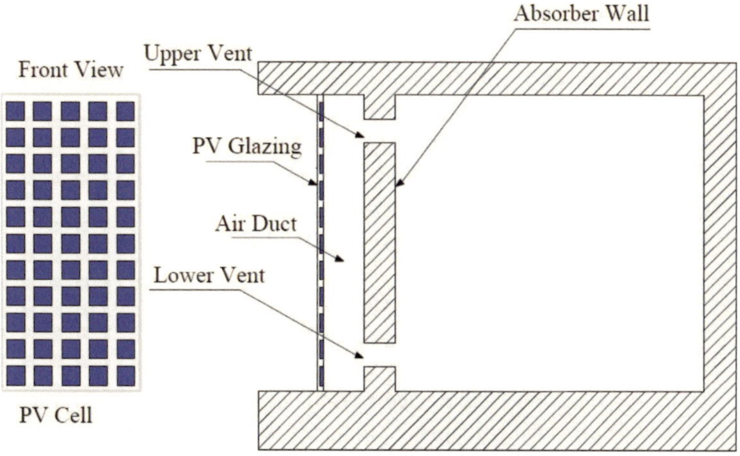

Figure 6.
Photovoltaic-Trombe wall.

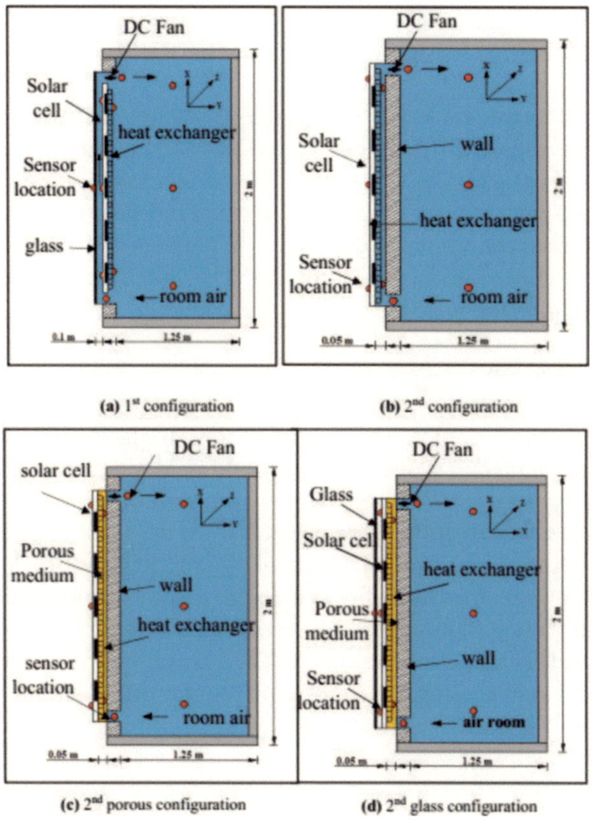

Figure 7.
Modified Photovoltaic-Trombe walls.

and electrical systems. Furthermore, the PV/TW system can contribute to the reduction in the consumption of fossil fuels for large combined production [20].

Several modifications were performed to improve the performance of PV/TW. Verifying the thermal and electrical performance of the PV/TW system has been done by Ahmed et al. [21, 22]. The base configurations were modified to improve the performance of the PV/TW system. These modifications are inserting the DC fan, heat exchanger, porous medium, and glass cover. Incorporating the DC fan and bi fluid (air and water) cooling circuit offered desirable features of the system performance. This study dealt with the design of two different configurations PV/TW system, as shown in **Figure 7** as follows:

1. The first configuration consists of a glass cover on the front, then the air duct, and the solar cell panel attached to the massive wall, as shown in **Figure 7a**.

2. The second configuration includes the solar cell panel in the front, then the air duct, and the massive wall, as shown in **Figure 7b**.

Also, various suggestions were performed to improve system performance. These modifications for both configurations are inserting a cooling coil and DC fan. Moreover, the second configuration, the porous medium, was added inside the gap, as shown in **Figure 7c**, and a glass cover was attached to the front of the solar cell, as shown in **Figure 7d**.

4. Variables affecting the performance of the PV/Trombe wall

PV/TW considers favorable architectural technology and utilizes solar energy for cooling and heating in different climate areas. Many factors affect the efficiency of PV/TW, and these factors should be considered when designing this system in buildings. These factors include operation parameters (mass flow rate, cooling, partial covering, southern window, air duct, and tilt angle of the solar cell) and design parameters [properties of glass, direct current (DC) fan, channel depth, thermal insulation, packing factor, and materials].

4.1 Effect of glass cover

During the design of PV/TW, the characteristics of glass, such as the number of glass, glazing thickness, and types of glass (single glass, double glass, and double glass filled with argon), significantly affect the amount of solar radiation absorbed and transmitted, as well as the heat transfer between the interior and the ambient. Irshad et al. [23] presented a simulated model of a room with PV/TW. The performance evaluated by varying the airflow velocity for three different PV/TW glazing types (i.e., single glazing, double glazing, and double glazing filled with argon). The results showed that double-glazing PV/TW loaded with argon demonstrated a significant reduction in cooling load and room temperature, while PV productivity increased.

4.2 Effect of DC fan

Jie et al. [14] introduced a novel system for PV/TW-solar cell connected to a DC fan and compared the performance of the system with a normal Trombe wall-solar cell, as shown in **Figure 8**. The results showed a significant increase in room temperature compared to the original room, as well as an increase in the electrical efficiency of the solar cell due to improved cooling. Yi et al. [24] presented a simulated system of novel PV/TW with a DC fan. The results showed that as fan speed increased, the electric and thermal efficiencies also increased.

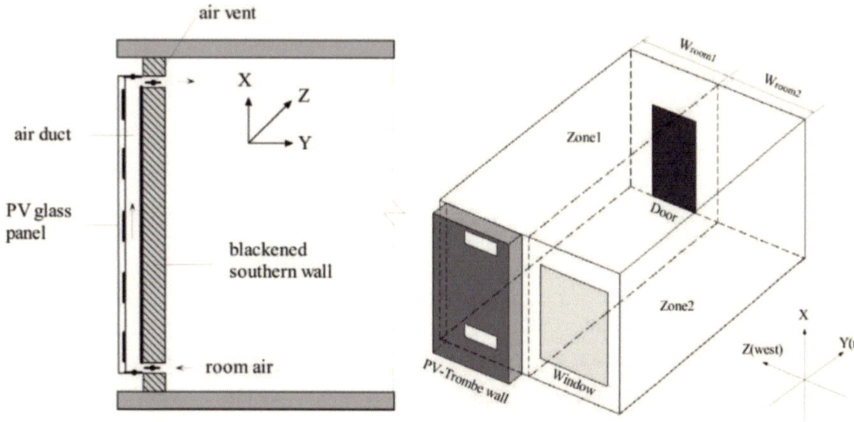

Figure 8.
Modified Photovoltaic-Trombe walls.

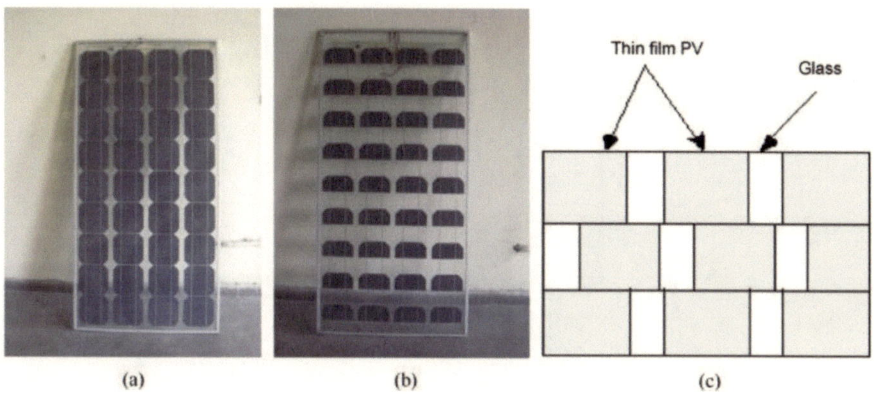

Figure 9.
(a) PV module with a cell area of 0.0139 m², (b) PV module with a cell area of 0.0069 m², and (c) the proposed design of thin-film PV with glass.

4.3 Effect of channel depth

The influence of channel depth, which represents the distance between a glass cover and a solar cell, was investigated. Ji et al. [25] studied the electrical and thermal performance of the PV/TW system utilized in Tibetan residential buildings, as shown in **Figure 8**. It was found that when the width of the PV/TW system increased, the room temperature also increased, although electrical effectiveness was almost constant.

Changing the air gap and different types of PV glazing (i.e., double glazing, single glazing, and double glazing filled with argon) was studied by Irshad et al. [26]. The results affirmed that double glazing PV/TW system filled with argon gives the maximum PV efficiency at a roof pitch angle of 20° and an air gap of 0.2 m. Ventilated PV/TW system and PV cell installed through the roof also reduced the cooling load of the room.

4.4 Effect of packing factor

Published studies have shown that reducing the packing factor leads to an increase in room temperature, as mentioned by Vats et al. [27]. The packing factor of a solar cell was dependent on changing the area of PV cells in a given area of the solar cell, as shown in **Figure 9**. The reduction in the temperature of the PV module was observed due to the reduction of the packing factor that enhances electrical effectiveness.

4.5 Effect of the coverage ratio

The coverage ratio of PV glazing is defined as:

$$e = A_{pv} / A_{glass} \tag{1}$$

where A_{pv} is the area of PV cells, and A_{glass} is the area of PV glazing. Jiang et al. [28] concluded that with the increase of coverage ratio, the thermal efficiency

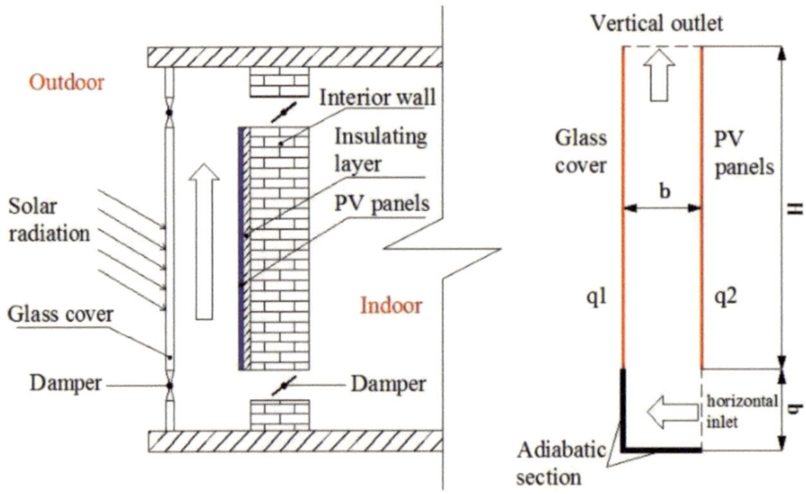

Figure 10.
Physical and simplified models of BIPV-TW.

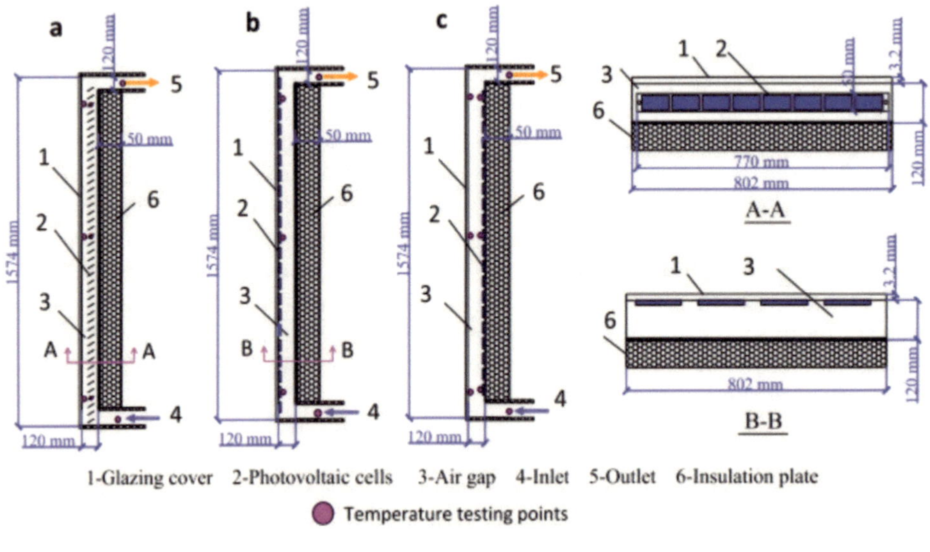

Figure 11.
PV-Trombe wall modules of (a) PV-BTW, (b) PV-GTW, and (c) PV-MTW.

and room temperature of the PV-TW decreased, but the electrical power and the total efficiency of PV/TW increased. Besides, the electrical efficiency of solar cells decreased, but the effect of the coverage ratio of electrical efficiency was small, less than 0.5%. If the glass were entirely covered with PVC, it would reduce solar radiation and might not be sufficient to provide heating. As a result, the air temperature in the channel was less than room temperature [29].

4.6 Effect of mass flow rate

Xu and Su [30] carried out a numerical simulation of airflow in BIPV-TW, as shown in **Figure 10**. The results demonstrated that the airflow rate increased linearly with duct height. Also, Hu et al. [31] investigated experimentally three models of PV/TW, as shown in **Figure 11**, i.e., PV-BTW, PV-GTW, and PV-MTW at the University of Science and Technology of China. The results showed that the angle of 50° and the entry air flow rate of approximately 0.45 m/s were the best parameters for the PV-BTW type. The results also showed that the PV-GTW model produced the highest electricity output compared to PV-MTW and PV-BTW.

5. Conclusions and recommendations

Previous works have shown that researchers are interested in finding a way to use solar energy as a substitute, useful, and clean energy. The researchers concluded the importance of studying design and operational information that led to an increase in the electrical and thermal efficiency of the PV/TW system. The following conclusions can be extracted from the previous studied:

1. The use of water to cool the solar cell reduces the temperature of the solar cell, which increases the electrical power produced that increases the efficiency of electricity generation. Also, it increases the overall thermal efficiency but reduces thermal air efficiency.

2. The use of a DC fan increases the produced electrical power and room temperature, which increases thermal and electrical efficiency.

3. The use of glass cover increases the temperature of the solar cell, which decreases the power and efficiency of the electricity generation. Also, the use of glass cover increases the room temperature and thermal efficiency.

4. The use of a porous medium increases the area of heat transfer, which increases the thermal and electrical efficiency and temperature of the room.

5. Thermal and electrical efficiencies have decreased in a dusty and cloudy climate due to a decrease in the amount of solar radiation reaching the solar cell of the PV/TW system.

Author details

Omer K. Ahmed
Northern Technical University, Kirkuk, Iraq

*Address all correspondence to: omerkalil@yahoo.com

IntechOpen

References

[1] Ahmed O, Mohammed Z. Influence of porous media on the performance of hybrid PV/thermal collector. Renewable Energy. 2017;**112**:378-387. DOI: 10.1016/j.renene.2017.05.061

[2] Taner T, Kürşat Demirci O. Energy and economic analysis of the wind turbine plant's draft for the Aksaray City. Applied Ecology and Environmental Sciences. 2014;**2**(3):82-85. DOI: 10.12691/aees-2-3-2

[3] Ahmed O, Hamada K, Salih A. A state of the art review of PV-Trombe wall system: Design and applications. Environmental Progress & Sustainable Energy. 2019:1-16. DOI: 10.1002/ep.13370

[4] Ahmed O, Bawa S. Reflective mirrors effect on the performance of the hybrid PV/thermal water collector. Energy for Sustainable Development. 2018;**43**:235-246. DOI: 10.1016/j.esd.2018.02.001

[5] Hussien A, Ahmed O. Assessment of the performance for a hybrid PV/solar chimney. International Journal of Engineering & Technology. 2018;**7**:114-120. DOI: 0.14419/ijet.v7i4.37.24085

[6] Abdullah A, Ahmed O, Ali Z. Performance analysis of the new design of photovoltaic/storage solar collector. Energy Storage. 2019;**1**(3):1-13. DOI: 10.1002/est2.79

[7] Ahmed O. A numerical and experimental investigation for a triangular storage collector. Solar Energy. 2018;**171**:884-892. DOI: 10.1016/j.solener.2018.06.097

[8] Ahmed O, Hussein A. New design of solar chimney (case study). Case Studies in Thermal Engineering. 2018;**11**:105-112. DOI: 10.1016/j.csite.2017.12.008

[9] Saadatian O, Sopian K, Lim C, Asim N, Sulaiman M. Trombe walls: A review of opportunities and challenges in research and development. Renewable and Sustainable Energy Reviews. 2012;**16**(8):6340-6351. DOI: 10.1016/j.rser.2012.06.032

[10] Fares A. The effect of changing Trombe wall component on the thermal load. Energy Procedia. 2012;**19**(1):47-54. DOI: 10.1016/j.egypro.2012.05.181

[11] Stazi F, Mastrucci A, Perna C. Trombe wall management in summer conditions: An experimental study. Solar Energy. 2012;**86**(9):2839-2851. DOI: 10.1016/j.solener.2012.06.025

[12] Hu Z, He W, Ji J, Zhang S. A review on the application of Trombe wall system in buildings. Renewable and Sustainable Energy Reviews. 2017;**70**(2017):976-987. DOI: 10.1016/j.rser.2016.12.003

[13] Mohammed F, Ahmed O, Emad A. Effect of climate and design parameters on the temperature distribution of a room. Journal of Building Engineering. 2018;**17**:115-124. DOI: 10.1016/j.jobe.2018.02.007

[14] Jie J, Hua Y, Gang P, Bin J, Wei H. Study of PV-Trombe wall assisted with DC fan. Building and Environment. 2007;**42**(10):3529-3539. DOI: 10.1016/j.buildenv.2006.10.038

[15] Taner T. A feasibility study of solar energy-techno economic analysis from Aksaray City, Turkey. Journal of Thermal Science and Engineering. 2018;**3**(5):25-30. DOI: 10.18186/thermal.505498

[16] Chow T, Hand J, Strachan P. Building-integrated photovoltaic and thermal applications in a subtropical hotel building. Applied Thermal Engineering. 2003;**23**(16):2035-2049. DOI: 10.1016/S1359-4311(03)00183-2

[17] Hu Z et al. Comparative study on the annual performance of three types of building integrated photovoltaic (BIPV) Trombe wall system. Applied Energy. 2017;**194**:81-93. DOI: 10.1016/j.apenergy.2017.02.018

[18] Nagano K, Mochida T, Shimakura K, Murashita K, Takeda S. Development of thermal-photovoltaic hybrid exterior wallboards incorporating PV cells in and their winter performances. Solar Energy Materials & Solar Cells. 2003;**77**(3):265-282. DOI: 10.1016/S0927-0248(02)00348-3

[19] Jie J, Hua Y, Wei H, Gang P, Jianping L, Bin J. Modeling of a novel Trombe wall with PV cells. Building and Environment. 2007;**42**(3):1544-1552. DOI: 10.1016/j.buildenv.2006.01.005

[20] Taffesse F, Verma A, Singh S, Tiwari G. Periodic modeling of semi-transparent photovoltaic thermal-Trombe wall (SPVT-TW). Solar Energy. 2016;**135**:265-273. DOI: 10.1016/j.solener.2016.05.044

[21] Ahmed O, Hamada K, Salih A. Enhancement of the performance of photovoltaic/Trombe wall system using the porous medium: Experimental and theoretical study. Energy. 2019;**171**:14-26. DOI: 10.1016/j.energy.2019.01.001

[22] Ahmed O, Hamada K, Salih A. Performance analysis of PV/Trombe with water and air heating system: An experimental and theoretical study. Energy Sources, Part A: Recovery, Utilization, and Environmental Effects. 2019;**86**:716-722. DOI: 10.1080/15567036.2019.1650139

[23] Irshad K, Habib K, Thirumalaiswamy N. Performance evaluation of PV-Trombe wall for sustainable building development. Procedia CIRP. 2015;**26**:624-629. DOI: 10.1016/j.procir.2014.07.116

[24] Jie J, Hua Y, Gang P, Bin J, Wei H. Optimized simulation for PV-TW system using DC fan. In: Proceedings of ISES World Congress 2007: Solar Energy and Human Settlement; 18-21 September 2007. pp. 1617-1622. DOI:10.1007/978-3-540-75,997-3_332

[25] Ji J, Yi H, Pei G, He H, Han C, Luo C. Numerical study of the use of photovoltaic—Trombe wall in residential buildings in Tibet. Proceedings of the Institution of Mechanical Engineers, Part A. 2007;**221**(A8):1131-1140. DOI: 10.1243/09576509JPE364

[26] Irshad K, Habib K, Kareem W. Effect of air gap on performance enhancement of building assisted with photo voltaic systems. ARPN. Journal of Engineering and Applied Science. 2016;**11**(20):12078-12083

[27] Vats K, Tomar V, Tiwari G. Effect of packing factor on the performance of a building integrated semitransparent photovoltaic thermal (BISPVT) system with air duct. Energy and Buildings. 2012;**53**:159-165. DOI: 10.1016/j.enbuild.2012.07.004

[28] Jiang B, Ji J, Yi H. The influence of PV coverage ratio on thermal and electrical performance of photovoltaic-Trombe wall. Renewable Energy. 2008;**33**(11):2491-2498. DOI: 10.1016/j.renene.2008.02.001

[29] Sun W, Ji J, Luo C, He W. Numerical study of performance of Trombe wall with PV cells. ISES Solar World Congress. 2007;**1**:397-400

[30] Xu X, Su Y. Numerical simulation of air flow in BiPV-Trombe wall. Advances in Materials Research. 2013;**860-863**:141-145. DOI: 10.4028/www.scientific.net/AMR.860-863.141

[31] Hu Z et al. Design, construction and performance testing of a PV blind-integrated Trombe wall module. Applied Energy. 2017;**203**:643-656. DOI: 10.1016/j.apenergy.2017.06.078

Chapter 5

Energetics and GHG Emission Mitigation Potential Estimation of Solar Water Heating System in India

Bhupendra Singh Rawat, Pradeep Chandra Pant, Poonam Negi and Bharti Ramola

Abstract

The aim of this study is focused on "energetics" or energy analysis and evaluation of greenhouse gas (GHG) mitigation potential of solar water heating system (SWHS) having 100 litre per day (lpd) capacity in 28 states of India. Different types of collectors are used in solar water heating system, which may affect on the feasibility of the system. So to analyze this factor, the present approach focused on three types of glazed flat plate collectors. In addition to this, the feasibility of the system also does depend upon the end user applications. So to incorporate more feasible analysis, three different scenarios were considered. For the analysis of environmental impact of the system, GHG emission mitigation potential has also been evaluated by assuming the replacement of the system with frequently used fuels, viz., natural gas, coal, diesel, etc. This was done for both seasonal and annual use of this renewable energy thermal device. The result of this study shows that energy payback period (EPBP) and energy yield ratio (EYR) values for selected locations are in the adoption level. In case of coal, the GHG-emission reduction was found to have a maximum value of 22.4 tCO_2 for the state of Rajasthan and a minimum value of 13.4 tCO_2, for Jammu & Kashmir, for 20 years of useful lifetime of the system.

Keywords: energy analysis, energy payback period, energy yield ratio, glazed flat plate collectors, GHG emission mitigation

1. Introduction

Energy security is a goal that many countries are pursuing to ensure that their economies function without interruption and that their peoples have access to adequate, reliable and affordable supplies of modern and clean energy. It is a pressing concern because the demand for energy is growing rapidly due to robust economic expansion, population growth, new uses of energy and income growth and yet the supplies of energy resources required to power these needs are finite and in most cases nonrenewable. Furthermore, the production, transportation and utilization of energy are a major source of greenhouse gases that cause global warming and climate change. In this perspective energy management may play vital role in the optimization of any renewable energy system via energy analysis or by means of some other methods. Further, renewable energy systems have been proven to have potential in the reduction of greenhouse gas emission [1–7].

The application of solar energy utilization in the form of thermal energy is for heating water for domestic, commercial and industrial uses and range from the temperature 80–240°C [8–10]. During the last two decades with the emerging technologies and increasing demand the potential use of solar water heating system has been widely expanded, particularly in the residential sector [11]. The estimated gross potential for solar water heating systems in residential sector of India is quite massive [12]. In a typical Indian house, the energy consumed for heating water for bathing purpose ranges from 2000 to 7000 GJ/annum depending upon the location/ weather conditions. In industries also, hot water requirement is quite high and therefore the energy consumption is many times higher than the above value. This energy requirement is being fulfilled by grid extension, coal, furnace oil, LPG/ natural gas, wood, etc. For development, dissemination and the promotion of solar water heating system in different sectors, significant efforts have been made by the government of India. Solar water heating system of 5.6 GWth, capacity was proposed to be added in the duration of 2012–2017 and achieve 14 GWth by 2022 [13]. For the promotion and dissemination of solar water heating system like other renewable energy systems among people, government of India has been implemented several programs. This includes the dissemination of the projects, financial support or subsidies, awareness programs and through research and development [14].

A significant energy input is expected in the installation of solar water heating systems worldwide. *Energy analysis* of solar water heating system for domestic or commercial applications, gives information about the embodied energy in its various components and their share during installation and its useful lifetime. The significance of energy analysis lies in the fact that present day energy requirements are being met mainly from non-renewable energy sources, which are available as finite stores and are bound to be exhausted. Webb and Pearce [15] attribute the genesis of energetic to the awareness that these sources need to be judiciously utilized. Identification of energy-intensive components and their replacement with lesser once is the main advantage of energy analysis. This minimizes the overall energy demand of the system and gives a chance of utilization or implementation of this energy. In addition to this, energy analysis of the system and its comparison with the systems of similar technology may reduce the energy demand in any particular selected area. Energy analysis may provide useful approach for evaluation of energy yield ratio (EYR), energy payback period (EPBP) and GHG-emission mitigation potential of solar water heating systems for selected locations. This may also provide a comparative analysis of systems by different manufacturers and installation making decision for any selected location [16].

The design and efficiency of solar water heating system or any renewable energy system does depend upon the end user applications and climatic conditions of that particular location. Though the output energy of solar water heating system does depends upon the type of solar collector used, its efficiency, overall top loss heat transfer coefficient, number of glass covers used, available solar intensity, wind velocity of the location of installation [17–34].

There are several methods and simulation-based tools available for the design and energy analysis purpose. RETScreen, TRNSYS and SOLCHIPS, etc., are some user friendly tools, with the help of these tools, by the execution of number of simulation; one may analyze the effect of the variation of different variables on energy output given by the system. RETScreen software which is used in present study is also helpful for the evaluation of greenhouse gas emission mitigation potential of any renewable energy system in any particular location [35–38]. In addition to these tools, linear and nonlinear optimization techniques and evolutionary search algorithms have also been reported by different researchers to design the solar water heating systems [39–48].

Significant works have been carried out by different researches worldwide for the energy analysis and for the estimation of greenhouse gas emission mitigation potential of solar water heating systems having different collector area [49–55]. A comparative energy analysis study for the evaluation of energy payback period of solar water heating system having 3.6 m^2 of collector area with electric heater and gas water heater has been carried out by Crawford and Treloar [56]. A detailed life cycle analysis of solar water thermosyphonic system having collector area 2.13 m^2 was carried by Ardente et al. [57], where the estimated energy payback period was less than 2 years. An estimated energy payback period in between 5 and 19 months was found by Battisti and Corrado [58] after a detailed life cycle analysis for a thermosyphonic system having a 1.7 m^2 flat plate collector. They evaluated energy payback period on the basis of energy substituted with an estimated yearly output of 1958 MJ of the system. Asif et al. [59] compared the life cycle performance of two laboratory prototypes solar collectors made of Aluminum and Stainless Steel with built-in storage. They considered the collector area of 1 m^2 and found energy payback period in between 6 and 7 months for different locations on the basis of estimated energy output of the systems. Kalogirou [60] made an analysis for pressurized and thermosyphonic system having collector area of 3.8 m^2. The estimated value was found to be 1.2 years for the pumped system and 1.1 years for the thermosyphonic system. A fairly comprehensive comparative study of solar water heating system with three systems, based on conventional fuels was also carried out by Kalogirou [61]. In India, the evaluation of energy payback period of solar water heating system for different low altitude locations on the basis of energy analysis has been carried out by Mathur and Bansal [62]. Cumulated energy demand for the systems of different technologies was estimated by Mathur and Bansal [16]. They compared the solar water heating system with electric water heating system on the basis of energy analysis. Comparative study of solar water heater heating system of capacity 100 lpd with electric water heating system by using energy analysis method has been reported by Marimuthu et al. [63]. However, the estimation of energy consumption of the electrical water heater system was for the average temperature of Chennai. They also estimated the environmental benefits of the system on the basis of life cycle assessment of the system.

In the present study, "energetics" or energy analysis exercise has been undertaken on a domestic solar water heating system of 100 lpd capacity. Energy embodied in such systems have been estimated for three most prominent combinations of materials viz. Copper-Copper (Cu-Cu), Copper-Steel (Cu-Steel) and Copper- Aluminum (Cu-Al) were used in manufacturing the flat plate absorber of the flat plate solar collector, the main component of the solar water heating system. Three scenarios viz. optimistic, most probable and pessimistic were considered for the estimation of energy requirement in each and every component of the system during its useful lifetime of 20 years. Estimation of energy output by the system for one representative city of the 28 states has been considered. The lifecycle energy output and GHG-emission mitigation by the replacement of the solar water heating system with frequently used fuels viz. natural gas, coal, diesel, etc., have been estimated.

2. Methodology

In the present study, solar water heating system having flat plate collector, specified by Bureau of Indian Standard (BIS) of capacity 100 lpd is taken [64]. The Description of Component of solar water heating systems may be found anywhere [65]. The energy analysis of renewable energy systems or any energy production systems can be performed by using process analysis, input–output analysis or net energy analysis method [66]. In this present study process analysis method has been considered for the evaluation of total energy embodied during 20 years of useful life time of the system. In this process, energy inputs in individual component are estimated. Energy embodied in individual components is estimated by the estimation of direct and indirect energy. This is taken into account for the entire process of manufacturing the finished component and maintenance during its useful lifetime. Monthly average daily radiation, temperature, relative humidity and the values of wind speed for different locations have been taken from the RET-Screen online weather database. The renewable energy outputs of the system and GHG-emission reduction for annual use of solar water heating system for different selected locations of India have been estimated using RETScreen–decision support software, developed and disseminated by renewable energy decision support centre, Canada [67]. The climate of India has been divided in six zones according to variation of its climatic conditions by Bansal and Minke [68]. However, this study has been undertaken for representative city of 28 states to analyze precise feasibility of installing of solar water heating system. Energy yield ratio and energy payback period has been estimated to judge the energetic viability of the system. Energy embodied in different components or the entire system and energy output are considered as primary energy to expedite a proper comparison and evaluation. For the sake of simplicity, estimation of energy embodied in solar water heating system some assumptions have been made. Firstly, while estimating energy input, manual labor is not taken into consideration because during manufacturing it does not have any impact on primary source of energy. In addition to this there may arise error and uncertainties while converting man hour into equivalent primary energy and process analysis has been carried out up to second level of regression. For evaluating the energetic feasibility of solar water heating system, two measures namely, energy yield ratio (EYR) and energy payback period (EPBP) have been considered.

2.1 Energy embodied analysis

Estimation of energy embodied under the considered method comprises several steps; Evaluation of quantity of primary material after the identification of different components of the system, followed by the estimation of the total energy input in individual component. Estimation of the total energy embodied in the system by the calculating useful life time of each of the individual component.

The life cycle embodied energy E_{emb} is calculated by the following expression:

$$E_{emb} = E_{direct} + E_{mate} + E_{om} \tag{1}$$

where, E_{direct} is the direct energy consumed during manufacturing process of different components and E_{mate} is the energy embodied for the materials in the system and is given by:

$$E_{mate} = \sum_{i=1}^{n} \xi_i m_i \tag{2}$$

E_{om} is the energy embodied in maintenance and periodic replacement the system and is given by:

$$E_{om} = \sum_{i=1}^{n} \left[\frac{UL_{dswh}}{FR_i} - 1\right]^{+} (\xi_i m_i) \tag{3}$$

The "+" sign signifies the next higher whole number of the quantity inside the bracket.

2.2 Energy output analysis

The annual energy output of a system depends on design and operational parameters which include solar radiation availability, thermal performance of the collector type and collector tilt, storage tank, operating temperature and ambient conditions [69–72]. In this work RET-Screen system support software has been used for evaluation of energy output analysis.

2.3 Measured parameters of energetic performance

2.3.1 Energy yield ratio (EYR)

It is the ratio of lifecycle primary energy output of the system to the lifecycle energy embodied in the system and is expressed as:

$$EYR_{dswh} = \frac{\left[\left(\frac{E_{o,u}}{\eta_f}\right)(1+\alpha)\right] UL_{dswh}}{lifecyle\ energy\ embodied} \tag{4}$$

2.3.2 Energy payback period ($EPBP_{dswh}$)

It is the time taken to recuperate the initial primary energy embodied in any system by the net annual primary energy saving by the use of the system. It is given by:

$$EPBP_{dswh} = \frac{(E_{direct} + E_{mat})}{\left[\left(\frac{E_{out}}{\eta_f}\right)(1+\alpha)\right] - \left(\frac{E_{om}}{UL_{dswh}}\right)} \tag{5}$$

3. Results and discussion

Table 1 presents the energy embodied in manufacturing a flat plate collector of different absorber plates, viz., (a) Aluminum absorber plate having copper headers and risers (b) Copper absorber plate having copper headers and risers and (c) Steel absorber plate having copper header and risers. From the present table one may predict, the type of material and quantity required, and the corresponding embodied energy values (ξi) for three different types of collectors in manufacturing a 100 lpd solar water heating system. The value of ξi has been obtained from the literature [73]. It is observed that the collector with Aluminum absorber plate and copper header and riser has maximum energy embodied system (6382 MJ), which is followed by collector made up of steel absorber plate and copper header and riser (5899 MJ). Least energy-embodied system is found in collector made up of copper absorber plate header and riser (5815 MJ).

Flat plate collector component	Material	Quantity	Unit	Energy density (MJ/unit)	Embodied energy (MJ)
section frame (100x25)	Aluminium	4.25	kg	254	1080
sheet for back of the box (24 swg)	Aluminium	4.50	kg	254	1143
angle for box (1"x1"x 16 swg)	Aluminium	1.50	kg	254	381
foil for heat reflection (0.5mm)	Aluminium	1.50	kg	254	381
neelpop revie ts (60nos)	Aluminium	0.03	kg	254	8
self-tapping screw	Steel	0.50	kg	50	25
insulation	Glass wool	0.12	m^3	114	14
glazing (4 mm thick)	Toughened	35.00	kg	30	1050
(a) absorber plate (0.71 mm thick)	Aluminium	4.46	kg	254	1133
(b) absorber plate (34 swg)	Copper	4.25	kg	133	565
(c) absorber plate (0.71 mm thick)	Steel	13	kg	50	650
riser tube (0.5" diameter,24 swg 10nos)	Copper	4.00	kg	133	532
header tube (1" diameter,22 swg 2 nos)	Copper	2.25	kg	133	299
flanges (4 nos, 80 mm outer diameter)	Brass	0.80	kg	133	106
beading for glass sealing	Rubber	1.50	kg	130	195
black paint		0.50	kg	72	36
Total (a) with aluminium absorber plate and copper headers and risers					6382
Total (b) with copper absorber plate, headers and risers					5815
Total (c) with steel absorber plate and copper headers and risers					5899

Table 1.
Energy embodied in manufacturing a flat plate collector of different absorber plates.

Total energy embodied in balance of the system was found to be 2924 MJ. Energy embodied estimated values of 625.16, 636 and 1219.16 MJ were found, respectively, for storage tank, stand and pipeline (**Table 2**).

From **Table 3**, one can predict the required energy of the system during its useful life time for three collector types and on the basis of different considered scenarios. The components of solar water heating system viz. storage tank, absorber plate, outer box, etc., have long life time. However, other subsidiary components may require replacement in the 20 years of useful life time of the system due to their short life time. Maximum energy requirement is found in case of glass whereas minimum for the insulation. Total estimated values of embodied energy in a complete system were found to be 9306, 8739 and 8823 MJ, respectively, in case of Cu-Cu, Al-Cu and Steel-Cu collector. One may also predict from the table that about 77,

System component	Material	Quantity	Unit	Energy density (MJ/unit)	Embodied energy (MJ)
Storage tank					
tank	Stainless Steel	12.50	kg	50	625
insulation	Glass wool	0.55	m3	114	62.7
cladding	Aluminium	1.50	kg	254	381
Stand					
angle	Steel	12.00	kg	50	600
paint		0.50	kg	72	36
Pipeline (20")	GI	20.00	kg	50	1000
insulation	Glass wool	0.14	m3	114	15.96
cladding	Aluminium	0.80	kg	254	203.2
					2924

Table 2.
Estimated energy input values in manufacturing the subsidiary components of the system.

Material replacement	Quantity	Unit	Energy Intensity (MJ/unit)	Optimistic scenario		Most probable		Pessimistic	
				FR_i (Years)	E_{om} (MJ)	FR_i (Years)	E_{om} (MJ)	FR_i (Years)	E_{om} (MJ)
Glass	35	Kg	30	10	1050	7	2100	5	3150
Rubber beading/gaskets	1.5	Kg	130	10	195	7	390	5	585
Paint	1	litre	72	7	144	5	216	3	432
Insulation	0.81	m3	114	10	92.34	7	184.68	5	277
Total					1481		2891		4444
Annual energy requirement in maintenance					74		145		222
Life cycle energy embodied	Cu-Cu		9306		10787		12197		13750
	Al-Cu		8739		10220		11629		13183
	Steel-Cu		8823		10305		11714		13267

FR_i= frequency of replacement
E_{om}=energy embodied in periodic replacement and maintenance

Table 3.
Lifecycle embodied energy in a 100 lpd solar water heating system (Useful life time = 20 years).

145 and 222 MJ energy is required for optimistic, most probable and pessimistic scenarios, respectively, for the annual maintenance of the system.

Energy output through solar water heating system at different locations of India for the annual use and use of 5.5 months per year of the system has been shown in **Figure 1**. Monthly average daily radiation in the plane of solar collector (kWh/m^2/d) has been evaluated on the basis of monthly average daily radiation on horizontal surface (kWh/m^2/d), monthly average temperature (°C), monthly average relative humidity (%) and monthly average wind speed (m/s). In addition to this, evaluation of energy output for specific location is based on different parameters viz. slope of collector of the solar water heating system was taken equivalent to latitude of the location. Solar collector of area 2 m^2, Fr ($\tau\alpha$) coefficient of 0.85, wind correction for Fr ($\tau\alpha$) of 0.040, Fr UL coefficient 11.56 (W/m^2)/°C, wind correction for Fr UL 4.37 (J/m^3)/°C, pipe diameter of 10 mm, piping and solar tank losses of 1%, losses due to snow and/or dirt of 2%, horizontal distance from mech. Room to collector of 5 m and horizontal distance of floors from mech. Room to collector of 2 m have been taken, heat exchanger/antifreeze protection was not considered, for the analysis of energy output.

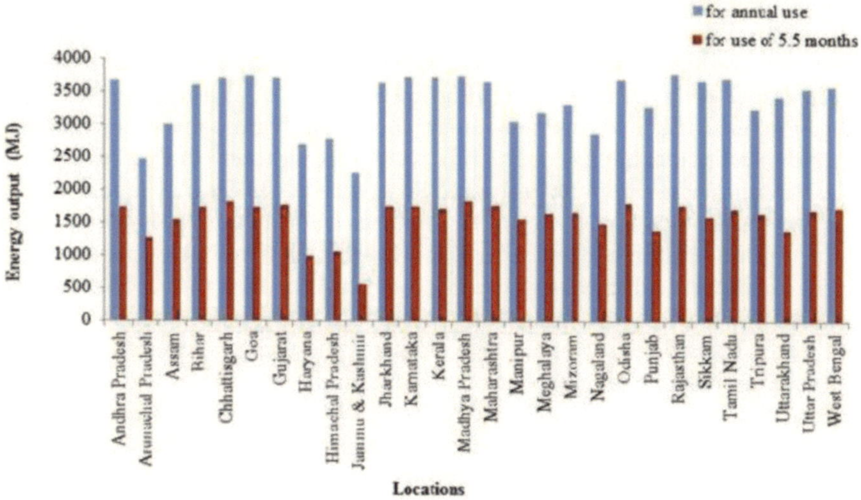

Figure 1.
Energy output through SWHS at different locations of India for annual use and of 5.5 months.

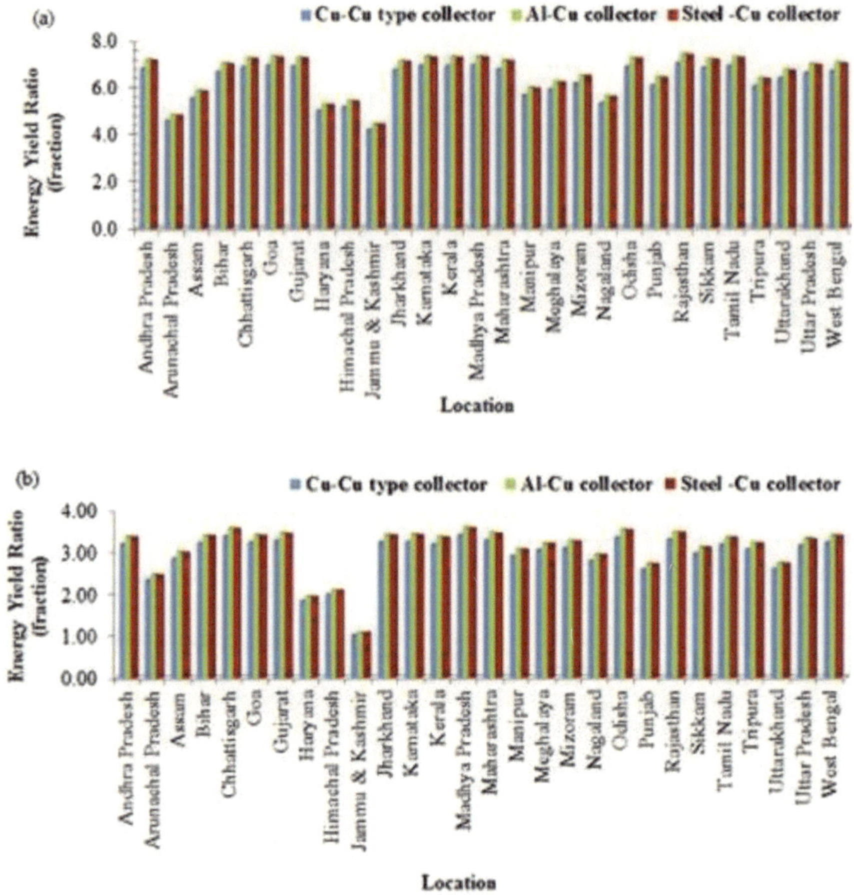

Figure 2.
EYR values for different locations of India of 20 years of useful lifetime of SWHS in case of Optimistic scenario for (a) annual use (b) use of 5.5 months per year.

The minimum value of energy output of 2258 MJ for Jammu Kashmir and maximum value of 3781 MJ was found in case of Rajasthan for the annual use of the solar water heating system. However, use of 5.5 months of the solar water heating system, the minimum and maximum value of energy output of 555 and 1833 MJ, respectively, has been found for the locations Jammu Kashmir and Madhya Pradesh. The lower value of energy output in Jammu and Kashmir as compared to other locations is due to relatively poor weather conditions.

The variation of energy yield ratio (EYR) in case of Cu-Cu, Al-Cu and Steel-Cu collector types for different locations of India has been shown in **Figures 2–4**, for 20 years of useful lifetime of the solar water heating system. In this perspective, on the basis of different adopted scenarios viz., optimistic, most probable and pessimistic, detailed location wise graphical representation has been considered for the feasibility analysis of the system of its annual use and use of 5.5 months in a year. From these figures, one may predicts that, for annual use of solar water heating system, in all locations for the selected collector types and for different scenarios, the energy yield ratio is greater than one. This signifies that the installation of solar water heating system in these locations is energetically feasible. The use of system for selected months in different locations of India has also been found feasible in terms of energy yield ratio expect Jammu and Kashmir. However, due to cold

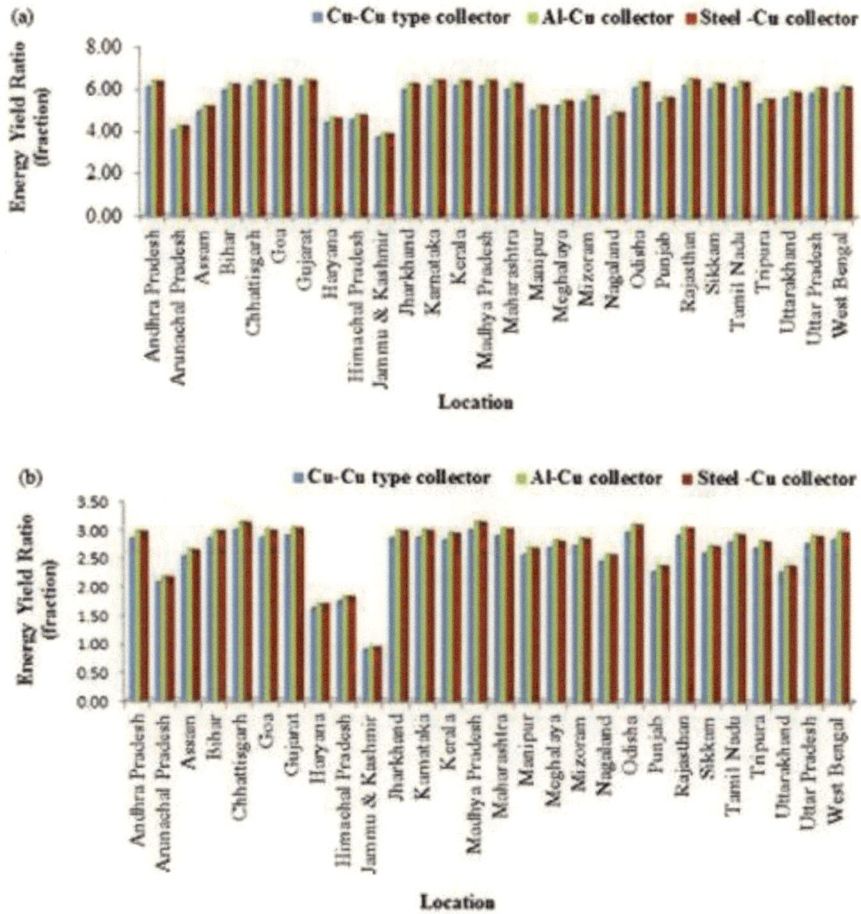

Figure 3.
EYR values for different locations of India of 20 years of useful lifetime of SWHS in case of Most Probable scenario for (a) annual use (b) use of 5.5 months per year.

climatic conditions of said location, there is requirement of annual use or for commercial applications of solar water heating system.

Figures 5–7 depicts the energy payback period for 28 locations in India for the two utilization conditions of the system, i.e., for the use of the system 5.5 months in a year and use of the system for entire year. Evaluation of energy payback period has also been taken for 20 years of useful life time of the system, collectors used and scenarios as considered in the estimation of EYR. From these figures, one may also observe that, the annual use of the system and use of 5.5 months in a year, the value of energy payback period is less than its useful lifetime, in case of different collector types and scenarios. However, the use of the system for 5.5 months in a year, energy payback period is found greater than its useful lifetime for Jammu and Kashmir.

Estimated values of GHG-emission reduction for annual use of solar water heating system has been shown in **Table 4** for different locations of India. GHG-emission mitigation potential of solar water heating system has been evaluated on the basis of energy delivered by the system, and the replacement of solar water heating system with different fuels based systems. Base case GHG-emission factor (tCO_2/MWh) in case of natural gas, coal, oil, diesel oil, biomass and propane has been taken 0.491, 0.491, 1.018, 0.975, 0.030 and 0.552, respectively. The value of

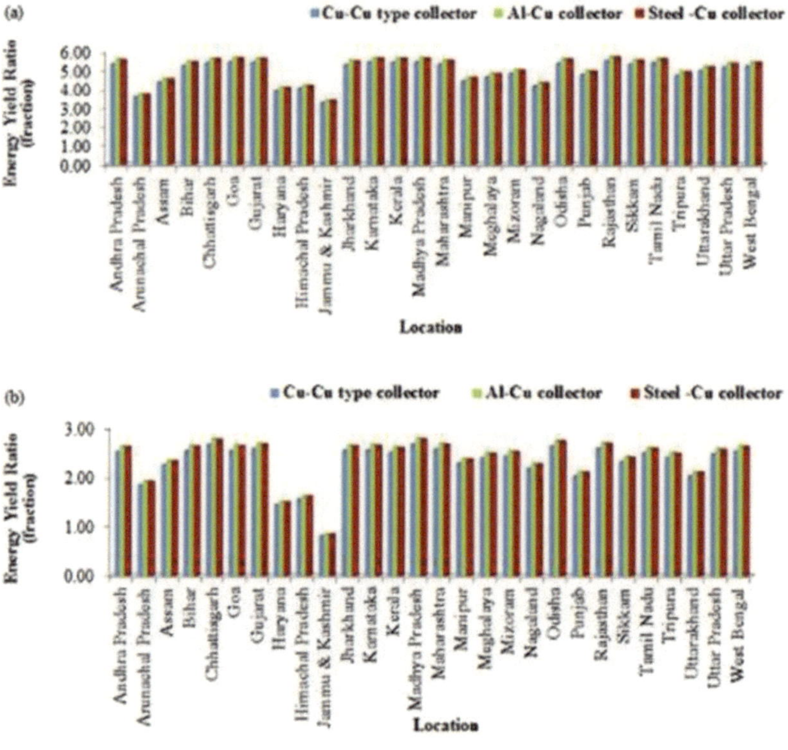

Figure 4.
EYR values for different locations of India of 20 years of useful lifetime of SWHS in case of Pessimistic scenario for (a) annual use (b) use of 5.5 months per year.

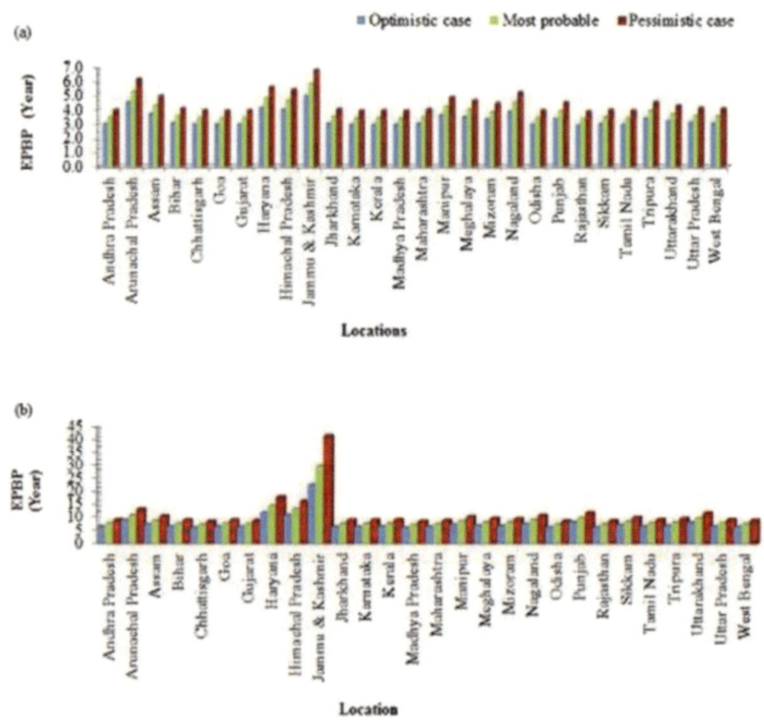

Figure 5.
EPBP values for 20 years of useful lifetime of SWHS at different locations of India in case of Cu-Cu collector material (a) for its annual use (b) for its use of selected months.

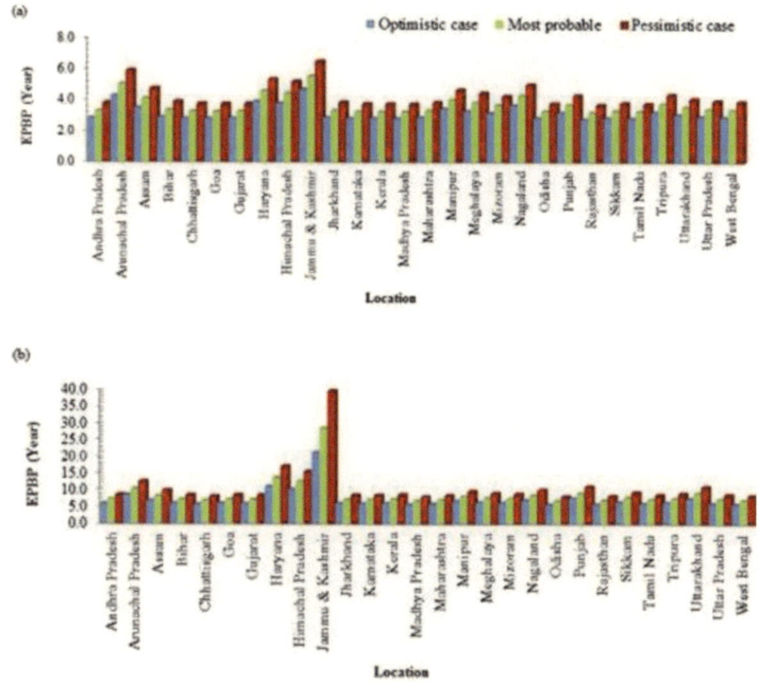

Figure 6.
EPBP values for 20 years of useful lifetime of SWHS at different locations of India in case of Al-Cu collector material (a) for its annual use (b) for its use of selected months.

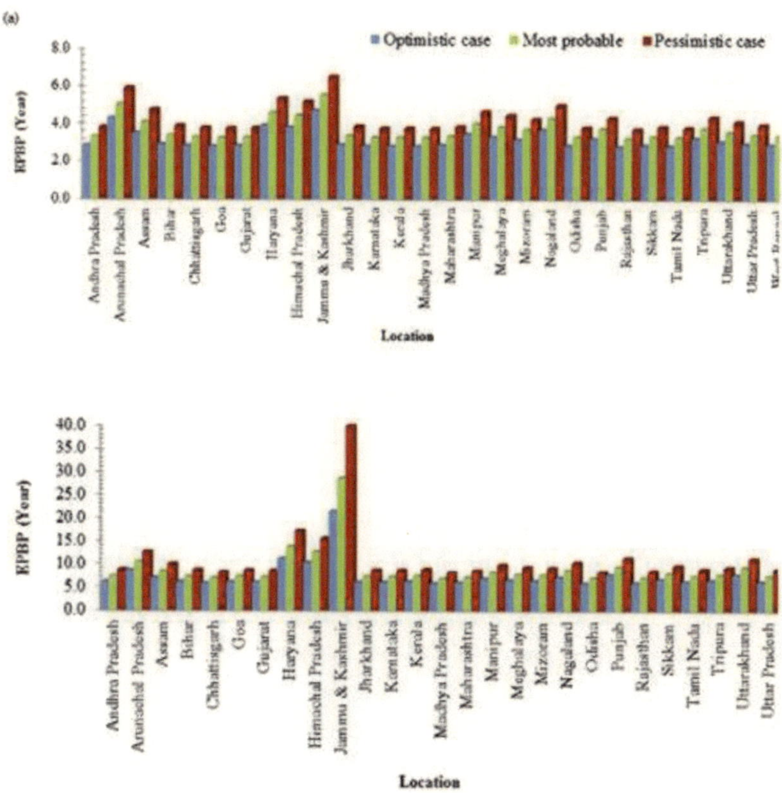

Figure 7.
EPBP values for 20 years of useful lifetime of SWHS at different locations of India in case of Steel-Cu collector material (a) for its annual use (b) for its use of selected months.

Location	GHG emission reduction (tCO2)= Base case GHG-emission factor (tCO2/MWh x End use energy delivered (MWh) + Proposed case GHG-emission factor (tCO2/MWh)						
	End use energy delivered (MWh)	Base case system (fuel type)					
		Natural gas	Coal	oil	Diesel oil	Biomass	Propane
		Base case GHG emission factor (tCO2/MWh)					
		0.491	1.069	1.018	0.975	0.030	0.552
		Annual GHG emission reduction (tCO2/year)					
Andhra Pradesh	1.02	0.50	1.09	1.04	1.00	0.03	0.56
Arunachal Pradesh	0.69	0.34	0.73	0.70	0.67	0.02	0.38
Assam	0.83	0.41	0.89	0.85	0.81	0.02	0.46
Bihar	1.00	0.49	1.07	1.02	0.98	0.03	0.55
Chhattisgarh	1.03	0.51	1.10	1.05	1.00	0.03	0.57
Goa	1.04	0.51	1.11	1.06	1.01	0.03	0.57
Gujarat	1.03	0.51	1.10	1.05	1.01	0.03	0.57
Haryana	0.75	0.37	0.80	0.76	0.73	0.02	0.41
Himachal Pradesh	0.77	0.38	0.82	0.78	0.75	0.02	0.42
Jammu & Kashmir	0.63	0.31	0.67	0.64	0.61	0.02	0.35
Jharkhand	1.01	0.50	1.08	1.03	0.98	0.03	0.56
Karnataka	1.04	0.51	1.11	1.06	1.01	0.03	0.57
Kerala	1.03	0.51	1.10	1.05	1.01	0.03	0.57
Madhya Pradesh	1.04	0.51	1.11	1.06	1.01	0.03	0.57
Maharashtra	1.01	0.50	1.08	1.03	0.99	0.03	0.56
Manipur	0.85	0.42	0.91	0.86	0.83	0.03	0.47
Meghalaya	0.88	0.43	0.95	0.90	0.86	0.03	0.49
Mizoram	0.92	0.45	0.98	0.94	0.90	0.03	0.51
Nagaland	0.80	0.39	0.85	0.81	0.78	0.02	0.44
Odisha	1.03	0.50	1.10	1.05	1.00	0.03	0.57
Punjab	0.91	0.45	0.97	0.93	0.89	0.03	0.50
Rajasthan	1.05	0.52	1.12	1.07	1.02	0.03	0.58
Sikkam	1.02	0.50	1.09	1.04	0.99	0.03	0.56
Tamil Nadu	1.03	0.51	1.10	1.05	1.01	0.03	0.57
Tripura	0.90	0.44	0.96	0.92	0.88	0.03	0.50
Uttarakhand	0.95	0.47	1.02	0.97	0.93	0.03	0.53
Uttar Pradesh	0.95	0.47	1.02	0.97	0.93	0.03	0.53
West Bengal	1.00	0.49	1.07	1.02	0.97	0.03	0.55

Table 4.
Estimated values of GHG-emission reduction for annual use of the system.

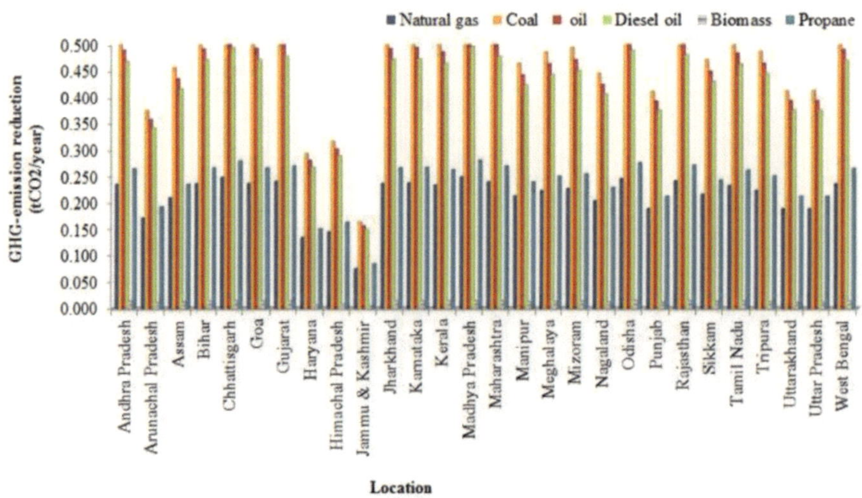

Figure 8.
Estimated values of GHG-emission reduction of using the system for selected months in a year.

proposed case emission factor has been taken as zero in case of solar water heating system, based on the fact that there is no pollution during the operation of the system. From **Table 4**, one may depict the annual variation range of 0.31 tCO_2 to 0.51 tCO_2, 0.67 tCO_2 to 1.12 tCO_2, 0.70 tCO_2 to 1.06 tCO_2, 0.02 tCO_2 to 0.03 tCO_2 and 0.35 tCO_2 to 0.58 tCO_2 in case of Natural gas, coal, oil, diesel oil, biomass and propane, respectively, for different selected location of India. Estimated values of GHG-emission reduction of using the system for 5.5 months in a year for different locations has been shown in **Figure 8**.

4. Conclusions

In solar water heating system having glazed flat collector, collector with copper absorber plate and headers and risers made of copper required minimum energy in compare to Steel-Copper and Aluminum-Cooper. Being the feasibility of the system does depending upon the availability of solar radiation of that particular location; it is also depending upon the utilization of the system by end user. So the present study on the basis of different scenarios viz. optimistic, most probable and pessimistic may give more classified information to the end users, researchers, academician and industrialist. Significant amount of GHG emission mitigation potential of the system have been estimated for different locations. Further the amount of GHG emission or mitigation does depend on the utilization of solar water heating system and use of fuel type by end users of those particular locations.

Acknowledgements

Authors are thankful to the Ministry of New and Renewable Energy Sources (MNRE), India, for funding the investigations and Solar Energy Centre, Gual Pahari, Gurgaon (SEC), India, for providing research facilities.

Nomenclature

ξ_i	energy intensity of the ith component of the material (in MJ/mass or MJ/volume)
m_i	mass per unit volume of the ith component
n	total number of components in the system
UL_{dswh}	expected useful life of the system
FR_i	frequency of replacement of the ith component
n_f	efficiency of utilization in the corresponding heating device
α	fraction of the process energy required to make the fuel available to the user

Author details

Bhupendra Singh Rawat[1], Pradeep Chandra Pant[2], Poonam Negi[3*] and Bharti Ramola[3]

1 Department of Physics, Faculty of Applied Sciences, Uttaranchal University, Dehradun, UK, India

2 Ministry of New and Renewable Energy (MNRE), New Delhi, India

3 Department of Chemistry, Faculty of Applied Sciences, Uttaranchal University, Dehradun, UK, India

*Address all correspondence to: poo.chm20@gmail.com

References

[1] Palensky P, Dietrich D. Demand side management: Demand response, intelligent energy systems, and smart loads. IEEE Transactions on Industrial Informatics. 2011;7(3):381-388

[2] Abdelaziz EA, Saidur R, Mekhilef S. A review on energy saving strategies in industrial sector. Renewable and Sustainable Energy Reviews. 2011;**15**: 150-168

[3] Petrecca G. Industrial Energy Management: Principles and Applications. Massachusetts: Kluwer Academic Publisher; 1993

[4] Gielen D, Taylor P. Indicators for industrial energy efficiency in India. Energy. 2009;**34**:962-969

[5] Taner T. Energy management for Turkish Industry (in Turkish) [MSc thesis]. Denizli: Pamukkale University Institute of Science; 2002

[6] Streimikiene D, Burbeikis J, Punys P. Review of renewable energy use in Lithuania. Renewable and Sustainable Energy Reviews. 2005;**9**:29-49

[7] Balusamy T, Sadhishkumar S. Performance improvement in solar water heating systems—A review. Renewable and Sustainable Energy Reviews. 2014;**37**:191-198

[8] Proctor D, Morse RN. Solar energy for the Australian food processing industry. Solar Energy. 1977;**19**:63-72

[9] Kalogirou SA. The potential of industrial process heat applications. Applied Energy. 2003;**76**:337-361

[10] Islam MR, Sumathy K, Khan SU. Solar water heating systems and their market trends. Renewable and Sustainable Energy Reviews. 2013;**17**:1-25

[11] Timilsina GR, Kurdgelashvili L, Narbel PA. A Review of Solar Energy—Markets, Economics and Policies. Policy Research Working Paper 5845. The World Bank, Development Research Group, Environment and Energy Team. Washington, DC: World Bank; 2011. Available from: http://documents.worldbank.org/curated/en/5460914 68178728029/A-review-of-solar-energy-markets-economics-and-policies

[12] Eléonore Q. Overview of Sustainable Renewable Energy Potential of India—2010. Report Updated by Avinash HN, Palak Timbadiya Research Associates. San Diego, California, USA: Global Energy Network Institute (GENI); 2010

[13] Annual Report. New Delhi: Ministry of Non-Conventional Energy Sources. 2002. Available from: http://mnes.nic.in

[14] Naidu BSK. Indian scenario of renewable energy for sustainable development. Energy Policy. 1996; **24**(6):575-581

[15] Webb M, Pearce D. The economics of energy analysis. Energy Policy. 1975; **3**:318-333

[16] Mathur J, Bansal N. Analysis of selected renewable energy options for India. Energy Sources. 2001;**23**:877-888

[17] Vettrivel H, Mathiazhagan P. Comparison study of solar flat plate collector with single and double glazing systems. International Journal of Renewable Energy Research. 2017;7(1): 266-274

[18] Hottel HC, Whillier A. Evaluation of Flat-Plate Collector Performance. Tucson, AZ, USA: University of Arizona Press; 1958

[19] Klein SA. Calculation of flat-plate collector utilizability. Solar Energy. 1978;**21**(8):393-402

[20] Klein SA, Beckman WA, Duffie JA. A design procedure for solar heating systems. Solar Energy. 1976;**18**:113-127

[21] Atipoang N, Chansena C, Kiatsiriroat T. Performance analysis of solar water heater combined with heat pump using refrigerant mixture. Applied Energy. 2009;**86**(5):748-756

[22] Ruchi S, Sumathy K, Erickson P, Gong J. Recent advances in the solar water heating systems: A review. Renewable and Sustainable Energy Reviews. 2013;**19**:173-190

[23] Matuska T, Vladimir Z, Juliane M. Detailed modelling of solar flat-plate collectors with design tool kolektor 2.2. In: 11th International IBPSA Conference, Scotland; 2009. pp. 2289-2296

[24] Agbo SN, Okoroigwe EC. Analysis of thermal loses in the flat-plate collector of a thermosyphon solar water heater. Research Journal of Physics. 2007;**1**:35-41

[25] Pillar PK, Agarwal RC. Factors influencing solar energy collector efficiency. Applied Energy. 1981;**8**: 205-213

[26] Buchberg H, Catton I, Edwards DK. Natural convection in enclosed spaces—A review of application to solar energy collection. Journal of Heat Transfer. 1976;**98**(2):182-188

[27] Sekhar YR, Sharmaand KV, Basaveswara RM. Evaluation of heat loss coefficients in solar flat plate collectors. ARPN Journal of Engineering and Applied Sciences. 2009;**4**(5):15-19

[28] Agbo SN, Unachukwu GO. Performance evaluation and optimization of the NCERD thermosyphon solar water heater. In: Proceedings of the World Renewable Energy Congress; August 19–25; Florence, Italy; 2006

[29] Duffie JA, Beckman WA. Solar Energy Thermal Processes. New York: John Wiley Inc.; 1974

[30] Eisenmann W, Vajen K, Ackermann H. On the correlations between collector efficiency factor and materials content of parallel flow flat plate solar collectors. Solar Energy. 2004;**76**:381-387

[31] Malhotra A, Garg HP, Palit A. Heat loss calculation of flat plate solar collectors. Journal of Thermal Engineering. 1981;**2**:59-62

[32] Mumah SN. Transient response of thermosyphon solar collectors. Solar Energy. 1995;**24**:55-61

[33] Morrison GL, Ranatunga DB. Sizing and performance evaluation of a flat-plate collector for a solar absorption refrigeration system. Nigerian Journal of Solar Energy. 1980;**13**:67-75

[34] Yeh HM, Ho CD, Yeh CW. Effect of aspect ratio on the collector efficiency of sheet and tube solar water heaters with the consideration of hydraulic dissipated energy. Renewable Energy. 2003;**28**:1575-1586

[35] Natural Resources Canada. RETScreen International. Available from: http://www.retscreen.net

[36] TRNSYS: Transient Systems Simulation Program. 2010. Available from: http://sel.me.wisc.edu/trnsys/

[37] POLYSUN: Simulation Software for Solar Thermal Systems. 2010. Available from: http://www.solarconsulting.us/polysun.html

[38] TSOL, Dynamic Simulation Programme for the Design and Optimisation of Solar Thermal Systems. 2010. Available from: http://www.valentin.de/

[39] Kalogirou SA. Optimization of solar systems using artificial neural-networks

and genetic algorithms. Applied Energy. 2004;**77**:383-405.17

[40] Kulkarni GN, Kedare SB, Bandyopadhyay S. Design of solar thermal systems utilizing pressurized hot water storage for industrial applications. Solar Energy. 2008;**82**(18): 686-699

[41] Kim YD, Thu K, Bhatia HK, Bhatia CS, Ng KC. Thermal analysis and performance optimization of a solar hot water plant with economic evaluation. Solar Energy. 2012;**86**(19):1378-1395

[42] Atia DM, Fahmy FH, Ahmed NM, Dorrah HT. Optimal sizing of a solar water heating system based on a genetic algorithm for an aquaculture system. Mathematical and Computer Modelling. 2012;**55**(20):1436-1449

[43] Bornatico R, Pfeiffer M, Witzig A, Guzzella L. Optimal sizing of a solar thermal building installation using particle swarm optimization. Energy. 2012;**41**(21):31-37

[44] Cheng Hin JN, Zmeureanu R. Optimization of a residential solar combisystem for minimum life cycle cost, energy use and exergy destroyed. Solar Energy. 2014;**100**(22):102-113

[45] Myeong JK. Analysis and optimization design of a solar water heating system based on life cycle cost using a genetic algorithm. Energies. 2015;**8**:11380-11403. DOI: 10.3390/en81011380

[46] Matrawy KK, Farkas I. New technique for short term storage sizing. Renewable Energy. 1997;**11**:129-141

[47] Loomans M, Visser H. Application of the genetic algorithm for optimization of large solar hot water systems. Solar Energy. 2002;**72**(5): 427-439

[48] Krause M, Vajen K, Wiese F, Ackermann H. Investigation on optimizing large solar thermal systems. Solar Energy. 2002;**73**:217-225

[49] Bailey RA. Net energy analysis of eight technologies to provide domestic hot water heat. Energy. 1981;**983**:6

[50] Lund PD, Kangas MT. Net energy analysis of district solar heating with seasonal heat storage. Energy. 1983;**8**: 813

[51] Hernandez P, Kenny P. Net energy analysis of domestic solar water heating installations in operation. Renewable and Sustainable Energy Reviews. 2012; **16**:170-177

[52] Shahzada M, Nawaz N, Heliyo SA. Energy security for socioeconomic and environmental sustainability in Pakistan. Heliyon. 2018;**4**:e00854. DOI: 10.1016/j.heliyon.2018.e00854

[53] Denholm P. The Technical Potential of Solar Water Heating to Reduce Fossil Fuel Use and Greenhouse Gas Emissions in the United States. Technical Report NREL/TP-640-41157. Battelle,USA: National Renewable Energy Laboratory; 2007

[54] Purohit P, Michaelowa A. CDM potential of solar water heating systems in India. Solar Energy. 2008;**82**:799-811

[55] Han J, Mol APJ, Lu Y. Solar water heaters in China: A new day dawning. Energy Policy. 2010;**38**:383-391

[56] Crawford RH, Treloar GJ. Net energy analysis of solar and conventional domestic hot water systems in Melbourne, Australia. Solar Energy. 2004;**76**:159-163

[57] Ardente F, Beccali G, Cellura M, Lo Brano V. Life cycle assessment of a solar thermal collector. Renewable Energy. 2005;**30**:1031-1054

[58] Battisti R, Corrado A. Environmental assessment of solar

thermal collectors with integrated water storage. Journal of Cleaner Production. 2005;**13**:1295-1300

[59] Asif M, Currie J, Muneer T. Comparison of aluminium and stainless steel built-in-storage solar water heater. Building Services Engineering Research and Technology. 2007;**28**:337-346

[60] Kalogirou S. Thermal performance, economic and environmental life cycle analysis of thermosyphon solar water heaters. Solar Energy. 2009;**83**:39-48

[61] Kalogirou S. Environmental benefits of domestic solar energy systems. Energy Conversion and Management. 2004;**45**:3075-3092

[62] Mathur J, Bansal N. Energy analysis of solar water heating systems in India. The International Journal of Life Cycle Assessment. 1999;**4**:113-116

[63] Marimuthu CL, Kirubakaran V. Carbon payback period and energy payback period for solar water heater V.2. International Research Journal of Environmental Sciences. 2014;**3**(2): 93-98

[64] BIS. Indian Standard-Solar Flat Plat Collector-Specification IS 12933 (Part 5). New Delhi, India: Bureau of Indian Standards; 1992

[65] Pant PC, Mishra AK, Kandpal TC. Energetics of solar water heating system. In: Proceeding of 3rd International Conference on Solar Radiation and Day Lighting (SOLARIS 2007) organized by Centre for Energy Studies, Indian Institute of Technology, Delhi; New Delhi, India; 2007. pp. 509-516

[66] Chapman PF. Energy costs: A review of methods. Energy Policy. 1974; **2**(2):91-103

[67] RET-Screen Software, Government of Canada, Natural Resources Canda's CANMET Energy Diversification Research Laboratory (CEDRL). 2000. Available from: http://rescreen.gc.ca

[68] Bansal NK, Minke G. Climatic Zones and Rural Housing in India. Germany: Scientific Series of International Bureau; 1995

[69] Morrison GL, Braun JE. System modeling and operation characteristics of thermosyphon solar water heaters. Solar Energy. 1985;**34**:389-405

[70] Fanney AH, Klein SA. Performance of solar domestic hot water systems at the National Bureau of Standards—Measurements and predictions. ASME Journal of Solar Energy Engineering. 1983;**105**:311-321

[71] Morrison GL, Tran HN. Simulation of the long term performance of thermosyphon solar water-heaters. Solar Energy. 1984;**33**:515-526

[72] Shariah A, Shalabi B. Optimal design for a thermosyphon solar water heater. Renewable Energy. 1997;**11**:351-361

[73] Boustead I, Hankok GF. Handbook of Industrial Energy Analysis. Chichester West Sussex, England: Ellis Horwood Limited; 1979

Chapter 6

Photo-Voltaic (PV) Monitoring System, Performance Analysis and Power Prediction Models in Doha, Qatar

Farid Touati, Amith Khandakar, Muhammad E.H. Chowdhury, Antonio Jr. S.P. Gonzales, Christian Kim Sorino and Kamel Benhmed

Abstract

This study aims developing customized novel data acquisition for photovoltaic systems under extreme climates by utilizing off-the-shelf components and enhanced with data analytics for performance evaluation and prediction. Microcontrollers and sensors are used to measure meteorological and electrical parameters. Customized signal conditioning, which can withstand high-temperature along with microcontrollers' development boards enhanced with appropriate interfacing shields and wireless data transmission to iCloud IoT platforms, is developed. In addition, an automatically controllable in-house electronic load of the PV system was developed to measure the maximum power possible from the system. LabVIEW™ program was used to allow ubiquitous access and processing of the recorded data over the used IoT. Furthermore, machine learning algorithms are utilized to predict the PV output power by utilizing data collected over a two-year span. The result of this study is the commissioning of original hardware for PV study under extreme climates. This study also shows how the use of specific ML algorithms such as Artificial Neural Network (ANN) can successfully provide accurate predictions with low root-mean-squared error (RMSE) between the predicted and actual power. The results support reliable integration of PV systems into smart-grids for efficient energy planning and management, especially for arid and semi-arid regions.

Keywords: PV, environmental parameters, sensors, data acquisition system, iCloud storage, PV power prediction, ML

1. Introduction

Qatar's rapid development over the past decade led to a remarkable growth on its economy and population. Hence, increasing the demands on food, water, electronics and services. All of which relies on electricity to power the industries such as desalinization plants, farms, commercial infrastructures, semiconductor factories and more. According to the Qatar Water and Electricity Corporation or QWEC, a foremost power generation plant in the country stated that the electricity demand in the country is increasing at an estimated yearly average growth rate of 6–7% in the coming years [1]. In order to address the increasing electricity demand, the state is considering a new energy strategy that would foster sustainability, but also contribute to the reduction of the greenhouse gas emission levels. Fortunately, the gulf region where the country resides, experiences 6 $kWh/m^2/day$ amounting to 4449 h/year where 70% comes from sunshine, thus, focusing on optimization of energy extraction from sunlight is a viable solution [2]. In fact, renewable energy sources such as those from photovoltaic cell (PV) plants are estimated to contribute 11% to the global demand by 2050 according to the International Energy Agency (IEA) [3].

Another possible source of renewable energy in Qatar can be harnessed from wind turbines. An assessment on wind energy potential in Qatar conducted by Qatar Petroleum [4] revealed that Qatar may employ use of small and medium wind turbines since 80% of the time wind speed over the country exceeds the critical speed of 3 m/s with annual mean speed over land and offshore of 4.3 and 5.7 m/s, respectively. It was estimated that 150 W/m^2 may be harnessed from a 5 m/s wind speed but the power generated from wind turbines may be 8% less compared to the gas fired electricity. The cost projected for an offshore wind turbine is 10% less than the gas-based counterpart. Although wind turbines sound promising as a potential source of renewable energy, it does present several disadvantages compared to PV plants such as: annual maintenance on the turbine's gear box in contrast to minimal maintenance for the PV, loud noise during operation for nearby inhabitants, and smaller life span of 20–25 years compared to 30 year life span of PV [5]. Qatar does not have immediate plans for installing wind turbines yet, instead it has been focusing on solar energy by allocating US $1 billion investment for the project which includes desalinization plants and a 200 MW power plant by Kahramaa [4]. With the upcoming 2022 FIFA cup, the country aims to be the first carbon neutral world cup utilizing solar energy to power air conditioning and fan zones. Since the state is leaning towards utilizing mostly solar energy to help power its industry, this study was conducted to primarily focus on PV alternative that was designed specifically for Qatar's environment to test and understand its performance through measurment, prediction and analysis that should provide possible references for its solar industry.

Large-scale PV farms are usually situated where maximum solar energy conversion can be generated which are either semi-arid lands or a desert. However, soaring temperatures reaching 50°C or more, high humidity and heavy sandstorms are some examples of environmental factors that may significantly reduce the efficiency in power generation of the PV systems. These issues are region-specific and may differ from one place to another even within the Gulf region, Hence, it is significant to investigate the modern PV technology under these harsh conditions that are specifically present in Qatar so that performance could be strongly correlated to it [6]. One apparent benefit from this is that the uncertainty of PV performance will be greatly reduced leading to a more predictable and profitable solar megaprojects that are planned to be constructed in the area [7–9]. The results could also cater to the interests of the manufacturers, researchers and technology enthusiasts in order to develop or innovate solutions.

Efficient energy management is among the benefits from understanding PV performance since some modern communities now use hybrid systems where they integrate renewable sources of energy such as solar PV to determine how it behaves in such systems. In [10], the authors discussed modeling and optimization of urban integrated energy systems to provide an energy plan or policy for a better energy efficiency aiming to mitigate energy crisis experienced in urban communities. In addition, Menetti et al. [11] proposed an efficient energy management that effectively use energy storage systems for renewable energy sources and the electric grid to reduce energy exchanged and power peaks on the grid. The data from the monitoring system becomes a necessary tool for conducting important analysis on the system for a region such as [12] to determine its costs and profit throughout its operation to assess its financial sustenance and feasibility for its possible application to other regions. In addition, it would also aid in contributing to the continuing development of efficient operations in industries to yield better results through exergy and energy analysis such as in [13, 14] and techno-economic analysis in [15, 16]. With increasing amount of studies being conducted centered on renewable energy especially on solar energy and PV, this study will prove useful to the scientific community and may serve as a significant reference to the ones conducted similarly in Qatar.

Several similar investigations in Qatar with same line of inquiries [17–26] were conducted but none has been able to provide a cost-effective yet reliable system that satisfies the requirement for accessing, monitoring and predicting PV yield. Another major concern is the data acquisition system (DAS); most available commercial DAS tend to be costly when implemented for large solar PV plants. In addition, commercial DAS are inflexible for reconfigurations and modifications for various scenarios, thus, limiting its use. Furthermore, numerous efforts have been conducted in designing and implementing PV monitoring systems that utilize several sensors and data acquisition [27]. The system in [28] included an off-shelf component of Agilent 24902A, wherein the data were transmitted to the wired general purpose instrumentation bus to a computer that is running a LabVIEW™ program to determine the impact of solar irradiance and ambient temperature. Haba [29] developed a designated monitoring system for several PV panels that utilizes three gateways intended for weather station, current and voltage readings and storm detection which were then sent and hosted to online cloud specifically freeboard.io. A readily available commercial DAS was used for investigating the impact of module temperature and solar irradiance on PV efficiency and transmits to a server through the use of GPIB bus and cloud service [30]. Study [31] used a system consisting of LM35 temperature sensor and LDRs (light dependent resistors) for measuring ambient temperature and solar irradiance of PV module, respectively. The data is then transmitted to the computer wirelessly via Wi-Fi by connecting the microcontroller with EGSR7150 modem through its serial interface.

Forecasting of PV performance were recently introduced to improve the quality of the systems such as providing dispatch management, control operations, power ramp and flicker prediction on hourly basis; and load consumption and production monitoring on daily basis [32]. Parametric models were also utilized for forecasting which are mostly affected by the execution of the component models and factors that are not readily available, thus, affects the accuracy of the system [33]. Recently, ML was introduced to overcome the above drawbacks; which is driven by the interactions between the input and output variables according to the data. Several studies were already conducted like in [34] were they determined the solar potential from rooftops in Switzerland by utilizing ML. Li et al. [35] used ML to predict solar irradiance to precisely determine the PV output utilizing Markov model and regression. Most of these forecasts were conducted on a specific environment, hence it would not be able to provide the same accruacy when used in another locations that exhibits different environmetal parameters like in Doha were it experiences unique intense heat and heavy dust storms that lasts year long. Therefore, we planned to deliberately harness ML for predicting the performance of PV systems from the various environmetal parameters that are present in Doha along the year for viability and bankability of PV energy source.

This study describes the development of an in-house customized DAS system that is viable for monitoring PV systems under Qatar's climate and which comprises of two parts: hardware and software. Also, the study is enhanced by describing the calibration tools that are necessary in such studies. The remainder of the study is as follows: Section 2 describes the hardware and signal acquisition. Section 3 depicts the ML used for the data gathered throughout the duration of the study. Section 4 discusses the results from the developed system and the ML results. Finally, the conclusion and future work is provided in Section 5.

2. Hardware and signal acquisition

The hardware and signal acquisition system were installed in the Solar Lab facility under the College of Engineering, Qatar University. The ground floor of the solar lab facility houses computer workstation and wireless access point while its rooftop emulates the PV panel remote site where PV panels and data acquisition hardware system are mounted along with all environmental sensors and transducers. Qatar, having an arid environment with extreme ambient temperature easily surpassing 38°C during summer and often approaches 50°C with a humidity of 90% [36].

The authors developed an in-house and customized DAS that acquires six environmental parameters and two electrical parameters enhanced by analog filters with gain and offset adjustments for calibration purposes. The in-house DAS was designed to allow flexibility in order to construct a customized signal conditioning circuit suitable for each sensor that are deemed appropropriate for the range of parameter values in an arid environment. The selected sensors along with the signal conditioning circuit and topology were chosen in order to implement a robust DAS that is appropriate to Doha's harsh weather condition.

Figure 1 depicts the overall data acquisition framework. Data acquisition starts from the PV panel remote site where the PV panels are installed to ensure maximum exposure to sun's irradiance, free from shadows due to obstructions. Selection of azimuth and tilt angle of PV panels are also important mounting details that needs to be considered. Two polycrystalline PV panels connected in series were installed in the remote site where the electrical and environmental parameters are needed to be monitored periodically in a specified sequence of steps as shown in the generalized flowchart in **Figure 2**. Periodic acquisition are normally spaced 15 minutes apart to ensure seamless wireless transmission between the PV panel remote site to the research lab site due to the considering the response time of the hardware. Information collected in the research lab site are stored locally and to the file hosting service of Dropbox™ along with the visualization facility of ThingSpeak™ through and iCloud™ server.

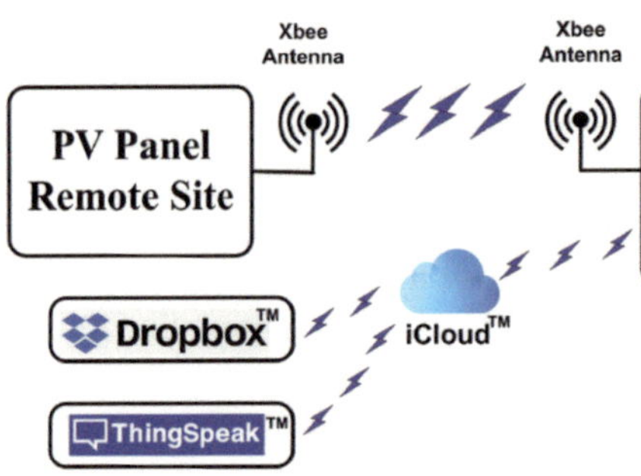

Figure 1.
Overall data acquisition system.

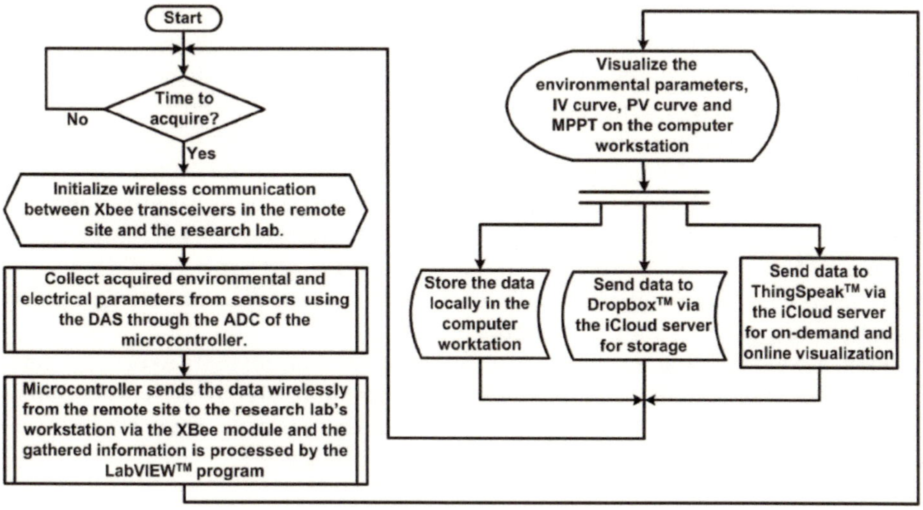

Figure 2.
Generalized flowchart of the PV monitoring system.

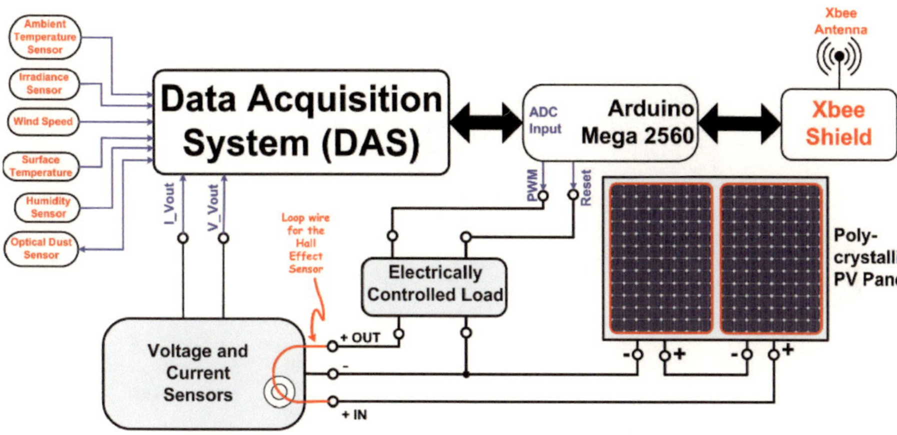

Figure 3.
System set-up of the PV panel remote site.

Parameter to be measured	Discrete sensor	Manufacturer	Measurement range
Ambient temperature	LM35	National Instruments	0–70°C
Irradiance	Pyranometer SP-110	Apogee	0 $\frac{W}{m^2}$ to 1500 $\frac{W}{m^2}$
Wind speed	Type 485 Wind sensor	QS-FS	0 $\frac{m}{s}$ to 32.4 $\frac{m}{s}$
Surface temperature	Platinum RTD PT100	Farnell	0–100°C
Humidity	HIH-4000-003	Honeywell	0–100%
Dust level	GP2Y1010AU0F Optical Sensor	Sharp	0 $\frac{mg}{m^3}$ to 0.6 $\frac{mg}{m^3}$
Voltage	Voltage transducer LV 25-P	LEM	0 V to 40 V
Current	Hall effect current transducer LA 100-P	LEM	0A to 5A

Table 1.
PV system specifications of each sensor.

A detailed illustration of the connection diagram exhibiting important components of the PV panel remote site is shown in **Figure 3**. Six environmental and two electrical parameters, namely; (1) ambient temperature, (2) irradiance level, (3) wind speed, (4) surface temperature, (5) relative humidity, (6) dust levels, along with PV voltage and current are carefully studied and chosen by the authors in [37, 38] in order to provide highest probable impact contributing to the correlation to PV panel performance and efficiency, thus, allowing higher reliability when applying ML algorithms in [37–39]. The system specifications of each sensor are enumerated in **Table 1** that includes actual part number of the off-the-shelf sensors along with the manufacturer and range of operation. The details of DAS design and operation were presented by the authors in [37–40].

Figure 4 exhibits the simplified connection of various elements to process the required signal for redundant storage and visualization in the research lab set-up. The computer workstation uses LabVIEW™ program to process data that allows visualization of recently acquired data as depicted in **Figure 5**.

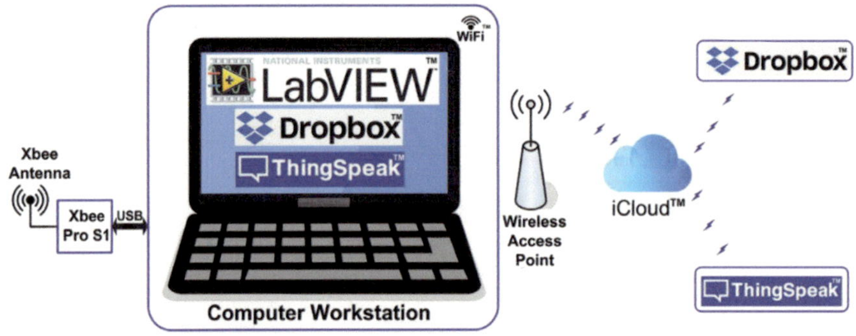

Figure 4.
System set-up of the research lab site.

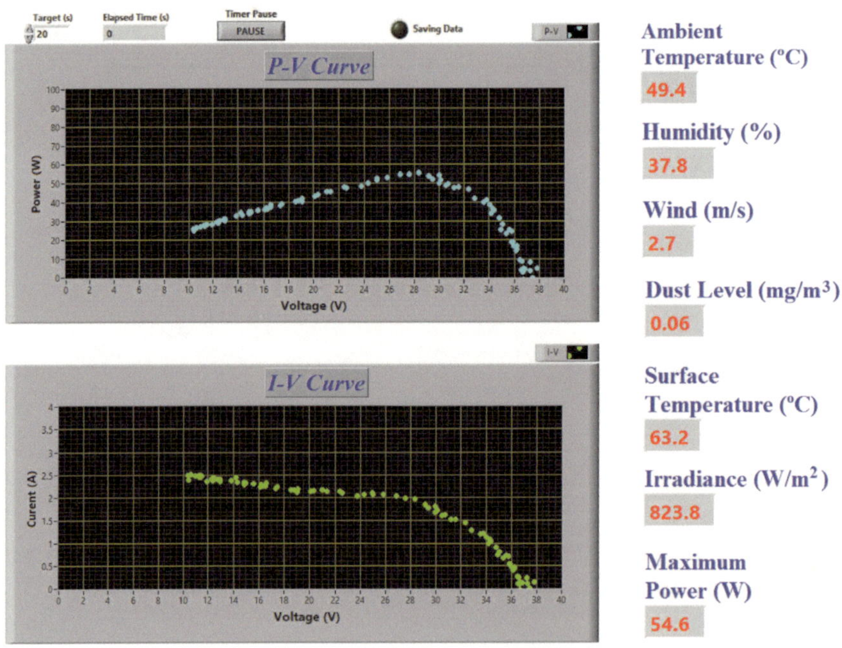

Figure 5.
Sample visualization of collected data using LabVIEW™ in the computer workstation.

3. Power prediction using machine learning

ML is the process of training a system to automatically predict output from given inputs. The system is trained using available set of inputs and their respective outputs. The concept of ML is useful in biomedical applications [41, 42], power prediction [43] and in general for any data processing and analysis studies. ML will be used to learn from the large amount of monitoring data collected from the setup discussed in the previous section and this phase is the training phase. During the training phase a part of the input data used for training is kept for validation purposes of the trained network. The validation accuracy is a metric used to determine how good or bad a trained ML network is. This ML trained network is then used for testing some data, which was unknown to the ML network, and is used to check if the ML trained network can actually predict the output correctly. The best performing ML network can later be used to predict the PV performance in the future based on the environmental and electrical inputs. The various stages that are involved in the ML are shown in **Figure 6** and will also be discussed in details in the sub sections below.

3.1 Pre processing

It is always important to make sure that the data given to the ML network for training is correctly formatted, making sure all outliers in the data or data which are incorrect and not trustable are removed. The data should be made in a format which is acceptable to the ML network in whichever platform it is being operated on. The ML Toolbox in Matlab 2019a version was used in the study. There are many other popular ML platforms available such as TensorFlow, Keras, Shogun, and RapidMiner.

3.2 Feature selection

Once the data (input and output) for the training and testing purpose is ready, it is important to select the inputs that can help in predicting the output better. Sometimes giving more input or options to help in prediction can lead to overfitting

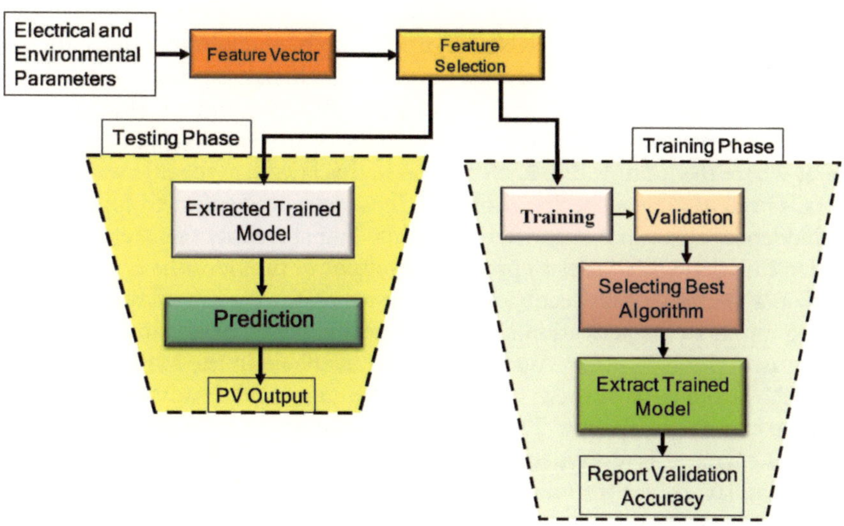

Figure 6.
Stages involved in ML training and testing phase.

problem. Overfitting is an issue where a ML network is trained to work the best for only the trained dataset and predicts mostly wrong outputs in the testing phase. This process of selecting the input data that can increase the testing accuracy is called feature selection. Selection of features is the process of selecting a subset of relevant, high-quality and non-redundant features to create learning models with better accuracy [44, 45]. Well known feature selection techniques – Correlation feature selection (CFS) and Relief feature selection (ReliefF) was used in this study. CFS technique selects feature sub-sets based on correlation-based heuristic evaluation function and ReliefF is an instance-based algorithm that assigns a relevance weight to each feature that reflects its ability to differentiate class values [43].

3.3 Prediction models

Once the data that will be given as input to the ML training phase is selected, then there are several ML techniques that can be used to see which techniques help in reaching better performance. The techniques used in this study can be broadly classified into two categories: Classical ML Technique and Artificial Neural Network. These techniques are compared in the performance in prediction during the testing phase and the best performing technique is archived for future use.

3.4 Classical machine learning

Several simple and popular regression and prediction models are stated in this work to estimate the PV output power. These are namely Simple Linear Regression [46], Gaussian Process Regression (GPR) [47] from the regression learner, and M5P regression tree [37, 48]. Simple linear regression model has a linear relationship between the output response and the input parameters. GPR involves a Gaussian process using lazy learning and a measure of the point similarity (kernel function) to predict the value from the training data for an unseen point. The M5P regression tree uses algorithm which contains if and else statements [48, 49]. In other words, predicted power will be the result of "if… then … else … " statements.

3.5 Artificial neural network

Artificial Neural Network (ANN) (**Figure** 7) can be thought of a replication of how the human nervous system works, but as it is artificial thus it gets its name [50]. ANN has three major layers: (1) Input Layer, Output Layer and the Hidden Layer. The input layer are the artificial neurons where the actual learning happens and is also the layer where the input is fed. Each neuron in this layer has specific weights, which are details used to solve a specific problem. These weighted summed inputs are used in the hidden layers or in the transfer functions. Transfer functions are then inputs to activation function which tries to predict the output or provides the error back to the network as a feedback. This feedback acts as learning for the input layers again to try providing inputs to the activation function to help in better prediction.

There are several Training Algorithms (TA) available in the Matlab implementation of ANN and each of them have their advantages and disadvantages and each application can have a specific TA giving better results than the others due to the nature of the data. It is always important to explore various combinations of number of hidden layers and training functions to find the best combination that predicts the PV power most accurately, as shown in **Figure 8**. The algorithm first varies the training algorithms, then the number of hidden layers and then does many tries

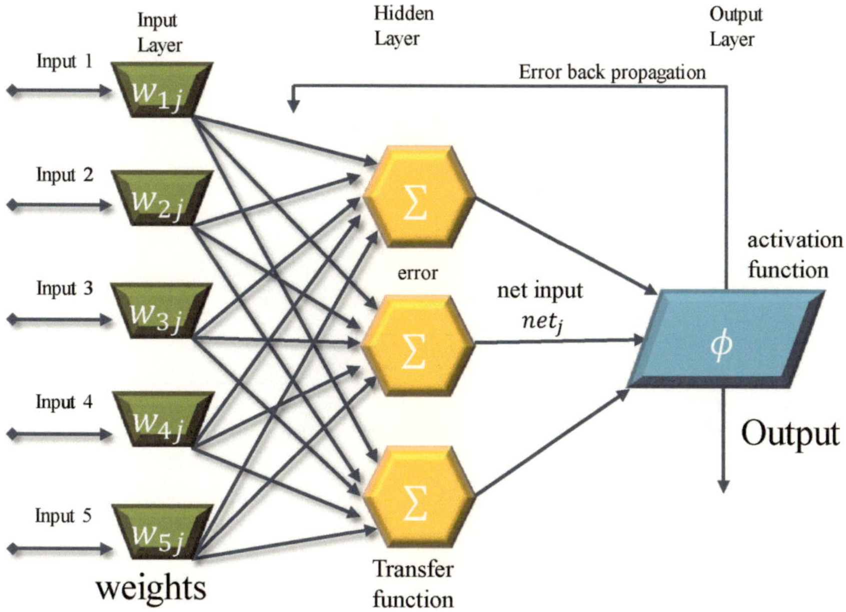

Figure 7.
ANN architecture and its main components.

using the combination. During each trial the algorithm stores the network with best performance for testing purpose. The final best network is used for predicting the PV power using the input variables.

Figure 9 summarizes the network settings for the ANN based PV power prediction. The optimum number of hidden layers providing the best model were different for all features (60), CFS technique (260) and ReliefF technique (180) and were found using the algorithm stated in **Figure 8**.

In order to compare between the various categories, techniques of ML and also the various feature selection techniques the below statistical parameters were used as performance metrics [51].

$$Correlation\ Coefficient, r = \frac{Con(X, Y)}{\sigma_x \sigma_y} \tag{1}$$

$$Mean\ absolute\ error, MAE = \frac{1}{n}\sum_{n}|X - Y| \tag{2}$$

$$\text{Mean Squared Error, MSE} = \frac{\sum |X - Y|^2}{n} \tag{3}$$

$$Root\ mean\ square\ error, RMSE = \sqrt{\frac{\sum |X - Y|^2}{n}} = \sqrt{MSE} \tag{4}$$

$$\text{Coefficient of determination, or } \mathrm{R}^2 = 1 - \frac{MSE\ (Model)}{MSE\ (Baseline)} \tag{5}$$

$$\text{MSE(Baseline) is calculated by } \frac{\sum |X - \overline{Y}|^2}{n} \tag{6}$$

where X is the actual data vector, Y and
$\overline{\mathrm{Y}}$ are the predicted data vector and mean of the predicted data vector.

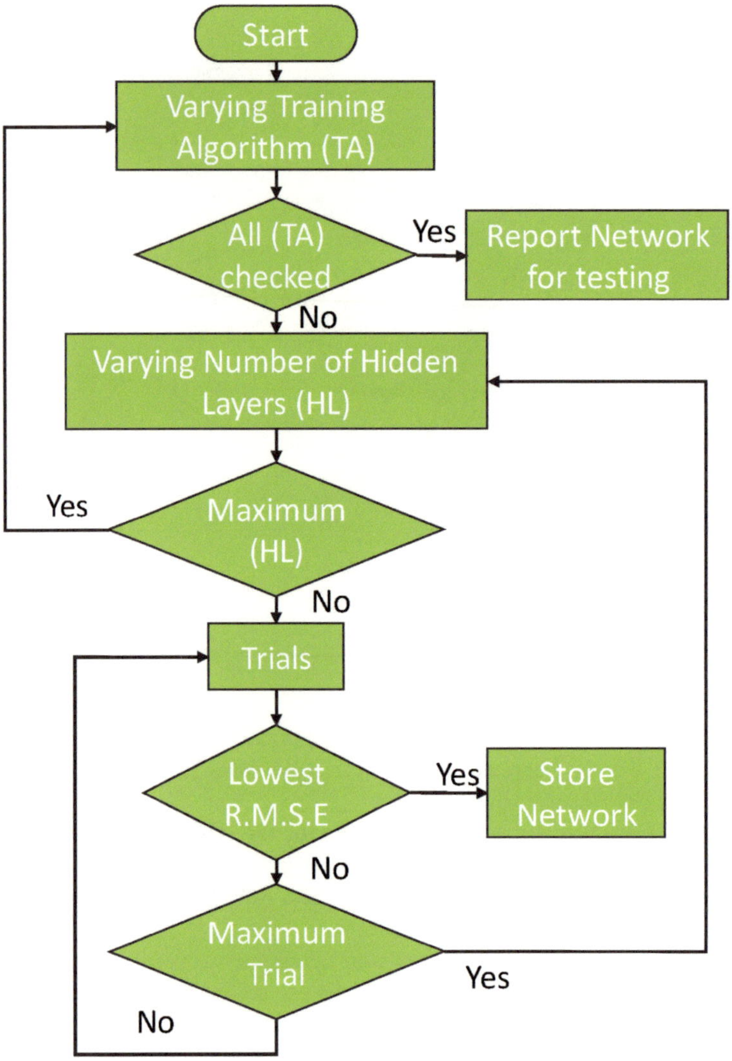

Figure 8.
Method to find the best ANN to predict PV power.

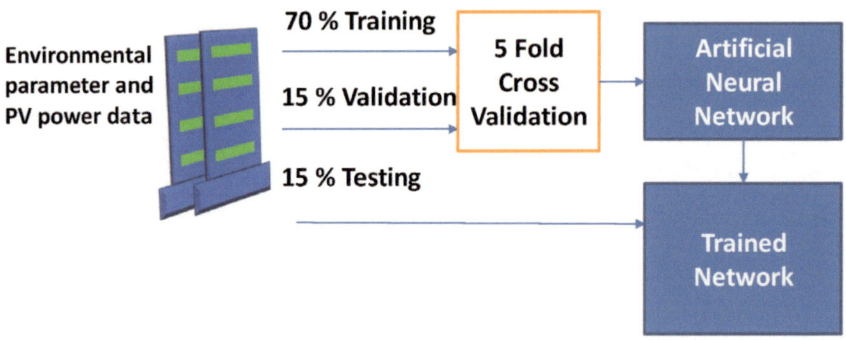

Figure 9.
Details of the ANN.

4. Results from the setup and machine learning

The prototype system (setup shown in **Figures 1** and **3**) was used for collecting the PV and environmental parameters and PV power output data from the period November 2014 until October 2016. Summary of the PV and environmental parameters and the data used for deriving the predictive model of the PV power is shown in **Table 2**.

Table 3 summarizes the parameters selected based on the feature selection techniques CFS and relief F.

Table 4 summarizes the performance of the different classical ML techniques with the different feature selection techniques. It shows both the Training and Testing Phase performance metrics. It can be clearly seen the best performance is the CFS feature selection technique using the GPR algorithm with RMSE of 12.7144 watts compared to the maximum power of 114.2017 watts generated from the setup, as shown in **Table 2**.

Table 5 summarizes the performance of the ANN best trained network found using the algorithm in **Figure 8** and with the different feature selection techniques.

Environmental Parameters	Max	Min	Unit
Temperature	61	14.6365	Degree Celsius
Relative Humidity	90.76345	27.8157	%
PV surface Temperature	74.3968	9.3037	Degree Celsius
Irradiance	1033.529	38.0076	W/m2
Dust Accumulation	1.1142	0.0553	mg/m3
Wind Speed	34.2437	0.5893	km/h
Power	114.2017	0.0368	W

Table 2.
Details of the environment parameters used for the predictive model.

Selection technique	Selected features	
Filters	CFS	Temperature
		PV Temperature
		Irradiance
		Dust Accumulation
	ReliefF (Ranked Features)	Irradiance
		Wind Speed
		PV Temperature
		Temperature

Table 3.
Selected features vector.

Selection criterion	Performance Metric	Training Phase			Testing Phase		
		Linear regression model	M5P tree model	GPR	Linear regression model	M5P tree model	GPR
Without feature Selection (all features are used)	r	0.9853	0.8908	0.9833	0.8550	0.7139	0.8662
	MAE	5.3592	6.7867	4.2601	13.1340	16.7506	12.6338
	MSE	59.1440	80.1867	44.8460	212.1000	418.5574	195.7792
	RMSE	7.6905	8.9547	6.6967	14.5639	20.4587	13.9921
	R^2	0.8100	0.9167	0.8600	0.7310	0.5096	0.7502
CFS	r	0.9814	0.9727	1.0000	0.8672	0.6809	0.8895
	MAE	5.7514	6.6892	5.1657	12.4394	27.3516	11.5247
	MSE	66.7100	92.8209	61.4970	194.3095	1050.1900	161.6557
	RMSE	8.1680	9.6344	5.1657	31.9395	32.4066	12.7144
	R^2	0.7900	0.7048	0.8100	0.7520	0.4636	0.7912
ReliefF	r	0.9837	0.9694	1.0000	0.8650	0.5797	0.8815
	MAE	5.3727	6.9110	4.2257	12.7339	18.8818	11.5482
	MSE	60.0210	103.9819	46.4300	197.5275	614.7668	173.3196
	RMSE	7.7473	10.1972	6.8139	14.0544	24.7945	13.1651
	R^2	0.8100	0.6693	0.8500	0.7482	0.3361	0.7771

Table 4.
Performance comparison between the various regression techniques.

Selection criterion	Performance Metric	Training Phase	Testing Phase
		ANN	ANN
Without feature Selection (all features are used)	r	0.9967	0.9856
	MAE	2.1275	3.2945
	MSE	4.5952	30.0134
	RMSE	2.1436	5.4784
	R^2	0.9641	0.9538
CFS	r	0.9852	0.9713
	MAE	4.8239	7.7453
	MSE	37.8900	130.3845
	RMSE	6.1555	11.4186
	R^2	0.8396	0.8213
ReliefF	r	0.9910	0.9804
	MAE	3.7225	5.0234
	MSE	30.6370	36.2345
	RMSE	5.5351	6.0145
	R^2	0.9032	0.9013

Table 5.
Performance comparison between the various ANN techniques.

It can be clearly seen that the ANN trained network outperforms the techniques in the classical ML techniques. In ANN, without feature selection techniques provides the best testing performance with RMSE of 5.48 watts compared to the maximum power of 114.20 watts generated from the setup, as shown in **Table 2**.

5. Conclusion and future work

A customized PV system was developed at Qatar University to monitor, analyze and evaluate the performance of PV using various weather factors. The study also showed details of how the data collected could be used for training different ML algorithms which were compared using different statistical analytical tools. Several feature selection techniques were also used to avoid the problem of overfitting. Comparison between the different ML techniques and different feature selection techniques helped in concluding an ANN model to be used for predicting PV performance using different environment and electrical parameters. The paper also showed the opportunity of tuning the ANN by varying the number of hidden layers and changing the training algorithm. This study describes the development of an in-house customized DAS system that is viable for monitoring PV systems under Qatar's climate and which comprises of two parts: hardware and software. Also, the study is enhanced by describing the calibration tools that are necessary in such studies. The remainder of the study is as follows: Section 2 describes the hardware and signal acquisition. Section 3 depicts the ML used for the data gathered throughout the duration of the study. Section 4 discusses the results from the developed system and the ML results. Finally, the conclusion and future work is provided in Section 5.

Acknowledgements

The authors would like to thank Qatar University for the financial, technical, and administrative support, without which this work would have not been achieved.

Conflict of interest

The authors declare no conflict of interest.

Author details

Farid Touati, Amith Khandakar*, Muhammad E.H. Chowdhury,
Antonio Jr. S.P. Gonzales, Christian Kim Sorino and Kamel Benhmed
Qatar University, Doha, Qatar

*Address all correspondence to: amitk@qu.edu.qa

References

[1] Qatar Electricity and Water Corporation. Qatar's Installed Power Capacity Doubles in 2 Years as Demand Rises [Internet]. 2013. Available from: https://www.qewc.com/qewc/en/index.php/qewc/77-gulf-times/125-qatar-s-installed-power-capacity-doubles-in-2-years-as-demand-rises [Accessed: 19 December 2019]

[2] Aksakal A, Rehman S. Global solar radiation in northeastern Saudi Arabia. Renewable Energy. 1999;**17**:461-472. DOI: 10.1016/S0960-1481(98)00769-1

[3] US Energy Information Administration. International Energy Outlook [Internet]. 2018. Available from: http://www.eia.gov/forecasts/ieo/pdf/0484(2013).pdf [Accessed: 20 October 2019]

[4] Assessment of solar and wind energy potential in Qatar [Internet]. 2013. Available from: http://www.jccp.or.jp/international/conference/docs/14assessment-of-solar and-wind-energy-potential-in.pdf [Accessed: 18 April 2020]

[5] Renewables comparison: wind vs. solar energy [Internet]. 2016. Available from: https://icap.sustainability.illinois.edu/files/projectupdate/4045/wind%20v%20solar.pdf [Accessed: 18 April 2020]

[6] Touati FA, Hitmi MAA, Bouchech HJ. Study of the effects of dust, relative humidity, and temperature on solar pv performance in Doha: Comparison between monocrystalline and amorphous PVS. International Journal of Green Energy. 2013;**10**:680-689. DOI: 10.1080/15435075.2012.692134

[7] Adnene C, Hamadi Z, Moncef J. Control and real time monitoring of less battery storage photovoltaic plants. International Journal of Sustainable Energy. 2005;**24**:87-98. DOI: 10.1080/14786450512331329574

[8] Tina GM, Grasso AD. Remote monitoring system for stand-alone photovoltaic power plants: The case study of a PV-powered outdoor refrigerator. Energy Conversion and Management. 2014;**78**:862-871. DOI: 10.1016/j.enconman.2013.08.065

[9] Fathi AE, Nkhaili L, Bennouna A, Outzourhit A. Performance parameters of a standalone PV plant. Energy Conversion and Management. 2014;**86**: 490-495. DOI: 10.1016/j.enconman.2014.05.045

[10] Li S, Guo L, Zhang P, Wang H, Cai Z, Zhu X, et al. Modeling and optimization on energy efficiency of urban integrated energy system. In: 2nd IEEE Conference on Energy Internet and Energy System Integration (EI2); 20–22 October 2018. Beijing, China: IEEE; 2018. pp. 1-6

[11] Menniti D, Pinnarelli A, Sorrentino N, Vizza P, Burgio A, Brusco G, et al. A real-life application of an efficient energy management method for a local energy system in presence of energy storage systems. In: IEEE International Conference on Environment and Electrical Engineering and 2018 IEEE Industrial and Commercial Power Systems Europe (EEEIC/I&CPS Europe); 12–15 June 2018. Palermo, Italy: IEEE; 2018. pp. 1-6

[12] Swarnkar NM, Gidwani L. Economic and financial assessment of integrated solar and wind energy system in Rajasthan, India. In: International Conference on Computation of Power, Energy Information and Commuincation (ICCPEIC); 22–23 March 2017. Melmaruvathur, India: IEEE; 2017. pp. 471-476

[13] Kumar M, Saini R. Energy and exergy analysis for heliostat based solar thermal power plant. In: 2nd IEEE

International Conference on Power Electronics, Intelligent Control and Energy Systems (ICPEICES); 22–24 October 2018. Delhi, India: IEEE; 2018. pp. 120-126

[14] Hamdi I, Kooli S. Exergy and energy analysis of the solar drying processes of tomatoes in Tunisia. In: 9th International Renewable Energy Congress (IREC); 20–22 March 2018. Hammamet, Tunisia: IEEE; 2018. pp. 1-6

[15] Šimunović J, Barbir F, Radica G, Klarin B. Techno-economic analysis of PV/wind turbine stand-alone energy system. In: 4th International Conference on Smart and Sustainable Technologies (SpliTech); 18–21 June 2019. Split, Croatia: IEEE; 2019. pp. 1-5

[16] Benmedjahed M, Maouedj R. Technical and economic analysis of wind turbine system for isolated location at Adrar in Algeria. In: 6th International Renewable and Sustainable Energy Conference (IRSEC); 5–8 December 2018. Rabat, Morocco: IEEE; 2018. pp. 1-4

[17] Jeevandoss CR, Kumaravel M, Kumar VJ. Sunlight based I-V characterization of solar PV cells. In: Instrumentation and Measurement Technology Conference; 10–12 May 2011. Binjiang, China: IEEE; 2011. pp. 1-4

[18] Ali MA, Emziane M. Performance analysis of rooftop PV systems in Abu Dhabi. Energy Procedia. 2013;**42**: 89-697. DOI: 10.1016/j.egypro.2013.11.071

[19] Atmaja TD. Façade and rooftop PV installation strategy for building integrated photo voltaic application. Energy Procedia. 2013;**32**:105-114. DOI: 10.1016/j.egypro.2013.05.014

[20] Congedo PM, Malvoni M, Mele M, Giorgi MGD. Performance measurements of monocrystalline silicon PV modules in South-Eastern Italy. Energy Conversion and Management. 2013;**68**:1-10. DOI: 10.1016/j.enconman.2012.12.017

[21] Dubey S, Sarvaiya JN, Seshadri B. Temperature dependent photovoltaic (PV) efficiency and its effect on PV production in the world - A review. Energy Procedia. 2013;**33**:311-321. DOI: 10.1016/j.egypro.2013.05.072

[22] Tebibel H, Labed S. Performance results and analysis of self-regulated PV system in Algerian Sahara. Renewable Energy. 2013;**60**:691-700. DOI: 10.1016/j.renene.2013.06.032

[23] Aste N, Pero CD, Leonforte F. PV technologies performance comparison in temperate climates. Solar Energy. 2014;**109**:1-10. DOI: 10.1016/j.solener.2014.08.015

[24] D'Orazio M, Perna CD, Giuseppe ED. Experimental operating cell temperature assessment of BIPV with different installation configurations on roofs under Mediterranean climate. Renewable Energy. 2014;**68**:378-396. DOI: 10.1016/j.renene.2014.02.009

[25] Saber Esmail M, Lee SE, Manthapuri S, Yi W, Deb C. PV (photovoltaics) performance evaluation and simulation-based energy yield prediction for tropical buildings. Energy. 2014;**71**:588-595. DOI: 10.1016/j.energy.2014.04.115

[26] Maturi L, Belluardo G, Moser D, Buono MD. BiPV system performance and efficiency drops: Overview on PV module temperature conditions of different module types. Energy Procedia. 2014;**48**:1311-1319. DOI: 10.1016/j.egypro.2014.02.148

[27] Amhani A, Attia HA. Online multi-parameters electronic monitoring system for solar photovoltaic panel

applications. In: International Conference on Electrical and Computing Technologies and Applications (ICECTA); 21–23 November 2017. Ras Al Khaimah, United Arab Emirates: IEEE; 2018. pp. 1-4

[28] Chouder A, Silvestre S, Taghezouit B, Karatepe E. Monitoring, modelling and simulation of PV systems using LabVIEW. Solar Energy. 2013;**91**: 337-349. DOI: 10.1016/j.solener.2012.09.016

[29] Haba CG. Monitoring photovoltaic parks for damage prevention and optimal operation. In: International Conference on Electromechanical and Power Systems (SIELMEN); 11–13 October 2017. Iasi, Romania: IEEE; 2017. pp. 321-326

[30] Torres M, Muñoz FJ, Muñoz JV, Rus C. Online monitoring system for stand-alone photovoltaic applications—Analysis of system performance from monitored data. Journal of Solar Energy Engineering. 2012;**134**: 1-9. DOI: 10.1115/1.4005448

[31] Anwari M, Hidayat A, Hamid MI. Wireless data acquisition for photovoltaic power system. In: INTELEC 2009 - 31st International Telecommunications Energy Conference; 18–22 October 2009. Incheon, South Korea: IEEE; 2009. pp. 1-4

[32] Theocharides S, Makrides G, Georghiou G, Kyprianou A. Machine learning algorithms for photovoltaic system power output prediction. In: IEEE International Energy Conference (ENERGYCON); 3–7 June 2018. Limassol, Cyprus: IEEE; 2018. pp. 1-6

[33] Lorenz E, Scheidsteger T, Hurka J, Heinemann D, Kurz C. Regional PV power prediction for improved grid integration. Progress in Photovoltaics: Research and Applications. 2011;**19**: 757-771. DOI: 10.1002/pip.1033

[34] Assouline D, Mohajeri N, Scatterzzini J. Quantifying rooftop photovoltaic solar energy potential: A machine learning approach. Solar Energy. 2016;**141**:278-296. DOI: 10.1016/j.solener.2016.11.045

[35] Li J, Ward J, Tong J, Collins L, Platt G. Machine learning for solar irradiance forecasting of photovoltaic system. Renewable Energy. 2016;**90**: 542-553. DOI: 10.1016/j.renene.2015.12.069

[36] Average Weather in Doha, Qatar, Year Round - Weather Spark [Internet]. 2019. Available from: https://weatherspark.com/y/105083/Average-Weather-in-Doha-Qatar-Year-Round [Accessed: 01 November 2019]

[37] Touati F, Chowdhury NA, Benhmed K, Gonzales AS, Al-Hitmi MA, Benammar M, et al. Long-term performance analysis and power prediction of PV technology in the State of Qatar. Renewable Energy. 2017;**113**: 952-965. DOI: 10.1016/j.renene.2017.06.078

[38] Touati F, Al-Hitmi MA, Chowdhury NA, Hamad JA, Gonzales AS. Investigation of solar PV performance under Doha weather using a customized measurement and monitoring system. Renewable Energy. 2016;**89**:564-577. DOI: 10.1016/j.renene.2015.12.046

[39] Benhmed K, Touati F, Al-Hitmi M, Chowdhury NA, Gonzales AS, Qiblawey Y, et al. PV power prediction in Qatar based on machine learning approach. In: IRSEC 2018 - 6th International Renewable and Sustainable Energy Conference (IRSEC); 05–08 December 2018. Rabat, Morocco: IEEE; 2018. pp. 174-177

[40] Touati F, Gonzales AS, Qiblawey Y, Benhmed K. A customized PV performance monitoring system in Qatar's harsh environment. In: IRSEC

2018 - 6th International Renewable and Sustainable Energy Conference (IRSEC); 05–08 December 2018. Rabat, Morocco: IEEE; 2018. pp. 134-139

[41] Chowdhury ME, Alzoubi K, Khandakar A, Khallifa R, Abouhasera R, Koubaa S, et al. Wearable real-time heart attack detection and warning system to reduce road accidents. Sensors. 2019;**19**:2780. DOI: 10.3390/s19122780

[42] Chowdhury ME, Khandakar A, Alzoubi K, Mansoor S, Tahir AM, Reaz MBI, et al. Real-time smart-digital stethoscope system for heart diseases monitoring. Sensors. 2019;**19**:2781. DOI: 10.3390/s19122781

[43] Khandakar A, Chowdhury MEH, Khoda Kazi M, Benhmed K, Touati F, Al-Hitmi M, et al. Machine learning based photovoltaics (PV) power prediction using different environmental parameters of Qatar. Energies. 2019;**12**:2782. DOI: 10.3390/en12142782

[44] Guyon I, Elisseeff A. An introduction to variable and feature selection. Journal of Machine Learning Research. 2003;**3**:1157-1182. DOI: 10.1162/153244303322753616

[45] Wang H, Khoshgoftaar TM, Gao K, Seliya N. High-dimensional software engineering data and feature selection. In: 21st IEEE International Conference on Tools with Artificial Intelligence; 2–4 November 2009. Newark, NJ, USA: IEEE; 2009. pp. 83-90

[46] Yan X, Su X. Linear regression analysis: Theory and computing. World Scientific. 2009;**1**:9-28. DOI: 10.1142/6986

[47] MacKay DJ, Mac Kay DJ. Information Theory, Inference and Learning Algorithms. Cambridge, United Kingdom: Cambridge University Press; 2003. DOI: 10.1017/S026357470426043X

[48] Quinlan JR. Learning with continuous classes. In: 5th Australian Joint Conference on Artificial Intelligence. 1992. DOI: 10.1142/9789814536271

[49] Wang Y, Witten IH. Induction of Model Trees for Predicting Continuous Classes. Hamilton, New Zealand: Department of Computer Science, University of Waikato; 1996

[50] Mellit A, Pavan AM. A 24-h forecast of solar irradiance using artificial neural network: Application for performance prediction of a grid-connected PV plant at Trieste, Italy. Solar Energy. 2010;**84**:807-821. DOI: 10.1016/j.solener.2010.02.006

[51] Sheiner LB, Beal SL. Some suggestions for measuring predictive performance. Journal of Pharmacokinetics and Biopharmaceutics. 1981;**9**:503-512. DOI: 10.1007/BF01060893

Chapter 7

Computing the Global Irradiation over the Plane of Photovoltaic Arrays: A Step-by-Step Methodology

Oswaldo A. Arraez-Cancelliere, Nicolás Muñoz-Galeano and Jesús M. López-Lezama

Abstract

The quality of solar resource data is critical for the economic and technical assessment of solar photovoltaic (PV) installations. Understanding uncertainty and managing weather-related risk are essential for successful planning and operating of solar electricity assets. The input information available for PV designers is usually restricted to 12 monthly mean values of global horizontal irradiation (GHI) and average temperature, which characterize solar climate of locations. However, for calculating the energy production of a photovoltaic system, the global irradiation over the plane of the PV array is necessary. For this reason, this book chapter presents a methodology to appropriately determine the global irradiation over the plane of photovoltaic arrays. The methodology describes step by step the necessary equations for processing the data. Examples with numerical results are included to better show the data processing.

Keywords: global horizontal irradiation (GHI), photovoltaic (PV), energy production, solar resource, data processing

1. Introduction

Renewable energy resources have become a promissory alternative to overcome the problems related to high pollution and limited sources of conventional energy. So, the analysis of energy resources and their economic feasibility is a concern topic for researchers around the world [1–4]. In this context, photovoltaic power plants have become one of the most important renewable sources of energy that have rapidly spread in the last decade. However, the assessment of the solar resource is not a topic usually approached by engineers and researchers due to the complexity in the process of computing the data, being extensive when the global horizontal irradiation is processed to obtain the global tilde irradiation. Therefore, this book chapter provides a step-by-step methodology for computing the global irradiation over the plane of photovoltaic arrays.

The quality of solar resource data is critical for economic and technical assessment of solar power installation. Understanding uncertainty and managing weather-related risk are essential for successful planning and operating of solar electricity assets. High-quality solar resource and meteorological data can be obtained by two approaches: high-accuracy instruments installed at a meteorological station and complex solar meteorological models, which are validated using high-quality ground instruments [5].

The input information available for PV designers is usually restricted to the 12-monthly mean values of global horizontal irradiation (GHI) and average temperature, which characterize solar climate of locations. However, the global irradiation over the plane of the PV array is necessary to calculate the energy production of a photovoltaic system.

The assessment of radiation arriving on an inclined surface, using global horizontal data as input, raises two main problems; first, to separate the GHI into their direct and diffuse components (decomposition) and, second, from them, to estimate the radiation components falling on an inclined surface (transposition) [6]. To solve these problems, it is important to describe in detail parameters such as declination angle, solar hour angle, solar zenith, solar altitude, solar azimuth, angle of incidence, solar constant, extraterrestrial irradiance over the horizontal surface, clearness index, and diffuse fraction index [6–8].

As regards decomposition models, in Ref. [9], authors reviewed and compared four decomposition models for monthly average of horizontal daily irradiation: a linear relationship proposed by Page [10], two polynomial equations defined by the authors of Refs. [11, 12], and a local correlation proposed by Macagnan et al. [13]. The authors of Ref. [9] stand out for the Page decomposition model of Ref. [10] in combination with the Perez transposition model [13] as a good performance combination used for passing from global horizontal irradiation to effective in-plane irradiance when it is started from monthly average of daily irradiation values. After applying the decomposition models, direct, diffuse, and albedo components of irradiation can be obtained.

For obtaining radiation components falling on an inclined surface, geometric considerations can be taken into account [9]. In Ref. [9], eight transposition models were reviewed, obtaining the best results using the Perez model [13]. Similar results were obtained in Ref. [14], where four transposition models were compared and validated with two-year data measured at site, and Liu and Jordan isotropic sky model is used in this book chapter due to its simplicity of implementation and good results reported in Ref. [15].

This book chapter describes in detail a methodology to determine the global irradiation over the plane of photovoltaic arrays. The chapter is organized as follows: Section 2 contains the proposed methodology, Section 3 includes experimental results that include data processing, and Section 4 presents the conclusions.

2. Proposed methodology

The proposed methodology includes the data processing and also the definitions to understand it. Therefore, some basic definitions are presented below:

2.1 Declination angle (δ)

It is the angle between the equatorial plane and a straight line drawn between the center of the Earth and the center of the sun. It may be considered as approximately constant over the course of any day. It can be calculated using Eq. (1), where d_n is the day number counted from the beginning of the year [7].

$$\delta = 23.45° \times sin\left[\frac{360 \times (d_n + 284)}{365}\right] \tag{1}$$

2.2 Solar hour angle (ω)

It is the difference between noon and the selected moment of the day in terms of a 360° rotation in 24 h. ω is equal to 0 at midday of each day, and it is counted as negative in the morning and positive in the afternoon. The solar hour angle is given as:

$$\omega = 15 \times (T_{solar}(h) - 12) \tag{2}$$

$$T_{solar}(h) = T_{local}(h) + \frac{4(L_{st} - L_{loc}) + EoT}{60} \tag{3}$$

$$EoT\ (min) = 9.87\, sin\, 2B - 7.53\, cos\, B - 1.5\, sin\, B \tag{4}$$

$$B_n = \frac{360}{365}(d_n - 81) \tag{5}$$

$$L_{st} = 15 \times \Delta T_{gmt} \tag{6}$$

where T_{solar} and T_{local} are the solar time and the local clock time, respectively. L_{st} and L_{loc} are the standard meridian for the local time zone and the longitude of the location (east positive and west negative). ΔT_{gmt} is the local time zone (e.g., Bogotá, -5). EoT stands for equation of time, which is the time difference between the apparent solar time for people and the real mean solar time, and takes into account the perturbation of the earth's rotation [7].

2.3 Solar zenith (θ_{zs}) and solar altitude (γ_s)

Solar zenith is the angle between the vertical and the incident solar beam, and it can also be described as the angle of incidence of beam radiation on a horizontal surface [6]. The complement of the zenith angle is called the solar altitude, γ_s. These angles can be calculated using Eq. (7) and are function of the declination angle (δ), the latitude (ϕ) (north positive and south negative), and the true solar time (ω).

$$cos\, \theta_{zs} = sin\delta\ sin\phi + cos\delta\ cos\delta\ cos\omega = sin\, \gamma_s \tag{7}$$

$$\gamma_s = 90 - \theta_{zs} \tag{8}$$

Eq. (7) can be used to find the sunrise angle (ω_s) since at sunrise $\gamma_s = 0$, hence.

$$\omega_s = -cos^{-1}(-tan\ \delta tan\phi) \tag{9}$$

In accordance with the sign convention, ω_s is always negative. The sunset angle is equal to $-\omega_s$, and the length of the day is equal to:

$$T_d(hour) = \frac{2 \times abs(\omega_s)}{15°} \tag{10}$$

2.4 Solar azimuth (ψ_s)

It is the angle between the meridians of the locations and the sun. It can also be described as the angular displacement from noon to the projection of beam radiation on the horizontal plane. The solar azimuth is given by:

$$\cos\psi_s = \left(\frac{\sin\gamma_s \times \sin\phi - \sin\delta}{\cos\gamma_s \cos\phi}\right) \tag{11}$$

In the Northern Hemisphere, true solar is the reference of the system, and it is defined as positive toward the west, that is, in the evening, and negative toward the east, that is, in the morning [6]. In **Figure 1**, the solar zenith, solar altitude, and solar azimuth are described.

2.5 Angle of incidence (θ_i)

Most practical applications require the position of the sun relative to an inclined plane to be determined. The angle of solar incidence between the sun's rays and the normal to the surface is given by:

$$\begin{aligned}\cos(\theta_i) = {} & \sin(\delta)\sin(\varnothing)\cos(\beta) - [sign(\varnothing)]\sin(\delta)\cos(\varnothing)\sin(\beta)\cos(\alpha) \\ & + \cos(\delta)\cos(\varnothing)\cos(\beta)\cos(\omega) + [sign(\varnothing)]\cos(\delta)\sin(\varnothing)\sin(\beta)\cos(\alpha)\cos(\omega) \\ & + \cos(\delta)\sin(\alpha)\sin(\omega)\sin(\beta)\end{aligned} \tag{12}$$

where β is the tilt of the inclined plane (the angle formed with the horizontal), and α is the surface azimuth angle conventionally measured clockwise from the south (See **Figure 2**) [8]. The $sign(\varnothing)$ function is 1 when the latitude is greater than 0 and is -1 otherwise.

2.6 Solar constant (B_0)

It is the amount of solar radiation received at the top of the atmosphere on a normal plane at the mean Earth-sun distance [6]. A good approximation of this value is:

$$B_0 = 1367\ W/m^2 \tag{13}$$

Extraterrestrial irradiance over the horizontal surface ($B_0(0)$): Extraterrestrial radiation over a horizontal surface varies over the day, and it is given by:

$$B_0(0) = B_0 \times \left[1 + 0.033\cos\left(\frac{360}{365}d_n\right)\right]\cos\theta_{zs} \tag{14}$$

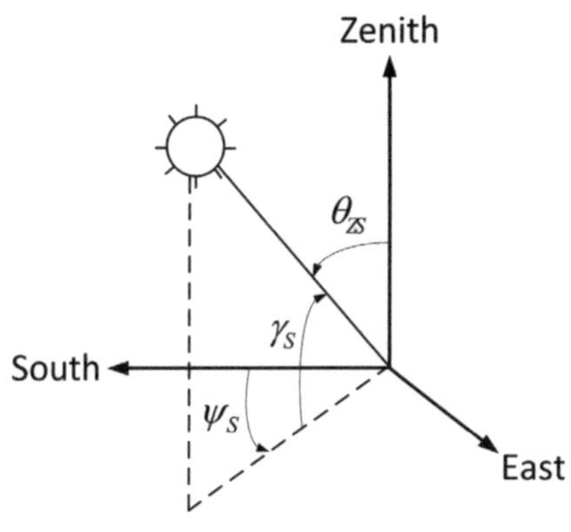

Figure 1.
Position of the sun relative to a fixed point on the earth defining the solar azimuth (ψ_s), the solar zenith (θ_{zs}), and the solar altitude (γ_s).

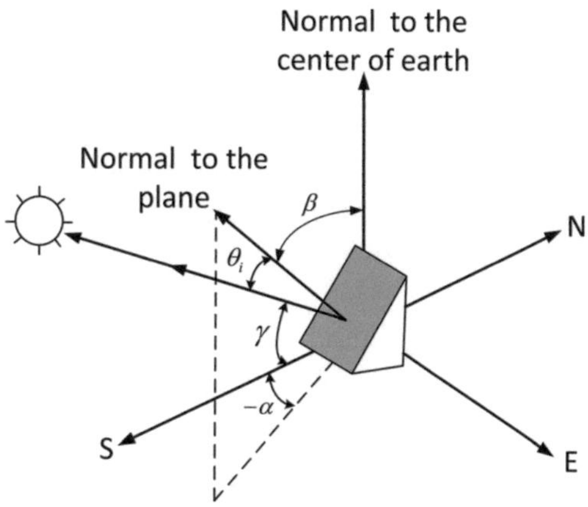

Figure 2.
Definition of angles used as coordinates for an element of sky radiation to an inclined plane of tilt β and oriented to α.

If Eq. (14) is integrated over the day, the following expression is obtained:

$$B_{0d}(0) = \frac{24}{\pi} B_0 \left[1 + 0.033 \cos\left(\frac{360}{365} d_n\right)\right] \left[\cos\phi \cos\delta \sin\omega_s + \frac{\pi}{180} \omega_s \sin\phi \sin\delta\right] \quad (15)$$

Hence, average daily extraterrestrial irradiation in a month over a surface is obtained by:

$$B_{0dm}(0) = \frac{1}{d_{n2} - d_{n1} + 1} \sum_{d_{n1}}^{d_{n2}} B_{0d}(0) \quad (16)$$

The value calculated in Eq. (16) is used to estimate the clearness index (K_{Tm}) [2].

2.7 Clearness index (K_{Tm})

It is the relation between the solar radiation at the Earth's surface and the extraterrestrial radiation over the horizontal plane. The clearness index K_{Tm} for each month is given by:

$$K_{Tm} = \frac{G_{dm}(0)}{B_{0dm}(0)} \quad (17)$$

where $G_{dm}(0)$ is the average daily horizontal global irradiation of a month, which is usually an input value [6].

2.8 Diffuse fraction index (K_d)

It is the relation between the diffuse radiation over the horizontal plane and the global radiation over the horizontal plane. This index is widely used on decomposition models to separate the global radiation into its direct and diffuse components.

$$K_{dm} = \frac{D_{dm}(0)}{G_{dm}(0)} \quad (18)$$

The modeling process for calculating the effective in-plane hourly irradiation when starting from monthly average of horizontal daily irradiation and using monthly average daily irradiance profiles is shown in **Figure 3** [9].

The daily irradiance profile can be defined in terms of irradiance divided by daily irradiation and on assuming that the profile of the extraterrestrial horizontal solar radiation translates directly into the profile of the diffuse component while an empirical correction is needed for global radiation [11]. The following equations describe the model to calculate the daily irradiance profile starting from daily average monthly values:

$$G(0) = r'_G \times G_{dm}(0) \tag{19}$$

$$D(0) = r'_D \times D_{dm}(0) \tag{20}$$

$$B(0) = G(0) - D(0) \tag{21}$$

with

$$r'_D = \frac{B_0(0)}{B_{0d}(0)} = \frac{\pi}{T} \times \left(\frac{\cos\omega - \cos\omega_s}{\frac{\pi}{180} \times \omega_s \times \cos\omega_s - \sin\omega_s} \right) \tag{22}$$

$$r'_G = r'_D \times (a + b \times \cos\omega) \tag{23}$$

$$a = 0.409 - 0.5016 \times \sin(\omega_s + 60) \tag{24}$$

$$b = 0.6609 + 0.4767 \times \sin(\omega_s + 60) \tag{25}$$

where ω and ω_s are expressed in degrees, and T is the day length, usually expressed in hours (24 h). The unit of indexes r_D and r_G is T^{-1}, and they can be used to calculate irradiation during short periods centered on the considered instant ω. Subscripts "d" and "m" refer to the daily and monthly average of daily values, respectively.

The diffuse component of the average daily irradiation, $D_{dm}(0)$, is derived from a decomposition model consisting of an empirical relationship between the clearness index, K_{Tm}, and the diffuse fraction, K_{dm}. In Ref. [9], the authors review and compare four decomposition models for monthly average of horizontal daily

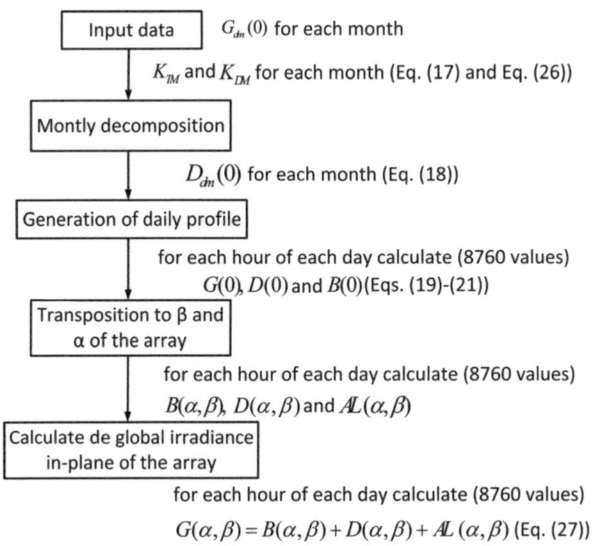

Figure 3.
Calculation of daily irradiation on an inclined surface.

irradiation: a linear relationship proposed by Page [10], two polynomial equations defined by the authors of Refs. [11, 12], and a local correlation proposed by Macagnan et al. [13]. The authors stand out the Page decomposition model in combination with the Perez transposition model [13] as a good performance combination used for passing from global horizontal irradiation to effective in-plane irradiance when it is started from monthly average of daily irradiation values.

The decomposition model proposed by Page consists of a linear equation that correlated the diffuse fraction index and the clearness index using data from locations situated between 40°N and 40°S, and it is given by:

$$K_{dm} = 1 - 1.13K_{Tm} \tag{26}$$

Once the global horizontal irradiance is separated into direct and diffuse components and the daily irradiance profile is obtained, it is necessary to calculate the effective irradiance on the plane of the array. The irradiance over the plane with a tilt β, in degrees, and oriented to angle α, conventionally measured clockwise from the south, can be obtained by:

$$G(\beta, \alpha) = B(\beta, \alpha) + D(\beta, \alpha) + AL(\beta, \alpha) \tag{27}$$

where G, B, D, and AL represent global, direct, diffuse, and albedo components, respectively. The irradiance components over the plane are given by:

$$B(\beta, \alpha) = B(0) \times r_B \tag{28}$$

$$D(\beta, \alpha) = D(0) \times r_D \tag{29}$$

$$AL(\beta, \alpha) = G(0) \times r_{AL} \tag{30}$$

The beam transposition factor is calculated straightforward from simple geometric considerations [9]:

$$r_B = \frac{max\ (0,\ cos\ \theta_i)}{cos\ \theta_{zs}} \tag{31}$$

Assuming isotropic albedo radiation, the corresponding transposition factor is given by:

$$r_{AL} = \rho \frac{1 - cos\ \beta}{2} \tag{32}$$

where ρ is the ground reflection factor. The albedo radiation is scarcely relevant and rarely measured. A general reflection value of 0.2 is considered since this is extendedly used on practice [9].

The diffuse transposition factor depends on the assumption made for the sky radiance distribution. In Ref. [9], eight transposition models are reviewed, obtaining the best results using the Perez model. Similar results are obtained in Ref. [14], where four transposition models are compared and validated with two-year data measured at site, and the most accurate results were obtained by the Hay and Davies transposition model and the Perez transposition model. In this work, the Liu and Jordan isotropic sky model is used due to its simplicity of implementation and good results as reported in Ref. [15].

In the transposition model proposed by Liu and Jordan, the diffuse radiation is given by an isotropic component coming from the entire celestial hemisphere. The diffuse transposition factor is given by:

$$r_D = \frac{1 + \cos\beta}{2} \tag{33}$$

In summary, the global irradiation over the tilted plane is calculated by:

$$G(\beta, \alpha) = B(0) \times r_B + D(0) \times r_D + G(0) \times r_{AL} \tag{34}$$

3. Experimental results: Data processing

Santa Cruz del Islote in Colombia was used as a location for the case study. Data that consist on the monthly average daily global horizontal irradiance (G_{dm}) and the monthly average ambient temperature ($T_{amb,m}$) were provided by Solargis through its pvPlanner platform, see **Table 1**.

A MATLAB routine was used to compute the monthly global irradiance over the horizontal and over the plane. Results were also compared with the data that can be processed from Solargis, and **Table 2** shows the results obtained. Comparison shows that the results are good enough for the purpose of this work.

Location	**Islote de Santa Cruz, Colombia**	
Latitude	9.79	
Longitude	−75.859167	
Time zone (GTM)	−5	
Tilt	10	Plane inclination
Azimuth	0	Plane orientation (South = 0°; West 90°)
GRF	0.2	Ground reflection factor (0–1) Default = 0.2
Month	Daily sum of global irradiation [Wh/m^2]	Average diurnal (24-h) air temperature [°C]
January	5922.6	27.8
February	6271.4	27.6
March	6267.7	27.6
April	5906.7	27.7
May	5367.7	27.9
June	5396.7	28.5
July	5587.1	28.7
August	5538.7	28.5
September	5363.3	28.2
October	5025.8	27.9
November	4970.0	27.8
December	5200.0	28.0
Average	5568.1	28.0

Table 1.
Meteorological input parameters [16].

	Global horizontal irradiation (kWh/m^2) Solargis	Global horizontal irradiation (kWh/m^2) calculated	Dev (%)	Global tilted irradiation (kWh/m^2) Solargis	Global tilted irradiation (kWh/m^2) calculated	Dev (%)
January	183.6	182.0	−0.88%	201.9	198.6	−1.65%
February	175.6	174.2	−0.81%	186.9	184.3	−1.41%
March	194.3	193.0	−0.68%	198.5	196.2	−1.14%
April	177.2	176.1	−0.65%	175	172.9	−1.17%
May	166.4	165.2	−0.70%	160.1	158.8	−0.83%
June	161.9	160.8	−0.71%	153.6	152.6	−0.65%
July	173.2	172.0	−0.69%	165.3	163.9	−0.84%
August	171.7	170.6	−0.65%	167.8	166.1	−1.03%
September	160.9	159.8	−0.70%	162	160.1	−1.17%
October	155.8	154.4	−0.91%	162.4	159.5	−1.79%
November	149.1	147.7	−0.96%	160.5	157.1	−2.13%
December	161.2	159.7	−0.93%	177.8	174.1	−2.09%
Year	2030.9	2015.3	−0.77%	2071.8	2044.1	−1.34%

Table 2.
Global horizontal and tilted irradiations.

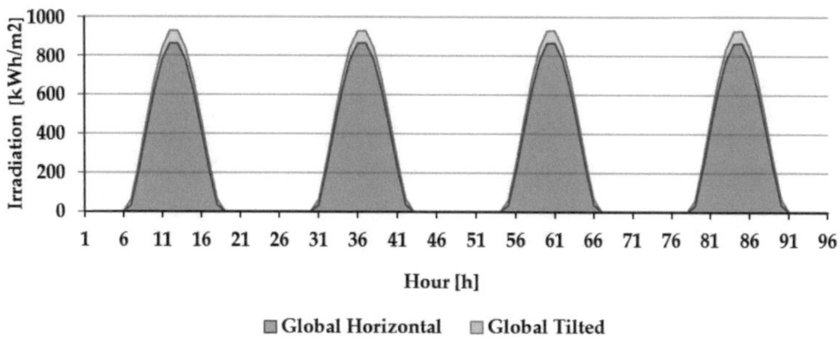

Figure 4.
Daily global irradiation calculated for "Islote de Santa Cruz."

In **Figure 4**, it is shown the daily global profile on the horizontal and on the tilted plane for the first 4 days of the year calculated for the selected location. As expected, global tilde irradiation is higher than the global horizontal irradiation.

4. Conclusions

This book chapter presented a methodology that describes in detail the data processing to obtain the global irradiation over the plane of photovoltaic arrays. The methodology includes definitions and equations necessary to perform the processing of the data. The following parameters were described and along with their equations: declination angle, solar hour angle, solar zenith, solar altitude, solar azimuth, angle of incidence, solar constant, extraterrestrial irradiance over the horizontal surface, clearness index, and diffuse fraction index.

The obtention of the global irradiance over a tilde plane requires decomposition models, which provide direct, diffuse, and albedo components of irradiation. This book chapter provides global horizontal irradiation to effective in-plane irradiance when it is started from monthly average of daily irradiation values.

This book chapter is extensive in the use of geometric considerations. This is due to the fact that input data are global horizontal irradiation and average temperature, while output data are the global irradiation over a tilde PV plane. Therefore, equations and schemes were provided for facilitating the explanation.

Acknowledgements

The authors gratefully acknowledge the financial support provided by the Colombia Scientific Program within the framework of the call "Ecosistema Científico" (Contract No. FP44842-218-2018). The authors also want to thank the "Programa de Sostenibilidad de la UdeA."

Author details

Oswaldo A. Arraez-Cancelliere, Nicolás Muñoz-Galeano* and Jesús M. López-Lezama
Facultad de Ingeniería, Grupo GIMEL, Universidad de Antioquia, Medellín, Colombia

*Address all correspondence to: nicolas.munoz@udea.edu.co

References

[1] Taner T, Demirci OK. Energy and economic analysis of the wind turbine plant's draft for the Aksaray City. Applied Ecology and Environmental Sciences. 2014;**2**(3):82-85. DOI: 10.12691/aees-2-3-2

[2] Taner T. Economic analysis of a wind power plant: A case study for the Cappadocia region. Journal of Mecha nical Science and Technology. 2018; **32**(3):1379-1389. DOI: 10.1007/ s12206-018-0241-6

[3] Taner T, Dalkilic ASA. Feasibility study of solar energy-techno economic analysis from Aksaray City, Turkey. Journal of Thermal Engineering. 2019;**3**(5):1-1. DOI: 10.18186/ thermal.505498

[4] Taner T, Naqvi SAH, Ozkaymak M. Techno-economic analysis of a more efficient hydrogen generation system prototype: A case study of PEM Electrolyzer with Cr-C coated SS304 bipolar plates. Fuel Cells. 2019;**19**:1-8. DOI: 10.1002/fuce.201700225

[5] Solargis. Solargis Solar Resource Database Description and Accuracy [Internet]. 2019. Available from: https:// solargis2-web-assets.s3.eu-west-1.ama zonaws.com/public/doc/d 143113beb/Solargis-database-desc ription-and-accuracy-v2.pdf. [Accessed: 15-11-2019]

[6] Luque A, Hegedus S. Handbook of Photovoltaic Science and Engineering. 2nd ed. Chichester: Wiley; 2011. p. 1132. DOI: 10.1002/9780470974704

[7] Khatib T, Elmenreich W. A model for hourly solar radiation data generation from daily solar radiation data using a generalized regression artificial neural network. International Journal of Photoenergy. 2015;**2015**:1-13. DOI: 10.1155/2015/ 968024

[8] Demain C, Journée M, Bertrand C. Evaluation of different models to estimate the global solar radiation on inclined surfaces. Renewable Energy. 2013;**50**:710-721. DOI: 10.1016/j. renene.2012.07.031

[9] Moretón R, Lorenzo E, Pinto A, Muñoz J, Narvarte L. From broadband horizontal to effective in-plane irradiation: A review of modelling and derived uncertainty for PV yield prediction. Renewable and Sustainable Energy Reviews. 2017;**78**:886-903. DOI: 10.1016/j.rser.2017.05.020

[10] Page JK. The Estimation of Monthly Mean Values of Daily Short Wave Irradiation on Vertical and Inclined Surfaces from Sunshine Records for Latitudes 60 N to 40 S [Thesis]. University of Sheffield; 1976

[11] Collares-Pereira M, Rabl A. The average distribution of solar radiation-correlations between diffuse and hemispherical and between daily and hourly insolation values. Solar Energy. 1979;**22**:155-164. DOI: 10.1016/ 0038-092X(79)90100-2

[12] Erbs DG, Klein SA, Duffie JA. Estimation of the diffuse radiation fraction for hourly, daily and monthly-average global radiation. Solar Energy. 1982;**28**(4):293-302. DOI: 10.1016/ 0038-092X(82)90302-4

[13] Macagnan MH, Lorenzo E, Jimenez C. Solar radiation in Madrid. International Journal of Solar Energy. 1994;**16**(1):1-14. DOI: 10.1080/ 01425919408914262

[14] SolarPower Europe. O&M Best Practices Guidelines [Internet]. 2019. Available from: https://www.researchga te.net/publication/304624410_Sola rPower_Europe_OM_Best_Practices_ Guidelines. [Accessed: 15-11-2019]

[15] Duffie JA, Beckman WA. Solar Engineering of Thermal Processes. 4th ed. Hoboken, New Jersey, United States: Wiley; 2013. p. 910. DOI: 10.1002/9781118671603

[16] Solargis. pvPlanner [Internet]. 2019. Available from: https://solargis.info/pvplanner/#tl=Google:hybrid&bm=satellite. [Accessed: 15-11-2019]

Chapter 8

Pulsed Electrical Discharge and Pulsed Electric Field Treatment during Sunflower Seed Processing

Ivan Shorstkii and Evgeny Koshevoi

Abstract

For the successful implementation of emerging electrical technologies in the oil pressing process, optimization of process parameters in combination with parameters from electrical process is crucial. The aim of this study was to evaluate the effect of the following pretreatments: pulsed electrical discharge (PED) and pulsed electric field (PEF) on rheological properties, morphological capillary-porous structure, and oil recovery of sunflower seed. FESEM analysis of the surface microstructure, pressing, and solvent extraction were used to obtain treatment efficiency after novel technologies. The results of this study show that PED and PEF treatments could be used as a pretreatment before sunflower seed processing to modify internal structure, increase the oil yield, or contribute to the mechanical destruction of oil globules and the release of free oil to the surface under gentle conditions.

Keywords: pulsed electric field, electrical discharge, crop, oil, oilseeds, processing, pressing, extraction, biofuel

1. Introduction

Sunflower oil production is rapidly emerging in Russia, Ukraine, Turkey, and other countries, filling an important niche of locally supplied protein and fat sources. Sunflower seeds are becoming of outmost importance to fulfill the requirements for safe products and find efficient ways to reduce potential chemical and technological hazards. Current methods of sunflower seed processing rely on well-developed and established food industry thermal treatment (roasting, drying, heating, and cooling), mechanical treatment (cleaning, flaking, grinding, and pressing), and fractionation methods (extraction, sedimentation, separation, centrifugation, etc.) [1–4]. Application of novel emerging processing technologies, which potentially can improve processing efficiency and quality of the products to the sunflower seeds, is rather limited.

Development of a novel scientific direction should be based on an active implementation of green technologies [5]. Green technologies include methods that enhance the efficiency of extraction of target components (oils, fats, phospholipids, polyphenols, pigments, etc.) from plant materials and can improve the traditional processing technology or combine traditional technology with novel emerging technologies [5].

Sunflower oil is a high-calorie product that is widely used in its natural form for food, canning, and cosmetic production purposes [6]. Sunflower oil is attracting increasing attention as a source of renewable raw materials for chemical and energy industries. The range of its use for chemical purposes is extremely wide—from use as a starting material for chemical synthesis to use as lubricants. Sunflower oil is more environmentally friendly: it decomposes after 7 days by 95%, whereas mineral oil decomposes only by 15–17%. As a biodiesel fuel, sunflower oil can partially replace natural oil reserves, reducing CO2 emission [7]. Sunflower is one of the main crops used for biofuel production. An additional raw material during sunflower processing is sunflower husk, which is actively introduced as fuel pellets in alternative energy [8, 9].

Recently some promising emerging technologies of electrical treatment such as pulsed electric field (PEF) and high-voltage electrical discharges (HVED) for oil production were reported by several authors [10, 11]. As a novel industry-scale technology, pulsed electric field has already been mentioned as an innovative solution for electroporation of oil cells [12, 13] during extraction process. Permeabilization of the oil cell membrane can be increased due to rectangular bipolar or monopolar electrical pulses (millisecond or even microsecond pulse width). During volumetric PEF treatment, micro- and nano-pores are formed [14]. Such gentle treatment with low-temperature effects is quite important for heat-sensitive and organic materials. For oilseeds this technology forms an intercellular component film on the surface of treated materials due to electroporation [15]. This effect is studied by our research group and has great potential in oil extraction process.

To release the oil from the solid seed matrix, at a preparation stage, flaking, crushing, and roasting are used. According to [16] more than 40% sunflower spherosomes remain undestroyed after crushing and roasting. This limits a residual oil yield in sunflower meal. PEF or PED pretreatment assists in releasing the oil from lipo-vacuoles of mesocarp cells that have not been disrupted. Depending on process conditions (possibility of soaking or wetting of treating mass), PEF or PED treatment is preferred.

In order to expand the scope of application of PED and PEF treatments on an industrial scale, a clear understanding of the changes in the internal material structure, rheological behavior, processing efficiency, and extraction kinetics is necessary.

2. Pulsed electrical discharge treatment

As a novel technology, pulsed electrical discharge (PED) treatment has already been mentioned as innovative solution for electroporation of oil cells [10, 11, 15] during extraction process. The oil cell membrane can be charged sufficiently using monopolar electrical pulses (millisecond or even microsecond pulse width) to cause the rapid dielectric breakdown of material tissue under focused discharge energy. The main result of PED treatment is micro- and nano-channel formation [17] with low-temperature effects, which is quite important for heat-sensitive materials. For oilseeds this technology forms an intercellular oil film on the surface of treated materials due to electroporation [15]. This effect is studied by our research group and has great potential in oil extraction process.

Pilot-scale PED treatment set-up is described in **Figure 1a**. Pulsed electrical discharges were distributed to the treating mass in a filamentary glow discharge mode. Treatment chamber was in a point-plane electrode configuration. A stainless steel sphere (10 mm in diameter) was connected to a permanent magnet to

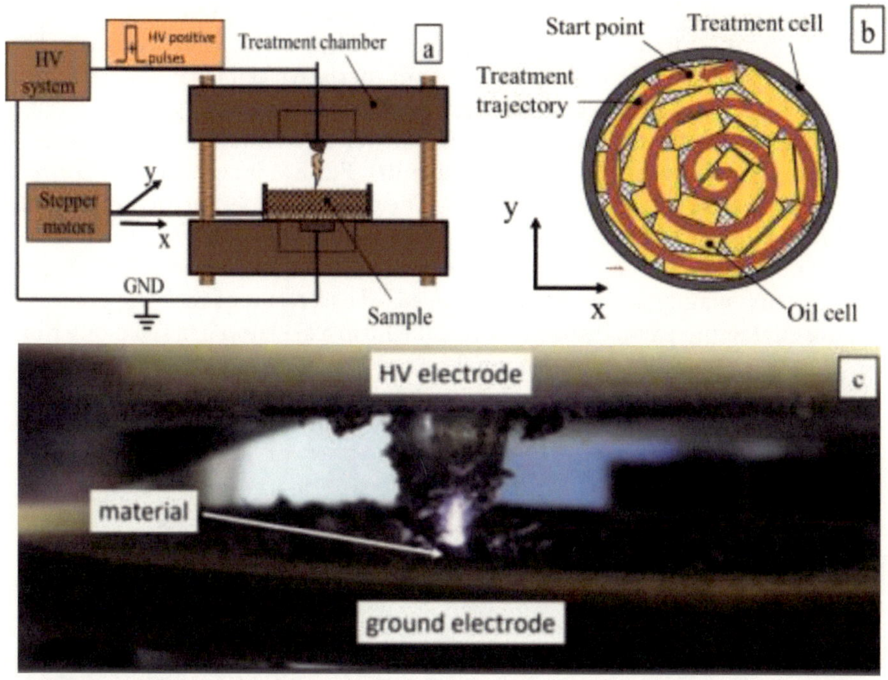

Figure 1.
Experimental set-up used to generate pulsed electrical discharges in air (a), schematic visualization of treatment trajectory in a treatment cell (b), and visualization of a treatment process (c) (adopted from [18]).

focus discharges. Top and bottom electrodes were placed in a dielectric holder. The electrode gap was set as 15 mm. Treatment chamber was set on positioning platform with two stepper motors. Sunflower mass treatment trajectory was controlled by the authors' software in accordance with **Figure 1b**. Immediately after PED treatment, sunflower mass was transfer for pressing step.

In Advanced Technologies Lab (Krasnodar, Russia), PED treatment was performed using the Matsusada power high-voltage system in combination with high-voltage amplifier [18]. In the set-up used in these experiments, the pulse or discharge duration was of 10 μs, and the frequency was 30 Hz. Each pulse or discharge applied provides a voltage of 30 kV. Specific input energy ranged from 10.0 J/kg up to 1.0 kJ/kg (**Figure 1c**).

The nature of the applied discharge in the treatment is shown in **Figure 1**. The high-voltage discharge oscillogram (**Figure 2**) is shown as two signals (purple and yellow). The yellow line characterizes the input signal coming to the amplifier and has distortion on the increasing front of the forward-angle pulse. The purple signal characterizes the output discharge on the electrodes. In this case, the microprobe on the falling edge of the rectangular signal in the form of repeated exponential pulses is clearly visible. These pulses characterize the presence of a breakdown in the air, visually observed during the experiment.

2.1 PED treatment effect on seed structure

Crushing and roasting are industrial technological steps assisting to release the oil from solid seed matrix, but not totally. To analyze the effect of PED treatment, treated sample was covered with gold and analyzed on FESEM. When analyzing the micrographs of the seed cake surface after roasting and PED treatment, an oil film (F) is clearly visible, and electrical holes (H), expressed in the form of convex craters, less than 2 microns in size are noticeable (**Figure 3c**). PED pretreatment assists in releasing the oil from lipo-vacuoles of mesocarp cells that have not been disrupted. After sunflower seed cake roasting operation (110°C), globules of oil (O) are clearly visible in the dark and light areas, as well as an oil film (F), expressed in the form of light homogeneous areas (**Figure 3b**). FESEM images of PED-treated sunflower seed cake are shown in **Figure 3**. The oil cell membrane (M) is presented in the form of looped light fibers surrounding the oil cell (**Figure 3a, b**). At rest the pulp there is a loose coagulation structure. The adhesion between the dispersed particles occurs mainly due to the free oil released as a result of roasting process. Thus, the contacting particles of the seed cake form a skeleton in a fixed oily film, adhered to the walls of the screw channel.

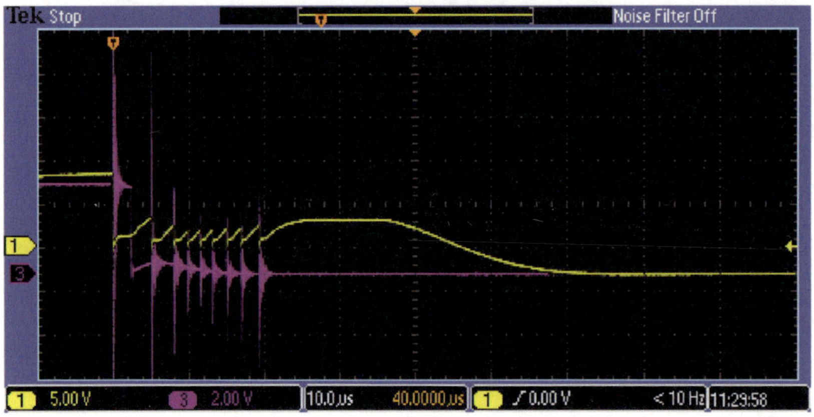

Figure 2.
Waveform of the input signal coming to the amplifier (yellow) and the output high-voltage signal coming to the electrodes (purple).

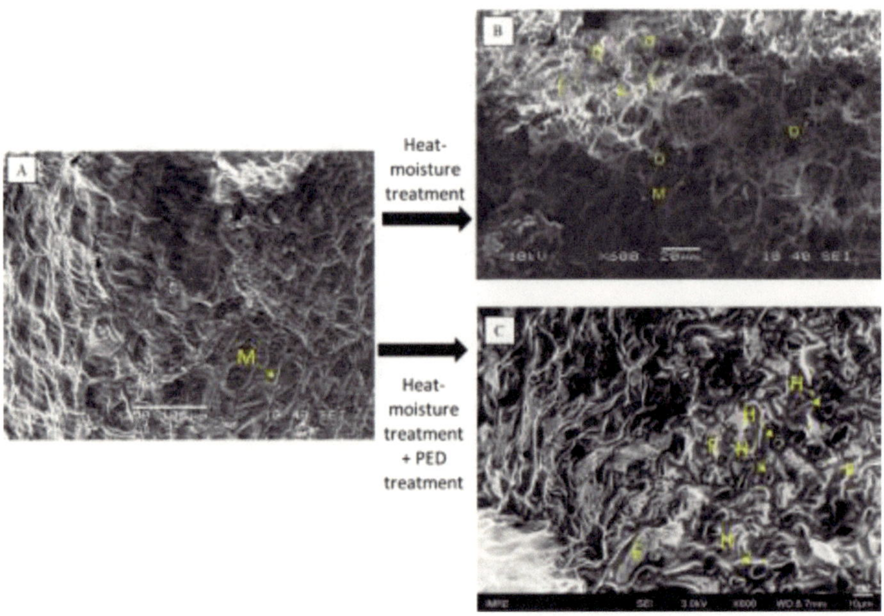

Figure 3.
SEM image of initial (a) and heat moisture-treated sunflower seed cake (b) and FESEM image of heat moisture-treated plus PED-treated sunflower seed cake at x250 and x600 magnification (adopted from [18]).

2.2 Rheological properties of sunflower seed mesh after PED treatment

Studying the pulsed electrical discharge effect on rheological behavior of sunflower seed cake can help to advance work and to develop projects for their industrial application. Few authors reported positive effects on rheological properties on some oil crops [19, 20]. Positive effects of electric discharges on rheological behavior of material with high viscosity are reported by several authors [21, 22]. It means that electrical treatment process can modify internal seed structure, increase the amount of destroyed spherosomes, and change rheological parameters of seed cake.

Structurally sunflower seed cake is a complicated dispersed system consisting of dispersed phase bubbles, oil globules with husk droplets, and dispersed medium as a protein shell (**Figure 3**). Such dispersed systems usually are non-Newtonian and characterized by complex rheological behavior.

Each individual pulsed discharge develops according to filamentary glow discharge mechanism. With a large surface resistance of the dielectric material (sunflower cake), a charge is created on its surface, created both during the charge drift from the discharge zone of the gas gap and as a result of the surface discharge. Electrical discharges penetrate the sunflower seed mass causing the crater appearance in a treatment zone. As a focused electrical discharge, energy forms a channel and damages cell membranes.

When analyzing SEM images and considering the supplied pulse, discrete traces of surface discharges can be noted. The size of the craters remaining on the test material averages 3–10 μm (**Figure 3c**). It is important to note that for discharge with a "positive" voltage, they do not merge with each other, which indicates the same sign of the electric charge distributed over their body.

Considering the duration of the development of plasma-chemical processes, the speed of ions in an electric field, and the diffusion of chemically active compounds, the processing time of the material was determined to be minimal in terms of product quality. The maximum number of destroyed cell membranes and the maximum release of free oil to the surface were observed when exposed to a maximum number of pulses.

Using a modified rotational viscometer Fungilab One Pro (Fungilab, Spain) for sunflower mass, an apparent viscosity via shear rate dependence was obtained. **Figure 4** shows the apparent viscosity for PED-treated and non-treated sunflower seed mass as a function of the shear rate. Non-Newtonian shear- thinning behavior from the flow curves sunflower seed cake (pretreated and non-treated) exhibit is observed. A similar shear-thinning behavior has also been observed for sesame seed [23] and sage seed solutions [24].

Pulsed electrical discharges contributed to the mechanical destruction of oil globules and the release of free oil to the surface. PED treatment decreases the initial shear stress from 24.5 to 22.9 Pa for samples after treatment with an $E = 16$ kV/cm field and number of discharges $n = 1800$.

Most of the oilseed materials have inhomogeneous loose structure in comparison with cellular structure of fruits and vegetables. PED effect can be visually observed for a first few cell layers on sunflower mass surface. This indicates a filtration or diffusion processes acting inside the internal material structure. As a result, there is no significant difference between viscosity curves for treated by PED and untreated materials (**Figure 4**).

2.3 PED treatment effect on sunflower oil yield during pressing

Initial oil content of heated sunflower meal according to the specification was 49.78 ± 0.5%. To analyze PED treatment effect on oil extraction, heated sunflower meal was treated and then pressed and extracted using n-hexane. **Figure 5** shows the extraction kinetic curves without and after PED treatment. PED treatment was performed at the following parameters: field strength 13.3 kV/cm, number of pulses 3600, and processing time 2 minutes. **Table 1** shows the oil yield data at various stages of the experiment. With the use of pretreatment with high-voltage discharges, the maximum oil yield was 15.7%, compared to the oil yield without treatment of 13.8% after 3 minutes of pressing (**Table 1**). It is worth noting that the improved oil yield later positively affected the residual oil content in sunflower meal, reducing the amount of oil by 0.58%.

Due to the preliminary destruction of the integrity of the oil cell membranes, the mass has a "sponge-like" structure. Like dielectric breakdown process, the flow of charged particles passes through the structure of the material and forms a channel. Due to the formation of numerous channels, mass transfer characteristics are improved.

Presented curves at **Figure 5** have a similar appearance to the extraction kinetics curve. It can be seen that due to the PED pretreatment, the integrity of the oil cells

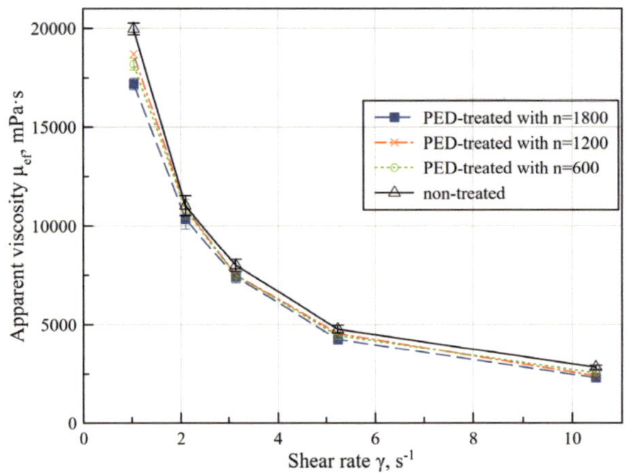

Figure 4.
Apparent viscosity and shear rate dependence of non-treated seed cake (Δ) and after PED treatment for protocol a with number of discharges n = 1800 (■), n = 1200 (x), and n = 600 (○).

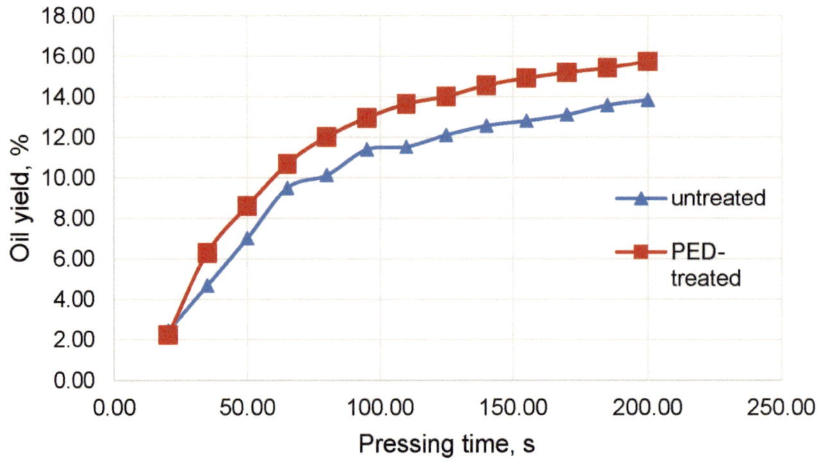

Figure 5.
The yield of oil versus time dependence during pressing on hydraulic press.

	Pressing		Solvent extraction	
	Without PED	After PED	Without PED	After PED
Oil yeild, %	13.8 ± 0.2	15.7 ± 0.2	34.18 moisture (3.01)	33.47 moisture (1.55)
Residual oil content of sunflower meal, %			1.19 ± 0.06	0.61 ± 0.05

Table 1.
Oil yield values at various stages for treated and untreated heated sunflower meal samples.

disintegrated, which as a result caused an improved oil yield. Previous data related to the application of a pulsed electric field to the oilseed material sunflower showed the possibility of destruction of oilseed cells using electric fields [11]. This is also confirmed by microstructural studies [17].

In sunflower oil production depending on the modernity of production, 0.5–5% of the oil remains in the sunflower oilseed meal [16]. Totally, it reaches up to 25 g/1 kg of sunflower oil as a monetary loss. Preliminary PED treatment could potentially help to extract 99.8% of sunflower oil during extraction. For a large-scale sunflower oil plant with 1200 tons/day capacity, PED technology could potentially increase the sunflower oil production by 6.96 tons per day. This additional revenue increases profit margins and pays back the investment in PED equipment.

2.4 Pulsed electrical discharge treatment effect on oil quality

Chemical parameters such as peroxide and acid value of extracted by pressing sunflower oil are shown in **Table 2**. The peroxide value characterizes the degree of oxidation of fats and oils and expressed in terms of the number of grams of iodine absorbed per gram of the sample. The peroxide value for oil for untreated and PED treated samples were 6.8 and 13.7, respectively. From the authors' point of view, the double roasting process could cause such high value of the peroxide value after PED treatment. Since to recreate the industrial process of heated sunflower meal processing it was heated for a long time to a temperature of 100°C, first for processing with high-voltage discharges, and as a preparation for pressing. As a result, the active action of oxygen during prolonged heating process contributed to an increase in this parameter by two times.

Acid value (the number of milligrams of potassium hydroxide required to neutralize free acids in 1 g of the sample) was used to check the purity of the oil and characterizes the degree of lipid hydrolysis. The acid number of the processed heated sunflower meal did not change and amounted to 1.43 KON/g.

For a more detailed analysis of the oil quality after treatment, IR spectrometry was used. Data from the obtained spectra for samples without and after processing are shown in **Figure 6**.

The spectrum obtained for pretreated oil had an additional absorption band in the region of 2300–2500 cm^{-1}. Several groups of RNH_3^+, $R_2C{=}NH^+$, and R_nPH_{3-n} phosphines can fluctuate in this region. Based on these data, it can be assumed that as a result of pulsed electrical discharge treatment, a new substance may be formed, which includes an amino or phospho group, due to the transition of molecules to an excited state. Sunflower seeds contain a whole complex of vitamins and microelements, including phosphorus, which are not present in the sunflower oil. Perhaps a new method of pretreatment allows you to extract nitrogen-containing vitamins such as B_6 and B_9. **Tables 3** and **4** provide a detailed breakdown of the spectrum for amino or phospho groups.

Sample	Peroxide value	Acid value
Untreated	6.8 mmol active O_2/kg	1.45 ml KOH/g
PED treated with E = 13.3 kv/cm and number of pulses n = 3600	13.7 mmol active O_2/kg	1.43 ml KOH/g

Table 2.
Oil quality indicators for samples treated and untreated by PED.

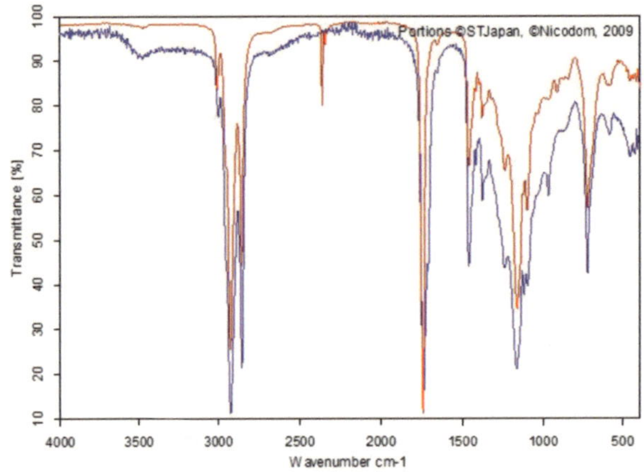

Figure 6.
IR spectrum of untreated extracted oil (blue line) and after PED treatment (red line).

Salts of amines			
NH_4^+	**3300–3030 (3.03–3,30)** **1430–1390 (7.00–7.20)**	**c.** **c.**	**υ_{N-H}, wide,** **δ_{N-H}, wide**
RNH_4^+	~3000 (3.33) ~2500 (4.00) ~2000 (5.00) 1600–1575 (6.25–6.35) 1500 (6.67)	δ_{N-H}, wide, c. δ_{N-H}, med. δ_{N-H}, med. δ_{N-H} δ_{N-H}	Few bands
$R_2NH_4^+$	2700–2250 (3.70–4.33)	c.	υ_{N-H}, sometimes it appears as a group of bands
$R_3NH_4^+$	2700–2250 (3.70–4.33)	c.	υ_{N-H}, sometimes it appears as a group of bands
R_4H^+	—	—	Does not have characteristic bands
$R_2C=NH^+$	2500–2300 (4.00–4.34) 2200–1800 (4.55–5.56)	cp	Wide band, sometimes manifested in the form of groups of strips

Table 3.
Decoding the IR spectrum by the amino group.

Thus, according to the results of oil quality parameters, it was shown that PED treatment at the stage of preparation of the material for subsequent processing has a nonsignificant effect. The detected deterioration of the peroxide number value is primarily associated with the setting of the experiment and requires a more detailed analysis.

Phosphines R_nPH_{3-n}	2275–2440 (4.40–4.01) 1080–1090 (9.26–9.17) 910–940 (10.00–10.64) 1405–1440 (7.12–6.94)	υ_{PH}, med. δ_{PH2}, med. δ_{PH2} δ_{CH2P}	
Phosphine oxides $R_3P^+ \rightarrow O^-$	1140–1300 (8.77–7.69) ~1150 (~8.70) ~1190 (~8.40)	c. c. c.	The frequency is affected by the electronegativity of substituents Aliphatic Aromatic
Phosphoric acid $R_n(HO)_{3-n}PO$	2550–2700 (3.92–3.70) 2100–2350 (4.76–4.26)	υ_{OH}, med. υ_{OH}, med.	Wide band

Table 4.
Decoding the IR spectrum for the phospho group.

According to the data obtained, a novel technology for sunflower oil production called pulsed electrical discharge pretreatment has been shown as an emerging technology to improve sunflower oil yield.

3. Pulsed electric field treatment

Pulsed electric field technology is foreseen as a rather flexible tool applicable for a range of cases from stressing of single cells or electroporation of plant cells as a preparation method before extraction or drying [25]. PEF treatment is a series of high-voltage electric pulses of short duration that creates the electric field strength from 1 to 80 kV/cm inside the treatment chamber. In a recent study on oil crops [11], it has been discovered that PEF can damage the structure of oil cells during the process, as well as facilitate the oil extraction without any temperature effects before extraction.

With respect to oil crops, continuous PEF treatment method is a sequence of the following operations: preparing the material (grinding, soaking), PEF treatment in continuous flow mode, drying, and extraction (**Figure 7**).

PEF treatment requires pre-shredding and soaking process of the material to produce a homogeneous, conductive medium. PEF treatment technology has been tested on such oil materials, such as olives [27], canola [28], and rapeseed [13]. As noted above, one of the conditions of PEF treatment is sufficient conductivity of the medium (humidity of material should be not less than 50%) [13]. Conductivity (electrical conductivity) also varies with temperature and dielectric characteristics of the material itself. Most oil-bearing materials are polar dielectrics, which creates certain difficulties for the industrial application of the PEF treatment method. The requirement to conduct the process of wetting of oil-bearing material is not always possible on production and can lead to a negative effect of the treatment [15]. Given the heterogeneity of the structure of the crushed seed, treatment according to the PEF treatment method in the works [13, 15] was conducted in the following sequence: grinding, moisturizing with the addition of water (approximately 1:1), PEF treatment, drying, and screw pressing.

Recent trend of PEF treatment application as a pretreatment prior to extraction of oil from oil-bearing materials. The greatest effect of PEF treatment as a preparatory stage was noted for olives [27]. The authors note that the increase in the yield of olive oil was 13.3 wt.% after processing with pulsed electric field intensity of 2 kV/cm and energy expenditure of 11.25 kJ/kg. The content of

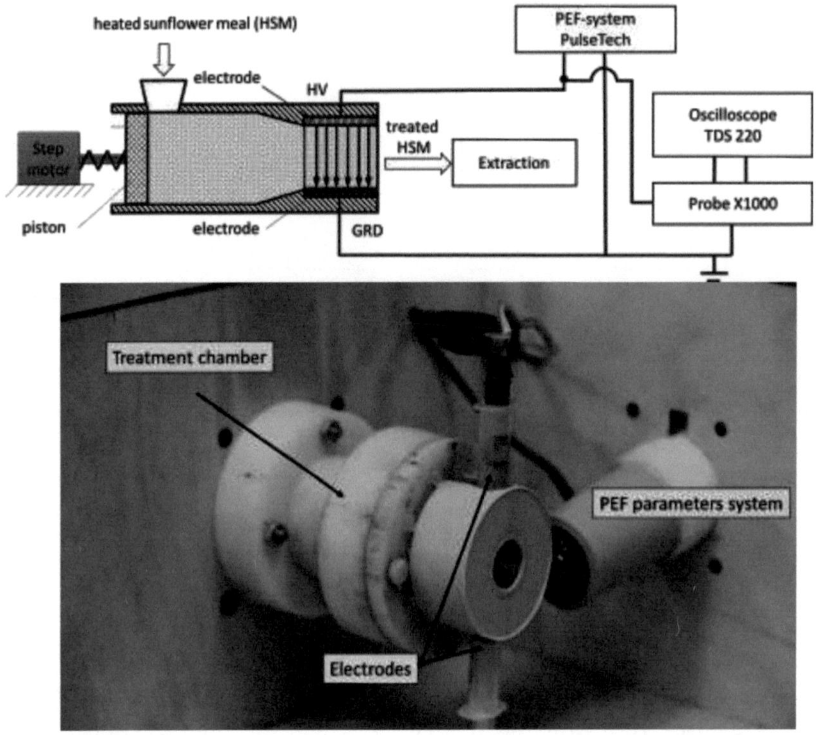

Figure 7.
Pilot PEF system with a continuous-flow chamber for sunflower oil production and continuous-flow treatment chamber with a parameter control system [26].

phenolic components, phytosterols, and tocopherols is significantly higher than in the samples without treatment.

PEF treatment with field strength E = 7 kv/cm and energy expenditure Q = 84 kJ/kg for rapeseeds allowed to increase the yield of oil by 2%, while the content of various associated useful components in the oil increased [13]. When black cumin seed was exposed to a field strength of E = 3.25 kV/cm, the yield of oil was increased by 37% [11]. However, the authors do not comment on such a significant increase in output of oil screw press after processing, limited only by the explanation of the effect of the opening of oil cells due to the treatment. [10] noted a 4.9 wt% increase in output of oil of sesame after PEF treatment with a field strength of E = 20 kV/cm, a pulse frequency of 0.5 Hz, and duration of 10 μs with preservation of quality characteristics.

PEF treatment method compared to the process of heating food products is more effective, since it allows to maintain the original quality characteristics of the food product. This treatment can be combined with other electro-physical methods of "green technologies" in the process of extracting vegetable oils [5]. With respect to oleaginous materials, data are available only on a pilot plant with a capacity of 2, 20, and 40 kW.

Based on the above, it can be concluded that PEF technology has a great potential of industrial application by combining the traditional processes of production of vegetable oils during the stages of preparation of the material for extraction and conducting extraction with an overlay of PEF and preparation prior to screw pressing. PEF treatment method regarding oil crops requires a number of preparatory operations such as soaking and subsequent drying, which creates a number of difficulties in the transition to large-scale production. However, this problem can be solved by soaking in solvent, as is done on the example of sunflower seeds [15].

3.1 Sunflower seed surface analysis after PEF treatment

Modification of the surface structure of a sunflower cotyledon after PEF treatment is shown in **Figure 8**. The possibility of sunflower oil cell destruction using short unipolar electrical pulses is clearly noticeable. It was experimentally defined that on the area of 1 mm^2, more than 25 convex craters appeared. According to average sunflower cell size of 100–200 μm, pulsed electric field was effecting on most of the cell membranes. Such kind of treatment could be used to create a new type of porous body structures [29].

In the case of pulse electric field application, sunflower body can be presented in the form of a modified capillary-porous model with additional microcapillaries. Such new model can be used to describe heat and mass transfer processes in sunflower seeds pretreated by pulsed electric field.

3.2 PEF treatment effect on the oil yield

Extraction yield obtained after PEF treatment is considered to be the main parameter to determine the economic efficiency and global performance of PEF-assisted sunflower oil production. In PEF application for the oil extraction processes from sunflower seeds, the main parameter is the applied energy. Sunflower seed preparation for PEF treatment and further extraction process were performed using ethanol solvent. It was determined that the increase in specific energy for a given electrical conductivity of samples has a monotonical increasing power dependence (**Figure 9**).

However, a further increase in the intensity index E does not significantly increase the oil yield index. The energy consumed (56.7 kJ/kg) for 7 kV/cm is the highest value of the oil yield of 48.5%, which was established experimentally and limited by temperature influence appearing during PEF treatment.

3.3 PEF treatment effect on the oil quality

From a chemical point of view, PEF treatment has no increased reaction of molecular oxygen with triacylglycerine, which negatively effects on chemical characteristics. PEF increases the content of human health-related compounds, such as polyphenols and tocopherols [15]. Sunflower oil spectra and color plot of untreated and PEF-treated sunflower seeds are shown in **Figure 10**.

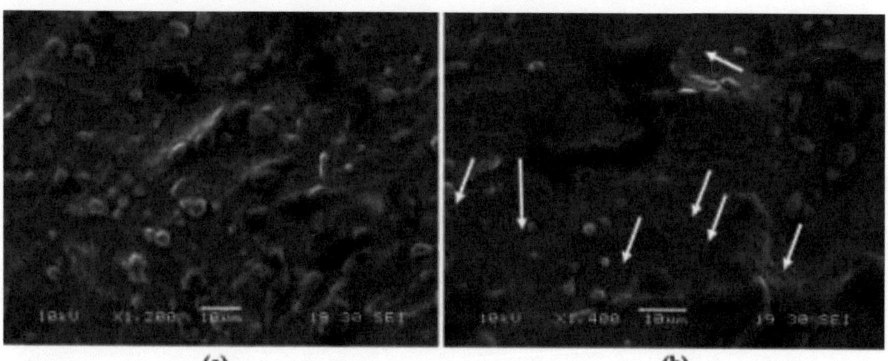

Figure 8.
Experimental set-up used to generate pulsed electrical discharges in air (a) and schematic visualization of treatment trajectory in a treatment cell (b).

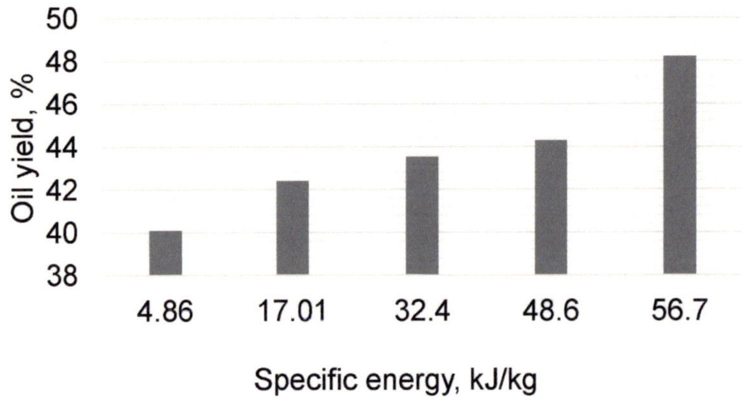

Figure 9.
The oil yield on specific energy of PEF treatment dependence.

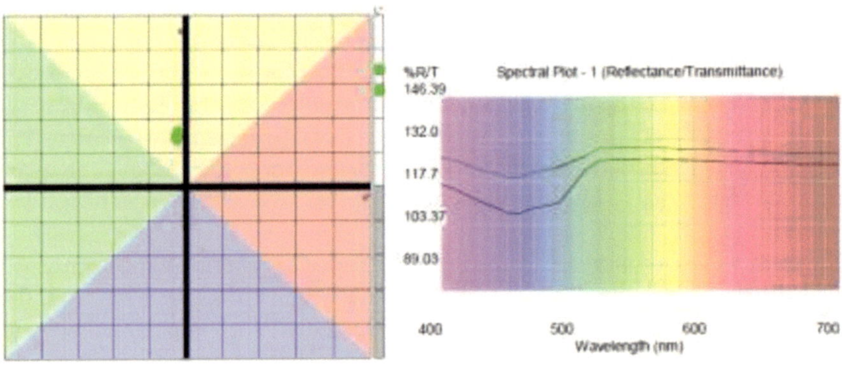

Figure 10.
Sunflower oil spectra plot and color plot for PEF-treated (1) marked as a black line and untreated samples (2) marked as a green line.

Non-treated oil obtained after ethanol extraction (**Figure 10**, black line) has a yellowish hue and lies in the second quadrant of the color spectrum. The same results of the oil yield were obtained for PEF-treated sunflower mass by ethanol (green line). Main color parameters such as L*, a*, and b* for extracted sunflower oil are listed in **Table 5**. PEF treatment affected lightness on sunflower oil, ranging from 91.68 in untreated to 92.06 in PEF-treated samples. Applying of PEF treatment allowed to increase the blue-yellow coordinate b* from −1.84 up to −1.46. For PEF pretreated sunflower mass, initial value of green-red coordinate, a*, decreased from 9.51 down to 7.06. From the authors' point of view, such difference in color parameters might be due to a chemical reaction of electrodes and plant material.

In summary, the initial sunflower oil after PEF treatment became more yellowish and greenish oil.

According to the data obtained in this chapter, a novel technology for sunflower oil production called pulsed electric field pretreatment has been shown as an emerging technology to improve sunflower oil yield up to 8.05% using ethanol as a solvent. This is not a maximum possible increase, since the disintegration index did not reach its maximal value for a completely destroyed amount of cell membranes. However, in production cycles, factories are using n-hexane which is more efficient for extraction processes. PEF treatment in a continuous flow of roasted sunflower seed cake material could help sunflower oil mills to increase profit margins.

Sample	L*	a*	b*
Untreated	91.68	-1.84	9.51
PEF-treated	92.06	-1.46	7.06

Table 5.
General color parameters of control and PEF sunflower oil.

4. Conclusions

Pulsed electrical discharge and pulsed electric field treatment were studied on physical, morphological, and oil recovery characteristics of sunflower seeds. The efficiency of PED technology for sunflower seed oil yield and oil quality, along with surface and rheological behavior, was studied. Solvent extraction efficiency and internal structure changes along with oil quality were studied for sunflower seeds pretreated by PEF. Our study demonstrates significance of both technologies for structural change and technological parameter efficiency.

The obtained effects are directly dependent on the processing method, preliminary preparation of the material, its initial component composition, and structure of the material. Moreover, as shown by literature data, the use of these electrical treatment methods allows to obtain products (oils), enriched with bioactive components. From the point of view of product quality, the proposed methods are able to preserve the original taste and organoleptic and qualitative indicators of oil-bearing materials and can be used to obtain high-quality functional products. Most of the research related to the study of methods of PED and PEF treatment should be carried out on pilot equipment, with technological parameters close to the industrial scale. It undoubtedly actualizes the obtained results and allows to predict the effects for the industrial scale.

Acknowledgements

The reported research was funded by internal grant of Kuban State University of Technology.

Conflict of interest

The authors declare no conflict of interest.

Author details

Ivan Shorstkii* and Evgeny Koshevoi
Kuban State University of Technology, Krasnodar, Russian Federation

*Address all correspondence to: i-shorstky@mail.ru

References

[1] Luciana J, Petrella I. Speculation in the oil market. Journal of Applied Econometrics. 2014;**30**:621-649. DOI: 10.2139/ssrn.2038977

[2] Ramadan MF. Healthy blends of high linoleic sunflower oil with selected cold pressed oils: Functionality, stability and antioxidative characteristics. Industrial Crops and Products. 2013;**43**:65-72. DOI: 10.1016/j.indcrop.2012.07.013

[3] Tasan M, Gecgel U, Demirci M. Effects of storage and industrial oilseed extraction methods on the quality and stability characteristics of crude sunflower oil (*Helianthus annuus* L.). Grasas y Aceites. 2011;**62**:389-398

[4] Temelli F. Perspectives on supercritical fluid processing of fats and oils. Journal of Supercritical Fluids. 2009;**47**:583-590. DOI: 10.1016/j.supflu.2008.10.014

[5] Chemat F, Vian MA, Cravotto G. Green extraction of natural products: Concept and principles. International Journal of Molecular Sciences. 2012;**13**:8615-8627. DOI: 10.3390/ijms13078615

[6] Borodin K, Salnikov S. Development of sunflower oil exports in Russia and the EEU: Main trends, prospects, and evaluations by the gravity model. International Economic Journal. 2018;**32**(3):418-437. DOI: 10.1080/10168737.2018.1520280

[7] Boumesbah I, Hachaïchi-Sadouk Z, Ahmia AC. Biofuel production from sunflower oil and determination of fuel properties. In: Dincer I, Colpan C, Kizilkan O, Ezan M, editors. Progress in Clean Energy. Vol. 2. Cham: Springer; 2018. DOI: 10.1007/978-3-319-17031-2_9

[8] Kułażyński M, Jabłoński S, Kaczmarczyk J, Świątek Ł, Pstrowska K, Łukaszewicz M. Technological aspects of sunflower biomass and brown coal co-firing. Journal of the Energy Institute. 2018;**91**:668-675. DOI: 10.1016/j.joei.2017.06.003

[9] Perea-Moreno MA, Manzano-Agugliaro F, Perea-Moreno AJ. Sustainable energy based on sunflower seed husk boiler for residential buildings. Sustainability. 2018;**10**:3407. DOI: 10.3390/su10103407

[10] Sarkis JR, Boussetta N, Blouet C, Tessaro IC, Ferreira Marczak LD, Vorobiev E. Effect of pulsed electric fields and high voltage electrical discharges on polyphenol and protein extraction from sesame cake. Innovative Food Science & Emerging Technologies. 2015;**29**:170-177. DOI: 10.1016/j.ifset.2015.02.011

[11] Bakhshabadi H, Mirzaei H, Ghodsvali A, Jafari SM, Ziaiifar AM. The influence of pulsed electric fields and microwave pretreatments on some selected physicochemical properties of oil extracted from black cumin seed. Food Science & Nutrition. 2018;**6**:111-118. DOI: 10.1002/fsn3.535

[12] Shorstkii I, Mirshekarloo MS, Koshevoi E. Application of pulsed electric field for oil extraction from sunflower seeds: electrical parameter effects on oil yield. Journal of Food Process Engineering. 2017;**40**:e12281. DOI: 10.1111/jfpe.12281

[13] Guderjan M, Elez-Martínez P, Knorr D. Application of pulsed electric fields at oil yield and content of functional food ingredients at the production of rapeseed oil. Innovative Food Science & Emerging Technologies. 2007;**8**:55-62. DOI: 10.1016/j.ifset.2006.07.001

[14] Boussetta N, Grimi N, Vorobiev E. Pulsed electrical technologies assisted polyphenols extraction from agricultural plants and bioresources: A review. International Journal of Food

Processing Technology. 2015;**2**:1-10. DOI: 10.15379/2408-9826.2015.02.01.1

[15] Shorstkii I, Khudyakov D, Mirshekarloo MS. Pulsed electric field assisted sunflower oil pilot production: Impact on oil yield, extraction kinetics and chemical parameters. Innovative Food Science & Emerging Technologies. 2020;**60**:102309. DOI: 10.1016/j.ifset.2020.102309

[16] Clef EL, Kemper T. Sunflower seed preparation and oil extraction. AOCS Press. 2015:187-226. DOI: 10.1016/B978-1-893997-94-3.50014-3

[17] Shorstkii IA, Zherlicin AG, Li P. Impact of pulsed electric field and pulsed microwave treatment on morphological and structural characteristics of sunflower seed. OCL. 2019;**26**:47. DOI: 10.1051/ocl/2019048

[18] Shorstkii I, Khudyakov D. Influence of pulsed electrical discharge, hydrostatic pressure and temperature on rheological properties of sunflower cake during oil pressing. Heliyon. 2020;**6**:e03046. DOI: 10.1016/j.heliyon.2019.e03046

[19] Han Z, Zeng X-A, Zhang B-S, Yu S. Effects of pulsed electric fields (PEF) treatment on the properties of corn starch. Journal of Food Engineering. 2009;**93**:318-323. DOI: 10.1016/j.jfoodeng.2009.01.040

[20] Pereira RN, Galindo FG, Vicente AA, Dejmek P. Food Biophysics. 2009;**4**:229-239. DOI: 10.1007/s11483-009-9120-0

[21] Ahmed J, Ramaswamy HS, Kasapis S, Boye JI. Novel Food Processing: Effects on Rheological and Functional Properties. Boca Raton: CRC; 2010. pp. 226-229

[22] Sizonenko ON, Kolmogorova RP, Iskimzhi AI, Taftay EI, Tkachenko A, Khvoshchan OV. The influence of surface-active agents treated by the electric discharge on rheological oil parameters. Neftyanoe Khozyaistvo (Petroleum Industry). 2003;**11**:79-81

[23] Akbulut M, Saricoban C, Ozcan MM. Determination of rheological behavior, emulsion stability, color, and sensory of sesame pastes (Tahin) blended with pine honey. Food and Bioprocess Technology. 2010;**5**:1832-1839. DOI: 10.1007/s11947-011-0668-6

[24] Salehi F, Kashaninejad M. Effect of drying methods on rheological and textural properties, and col-or changes of wild sage seed gum. Journal of Food Science and Technology. 2015;**52**:7361-7368. DOI: 10.1007/s13197-015-1849-5

[25] Alles MC, Smetana S, Parniakov O, Shorstkii I, Toepfl S, Aganovic K, et al. Bio-refinery of insects with pulsed electric field pre-treatment. Innovative Food Science & Emerging Technologies. 2020;**64**:102403. DOI: 10.1016/j.ifset.2020.102403

[26] Shorstkii I. Application of Pulsed Electric Field Treatment as a Biomaterials Preparation for Drying Process. Krasnodar: KubSTU; 2020. p. 172

[27] Abenoza M, Benito M, Saldaña G, Álvarez I, Raso J, Sánchez-Gimeno AC. Effects of pulsed electric field on yield extraction and quality of olive oil. Food and Bioprocess Technology. 2013;**6**:1367-1373. DOI: 10.1007/s11947-012-0817-6

[28] Teh SS, Niven BE, Bekhit AEDA, Carne A, Birch EJ. Microwave and pulsed electric field assisted extractions of polyphenols from defatted canola seed cake. International Journal of Food Science and Technology. 2015;**50**:1109-1115. DOI: 10.1111/ijfs.12749

[29] Tovbin YK. The Molecular Theory of Adsorption in Porous Solids. USA: CRC Press; 2017. p. 780. DOI: 10.1201/9781315116297

Section 3

Biomass Energy

Chapter 9

Bioenergy Potential of Turkey's Forest Sources, Biomass Energy Conversion Methods, Products, and Applications

Selçuk Sarıkoç

Abstract

Forests have been an important bioenergy source for mankind through the long ages, and they will continue as biomass feedstock sources in the future. This study aims to investigate Turkey's forest source, biomass resource, fuel wood, and forest residue potential to discover the bioenergy potential of Turkey. How to convert this potential to energy was evaluated in terms of applications and products. Thus, the most common biomass conversion methods such as thermal processes, pyrolysis, gasification, and combustion, and biological processes, fermentation, anaerobic digestion, and biophotolysis processes, have been explained as biomass energy conversion methods. Besides, the products of biomass are explained by its energy application fields. Overall, the bioenergy potential of Turkey's forest sources and biomass energy conversion methods will be overviewed by this study. Thus, this study will be attracted attention to forests' biomass source the effects on economic, ecological, and socio-economic respects.

Keywords: Turkey's forest, biomass, bioenergy potential, biomass energy conversion methods, alternative biofuels

1. Introduction

Rapidly growing population and industrialization have increased the energy request. It is very important that this increased energy requirement is from sustainable and environmentally friendly resources. At this point, biomass energy stands out with it being sustainable, environmentally friendly, and an inexhaustible resource that can be obtained anywhere. Especially in rural areas, it is becoming the most promising energy source due to its positive effects on socioeconomic developments [1]. In addition to this, woody biomass is estimated to meet approximately 2–18% of primary energy consumption in 2050 [2].

Biomass is vegetative organisms that plants produce and store from organic matter using photosynthesis using solar energy. Bioenergy is used to define energy and energy-related products produced from biomass. Biomass is formed as a result of the combination of sunlight and carbon dioxide and water in the atmosphere with photosynthesis reaction [1, 3].

$$CO_2 + 2H_2O \xrightarrow[\text{light+heat}]{} (|CH_2O| + H_2O) + O_2 \tag{1}$$

where $|CH_2O|$ refers to biomass as carbohydrate [3].

Biomass has been the most common and crucial energy source that is used in heating and cooking for thousands of years. Thus, wood is still the most widely used and richest biomass energy source. Today, plants, agriculture and forest residues, organic household waste, industrial waste, and algae are used as biomass sources. The biomass sources can be used in wide areas such as producing heat, electricity, fuel, and some chemicals [4].

When technical, economic, environmental, and social effects of alternative energy sources such as biomass, wind, hydroelectric, solar, and geothermal energy are evaluated, it is concluded that the most suitable alternative energy is biomass energy. The most important reason for this is that its social benefit is the highest among others [5]. In addition to this, the use of bioenergy has considerably the potential to reduce emissions of greenhouse gases. Bioenergy produces approximately the same amount of carbon dioxide as fossil energy sources, but net carbon emission is zero since the plant uses carbon dioxide by photosynthesis during the day [4].

Three sides of Turkey are surrounded by seas so that it has different climates. Besides, it is located in the center of the triangle connecting the continents of Asia, Europe, and Africa [6]. In 2015 the amount of carbon absorbed by forests in Turkey is 1.9 billion tons. In addition to this, oxygen production was annually calculated as 42 million tons [7]. Turkey has a very rich fauna and plant species source due to moderate climatic conditions. For this reason, it is among the countries rich in biodiversity. Turkey's territory is covered with 27.6% of forests, 31.1% of agricultural land, 18.6% of pasture, 21.3% of other areas, and 1.4% of water. The distribution ratios of the land situation in Turkey are shown in **Figure 1** [8].

The objective of this study is not only to address the current situation of Turkey's forest sources and their bioenergy potential but also to present the recent methods of biomass utilization in the applications. This book section exhibits forest bioenergy potential of Turkey and discusses the biomass conversion methods, products, and applications in terms of the production process and usage of the products in the field. This section aims to attract attention to the forests' bioenergy source and help to seek proper investments for the government and investors regarding forest biomass energy potential.

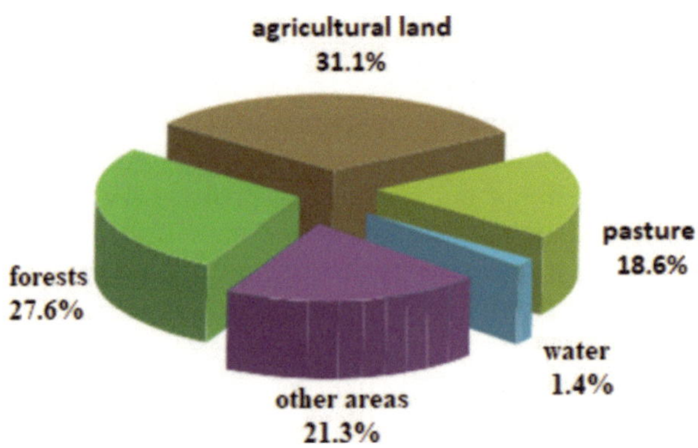

Figure 1.
The distribution ratios of the land situation in Turkey [8].

2. Turkey's forest biomass resources and distribution

Turkey is a transit point that connects the Asian and European continents. It is at the center of the triangle formed by the continents of Asia, Europe, and Africa. Besides, Turkey is surrounded by seas on three sides so it has a different climate. As a result, this situation makes it a rich country in terms of animal and plant diversity. Turkey's forests cover about 30% of the land area and have an equivalent of 11,000 plant species to plant diversity. Furthermore, 3708 of these plant species consist of endemic plant species. When we evaluate the forests of our country as tree species and the area they cover, the first three ranks are 18 species and 6,476,277 ha of oak (*Quercus* spp.), 5,420,524 ha of red pine (***Pinus brutia***), and 4,202,298 ha of larch (***Pinus nigra***) takes the forests. These are followed by beech (*Fagus orientalis* and ***Fagus sylvatica***), scots pine (***Pinus sylvestris***), and fir (*Abies nordmanniana* and *Abies cilicica*) forests. The classification of Turkey's forests is as follows: the Black Sea Region, the North Anatolia Forests which constitute 25% of the forests in Turkey, is the most forested area in Turkey, followed by Thrace, West and Middle Black Sea Forests, Eastern Black Sea Forests, Mediterranean Forests, and Central, East, and South East Anatolia Forests [6]. Turkey's forest asset map and the distribution of the forests are given in **Figure 2**.

2.1 Distribution of forest land in Turkey

Turkey has an ecologically rich diversity due to its geographical location and climatic diversity. The effects of forests on this ecological diversity and wealth are very important. Turkey has 78 million hectares of surface area. In addition to this, forest areas cover by 28.6% percentage except for treeless forest areas [7]. The ratio of land area and the amount of woodland in Turkey is given in **Figure 3**.

Forest areas can be divided into two classes as grove and coppice according to their operation types. Turkey's forests are composed of 88% of grove forest areas (19.6 million hectares) and 12% of coppice forest areas (2.7 million hectares) [7]. The rates of the forest areas according to the operation types are given in **Figure 4**.

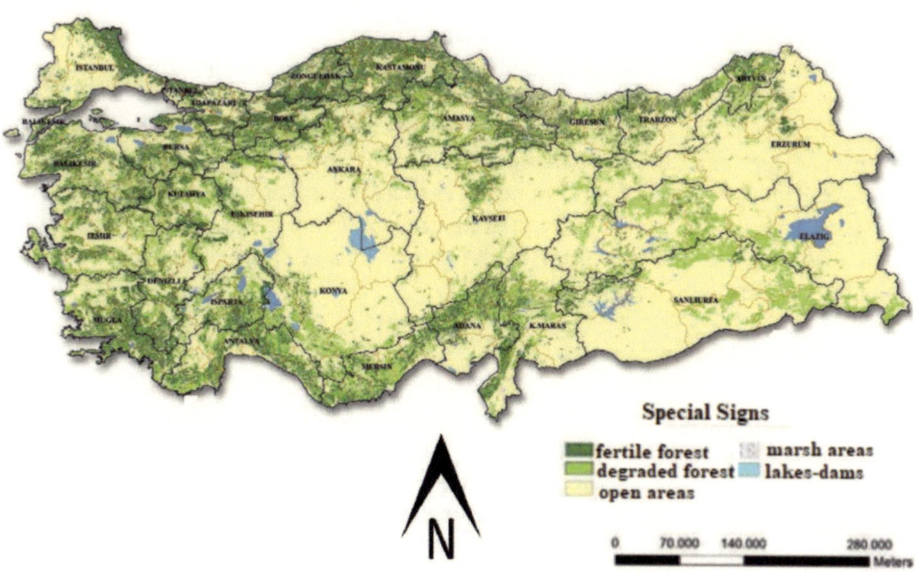

Figure 2.
Map of Turkey's forest assets [8].

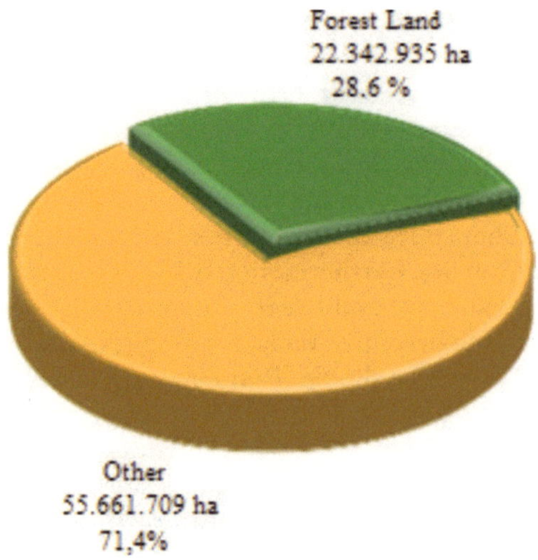

Figure 3.
The ratio of land area and the amount of woodland in Turkey [7].

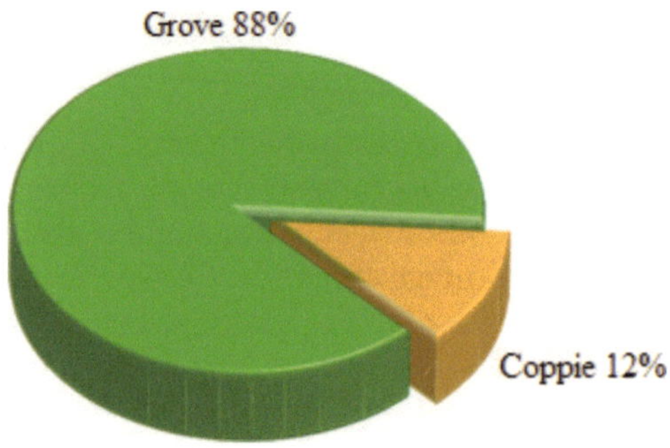

Figure 4.
The rates of the forest areas according to the operation types [7].

Turkey's forest lands' main function distribution is composed of 50% economic, 42% ecological, and 8% sociocultural [7]. Distribution rates according to the main functions of forest areas are given in **Figure 5**.

According to Turkey's forest lands taken into consideration for 42 years, the field of forest area size and change of forest wealth have increased through the years. Forest areas increased by 2.1 million hectares in 42 years. Activities such as protection, development, afforestation, and precautions for forests have been effective on this increase [7]. The amount of the forest area of Turkey through the years and the rate of the country land are given in **Table 1**.

2.2 Distribution of forest wealth

Turkey's forest assets are 20.2 million hectares in 1972 and reached 22.3 million hectares in 2015. In parallel with this, the wood wealth in forests increased from 0.9 billion m^3 in 1972 to 1.2 billion m^3 in 2003, to 1.6 billion m^3 in 2015. In respect to this, between 1973 and 2015, there has been an increase of 700 million m^3 in the tree wealth of the country's forests. In this increase, afforestation studies, migration of citizens living around the forest, and improvement of forest areas have been very effective [7]. The amount of coniferous, broad-leaved, mixed grove, and coppice forest areas of the forest asset in 2012 is given in **Table 2**.

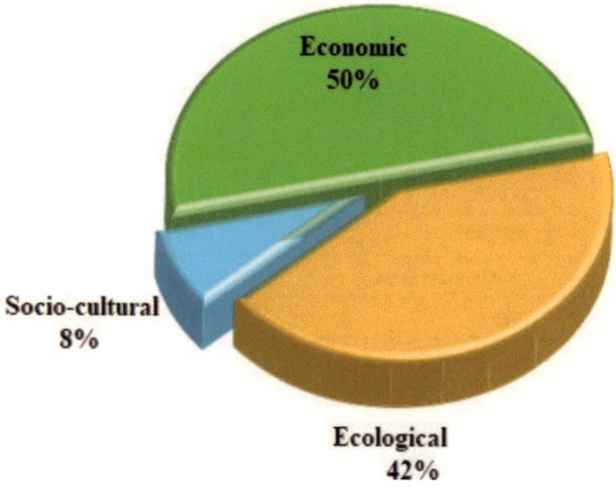

Figure 5.
Distribution rates according to the main functions of forest areas [7].

Years	Field (ha)	Rate (%)
1973	20.199.296	26.1
1999	20.763.248	26.7
2004	21.188.747	27.2
2012	21.678.134	27.7
2015	22.342.935	28.6

Table 1.
Forest area change by years [7].

According to the forest renovation plan in 2013–2015 in Turkey, the amount of forest area in 2015 was estimated to be 22.3 million hectares. The amount and rates of the distribution of the forest areas according to the operation types, the forest area, tree wealth, and annual current increase status are given in **Table 3**. Turkey's average annual amount of revenue derived from forests planted in 2015 is calculated as the volume-shelled body. This value was calculated as approximately 15.94 million m^3 from grove forests and 2.37 million m^3 from coppice forests. As a result, it was calculated as 18.31 million m^3 of total forest area [7].

Turkey's 13.9 million hectares of forest area (62%) is pure forest. In this amount, the rate of tree species entering the mixture is less than 10%. Besides, approximately 8.4 million hectares of forest (38%) is mixed forest [7]. The distribution of forest ratio of species to general forest area is given in **Figure 6**.

Qualification	Coniferous tree (ha)	Broad-leaved tree (ha)	Mixed grove (ha)	Total grove (ha)	Coppice forest (ha)	Total forest (ha)
Productive	6.792.336	2.156.746	1.332.464	10.281.728	1.276.940	11.558.668
Degraded	4.983.059	950.319	1.045.486	6.978.864	3.140.602	10.119.466
Total	11.775.395	3.107.066	2.378.131	17.260.592	4.417.542	21.678.135

Table 2.
Turkey's amount of forest assets in 2012 [8].

Operation types	Normal forest		Degraded forest		Total	
	ha	%	ha	%	ha	%
Forest area distribution						
Grove	11.919.061	54	7.700.657	34	19.619.718	88
Coppice	785.087	3	1.938.130	9	2.723.217	12
Total	12.704.148	57	9.638.787	43	22.342.935	100
Operation types	**Normal forest**		**Degraded forest**		**Total**	
	m3	%	m3	%	m3	%
Distribution of tree wealth						
Grove	1.506.131.410	93	59.996.731	4	1.566.128.141	97
Coppice	33.692.118	2	11.953.934	1	45.646.052	3
Total	1.539.823.528	95	71.950.665	5	1.611.774.193	100
Distribution of annual current increase						
Grove	42.322.876	92	1.484.455	3	43.807.331	95
Coppice	1.511.561	3	585.191	2	2.096.752	5
Total	43.834.437	95	2.069.646	5	45.904.083	100

Table 3.
The situation of forest areas according to their operation types [7].

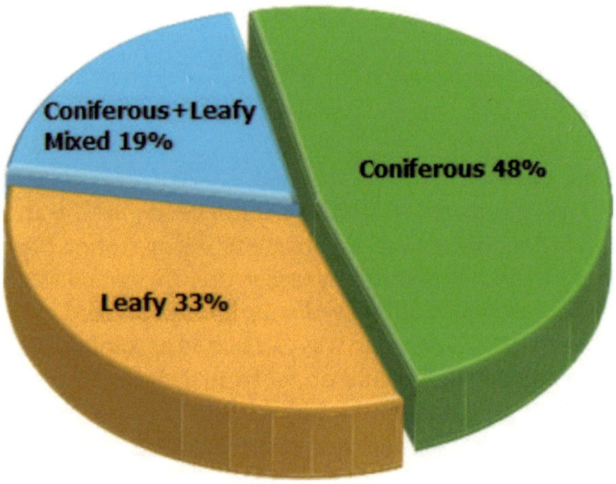

Figure 6.
Proportion of forest areas by tree type [7].

Turkey's forest areas consist of 33% broad-leaved forests (oak, beech, alder, chestnut tree species such as beech), 48% coniferous forests (tree species such as Turkish pine, crimean pine, scots pine, fir, spruce, cedar), 19% coniferous + broad-leaved mixed forests. Oak occupies the largest area in the forests (5.9 million ha), followed by Turkish pine, crimean pine, beech, scots pine, juniper, fir, cedar, spruce, stone pine, alder, chestnut, hornbeam, poplar, lime tree, ash tree, and eucalyptus [7]. Distribution values of forest areas by tree species are given for 2018 in **Table 4**.

Turkey's forest wealth distribution, the current value increment distribution, and distribution of forest areas by the year 2005–2018 are given as follows. **Figure 7** shows forest wealth distribution in 2018. The forest was composed of 95%

Total forest area by tree species (2018)		**Forest form**	
Tree type groups	**Total (ha)**	**Productive (ha)**	**Degraded (ha)**
Oak (*Quercus* spp.)	5,938,527	2,435,265	3,503,262
Turkish pine (*Pinus brutia*)	5,686,009	3,527,063	2,158,946
Crimean pine (*Pinus nigra*)	4,304,821	2,787,424	1,517,397
Beech (*Fagus orientalis*)	1,935,730	1,665,997	269,733
Scots pine (*Pinus sylvestris*)	1,538,304	901,606	636,698
Juniper (*Juniperus*)	963,217	223,097	740,120
Fir (*Abies* spp.)	593,201	391,842	201,359
Turkish cedar (*Cedrus libani*)	487,819	252,590	235,229
Oriental spruce (*Picea orientalis*)	327,890	234,224	93,666
Stone pine (*Pinus pinea*)	164,798	131,548	33,250
Alder (*Alnus* spp.)	149,215	115,646	33,569
Chestnut (*Castanea sativa*)	89,941	69,727	20,214
Hornbeam (*Carpinus* spp.)	35,609	28,872	6737
Poplar (*Populus* spp.)	16,430	6587	9843
Lime tree (*Tilia* spp.)	12,803	10,637	2166
Ash tree (*Populus* spp.)	7359	6854	505
Eucalyptus (*Eucalyptus* spp.)	1434	1383	51
Other species	368,826	192,784	176,042
Total	22,621,935	12,983,148	9,638,787

Table 4.
Distribution of forest areas by tree species 2018 [9].

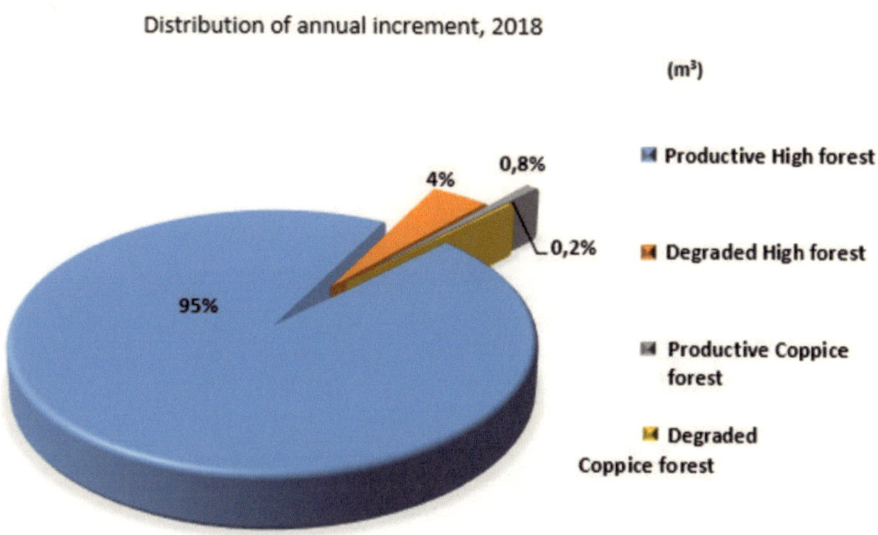

Figure 7.
Forest wealth distribution in 2018 [9].

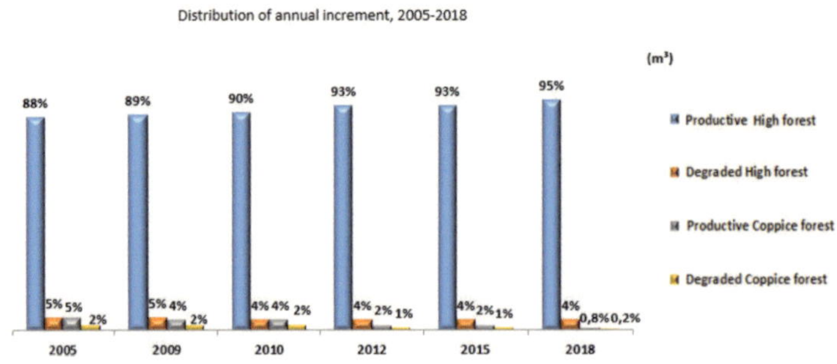

Figure 8.
Distribution of forest wealth between 2005 and 2018 [9].

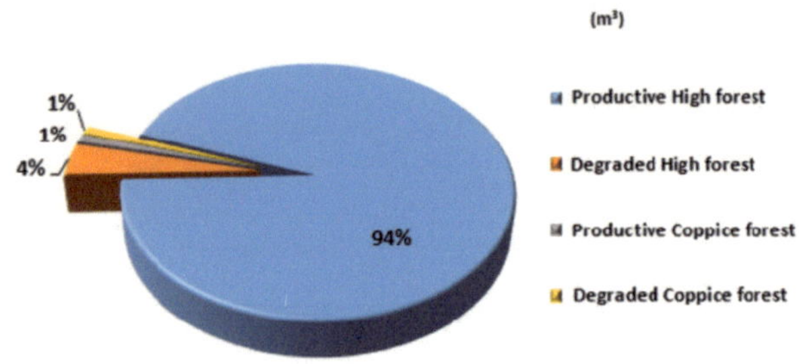

Figure 9.
Distribution of increment in 2018 [9].

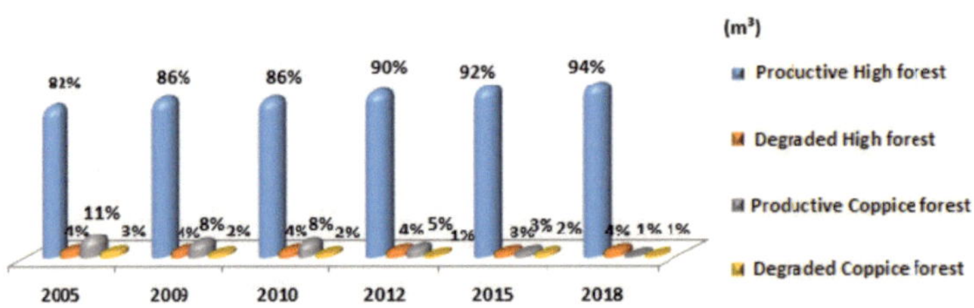

Figure 10.
Distribution of increment between 2005 and 2018 [9].

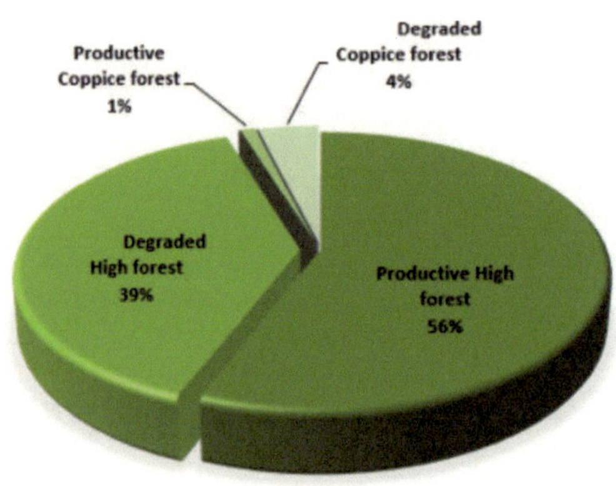

Figure 11.
Distribution of forest areas in 2018 [9].

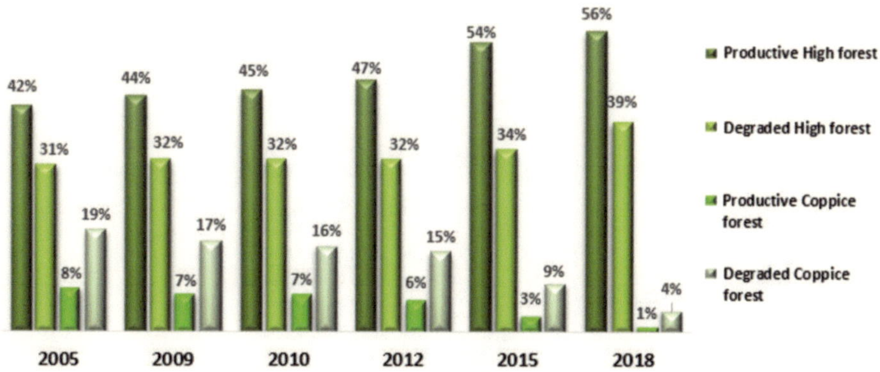

Figure 12.
Change of forest areas by years 2005–2018 [9].

productive high forest. **Figure 8** shows distribution of forest wealth between 2005 and 2018. The productive high forest percentage increased by 88–95% in 13 years. **Figure 9** indicates distribution of increment in 2018. **Figure 10** indicates distribution of increment between 2005 and 2018. **Figure 11** shows distribution of forest areas in 2018. **Figure 12** shows change of forest areas by years 2005–2018 [9].

In the last 30 years, an increase of approximately 990.000 ha has been achieved in forest areas with afforestation studies and increasing environmental awareness. Thus, not only superficial increase but also quality increase in forest areas was observed [6].

3. Turkey's forest bioenergy potential

Annual increment in volume of forests can be explained by the increase in total height and diameter of the tree in a cubic meter (m^3) during the growth period of the trees. Thus, the annual current increment was 28.1 million m^3 in total and 1.4 m^3 in a hectare in 1973. In addition to this, the annual current increment was calculated as 45.9 million m^3 in total and 2.1 m^3 in hectare in 2015. The reason for this increase is due to the increment in tree wealth and forest areas with the maintenance to forests [7]. In **Figure 13**, the wood biomass source that can be obtained from forests is given as a model.

Revenue in forestry is the annual revenue amount and is calculated in m^3. The amount of revenue in 2015 was determined as 15.942.459 m^3 in grove forests and 2.372.162 m^3 from coppice forests, with a total of 18.314.621 m^3 [7]. The change in forest revenue amounts by years is given in **Figure 14**.

3.1 Production amounts of fuel wood

The total amount of trees and annual revenue growth of forest areas can be considered as the biomass potential of forests. Thus, the amount of production and unprocessed wood production according to the tree species shows the potential of the production of firewood. **Tables 5** and **6** show the amount of wood that can be produced as fuel according to the forest area and tree types.

3.2 Forest waste bioenergy potential

It is estimated that available biomass energy potential from waste is about 8.6 million tons of oil equivalent (toe) in Turkey. Furthermore, it is anticipated that these waste biomass have a biogas potential of 1.5–2 Mtoe [10].

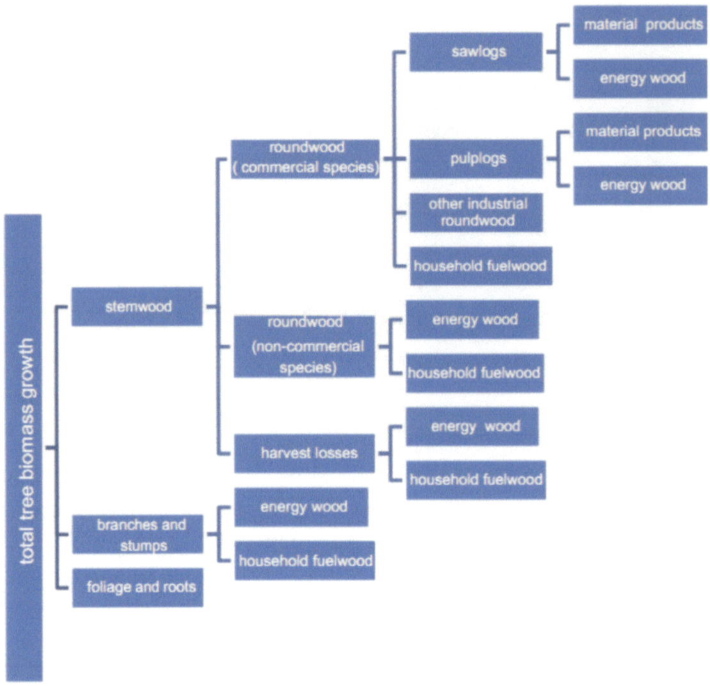

Figure 13.
Model of wood biomass source that can be obtained from forests [2].

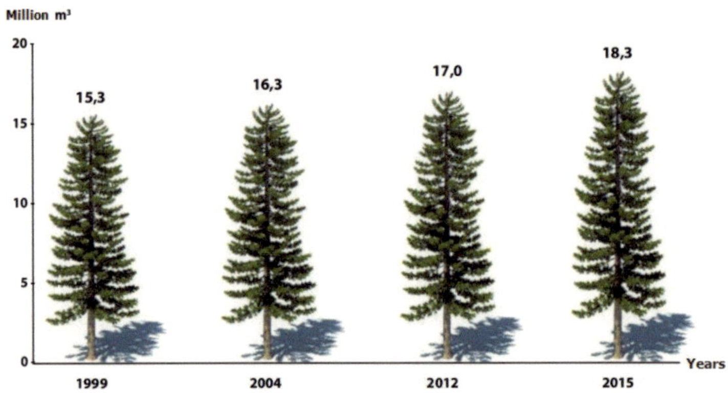

Figure 14.
Forest revenue by years [7].

Description	Unit	2014	2015	2016	2017	2018
Turkey (industrial wood)	m^3	14,923,209	16,637,598	17,009,998	15,521,622	19,080,137
Turkey (fuel wood)	m^3	5,257,994	5,022,986	4,877,067	4,359,646	4,890,455
Fuel wood (coniferous)	m^3	2,120,632	2,176,826	2,203,385	1,926,629	2,442,696
Fuel wood (non-coniferous)	m^3	3,137,362	2,846,160	2,673,682	2,433,017	2,447,759

Table 5.
Volume of wood between 2014–2018 [9].

Range of products		Coniferous							
		Cedrus	*Juniperus*	*Pinus brutia*	Other pinus	*Picea*	*Abies*	Other coniferous	Total
Sewn shell body volume (m^3)		177,316	29,959	6,541,644	7,443,431	636,806	2,067,953	408,451	17,305,560
Fuel wood	High forest (from allowable cut)	33,739	12,092	717,120	400,748	57,043	27,731	31,358	1,279,831
	Coppice (from allowable cut)	4651	318	8893	8926		5155	1141	29,084
	Site clearance, wreck, etc.	5491	754	292,649	662,736	43,024	90,171	38,959	1,133,781
	Total	43,881	13,164	1,018,659	1,072,410	100,067	123,057	71,458	2,442,696
Range of products		**Non-coniferous**							
		Quercus	*Carpinus*	*Fagus*	*Populus*	*Alnus*	Other non-coniferous	Total	Final total
Sewn shell body volume(m^3)		2,286,005	301,162	4,032,484	138,236	49,932	324,418	7,132,237	24,437,797
Fuel wood	High forest (from allowable cut)	506,016	57,851	358,467	4559	8200	139,551	1,074,608	2,354,439
	Site clearance, wreck, etc.	209,091	5334	332,170	7432	1605	38,958	594,590	1,728,371
	Total	1,323,625	69,497	793,986	12,373	9805	238,473	2,447,759	4,890,455

Table 6.
Production amounts of the tree species for volume of wood per species in 2018 [9].

Figure 15.
Distribution of Turkey's forest waste amount [11].

The total amount of waste originating from forests was calculated as 4.8 million tons (1.5 Mtoe) in Turkey. The gasification plant capacity that can be installed is estimated to be 600 MW [11]. The energy value of forest waste is estimated to be 859.899 toe/year in Turkey. The number of biomass electricity generation plant in Turkey is 128 units [12]. Wood biomass potential in forests depends on factors such as forest biomass increase, forest area, and forest growth [2]. Therefore, the bioenergy potential is also highly dependent on factors such as forest biomass increase, obtained from forest wastes, forest area, and growth of the forest. **Figure 15** shows the distribution of the amount of Turkey's forest wastes based on the amount of biomass.

4. Biomass energy conversion methods, products, and applications

The majority of biomass energy is used for cooking and heating in households. Approximately 6.5 million houses use wood as the main fuel for heating purposes in Turkey. Moreover, in the paper industry, approximately 60% of the factories' energy needs are obtained from waste wood [5].

There are main processes such as direct combustion, gasification, alcoholic fermentation, pyrolysis, liquefaction, anaerobic digestion, hydrogasification, and transesterification where energy is obtained from biomass. These processes have their own advantages according to the biomass source and the type of energy obtained. If biomass is converted using modern technologies and energy conversion efficiency is ensured, biomass energy could be a primary energy source in the future [4].

Demirbaş [13] classified wood as a second-generation biofuel in the study. Besides, examples of these biofuels are bio-alcohols, bio-oil, bio hydrogen, and bio Fischer-Tropsch diesel. In addition to these, using alternative fuels from biomass as fuel additives can improve fuel properties such as cetane and octane number, viscosity, and density in diesel and gasoline engines. Thus, fuels produced from biomass can be used as alternative fuels in internal combustion engines [14].

Biomass conversion techniques can be applied on biomass materials to obtain solid, liquid, and gaseous fuels. After the conversion process, fuels can be produced with the main products such as biodiesel, biogas, bioethanol, and pyrolytic gas. Besides, by-products such as fertilizer and hydrogen can be also obtained [15].

Biomass	Conversion method	Fuels	Application fields
Forest wastes	Anaerobic digestion	Biogas	Electric power production, heating
Agricultural wastes	Pyrolysis	Ethanol	Heating, transport vehicles
Energy crops	Direct combustion	Hydrogen	Heating
Animal waste	Fermentation and anaerobic digestion	Methane	Transport vehicles, heating
Garbage (organic)	Gasification	Methanol	Jet engines
Algae	Hydrolysis		Synthetic oil, rockets
Energy forests	Biophotolysis	Automotive gas oil	Drying
Vegetable and animal oils	Esterification reaction	Diesel fuel	Transport vehicles, heating, greenhouse cultivation

Table 7.
Biomass, biomass conversion methods, fuels, and application areas [15].

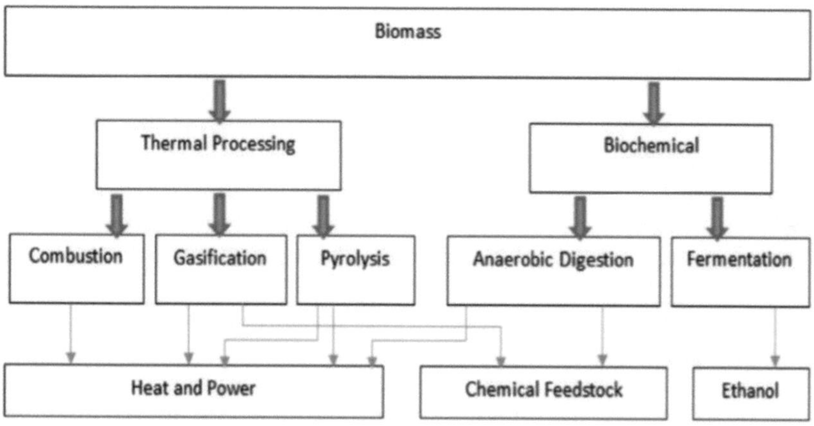

Figure 16.
Biomass processing method and products [16].

Alternative biofuels such as biogas and bioethanol fuels can be obtained by the biomass conversion process. Bioethanol can be used instead of oil, and biogas can be used instead of natural gas [16]. Conversion techniques using biomass sources, fuels obtained using these techniques, and application areas are given in **Table 7**.

The biomass processing process can be divided into two classes: thermal and biochemical. It can be divided into three subtitles as direct combustion, gasification, and pyrolysis in the thermal process. The biochemical process can be classified under two subtitles as fermentation and anaerobic digestion. **Figure 16** shows the methods and products obtained in the processing processes of biomass [16].

4.1 Thermal process

The majority of modern bioenergy plants are use biomass for obtaining heat and power. Developing gasification and pyrolysis bio-oil technology offers much more efficient energy conversion with turbine and combined cycle technologies [3]. Thermal process can be examined under three main headings: burning, gasification, and pyrolysis (**Figure 17**).

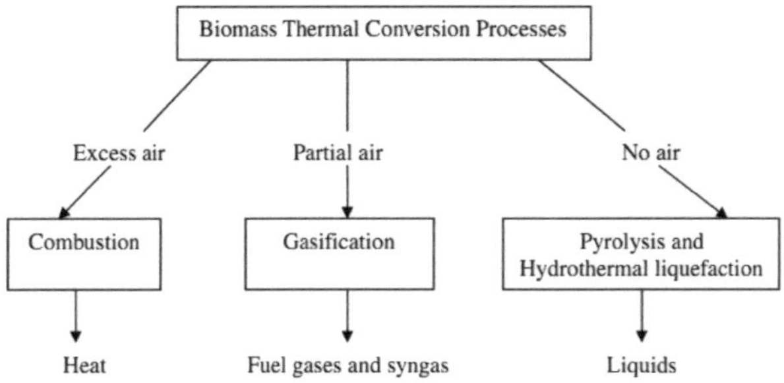

Figure 17.
Transformation of biomass in thermal process [17].

4.1.1 Pyrolysis

Pyrolysis process is the simplest and oldest method for biomass to gas from. Pyrolysis process is a physical and chemical situation that occurs by heating organic substances up to 500–600°C without oxygen. In this process, gas components, volatile condensates, charcoal, and ash are released. When it rises to high temperature, wood gas and components gas are released by heating the wood up to 900°C in an oxygen-free environment. As a result of pyrolysis, substances such as gases, water, organic compounds, tar, and charcoal are obtained [15, 18].

Pyrolysis is the method of obtaining solid, liquid, and gas products by breaking down the biomass with heat. Slow pyrolysis is a well-known method widely used in the production of charcoal. Fast pyrolysis is the method in which biomass converts more than 75% of liquid bio-oil at high temperatures. This bio-oil chemical composition obtained is very similar to biomass. This bio-oil can be used as renewable fuel in gas turbines, diesel engines, or boilers. Bio-oil has about 60% calorific value of conventional fuel oils by volume [3].

Bio-oil is a liquid fuel obtained by the thermochemical process of biomass. Bio-oil obtained from wood is liquid and dark brown-colored. Its density is 1200 kg/m^3 and it is more than the density of biomass and fuel oil. Bio-oil water content is 14–33 wt% by mass and cannot be removed by traditional methods such as distillation. Higher heating value (HHV) is 27 MJ/kg, and it is lower than traditional fuel oil (43–46 MJ/kg) [13]. The conversion of biomass into product by pyrolysis and the process steps of the products obtained are given in **Figure 18**.

4.1.2 Gasification

Gasification technology is one of the oldest conversion processes, and it has been used for more than 200 years [19]. The gasification process is the method for achieved combustible gas by dissolving solids like carbon-containing biomass at high temperatures. The process up to approximately 500°C in the gasification of organic substances is the pyrolysis phase. Here, carbon, gases (calorific value can be up to 20 MJ/m^3), and tar are obtained. When heating up to 1000°C, carbon reacts with water vapor to produce CO and H_2. Depending on the variable oxygen rate in the raw material, additional oxygen input may not be required for the gasification process. Gasification takes place in a reducing atmosphere with low air oxygen or steam injection. During this process, biomass is burned with the air supplied to the fuel cell under control, and the resulting products include combustible gases such as hydrogen, methane, as well as carbon monoxide, carbon dioxide, and nitrogen. Thus, combustible gases such as carbon monoxide, hydrogen, methane, and low

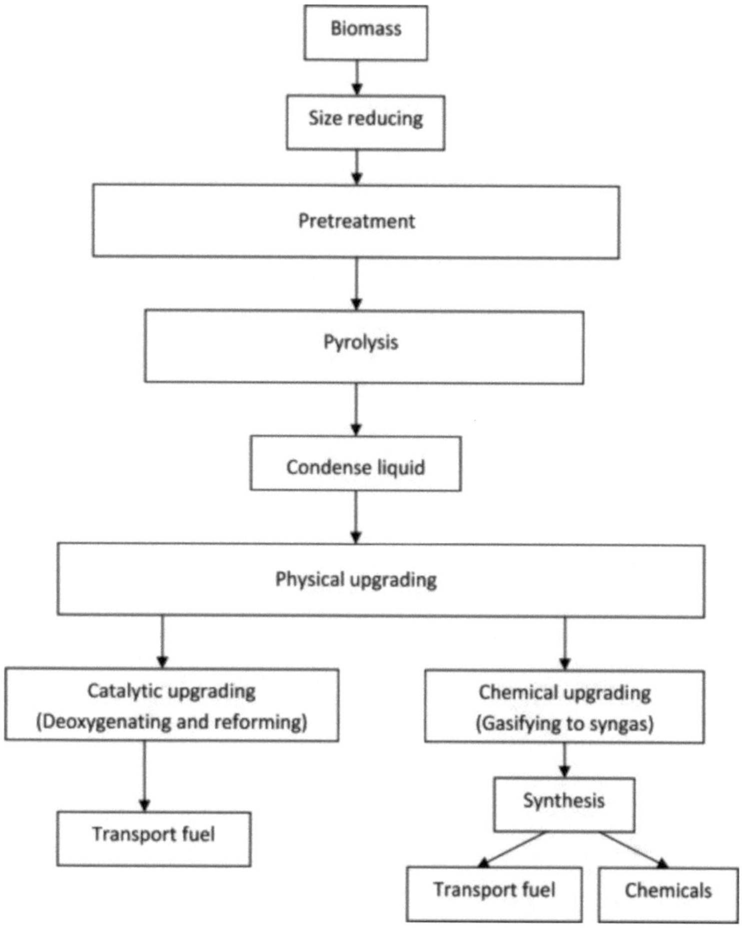

Figure 18.
Conversion of biomass to products by pyrolysis [13].

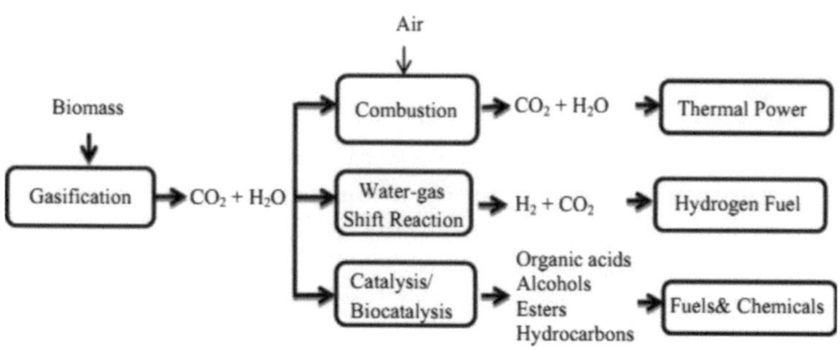

Figure 19.
Transformation processes of biomass into products by gasification [19].

amounts of other gases with low or medium calorific value are obtained. After cleaning the gas, it can be used as fuel in gas turbines, gasoline engines, dual fuel diesel engines, or in fuel cells after purification [3, 15, 18] (**Figure 19**).

Biomass gasification technology provides the opportunity to convert renewable biomass resources into clean gaseous fuels or synthesis gases. Heat or electricity is produced from these produced gases. In addition to these, there is the potential to produce liquid transportable fuel, hydrogen, or chemicals from them. Gasification

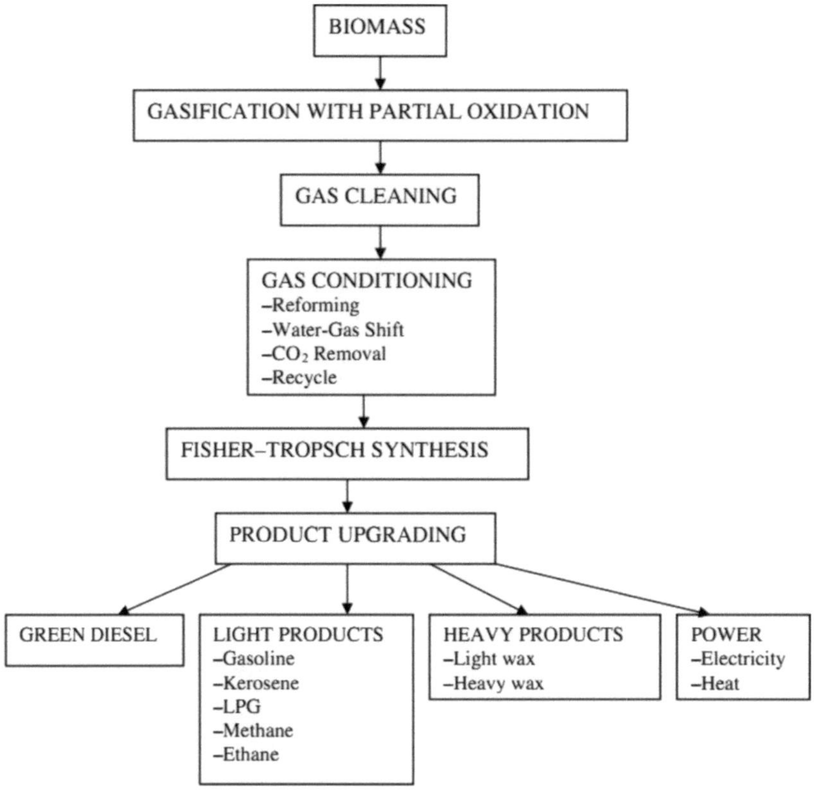

Figure 20.
Products obtained from biomass by Fischer-Tropsch synthesis [20].

is a promising energy conversion technology with its flexible, efficient, and environmentally adaptable features [17]. Besides, the most important feature of gasification is its high electrical efficiency. In the future, it is expected to be used instead of natural gas or diesel fuel in gas turbines or fuel cells, industrial boilers, and furnaces, to replace gasoline or diesel in internal combustion engines [19] (**Figure 20**).

4.1.3 Combustion

The process of biomass giving a fast chemical reaction with oxygen is called burning. As a result of combustion, heat, carbon dioxide, water vapor and some metal oxides are given to the environment [15]. The biomass and full combustion components are given in **Figure 21**.

Industrial and commercial combustion plants can burn a wide variety of fuels, from tree biomass to urban solid waste. Furnaces are the simplest combustion technology, and biomass burns in a combustion chamber. Combustion technology can be divided into two main categories as grate burner and fluid bed burner. In biomass combustion plants, a high-temperature and high-pressure steam is obtained as a result of combustion. This steam is passed through the turbine and converted into electrical energy with efficiency in the range of 17–25%. It can be increased up to 85% with efficient cogeneration systems [3, 19].

Pellets are generally solid wood particles with a cylindrical diameter of 10 mm and a length of less than 35 mm. Pellets produced from wood or waste wood are used to generate electricity in cogeneration systems, for heating in residences and industry. Wood pellets are the fuel with the highest thermal value after coal [22]. A comparison of the higher heat values of biomass and coal are given in **Table 8**.

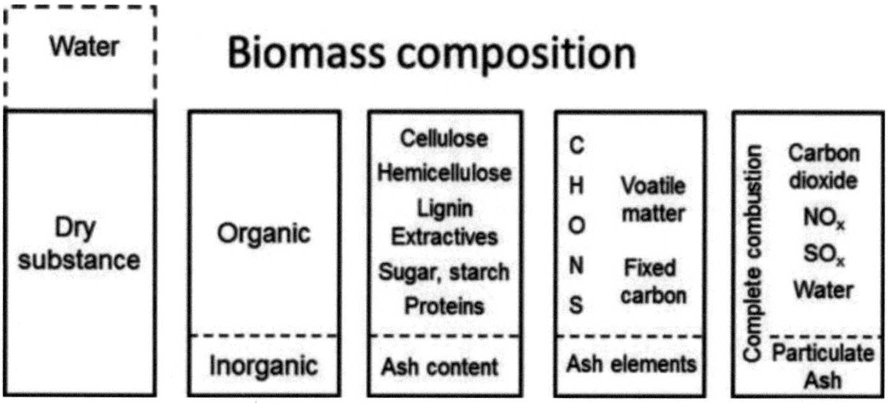

Figure 21.
The components of the biomass [21].

Fuel form	HHV (MJ/kg)
Wood	10–20
Vineyard pruning	14–18
Rice husk	12–14
Sawdust	12
Wood pellets	20
Coal	28

Table 8.
A comparison of the higher heat values of biomass and coal [22].

4.2 Biological process

4.2.1 Fermentation

It contains hemicellulose and lignin in different amounts in the biomass. Glucose can be obtained from cellulose using enzymes with chemical hydrolysis or after enzymatic hydrolysis with chemical processes. This process must be done with extreme care, as glucose can sometimes degrade during chemical hydrolysis. By fermentation of glucose, many chemical products can be obtained such as ethanol, acetone, and butanol which are equivalent to products from crude oil [15].

4.2.2 Digestion

Anaerobic digestion is a biological process and can take place in a completely oxygen-free environment. It is done by microorganisms that can live in an oxygen-free environment. The process is given in Eq. (2):

$$\begin{aligned}&\text{Organic matter} + \text{bacteria} + \text{water} \xrightarrow{\text{heat}} \text{methane}\\&\quad + \text{carbon dioxide} + \text{hydrogen, sulfur} + \text{stable, fertilizer} + \text{bacteria}\end{aligned} \tag{2}$$

Biomass can be separated by microorganisms through fermentation in an oxygen-free environment. End of the fermentation process, a valuable fertilizer, and gases such as methane and carbon dioxide products can be obtained [15].

Anaerobic digestion (AD) is a process of producing flammable gas, consisting of methane and carbon dioxide at a rate of 60:40 using microbes in an oxygen-free environment. Therefore, the biogas production process is a complex and sensitive process that contains many microorganism groups. Biogas is a flammable gas formed by decomposing biological wastes in an oxygen-free environment. Biogas approximately contains 50–60% methane gas. Biogas is a colorless, flammable gas. In addition to this, biogas consists of its main components such as methane and carbon dioxide. Besides, it contains a small amount of hydrogen sulfide, nitrogen, oxygen, and carbon monoxide. Generally, 40–60% of organic matter is converted to biogas. The general composition of biogas consists of 60% CH_4 and 40% CO_2, and its thermal value is 17–25 MJ/m^3. The remaining waste is an odorless solid or liquid waste suitable for use as fertilizer. After producing methane gas, methane gas can be used instead of LPG with very small changes. This gas can be used in spark ignition engines, gas turbines, and fuel cells [3, 16, 23, 24]. The components of biogas are given in **Table 9**.

Methane, CH_4	55–75%
Carbon dioxide, CO_2	25–45%
Carbon monoxide, CO	0–0.3%
Nitrogen, N_2	1–5%
Hydrogen, H_2	0–3%
Hydrogen sulfide, H_2S	0.1–0.5%
Oxygen, O_2	Traces %

Table 9.
Composition of biogas [24].

Biogas is a gaseous fuel as an alternative to natural gas. Thus, it can be used in the following fields: direct heating, motor fuel, turbine fuel power generation, fuel cells, additives for natural gas, and in the production of chemicals [23]. Flow diagrams of biogas production facilities are given in **Figures 22** and **23**.

4.2.3 Biophotolysis

Hydrogen and oxygen can be obtained by the biophotolysis process using some microscopic algae. These algae use solar energy in seawater, so they can work as a kind of solar cell. Thus, the microscopic algae can separate seawater photosynthetically to hydrogen and oxygen [15].

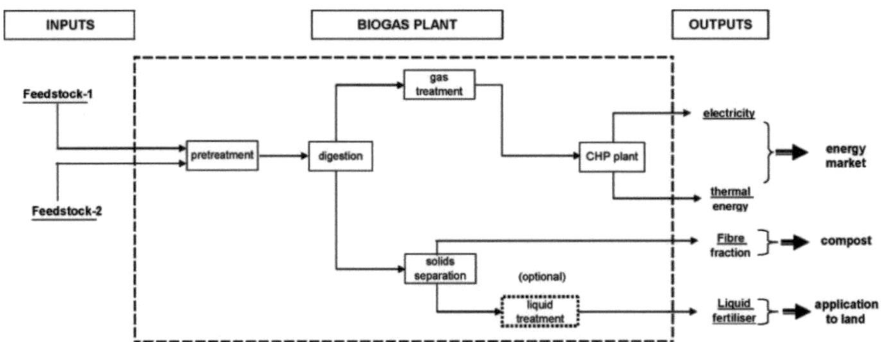

Figure 22.
Flow diagram of the biogas production facility [24].

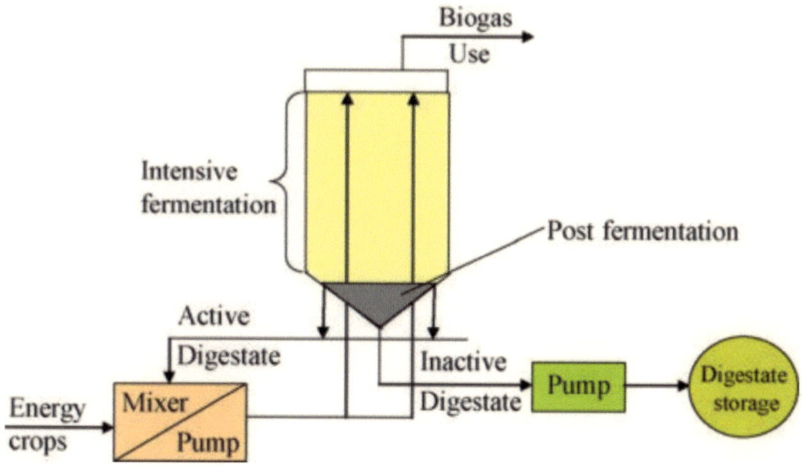

Figure 23.
Flow diagram of dry fermentation [24].

5. Conclusions

The study concludes that biomass energy in Turkey is seen as one of the most sustainable and promising renewable energy sources. Forest bioenergy potential can be converted to alternative biofuels. This process consists of the most common biomass conversion methods such as thermal processes, biological processes, and biophotolysis processes. The thermal processes consist of pyrolysis, gasification, and combustion, while the biological processes are fermentation and anaerobic digestion. Thus, forest bioenergy potential can be used for producing energy. In this regard, forest wastes or forest biomass can be turned into pellets and used in electricity generation in power plants. In addition to this, pyrolysis, gasification, fermentation and anaerobic digestion methods, alcohol, and biogas can be produced from forest wastes and used in the residential industry and transportation. Especially, bio-oils and bio-alcohols can be used in internal combustion engines, furnaces, or boilers as fuel. Besides, biogas also is used as a fuel in households or industry. Thus, Turkey can be reduced to its dependence on foreign energy demand due to the advantage of the rich forest resources. Besides, it is obvious that rich forest resources will contribute to both the ecological and socioeconomic structures of countries. Overall, the rich forest biomass potential is not only contributed to countries' economic field but also the ecological and socio-economic.

Acknowledgements

The author would like to thank Amasya University.

Abbreviations

CH_2O	carbohydrate
CH_4	methane
HHV	higher heating value (MJ/kg)
LPG	liquefied petroleum gas
Mtoe	million tons of oil equivalent
MW	megawatt

Author details

Selçuk Sarıkoç
Taşova Vocational School, Amasya University, Amasya, Turkey

*Address all correspondence to: sarikocselcuk@gmail.com

References

[1] Republic of Turkey Energy and Natural Resources Ministry and General Directorate of Energy Affairs. Home\Renewable Energy\Biomass\What Is the Biomass Energy?, 2020. Available from: http://www.yegm.gov.tr/yenilenebilir/biyokutle_enerjisi.aspx [Accessed: 14 March 2020]

[2] Lauri P et al. Woody biomass energy potential in 2050. Energy Policy. 2014;**66**:19-31

[3] Schuck S. Bioenergy as a sustainable energy source. Australian Journal of Multi-Disciplinary Engineering. 2015;5(1):69-74

[4] Chang J et al. A review on the energy production, consumption, and prospect of renewable energy in China. Renewable and Sustainable Energy Reviews. 2003;7(5):453-468

[5] Baris K, Kucukali S. Availability of renewable energy sources in Turkey: Current situation, potential, government policies and the EU perspective. Energy Policy. 2012;**42**:377-391

[6] Republic of Turkey General Directorate of Forestry, Republic of Turkey Ministry of Forestry and Water Affairs. General Directorate of Forestry—E-Library Publications, Forest of Turkey, 2019. Available from: https://www.ogm.gov.tr/ekutuphane/Yayinlar/Forests%20of%20TURKEY.pdf [Accessed: 27 December 2019]

[7] Republic of Turkey General Directorate of Forestry, Republic of Turkey Ministry of Forestry and Water Affairs. General Directorate of Forestry—E-Library Publications, Turkey Forest Wealth-2015, 2019. Available from: https://www.ogm.gov.tr/ekutuphane/Yayinlar/T%C3%BCrkiye%20Orman%20Varl%C4%B1%C4%9F%C4%B1-2016-2017.pdf [Accessed: 27 December 2019]

[8] Republic of Turkey General Directorate of Forestry, Republic of Turkey Ministry of Forestry and Water Affairs. General Directorate of Forestry—E-Library Publications, Forest Atlas, 2019. Available from: https://www.ogm.gov.tr/ekutuphane/Yayinlar/Orman%20Atlasi.pdf [Accessed: 27 December 2019]

[9] Republic of Turkey Ministry of Agriculture and Forestry General Directorate for Forest Management. General Directorate of Forestry—E-Library Publications, Statistics-Forestry Statistics 2018, 2019. Available from: https://www.ogm.gov.tr/ekutuphane/Sayfalar/Istatistikler.aspx?RootFolder=%2Fekutuphane%2FIstatistikler%2FOrmanc%C4%B1l%C4%B1k%20%C4%B0statistikleri&FolderCTID=0x012000301D182F8CB9FC49963274E712A2DC00&View={4B3B693B-B532-4C7F-A2D0-732F715C89CC [Accessed: 30 December 2019]

[10] Republic of Turkey Ministry of Energy and Natural Resources. Info Bank, Energy, Biomass, 2019. Available from: https://www.enerji.gov.tr/tr-TR/Sayfalar/Biyokutle [Accessed: 26 December 2019]

[11] Republic of Turkey Energy and Natural Resources Ministry and General Directorate of Energy Affairs. Home\Renewable Energy\Biomass\Turkey Forest Biomass Potential Source, 2019. Available from: http://www.yegm.gov.tr/yenilenebilir/tur_or_kay_biyo_pot.aspx [Accessed: 30 December 2019]

[12] Republic of Turkey Energy and Natural Resources Ministry and General Directorate of Renewable Energy. Biomass Energy Potential Atlas of Turkey. 2019. Available from: http://bepa.yegm.gov.tr/ [Accessed: 30 December 2019]

[13] Demirbas A. Competitive liquid biofuels from biomass. Applied Energy. 2011;**88**(1):17-28

[14] Sarıkoç S. Chapter 2: Fuels of the diesel-gasoline engines and their properties. In: Viskup R, editor. Diesel and Gasoline Engines. London: IntechOpen; 2020. pp. 1-17

[15] Republic of Turkey Energy and Natural Resources Ministry and General Directorate of Energy Affairs. Home\Renewable Energy\Biomass\Biomass Cycle Technologies, 2019. Available from: http://www.yegm.gov.tr/yenilenebilir/biyokutle_cevrim_tekno.aspx [Accessed: 30 December 2019]

[16] Hamawand I et al. Bioenergy from cotton industry wastes: A review and potential. Renewable and Sustainable Energy Reviews. 2016;**66**:435-448

[17] Demirbas A. Biofuels securing the planet's future energy needs. Energy Conversion and Management. 2009;**50**(9):2239-2249

[18] Republic of Turkey Energy and Natural Resources Ministry and General Directorate of Energy Affairs. Home\Renewable Energy\Biomass\Gasification, 2020. Available from: http://www.yegm.gov.tr/yenilenebilir/biyo_gazlastirme.aspx [Accessed: 14 March 2020]

[19] Bilgen S et al. A perspective for potential and technology of bioenergy in Turkey: Present case and future view. Renewable and Sustainable Energy Reviews. 2015;**48**:228-239

[20] Demirbas A. Biofuels sources, biofuel policy, biofuel economy and global biofuel projections. Energy Conversion and Management. 2008;**49**(8):2106-2116

[21] Dahlquist E. Biomass as Energy Source—Resources, Systems and Applications. Boca Raton, FL: Taylor & Francis; 2012. p. 287

[22] Nulocnes LJR, Matias JCO, Catalão JPS. Mixed biomass pellets for thermal energy production: A review of combustion models. Applied Energy. 2014;**127**:135-140

[23] Küçükçalı R. Yenilenebilir Enerjiler Alternatif Sistemler Isısan Çalışmaları No: 375. İstanbul, Türkiye: Isısan Akademi; 2008. p. 704

[24] Karellas S, Boukis I, Kontopoulos G. Development of an investment decision tool for biogas production from agricultural waste. Renewable and Sustainable Energy Reviews. 2010;**14**(4):1273-1282

Chapter 10

Combustion Characteristics and Behaviour of Agricultural Biomass: A Short Review

Swapan Suman, Anand Mohan Yadav, Nomendra Tomar and Awani Bhushan

Abstract

Biomass energy is one of the alternative sources of energy, which is particularly accessible in huge quantity worldwide in rural areas. Globally, solid biomass waste is the fourth as an energy resource after coil, oil and gas, which was providing approximately 14% of the world's energy needs. The potential of biomass materials depends on feedstock quantities and their composition. The use of biomass materials as energy source provides extensive benefits as far as the environment is concerned. The agricultural biomass materials absorb carbon dioxide (CO_2) during growth and emit it during combustion. Utilization of these types of wastes in various applications is in the form of a renewable and CO_2-neutral fuel. The physicochemical and structural analyses of agricultural biomass differ significantly with the feedstock types. This review study provides an alternative approach and better understanding to utilize huge amount of energy stored in biomass as the substitute of fossil fuels and also it should play an important role in sustainable energy systems as a component of a renewable energy mix.

Keywords: biomass, combustion characteristics, physicochemical properties, renewable energy, bio-energy

1. Introduction

Energy is most imperative need of human life. Energy consumption pattern indicates the social and economic development of any country [1]. Primary energy sources (natural gas, oil, coal) are considered as the main energy sources in the world [2]. **Table 1** shows the world's primary energy demand that is projected until 2035. As clearly can be seen, as the total worldwide demands for energy keep increasing year-by-year, biomass and other renewables are expected to gain significant contributions in meeting these demands. The world's depleting fossil fuels and increasing Green House Gas (GHG) emissions have given rise to much research into renewable and cleaner energy. In 2010, 76% of total GHG emissions, CO_2 remain the major anthropogenic GHG with the increasing fossil CO_2 emissions more than trebling from 420 $GtCO_2$ in 1970–1300 $GtCO_2$ in 2010 [3]. Since 2000, emissions of anthropogenic CO_2 have risen by more than 3% per year with the net addition likely to rise to 8–12 GtC by 2020 and as much as 6–23 GtC by 2050 [4, 5].

The concern over global warming and climate changes has stimulated a search for alternatives of energy that are renewable and environment friendly. There are various forms of energy sources that are abundantly available which includes nuclear, wind, solar, biomass, waste materials, geothermal, tidal, hydro, etc. The exploitation of these energy sources are growing in different parts of the world and its potential depends on various aspects such as, energy policy target, renewable energy market, technology development and topographical regions [6, 7].

Renewable energy resources that use domestic resources have the potential to provide energy services with zero emissions of both air pollutants and greenhouse gases [8]. Among the various renewable energy sources, biomass has the potential to be used as alternative source of energy with CO_2 neutral [9]. Apart from this, less content of N and S as compared to the fossil fuel makes it environment friendly and does not promote acid rain or greenhouse gas emission [10]. Biomass is a potentially important source of renewable energy in agricultural countries because of abundant supply and its low prices (**Table 1**).

1.1 Bio-waste sources

Bio-waste materials includes woody crops and wastes, agricultural wastes, bagasse, waste paper, sawdust, municipal solid waste, waste from food processing, and animal or cattle wastes. These wastes are significant potential resource for electricity generation, and like crop residues have many applications, especially in developing or developed countries [11]. Bio-waste or biomass is only substitute of fossil fuels which is renewable. Bio-waste contributes greater than 6% of global non-food energy consumption, which primitive low efficiency and highly polluting combustion in poorly controlled heating and cooking fires. Bio-waste offers important advantages as a combustion feedstock due to the high volatility of the fuel and the high reactivity [11]. Bio-waste contains much less carbon and more oxygen and has a low heating value than solid fossil fuels. The burning velocity of bio-waste (pulverized) is substantially higher than that of solid fossil fuels [12].

1.2 Forms of combustion and methodology

1.2.1 Forms of combustion

Combustion is categorised in different forms such as, direct combustion, evaporation combustion, decomposition combustion, surface combustion, and smouldering combustion [13]. In evaporation combustion, the sample containing

Primary energy	Years				
	2008	2015	2020	2030	2035
Coal	3315	3892	3966	3984	3934
Oil	4059	4252	4346	4550	4662
Gas	2596	2919	3132	3550	3748
Nuclear	712	818	968	1178	1273
Hydro	276	331	376	450	476
Biomass	1225	1385	1501	1780	1957
Other renewables	89	178	268	521	699
Total	12,271	13,776	14,556	16,014	16,748

Source: www.eia.org.

Table 1.
World's primary energy demands in MTOE (metric tonne of oil equivalent).

molecular structure with comparatively low fusing temperature evaporates by heating, and reacts with oxygen in gas phase. In decomposition combustion, gases (such as H_2, CO, C_mH_n, H_2O, and CO_2) produced from thermal decomposition reacts with oxygen in gas phase, and produces flame. Usually, char or say bio-char remains after these reaction (forms of combustion) and burns by surface combustion. Smouldering combustion is the thermal decomposition at temperature lower and then the ignition temperature of volatile component of the reactive biomass samples. In industrial direct combustion of biomass, decomposition combustion and surface combustion are the main forms of the combustion [14].

1.2.2 Methodology

The following two methodologies are used for combustion:

- Qualitative comparison and
- Quantitative comparison.

The qualitative comparison is based on literature available in particular field. Combustion of different biomass materials are distinguished on the basis of the type of combustion process used.

The quantitative comparison is based on description of individually built or planned plants or industries information that suppliers and owners of biomass combustion plants have given about their efficiencies, investment costs and emissions. In quantitative comparison we prepare a model or rough information about various biomass combustion plants and further we analyse their efficiency and other characteristics.

The features of some combustion methods are shown in **Table 2**.

Combustion method	Combustion type	Features
Fixed bed combustion	Horizontal/inclined grate Water-cooling grate Dumping grate	Grate is level or sloping. Ignites and burns as surface combustion of biomass supplied to grate. Used in small-scale batch furnace for biomass containing little ash
Moving bed combustion	Forward moving grate Reverse moving grate Step grate Louver grate	Grate moves gradually and is divided into combustion zone and after-combustion zone. Due to continuous ash discharge, grate load is large. The combustion obstruction caused by ash can be avoided. Can be applied to wide range of fuels from chip type to block type
Fluidized bed combustion	Bubbling fluidized bed combustion Circulation fluidized bed combustion	Uses sand for bed material, keeps fuel and sand in furnace in boiling state with high-pressure combustion air, and burns through thermal storage and heat transmission effect of sand. Suitable for high moisture fuel or low grade fuel
Rotary hearth furnace combustion	Kiln furnace	Used for combustion of high moisture fuel such as liquid organic sludge and food residue, or large waste etc. Restricted to fuel size on its fluidity
Burner combustion	Burner	Burns wood powder and fine powder such as bagasse pith by burners, same as that for liquid fuel

Table 2.
Combustion type and feature of biomass.

1.3 Bio-waste conversion technologies

Biomass can be converted to useful products by two main processes:

I. Thermochemical process

II. Bio-chemical process

1.3.1 Thermochemical process

Thermochemical process is one of the bioenergy conversion technologies, which are used to extract energy from the biomass. Thermo-chemical conversion process can be categorized as combustion, pyrolysis, gasification and liquefaction. These processes convert the solid waste biomass into energy rich valuable products. Selection of conversion process depends upon the feedstock type and quantity of biomass, desired form of the energy, i.e., end use requirements, environmental standards and economic conditions [15, 16]. The combustion of agricultural biomass produces (800–1600°C) heat energy for electricity generation. Gasification process generates (700–1000°C) heat energy with a combustible gas mixture, commonly known as producer gas or syngas, which can be used to make synthesize fuels or other chemicals using catalysts [17]. Pyrolysis also occurs at moderate to high temperatures (450–1000°C) in the absence of oxygen to produce energy-rich liquid known as bio-oil and solid char or sometimes biochar [18]. Liquefaction processing occurs at pressure of (5–25 MPa) to prevent boiling of water in the slurry and at temperatures ranging from 200 to 500°C, depending upon whether the desired products are fractionated plant polymers [19], a partially deoxygenated liquid product known as bio-crude [20]. Liquefaction processing at modest temperatures fractionates biomass into cellulose fibres, hemicellulose dehydration products, and lignin [19].

Thus, thermochemical processing offers opportunities for rapid processing of diverse feedstocks, including recalcitrant materials, for production of fuels, chemicals, and power.

Technology	Sub-categories of Technology	Products
Combustion	• Direct combustion to produce heat	• Heat
	• Direct combustion to produce steam	• Radiant heat, hot gas
	• Co-firing	• Electrical energy
	• Co-generation	• Heat (steam), electrical energy
Pyrolysis	• Fast pyrolysis	• Bio-oil, tar, gas, solid char
	• Slow pyrolysis (carbonization)	• Char, gas
	• Flash pyrolysis	• Bio-oil, char
	• Vacuum pyrolysis	• Bio-oil
	• Intermediate pyrolysis	• Tar, gas, char, bio-oil
	• Hydro-pyrolysis	• Bio-oil
Gasification	• Steam gasification	• Gas, char, liquid residue
Densification	• Briquetting	• Briquettes
	• Pelleting	• Pellets
Liquefaction	• Hydro-thermal	• Bio-oil

Table 3.
Conversion technologies for transforming biomass into energy [22, 23].

Ravandranath and Hall [21] and Amigun et al. [22] reported that the biomass materials are converted to biofuels from which energy is extracted from it for suitable utilization. They found that there are at-least five different forms of biofuels which are in use nowadays. These are bioethanol, biodiesel, biogas, bio-methanol and biochar. Traditional technologies depend on mainly on its efficient systems such as open fires for cooking and space heating. Improved technologies are tried to increase in efficiency. Additional details on these technologies are given in **Table 3**.

1.3.2. Bio-chemical process

Gumisiriza et al. [24] reported that in bio-chemical conversion processes two main processes are used, fermentation and anaerobic digestion, together with a lesser-used process based on mechanical extraction/chemical conversion.

- Fermentation is used on a large scale in various countries to produce ethanol from sugar crops and starch crops. The biomass is ground down and then the starch is converted by enzymes to sugars; after that, yeast converts the sugars into ethanol.

- Anaerobic digestion is the conversion of organic material directly into gas, termed biogas, a mixture of mainly methane and carbon dioxide with small quantities of other gases such as hydrogen sulphide. The biomass is converted by bacteria to produce gas with an energy content of about 20–40% of the lower heating value of the feedstock.

2. Combustion properties of biomass

Combustion properties of biomass can be classified as two following properties first is macroscopic properties and second is microscopic properties [25–27]. The macroscopic properties of agricultural biomass includes ultimate analysis, heating value, moisture content, particle size analysis, bulk density, and ash fusion temperature (AFT). And microscopic analysis of agricultural biomass includes thermal analysis, chemical kinetics, and mineral data. Fuel characteristics of biomass mention above have been reported by Bushnell et al. [26]. Fuel combustion properties of biomass can be conveniently grouped into physical, chemical, thermal, and mineral properties [27].

Physical properties for combustion such as porosity, bulk density, particle size, and shape distribution are related to fuel preparation methods. Chemical properties for combustion are the ultimate and proximate analysis, gross calorific value, and heating value of the volatiles [28]. Thermal properties of combustion such as rate of combustion with burning profiles, thermal analysis (TG/DTG), and emissivity vary with moisture content, temperature, and degree of thermal degradation [29]. Thermal degradation products of biomass consist of moisture, volatiles, char and ash. The yields depend on the pyrolysis time & temperature, heating rate, feedstock type, particle size, vapour residence time, sweeping gas flow rate, atmospheric gas flow rate, and types of reactors [30]. The standard methods for analyses of biomass fuel are given in **Table 4**.

Proximate analysis has long been established for assessing the quality of coal, biomass and biochar fuels through quantifying the concentrations of moisture, 'volatile matter', 'fixed C' and ash. Proximate analysis was performed on biomass waste samples for the determination of ash, moisture, volatile matter and fixed carbon contents [11]. To determine the basic elemental composition (carbon, hydrogen, nitrogen and sulphur content) of the biomass waste samples using CHNS

Property	Standard methods
Proximate analysis	
Ash	ASTM D1102 (873 K), ASTM E830 (848 K)
Moisture	ASTM E871
Volatile matter	ASTM E 872, ASTM E 897
Fixed carbon	By difference
Ultimate analysis	
Carbon, hydrogen	ASTM E 777
Nitrogen	ASTM E 778
Sulphur	ASTM E 775
Oxygen	By difference
Heating value (gross calorific value)	ASTM D 2015, E 711
Ash elemental	ASTM D3682, ASTM D2795, ASTM D4278, AOAC 14.7

Table 4.
Standard methods of biomass fuel [11].

analyser. Oxygen content was calculated by the difference. The gross calorific value (GCV) of all studied agricultural biomass samples was determined by bomb calorific measurement.

Ash or inorganic materials in agricultural biomass depend on the type of the feedstocks and the soil contamination in which the plant grows. Ash content is an important parameter directly affecting the heating value. High ash content of a plant part makes it less desirable as fuel [31, 32]. The composition of mineral matter can vary between and within each biomass sample. The higher GCV of WC, WD and CS are owing to lower content of incombustible mineral matter (ash) and higher amount of combustible components (VM, FC, C and H) [33].

The components of biomass include cellulose, hemicelluloses, lignin, lipids, proteins, simple sugars, starches, water, and other compounds [29]. The concentration of each class of compounds varies depending on different feedstocks. Due to the carbohydrate structure, biomass is highly oxygenated with respect to conventional fossil fuels including HC liquids and coals. Typically, 30–40 wt.% of the dry matter in biomass is oxygen [30]. The lignin value in case of woody and coconut shell biomass is higher than herbaceous and agricultural biomass [34]. Heating value, which is a very important factor affecting utilization of any biomass material as a fuel, is affected by the proportion of combustible organic components (called as extractives) present in it. The heating values of the extractive-free were found to be lower than those of the extracted parts, which indicate a likely positive contribution of extractives towards the increase of heating values [29]. Lignin has higher energy content (about 30%) than cellulose and hemicellulose, because of its higher degree of oxidation [35].

As we know that some of agricultural biomasses have high contents of alkali oxides and salts, the low melting points of which may lead to various problems during combustion [36, 37]. These agricultural residues include husks, straws, stalks etc. produced after harvesting the crop. As the crop residues are normally cultivated with the aid of chemical fertilizers, these residues generally contain higher amounts of sodium/potassium compounds presents in their ash. Higher concentration of these alkali compounds result in problems like bed agglomeration, slagging on furnace walls/super heater tubes and fouling of heat transfer surfaces [38]. The successful design and operation of a fluidized bed combustor depends on the ability to control and mitigate these ash related problems [39].

3. Combustion study for biomass materials

The use of agricultural biomass as a fuel provides significant benefits in various fields as energy source as far as the environment is concerned [40]. Agricultural biomass absorbs carbon dioxide (CO_2) during growth, and emits it during combustion. Therefore, it helps the atmospheric carbon dioxide (CO_2) recycling but, it does not contribute in the greenhouse gas (GHG) effect [41]. Agricultural biomass differs from coal in many important aspects, such as the physical and chemical properties. Biomass has less carbon, more oxygen, more silica and potassium, less aluminium and iron, lower heating value, higher moisture content, and lower density and friability than coal.

At low temperatures biomass materials would be instantly ignited, when they are feed in to the high temperature furnace. Here, the pivotal factor is the moisture content is low in most biomass materials, and also can be high for some biomass materials, e.g., bagasse. Basically in the all biomass materials found the high VM contents [42]. The volatiles comprise mainly of the combustibles like as CO, H_2 and C_xH_y. These aspects show that the combustion of the volatiles would be the dominant step during the combustion process.

Kaeferstein et al. [43] state that during the batch experiments in combustion process of biomass in a bubbling fluidized bed using oxygen concentration profiles measured directly over the bed with solid electrolyte sensor. During the combustion process, he observed that there was a rapid ingestion of oxygen, which took place in one phase. Whereas, for coal, the oxygen ingestion profile exhibited in two phases; a short phase for volatile combustion and a long phase for char combustion. The combustion process of biomass was almost complete after the complete combustion of volatiles. Heat distribution analysis during the combustion of biomass showed that over 67% of their calorific values were released through the combustion of the volatiles [44].

If we talking about designing a combustor for the precise use of agricultural biomass waste, we consider the suitable parameters likes, the variability of moisture, the volatile matter content, ash content, ash composition, and the energy content of the fuel [45, 46]. Accordingly follow above statement the requirements for designing a combustor such as low moisture content (about 3–9%) and high volatile matter (about 60–80%) for high burning rate reactivity, low ash content (about

Sl. no.	Property	Biomass samples	Coal
1.	M content (wt% of dry fuel)	1–10	2–2.5
2.	Ash content (wt% of dry fuel)	0.5–22	15–18
3.	C content (wt% of dry fuel)	30–50	70–75
4.	O content (wt% of dry fuel)	30–60	5–15
5.	S content (wt% of dry fuel)	<0.5	0.5–0.9
6.	SiO_2 content (wt% of dry ash)	3.78–78.20	8–15
7.	K_2O content (wt% of dry ash)	5.34–24.70	0.2–1.0
8.	Al_2O_3 content (wt% of dry ash)	0.71–11.69	4–6
9.	Fe_2O_3 content (wt% of dry ash)	0.14–16.69	0.01–2.0
10.	Ignition temperature (°C)	155–250	490–595
11	Peak temperature (°C)	250–320	—
12.	Heating value (MJ/kg)	14–20	23–28

Source: Refs. [11, 30].

Table 5.
Comparison on physical, chemical and combustion properties of biomass materials and coal.

0.5–6%, except rice and husk materials) for less in erosion, corrosion and ash fouling problems [45]. Due to high volatile matter of studied agricultural biomass materials, they have well in heating energy.

Table 5 shows the physical, chemical and combustion properties of biomass, coal and studied biomass samples. Biomass has significantly lower heating values than most of the coals. This may be caused, due to higher moisture and oxygen content. It was also to be seen that the lower heating values lead to lower burning temperatures. Agricultural biomass also has higher volatile matter [47] than coals. Agricultural biomass usually consists of 70–80% VM whereas coal consists of 10–50% VM [48].

4. Conclusions

From above discussion, it can be concluded that the high volatile matter contents of agricultural biomass have a significant effect on the combustion mechanisms. The volatiles consist mainly of combustibles and a significant amount of energy is released during their combustion. Biomass samples have ash content less than 5%, which is much less than any other fossil fuel used for combustion processes. Range obtained for VM content i.e., 60–85% in studied biomass is as high as any of the fossil fuels [49, 50], which can initiate ignition even at lower temperature (150–250°C) and support combustion processes, whereas in other fossil fuels like coal for initiating any combustion process the required ignition temperature ranges 260–450°C. Biomass also have less in nitrogen and sulphur content than any other fossil fuels (coal), which indicates less evaluation of NOx and SOx during combustion processes [48]. These above characteristics of biomass are importance with respect to the design and operation of combustion systems for agricultural biomass. The combustion significances of the biomass composition, particularly the fuel volatility, involve changing the process of combustion [51].

An attempt has also being made in this chapter to overview and understanding the different characteristics and properties of biomass samples and to evaluate their effect on the combustion characteristics.

Author details

Swapan Suman[1*], Anand Mohan Yadav[2], Nomendra Tomar[1] and Awani Bhushan[1]

1 Department of Mechanical Engineering, Meerut Institute of Engineering and Technology, Meerut, Uttar Pradesh, India

2 Department of Chemical Engineering, Meerut Institute of Engineering and Technology, Meerut, Uttar Pradesh, India

*Address all correspondence to: er.ssuman@gmail.com

References

[1] Halder PK, Paul N, Beg MRA. Assessment of biomass energy resources and related technologies practice in Bangladesh. Renewable and Sustainable Energy Reviews. 2014;**39**:444-460

[2] Yin CY. Prediction of higher heating values of biomass from proximate and ultimate analyses. Fuel. 2011;**90**:1128-1132

[3] IPCC. Climate change assessments Review of the processes and procedures of the IPCC. The Netherland: Inter Academy Council; 2014. Available from: https://www.ipcc.ch/pdf/IAC_report/IAC%20Report.pdf

[4] IPCC. Special report on carbon dioxide capture and storage. In: Metz B, Davidson O, de Coninck HC, Loos M, Meyer LJ, editors. Prepared by Working Group III of the Intergovernmental Panel on Climate Change. Cambridge, United Kingdom/New York, NY, USA: Cambridge University Press; 2005. p. 442

[5] IEA. IEA Technology Essentials-biofuel Production. International Energy Agency; 2007. pp. 1-4. Available from: http://www.iea.org/techno/essentials2.pdf

[6] REN21. Renewables Global Status Report: 2009 Update. Deutsche Gesellschaft für Technische Zusammenarbeit (GTZ) GmbH; 2009. Available from: http://www.ren21.net

[7] Energy Information Administration. The Annual Energy Outlook 2009 (AEO2009), Prepared by the Energy Information Administration (EIA), Presents Long-term Projections of Energy Supply, Demand, and Prices Through 2030, Based on Results from EIA's National Energy Modeling System (NEMS). EIA Published an "Early Release" Version of the AEO2009 Reference Case in December 2008. 2009

[8] Demirbas A. Potential applications of renewable energy sources, biomass combustion problems in boiler power systems and combustion related environmental issues. Progress in Energy and Combustion Science. 2005;**31**:171-192

[9] Gonzalez JF, Garcia CMG, Ramiro A, Gonzalez J, Sabio E, Ganan J, et al. Combustion optimisation of biomass residue pellets for domestic heating with a mural boiler. Biomass and Bioenergy. 2004;**27**(2):145-154

[10] Energy for cooking in developing countries. In: World Energy Outlook. Int Energy Agency (IEA); 2006. pp. 419-445. Available from: https://www.iea.org/publications/freepublications/publication/cooking.pdf [Accessed: 23 May 2016]

[11] Demirbas A. Combustion characteristics of different biomass fuels. Progress in Energy and Combustion Science. 2004;**30**:219-230

[12] Marks J. Wood powder: An upgraded wood fuel. Forest Products Journal. 1992;**42**:52-56

[13] Fujii S. Baiomasu Enerugino Riyo, Kenchiku, Toshi Enerugi Sisutemuno Shingijutsu. Kuuki Chowa Eisei Kogakkai Ed; 2007. pp. 212-218 [in Japanese]

[14] Mizutani Y. Nensho Kogaku. 3rd ed. Morikita Shuppan; 2002. pp. 169-181 [in Japanese]

[15] McKendry P. Energy production from biomass (part 1): Overview of biomass. Bioresource Technology. 2002;**83**(1):37-46

[16] Goyal HB, Seal D, Saxena RC. Bio-fuels from thermochemical conversion of renewable resources: A review. Renewable and Sustainable Energy Reviews. 2008;**12**:504-517

[17] Brown RC. Biomass refineries based on hybrid thermochemical/ biological processing—An overview. In: Kamm B, Gruber PR, Kamm M, editors. Biorefineries, Biobased Industrial Processes and Products. Weinheim: Wiley-VCH Verlag GmbH; 2005

[18] Bridgwater AV, Peacocke GVC. Fast pyrolysis processes for biomass. Renewable and Sustainable Energy Reviews. 2000;**4**:1-73

[19] Allen SG, Kam LC, Zemann AJ, Antal MJ Jr. Fractionation of sugar cane with hot, compressed, liquid water. Industrial & Engineering Chemistry Research. 1996;**35**:2709-2715

[20] Elliott DC, Beckman D, Bridgwater AV. Developments in direct thermochemical liquefaction of biomass: 1983-1990. Energy and Fuels. 1991;**5**(3):399-410

[21] Ravandranath NH, Hall DO. Biomass, Energy and Environment—A Developing Country Perspective from India. New York, USA: Oxford University Press; 1995

[22] Amigun B, Musango JT, Stafford W. Biofuels and sustainability in Africa. Renewable and Sustainable Energy Reviews. 2011;**15**:1360-1372

[23] National Association of Forest Industries (NAFI). Report 4 Converting Wood Waste into Renewable Energy: A Summary of Biomass Energy Conversion Technologies. Australia: Forest and Wood Products Research and Development Corporation; 2005

[24] Gumisiriza R, Hawumba JF, Okure M, Hensel O. Biomass waste-to-energy valorisation technologies: A review case for banana processing in Uganda. Biotechnology for Biofuels. 2017;**10**:11

[25] Demirbas A. Mechanism of liquefaction and pyrolysis reactions of biomass. Energy Conversion and Management. 2000;**41**:633-646

[26] Bushnell DJ, Haluzok C, Nikoo AD. Biomass Fuel Characterization Testing and Evaluating the Combustion Characteristics of Selected Biomass Fuels. Corvallis, OR: Bonneville Power Administration; 1989

[27] Ragland KW, Aerts DJ, Baker AJ. Properties of wood for combustion analysis. Bioresource Technology. 1991;**37**:161-168

[28] Yılmaz S, Selim H. A review on the methods for biomass to energy conversion systems design. Renewable and Sustainable Energy Reviews. 2013;**25**:420-430

[29] Gasparovic L, Korenova Z, Jelemensky L. Kinetic study of wood chips decomposition by TGA. Chemical Papers. 2010;**64**(2):174-181. DOI: 10.2478/s11696-009-0109-4

[30] Jones JL, Radding SB, Takaoka S, Buekens AG, Hiraoka M, Overend R. Thermal conversion of solid wastes and biomass. In: Symposium Series 130. Washington, DC: American Chemical Society; 1980. pp. 209-603

[31] Demirbaş A. Relationships between heating value and lignin, moisture, ash and extractive contents of biomass fuels. Energy Exploration and Exploitation. 2002;**20**:105-111

[32] Vassilev SV, Baxter D, Vassileva CG. An overview of the behaviour of biomass during combustion: Part I. Phase-mineral transformations of organic and inorganic matter. Fuel. 2013;**112**:391-449

[33] Chiang K-Y, Chien K-L, Lu C-H. Characterization and comparison of biomass produced from various sources: Suggestions for selection of pretreatment technologies in biomass-to-energy. Applied Energy. 2012;**100**:164-171

[34] Vassilev SV, Baxter D, Andersen LK, Vassileva CG. An overview of the chemical composition of biomass. Fuel. 2010;**89**:913-933

[35] Demirbas A. Relationships between lignin contents and fixed carbons of biomass samples. Energy Conversion and Management. 2003;**44**:1481-1486

[36] Vassilev SV, Baxter D, Vassileva CG. An overview of the behaviour of biomass during combustion: Part II. Ash fusion and ash formation mechanisms of biomass types. Fuel. 2014;**117**:152-183

[37] Reid JS, Koppmann R, Eck TF, Eleuterio DP. A review of biomass burning emissions part II: Intensive physical properties of biomass burning particles. Atmospheric Chemistry and Physics. 2005;5:799-825

[38] Bapat DW, Kulkarni SV, Bhandarkar VP. Design and operating experience on fluidized bed boiler burning biomass fuels with high alkali ash. In: Preto FDS, editor. Proceedings of the 14th International Conference on Fluidized Bed Combustion. Vancouver, New York, NY: ASME; 1997. pp. 165-174

[39] Zhang FS, Yamasaki S, Nanzyo M. Waste ashes for use in agricultural production: I. Liming effect, contents of plant nutrients and chemical characteristics of some metals. Science of the Total Environment. 2002;**284**: 215-225

[40] Southern Centre for Energy Environment. Implementation of renewable energy technologies—opportunities and Barriers: Zimbabwe Country Study. Denmark: UNEP Collaborating Centreon Energy and Environment; 2001

[41] Gadde B, Bonnet S, Menke C, Garivait S. Air pollutant emissions from rice straw open field burning in India, Thailand and the Philippines. Environmental Pollution. 2009;**157**:1554-1558

[42] Acma HH. Combustion characteristics of different biomass materials. Energy Conversion and Management. 2003;**44**:155-162

[43] Kaferstein P, Gohla M, Tepper H, Reimer H. Fluidization: Combustion and emission behaviour of biomass in fluidized bed combustion units. In: Preto FDS, editor. Proceedings of the 14th International Conference on Fluidized Bed Combustion. Vancouver, Canada, New York: ASME; 1997. pp. 15-27

[44] Hellwig G. Basic of the combustion of wood and straw. In: Palz W, Coombs J, Hall DO, editors. Energy from Biomass: 3rd E.C. Conference. London, UK: Elsevier Applied Science; 1985. pp. 793-798

[45] Jenkins BM, Baxter LL, Miles TR Jr, Miles TR. Combustion properties of biomass. Fuel Processing Technology. 1998;**54**:17-46

[46] Tian J et al. A biomass combustion chamber: Design, evaluation, and a case study of wheat straw combustion emission tests. Aerosol and Air Quality Research. 2015;**15**:2104-2114

[47] Suman S, Gautam S. Effect of pyrolysis time and temperature on the characterization of biochars derived from biomass. Energy Sources Part A: Recovery, Utilization, and Environmental Effects. 2017;**39**(9):933-940

[48] Suman S, Gautam S. Biochar Derived from Agricultural Waste Biomass Act as a Clean and Alternative Energy Source of Fossil Fuel. Energy System and Environment. Rijeka: InTechOpen; 2018. pp. 207-220. DOI: 10.5772/intechopen.73833

[49] Suman S, Gautam S. Pyrolysis of coconut husk biomass: Analysis of its biochar properties. Energy Sources, Part A: Recovery, Utilization, and Environmental Effects. 2017;**39**(8):761-767. DOI: 10.1080/15567036.2016.1263252

[50] Vassilev SV, Vassileva CG, Vassilev VS. Advantages and disadvantages of composition and properties of biomass in comparison with coal: An overview. Fuel. 2015;**158**:330-350

[51] Hroncova E, Ladomersky J, Valicek J, Dzurenda L. Combustion of Biomass Fuel and Residues: Emissions Production Perspective. Developments in Combustion Technology. Rijeka: InTechOpen; 2016. DOI: 10.5772/63793

Chapter 11

Potential of Microalgal Biodiesel: Challenges and Applications

Ashokkumar Veeramuthu and Chawalit Ngamcharussrivichai

Abstract

In the present scenario, rapid industrialization and urbanization have led to a dramatic increase in the levels of various hazardous pollutants in the environment, and this creates a serious threat to humankind. Today, most of the energy production comes from fossil fuel combustion, which is the key source of CO_2 emissions. Research studies show that the utilization of microalgae could be the best option for the production of renewable and sustainable energy and for the mitigation of CO_2 emission. Production of biofuels from microalgae can be classified as solid (biochar), liquid (bioethanol, biodiesel, bio-kerosene), and gaseous (biogas, bio-syngas, biohydrogen) fuels. Among these biofuels, biodiesel garners a lot of interest and attention because of its high accumulation of lipids (20–75%), which could be a potential alternative fuel for diesel engines. Algal lipids usually have a higher viscosity than petro-diesel; therefore, the transesterification process is required to decrease the viscosity of microalgal lipids before they can be combusted in the engines. However, microalgae are considered as a potential resource in the current biofuel industries; still, it fails at the commercial level. Thus, in this book chapter, we have discussed the microalgal biofuel production and the challenges behind and the future prospects.

Keywords: microalgae, cultivation, biomass, biodiesel, challenges

1. Introduction

Today's scientific reports revealed that the world's commercial primary energy needs are mostly coming from fossil fuel sources. It is forecasted that the primary energy demand by 2035 will increase to 54% and still fossil fuels contribute 82% of the global need [1]. Besides, an increase in the world population and their anthropogenic activities, such as transportation, land use, deforestation, industrialization, waste generation, etc., has been changing the natural structure of the earth. These activities lead to severe global climate problems in the present scenario. A remarkable change in the lifestyle of human beings is also building extra pressure on the production market to fulfill the demands and desires of society.

Nevertheless, the recent production and consumption models mainly rely on fossil fuel resources, which are affecting the environment and natural resources adversely and irreversibly. It has been reported that the majority of the global CO_2 emissions are due to the transport sector and the number of light motor vehicles is estimated to increase to over 2 billion by 2050 [2]. Hence, to address the significant

IntechOpen

issues such as energy depletion and hazardous gas emission, there is an urgent need for a substantial displacement of fossil fuel usage.

In this context, biofuels produced from biomass rather than fossil source are considered as a potential solution to address these challenges. Currently, researchers have been exploring various types of biofuel production such as solid (biochar), liquid (ethanol, vegetable oil, and biodiesel), and gaseous (biogas, bio-syngas, and biohydrogen), and they were categorized based on the type of feedstock used. It has been reported that the first-generation liquid biofuels which are produced using edible feedstock such as corn, soybean, sugarcane, and rapeseed have directly competed with food production. Meanwhile, food production is the other most critical challenge to society [3]. The second-generation biofuel production has been developed using the nonedible feedstock such as Jatropha, Switchgrass, etc.; however, these nonedible feedstocks also compete with food production [4]. Mainly the second-generation biofuel production relies on arable land, freshwater, and nutrients for their cultivation.

Therefore, to solve these critical issues, researchers have explored a third and fourth generation of biofuels using microalgae and macroalgae (third generation) and metabolic engineering of photosynthetic organisms to produce biofuels (fourth generation) [5]. In this regard, algae have received considerable attention in recent years because of their robust growth and potential to accumulate a high amount of lipid, carbohydrate, and protein, and these can be easily converted into various biofuels (biodiesel, bioethanol, and biogas). **Table 1** shows the potential applications of microalgal species for biodiesel production compared with other biomass sources. Besides, microalgae are the potential candidate for CO_2 sequestration, self-purification, and effective land utilization with high environmental benefits. Also, the cultivation and utilization of microalgae do not compete with food production for land, freshwater, and nutrient sources. Many studies have reported that microalgae offer a wide variety of bioproducts that can be utilized by various sectors, such as energy (biodiesel, biohydrogen, and bioethanol), pharmaceuticals, nutraceuticals, and feed and food supplements [6]. In the past few years, a large number of research work have been focused on microalgae for the potential biofuel production.

Feedstocks	Lipid content (% dry weight basis)	Biodiesel productivity (tons/year/ha)
Corn/maize (*Zea mays*)	44	0.152
Hemp (*Cannabis sativa*)	33	0.321
Soybean (*Glycine max*)	18	0.562
Jatropha (*Jatropha curcas*)	28	0.656
Camelina (*Camelina sativa*)	42	0.809
Canola/rapeseed (*Brassica napus*)	41	0.862
Sunflower (*Helianthus annuus*)	40	0.946
Castor (*Ricinus communis*)	48	0.156
Palm oil (*Elaeis guineensis*)	36	4.747
Microalgae (low lipid-yielding strains)	30	51.927
Microalgae (high lipid-yielding strains)	70	12.110

Table 1.
Lipid and biodiesel productivity of various feedstocks [7].

Nevertheless, the life cycle and techno-economic analysis have revealed that the biofuels derived from microalgae are not cost-competitive in comparison with conventional petrochemical fuels. Mostly 70% of the cost will be invested for cultivation and biomass harvesting [8]. However, a possible way to reduce the cost is to integrate the microalgal cultivation system with wastewater treatment. Generally, wastewater is rich in nutrients and other bioresources. It was reported that wastewater can produce 6.5 MJ/kL of energy, which constitutes 1% of the total world energy [9].

The nutrients presenting in the form of carbon, nitrogen, and phosphorous can be turned into an economic opportunity by feeding them to microalgae [10]. Microalgae can utilize the organic carbon in wastewater and tailor it into biomass. The use of wastewater for microalgal cultivation in mixotrophy and heterotrophy cultivation mode can balance the respiratory losses, improve energy budget, and give a boost to the biomass productivity. Earlier reports have shown that the utilization of nutrients and water from wastewater and industrial flue gas (CO_2) helps to decrease the cost and makes the algal biofuels commercially viable [11]. In addition, the recovery of other value-added bioproducts, rare earth metals, etc. can compensate for the cost involved during algal cultivation. Besides, the wastewater used for algal cultivation does not require any additional treatment to meet the ecological and environmental regulations. Therefore, the utilization of a large quantity of wastewater for microalgal cultivation could promote waste-free, carbon-neutral, and environmentally sustainable technology.

The previous studies have explored that the microalgal species, such as *Botryococcus braunii*, *Scenedesmus obliquus*, *Chlorella vulgaris*, *Chlamydomonas reinhardtii*, and *Nannochloropsis oculata*, are recognized as promising species for biofuel production [12, 13]. Nevertheless, the biofuels produced from highly potential microalgal feedstock need a powerful downstream processing technology. This book chapter is aimed to provide knowledge about the latest research and development of microalga-based biofuel and its challenges.

2. Microalgae and growth condition

Microalgae are simple microscopic heterotrophic or autotrophic photosynthetic organisms, and these organisms are also named phytoplankton. Generally, they are found in fresh, marine, and brackish water, and they utilize photonic energy (light sources), carbon dioxide (CO_2), and water for their growth. Microalgae are classified as green algae (Chlorophyceae), blue-green algae (Cyanophyceae), red algae (Rhodophyceae), brown algae (Phaeophyceae), and diatoms (Bacillariophyceae). The microalgal growth contains five different phases: (1) lag phase, initial growth period, where the microalgae take time to adapt themselves into a new environment; (2) log/exponential phase, here rapid cell division occurs, and growth is faster; (3) decline phase, this phase contains limiting cell division; (4) stationary phase, the cell density of microalgae is stable because of the limiting factors; and (5) death phase, in this stage almost the cell growth is stopped due to lack of nutrients. Besides, microalgae are easy to be cultivated because they can tolerate a broad range of pH, salinity, and temperature. Some researchers have reported that a lipid content of microalgae is usually between 20 and 50% on a dry weight basis [14], whereas in some microalgal species (e.g., *Botryococcus braunii*), the lipid production can be reached up to 75% on a dry weight basis. Microalgae are not only a good source of lipids, it is also a vital source of producing the bioproducts, such as polysaccharides, pigments, proteins, vitamins, bioactive compounds, and antioxidants.

2.1 Microalgal cultivation methods

Generally, according to the growth nutrient modes, the microalgal cultivation can be divided into three categories, which are autotrophic, heterotrophic, and mixotrophic cultivation. In autotrophic mode, the inorganic carbon and light/solar are the primary sources of energy for microalgal growth. The heterotrophic cultivation mode mainly uses organic carbon and energy from the Krebs cycle. On the other hand, in mixotrophic cultivation mode, the carbon sources for microalgal growth can be supplied by both inorganic and organic forms.

2.2 Large-scale microalgal cultivation and biomass production

In microalgal cultivation, large-scale biomass production can be performed by two major methods, such as open raceway and closed photobioreactor (PBR). The open pond is the traditionally used system for large cultivation of microalgal species, which is economically superior in comparison to PBR. In an open cultivation system, the sunlight and atmospheric CO_2 are used as sources for carbon production to achieve higher biomass productivities. On the other hand, the PBR method is suitable for axenic cultures, and this type of cultivation is widely used for the production of high value-added bioproducts (e.g., pharmaceuticals). Basically, an open raceway pond cultivation system consists of a simple water tank or bigger earthen pond in which the nutrients are added from outsourcing. Besides, the open pond is usually designed in a raceway or track configuration attached with paddle wheel to provide circulation and mixing of the algal cells and growth nutrients. The low-cost open raceway pond is typically made from poured concrete, or they are dug into the earth and lined with a plastic liner to avoid the groundwater. The growth medium is added in front of the paddle wheel for proper mixing and absorption.

PBR is the most common system used for the closed cultivation method. It is designed based on several basic features, including liquid circulation, illumination surface area, and gas exchange to supply CO_2 to PBR [15]. Generally, researchers use the PBR cultivation method for the production of high biomass with a controlled environment and to avoid contamination. This type of cultivation makes it easier to optimize the biomass productivity of selected algal species. Usually, PBR is designed by using glass or plastic, coupled with a gas exchanger to pass the nutrients and CO_2. This system contains an airlift pump usually used to circulate the microalgal culture grown in PBR, which helps to keep the culture in suspension state and improve the CO_2 dissolution. Researchers have identified various types of PBR design, such as polyethylene bags, glass fiber cylinder, tubular inclined, segmented glass plate, flat modular photobioreactor, and annular photobioreactor. In the PBR cultivation method, several studies have been carried out on catalyst improvement, shaping of the PBR, controlling environmental parameters, and axenic culture. During the cultivation period, the operational parameters, such as pH, temperature, and gas diffusion, are a crucial issue, and it should be adequately addressed in PBR [11].

Some studies have been performed to recycle the nutrients from wastewater sources, which is considered as a step-in treatment of industrial wastewater using microalgal species. As the world's population continuously increases day by day, wastewater discharge also increased. Thus, the utilization of this harmful wastewater as a source of microalgal growth nutrients is highly recommended for the environmentally friendly high production of biomass and lipid. Nevertheless, to make microalgal biodiesel production at commercial scale, an integrated biorefinery approach of wastewater utilization will strongly influence the future sustainability by addressing high-energy production, reducing greenhouse gas (GHG) emission, and lowering the production cost.

2.3 Harvesting technologies

Generally, two types of cultivation techniques are followed, i.e., batch and semi-continuous or continuous cultivation mode. During the cultivation period, biomass will be harvested and processed. It has been reported that the biomass harvesting accounts for approximately 20 and 30% of the total cost of microalgal down-stream processes [16]. Therefore, the harvesting cost is one of the major hurdles, which makes the algal cultivation unsuccessful at commercial scale. Hence, many researchers are finding more effective techniques for microalgal cell harvesting to overcome this issue. The microalgal harvesting process is expensive and energy-consuming because the density of algal cells in the culture medium is generally low and most of the microalgal cells carry a negative charge, which makes the cells in a suspension state. To achieve a maximum biomass production during the harvesting process, researchers explored several types of harvesting methods, such as filtration, centrifugation, sedimentation ultrasound, and floatation [14]. Nevertheless, these methods are not as efficient as flocculation because of their high cost and lower efficiency. The flocculation harvesting method is much more comfortable, with higher efficiency, than other methods; however, still, a lot of challenges are needed to be addressed. On the other hand, the harvesting of microalgal biomass using flocculants can contaminate the slurry concentrate; thus, it reduces the algal biomass market value, lipid conversion into biodiesel via transesterification process, and the application of this biomass for food industry and animal feeds. Therefore, the feasible way to minimize harvesting costs is only by improving the harvesting technologies. Besides, the suitable method for biomass harvesting mostly depends on the algal species.

2.4 Microalgal biomass and biodiesel production

The lipid accumulation differs from species to species; besides, it depends on the algal cultivation methods. It has been reported that the microalgal species, such as *Dunaliella*, *Chlorella*, *Isochrysis*, *Nannochloris*, *Scenedesmus*, *Tetraselmis*, *and Nannochloropsis*, accumulate the average lipid content of 15–60% on a dry weight basis (**Table 2**). Due to their relatively high lipid accumulation, the microalgal species are considered as a promising feedstock for biodiesel production [17]. The lipids obtained from microalgae are chemically similar to the conventional vegetable oils and so have been considered as a promising source for biodiesel [18]. The microalgal triglycerides can easily be converted into biodiesel, which is renewable, biodegradable, and environmentally friendly when compared to fossil fuel sources [19]. Besides, the microalga-derived biodiesel has a higher heating value (HHV) of 39–41 MJ/kg; thus, it is considered as a potential alternative for displacement of liquid transport fuels derived from petroleum crude [20]. Generally, microalgae produce a high level of triacylglycerols (TAG), which are accumulated in the plastids or found in the cytoplasm in the form of lipid bodies. It was observed that in algal species, during cultivation, the nitrogen starvation and other stressful situations, such as salinity, temperature, CO_2 concentration, and light intensity, stimulate the lipid biosynthesis, resulting in an enhanced lipid production [21, 22]. A study reported that the microalga *Haematococcus pluvialis* produces a high amount of neutral lipids when it is under a stressful environment (i.e., high light intensity and nitrogen starvation) [23].

Division	Microalgae	Volumetric productivity of biomass (g/L/day)	Lipid content (% dry weight biomass)
	Freshwater microalgae		
Green microalgae	*Ankistrodesmus* sp.	—	24–31
	Botryococcus braunii	0.02	20–75
	Chlorella emersonii	0.036	25–63
	Chlorella protothecoides	2	14–57
	Chlorella sorokiniana	0.23–1.7	19–22
	Chlorella vulgaris	0.02–0.2	5–58
	Chlorella pyrenoidosa	2.9	2
	Neochloris	29–65	—
	Chlorococcum sp.	0.28	19.3
	Haematococcus pluvialis	0.06	25
	Scenedesmus obliquus	0.74	11–55
	Scenedesmus quadricauda	0.19	1.9–18
	Scenedesmus sp.	0.26	19–21
Blue-green microalgae	*Spirulina platensis*	0.6–4	4–16
	Spirulina maxima	0.21–0.25	4–9
Red microalgae	*Porphyridium cruentum*	0.36–1.5	9–18.8
	Marine microalgae		
Green microalgae	*Dunaliella salina*	0.22–0.34	6–25
	Dunaliella primolecta	0.09	23
	Dunaliella tertiolecta	0.12	16.7–70
	Nannochloropsis sp.	0.17–1.4	12–53
	Tetraselmis suecica	0.12–32	8.5–23
	Tetraselmis sp.	0.3	12.6–14.7
	Pavlova salina	0.16	30.9
	Pavlova lutheri	0.14	35.4
	Isochrysis sp.	0.08–0.17	7.1–33
		Diatoms	
	Nitzschia sp.	—	16–47
	Skeletonema sp.	0.9	13.3–31
	Skeletonema costatum	0.8	13.5–51

Table 2.
Microalgal species and its productivity [25].

Thus, the microalgae accumulate a higher amount of lipids, especially TAG, which enhance a potential production of biodiesel. In microalgae, however, the biomass and lipid production plays a crucial role in biodiesel production at commercial scale; the quality of the lipids depends on algal species which greatly influences the biodiesel property. Nascimento et al. [24] investigated 12 algal species for potential biodiesel production, and results revealed that *Chlorella* and *Botryococcus* species accumulated high-level lipid content, which can be easily converted into biodiesel.

Some studies showed that the biodiesel obtained from the microalgae *Scenedesmus obliquus* and *Chlamydomonas* sp. contains a high level of saturated fatty acids (SFA) and has the highest cetane number (CN) of 63 along with enhanced oxidation stability. The lipids of some microalgal species, such as *Ankistrodesmus fusiformis*, *Kirchneriella lunaris*, *Ankistrodesmus falcatus*, and *Chlamydocapsa bacillus*, are rich in *polyunsaturated fatty acids* (PUFA), and the obtained biodiesel has low oxidation stability, high iodine value (IV), and low CN (42.5). The microalgal lipids with a high content of SFA and monounsaturated fatty acids (MUFA) give biodiesel with improved quality. Besides, the CN value is significantly correlated with a ratio of SFA to PUFA, which can be used to evaluate the delay between compression and ignition. Mostly, the algal biodiesel has higher CN value than that of the fossil fuels, which helps to shorten the delay in the ignition and complete combustion of the algal biodiesel.

An earlier study has reported that *Amphora* sp., a marine and freshwater diatom, produces a significant amount of MUFA, which can be considered as a potential feedstock to produce high-quality biodiesel [26]. Among several algal strains, such as *Phormidium* sp., *S. obliquus*, *C. vulgaris*, and *Dunaliella tertiolecta*, grown in a bubble PBR for biodiesel production, *Chlorella vulgaris* produces a significant amount of biomass and lipid. Besides, it is considered as the best choice for CO_2 sequestration at a rate of 17.8 $mgL^{-1}min^{-1}$. In addition, the alga *Chlorella vulgaris* biodiesel meets the ASTM 675 and EN 14214 standards because the lipids of these strains are rich in SFA of 43.5% and MUFA of 41.5% [27]. The microalga *Scenedesmus abundans* lipid is a rich source of MUFA (76%), which enhances the quality of resulting biodiesel to meet the European biodiesel standard (EN 14214), South African standard (SANS1935), and Germany's standard (DIN 51606) [28]. Some recent studies investigated different microalgal groups, such as green algae *Chlorella* sp., *Scenedesmus* sp., and *Selenastrum* sp.; red algae *Batrachospermum* sp.; diatoms *Navicula* sp. and *Phaeodactylum* sp.; and blue-green algal species (*Lyngbya* sp., *Isochrysis* sp., and *Prymnesiophytes* sp.), and observed that *Scenedesmus* sp., *Chlorella* sp., and *Isochrysis* sp. produce a significant amount of lipids, which are considered as a promising feedstock for biodiesel production [29]. A previous study isolated 96 microalgal strains from Singapore coastal area, which was then well screened for growth, biomass, and lipid productivity. The results revealed that the marine microalga *Nannochloropsis* sp. was found to be the most promising biomass material for biodiesel production because of its high lipid accumulation of 45% on a dry weight basis [29]. Islam et al. [30] investigated several microalgal species, such as *Ankistrodesmus* sp., *Botryococcus* sp., *Chlorella* sp., *Chlamydomonas* sp., *Coelastrum* sp., *Desmodesmus* sp., and *Scenedesmus* sp., for biodiesel production. The author observed that *Botryococcus* sp. produced a high lipid yield and the quality of the resulting biodiesel met the ASTM 6751-02 and EN 14214 standards. The quality of biodiesel obtained from different microalgal species, such as *Nannochloropsis* sp., *Scenedesmus* sp., and *Dinoflagellate*, was explored [31]. The biodiesel properties included density, kinematic viscosity, acid value, phosphorous content, sulfated ash content, and sulfur content, according to the Chinese National Standards (CNS). The results showed that the biodiesel characteristics were almost in the same range as fossil oil; nevertheless, the oxidative stability of those microalgal biodiesel was lower than that of the CNS standard. Also, they demonstrated that the oxidative stability of microalga-derived biodiesel could be improved by hydrogenation catalyzed over carbon-supported palladium (Pd/C). Some researchers reported that the microalga *Botryococcus* sp. produces lower biomass yield and higher lipid productivity, up to 75% on a dry weight basis, than other microalgal species, such as *Nannochloropsis* sp., *Nitzschia* sp., *Neochloris* sp., *Porphyridium* sp., *Dunaliella* sp., *Isochrysis*, and *Chlorella* sp., with a lipid content ranging from 20 to 50% but appropriate biomass yield.

3. Current challenges in microalgal biofuels

In the present scenario, rapid population growth and extensive fossil fuel usage increase the energy demand and significant environmental-related problems, which lead to global warming. Therefore, researchers have seriously searched for an alternative and sustainable solution to overcome those issues. In this context, microalgae are considered a promising candidate for an alternative fuel source and an excellent option for cleansing the environment. The previous studies have shown that the cost of biodiesel produced from microalgae is estimated at \$20.53 and \$9.84 per gallon using a PBR and open raceway pond cultivation method, respectively. This shows that microalgal biodiesel is a promising avenue for sustainable energy production. From the literature survey, it is clearly noted that even though several microalgal species are available for biodiesel production, only a few algal species are considered as the best choice because of its quality and quantity of lipid accumulation. Raja et al. [32] reported that, on earth, more than 25,000 microalgal species are available; however, only a few species are in use.

At present, the utilization of microalgae as a feedstock for the production of bioenergy and bioproducts still faces a lot of limitations and challenges, and we must be addressing these issues by improving the technologies from laboratory scale to commercial scale. The most critical problems are to improve the algal biomass productivity, dewatering and biomass productivity, pretreatment and extraction, and biodiesel production. Despite several advanced technologies are available for a large-scale biomass production and lipid conversion into biodiesel, still, microalgal biodiesel is too costly since the cultivation system design requires temperature and growth limiting condition control (viz., CO_2, water sources, nutrient source, and optimization). The other most crucial obstacle is biomass dewatering because this process is energy-intensive and so costly.

Generally, in a large-scale algal cultivation, the closed PBR system is more expensive than the open raceway ponds. The PBR system also faces major operating challenges, such as overheating and fouling, due to gas exchange limitation. In microalgal cultivation, open ponds, especially mixed raceway ponds, are much cheaper to be built and operated and are easily scaled up to several hectares, which make them the right choice for commercial-scale biomass production. About 95% of commercial microalgal biomass production is performed using open raceway ponds even for high value-added bioproducts, which sell for prices over a hundred/thousand dollars. Nevertheless, the open cultivation methods meet several limitations, mainly due to contaminations by other microalgal species, algal grazers, fungi, amoeba, etc., and temperature. A literature survey revealed that though hundreds of research papers were published, still now, no proper information is available on cultivation designs, operations, yields, and other important aspects at the commercial level [33]. A major bottleneck in microalgal biofuel production is the high capital and operating costs. However, several research studies have focused on microalga-based biofuels, still, there is a vast technological gap that was found during commercialization. In a large-scale biomass production, there is a large demand for water, CO_2, nitrogen, and phosphorous, which is believed as another major hurdle. The wastewater can be utilized as a source of nutrients; nonetheless, there is a serious concern on contamination by bacteria, pathogens, and chemical compounds presenting in wastewater. Earlier studies reported that 0.16 kg of nitrogen and 0.022 kg of phosphorous are required for producing 1 liter of algal oil [34]. Besides, for producing 1-liter algal oil, the microalgae need 3.5–9.3 kg of CO_2, which implies that algae utilize a large amount of CO_2 for its growth and biomass production.

Another major challenge is algal lipid extraction prior to biodiesel production. In this part, after biomass drying, the lipid extraction using expensive solvents significantly increases the production cost. Many researchers are searching for significantly advanced technologies without drying or solvent extraction of the algal slurry in order to reduce the biomass pretreatment cost.

The biodiesel production based on current methods is expensive since it requires neat lipid feedstock, free from free fatty acids (FFA), and water. For this kind of extraction technique, the biomass must be dried; however, biomass drying is another important process, and it requires a higher cost. To reduce the FFA content of lipid feedstock, the esterification process is carried out via either acidic or enzymatic route; however, this process is still at the research stage. The esterification through the enzymatic process using lipases may be considered as the best choice because it has added advantage of running even at low temperatures. Nonetheless, the primary issue in this method is glycerol formed as a by-product, which can inhibit the lipase activity. Some researchers demonstrated that using methyl acetate, as a substrate instead of methanol, avoids glycerol formation and lipase inhibition since triacetin is generated as a by-product [35].

3.1 Microalgal biofuels and future prospects

It has been reported that the current costs in biofuel produced from microalgal biomass are approximately estimated up to 50 $/L, and thus this makes algal biofuel unsuccessful at commercial scale [36]. Nevertheless, to reduce the cost and make the algal biofuel production at commercial success, research works are still ongoing. However, the most promising and sustainable way for biofuel production is to reduce the cultivation cost, particularly growth nutrient cost. The utilization of wastewater for large-scale cultivation of algal biomass is attracted between the researchers. It is possible to grow the algae at zero nutrient cost using wastewater obtained from various sources, such as industrial, municipal, and agricultural [17]. Recent researchers are extensively investigating an integrated biorefinery concept for producing the algal biomass at zero nutrient cost; besides, it is a possible way for treating the wastewater using zero-cost technology. Some studies focused on commercial interests in a large-scale microalgal culturing using coal-fired power plants or sewage treatment facilities. This approach not only provides the raw materials for the system, such as CO_2 and nutrients; besides it also produces valuable biofuels with a cleansed environment (**Figure 1**). Today, the price of crude oil is lower, so the biodiesel produced from microalgae is economically uncompetitive with fossil diesel [33]. From the above study, it clearly shows that low-cost biomass production

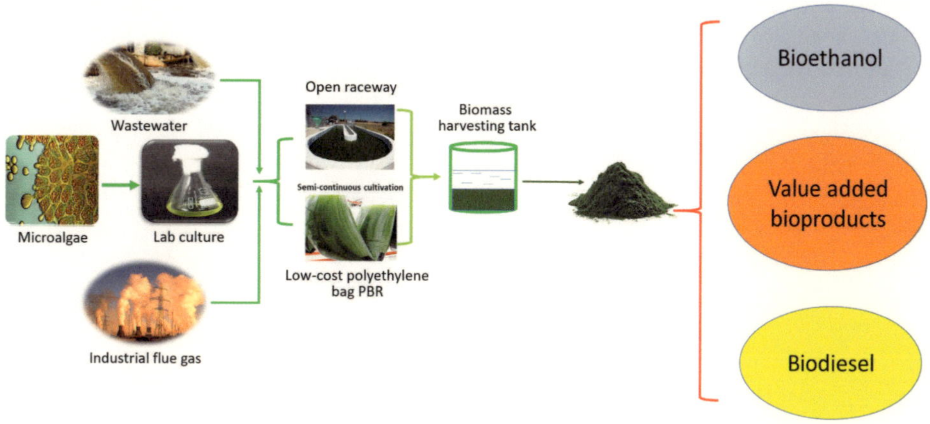

Figure 1.
Integrated microalgal biorefinery approach.

is a key issue towards the commercial production of algal biodiesel. Research efforts have been devoted to address the following problems: (1) to enhance the photosynthesis efficiency, biomass, and lipid production through genetic and metabolic engineering; (2) using high-efficient and low-cost biomass production system (open raceway pond or low-cost designed PBR); (3) cultivation mode (batch or semicontinuous); (4) utilization of wastewater and industrial flue gas in algal cultivation; (5) introducing novel microalgal harvesting methods; and (6) implementing low-cost with high-efficient oil extraction and transesterification methods (e.g., simultaneous oil extraction and transesterification, using novel heterogeneous acid catalysts, etc.).

4. Conclusion

Today, microalgae offer interesting characteristic features to qualify them as promising alternative feedstocks for various industrial and environmental applications. Nevertheless, many efforts are required to address different challenges particularly on low cost with high-efficiency biofuels, wastewater treatment, and CO_2 mitigation. Based on the research data available, it is clearly found that so far, microalgal biodiesel production is mainly stuck in a cost factor. It is clearly noticed that zero nutrient cost technology for biomass production, inexpensive large-scale harvesting, and biodiesel conversion process are yet to be improved through a detailed investigation. In this point of view, this review shows a clear outlook on algal cultivation for biofuel production and what are the challenges behind with future prospects. The production of microalgal biofuel at commercial scale can play a key role in the present global energy scenario and concern towards the related environmental issues. Researchers believe that microalgal as a third-generation candidate will be satiating the energy demand and its challenges in the future. The paramount challenges in algal biodiesel production are cultivation and harvesting techniques, and the limitations with an emphasis on the cost factor are discussed. Therefore, establishing a new and innovative biorefinery-based low-cost technology should be developed to overcome these problems. Recently, the microalga-based biorefinery is the emerging technology, which is aiming to address the above severe issues and make the algal biofuels sustainable and alternative. Besides, it helps wastewater treatment at zero-cost technology, CO_2 mitigation, and attractive value-added bioproducts. These economic processes could be improved by adapting various cost-cutting activities, such as utilizing wastewater and industrial flue gas as nutrient and carbon sources, respectively. Finally, the microalga-based biorefinery process seems to be the most feasible approach in the forthcoming years to compete with fossil fuels and to develop a sustainable and renewable bioenergy source.

Author details

Ashokkumar Veeramuthu[1] and Chawalit Ngamcharussrivichai[1,2*]

1 Center of Excellence in Catalysis for Bioenergy and Renewable Chemicals (CBRC), Faculty of Science, Chulalongkorn University, Bangkok, Thailand

2 Center of Excellence on Petrochemical and Materials Technology (PETROMAT), Chulalongkorn University, Bangkok, Thailand

*Address all correspondence to: chawalit.ng@chula.ac.th

References

[1] U.S. Energy Information Administration (EIA). Annual Energy Outlook 2015 with Projections to 2040; 2015

[2] Balat M, Balat H. Progress in biodiesel processing. Applied Energy. 2010;**87**(6):1815-1835

[3] HCJ G, Beddington JR, Crute IR, Haddad L, Lawrence D, Muir JF, et al. Food security: The challenge of feeding 9 billion people. Science. 2010;**327**(5967):812

[4] Mohr A, Raman S. Lessons from first generation biofuels and implications for the sustainability appraisal of second generation biofuels. Energy Policy. 2013;**63**:114-122

[5] Daroch M, Geng S, Wang G. Recent advances in liquid biofuel production from algal feedstocks. Applied Energy. 2013;**102**:1371-1381

[6] Bahadar A, Bilal Khan M. Progress in energy from microalgae: A review. Renewable and Sustainable Energy Reviews. 2013;**27**:128-148

[7] Ahmad AL, Yasin NHM, Derek CJC, Lim JK. Microalgae as a sustainable energy source for biodiesel production: A review. Renewable and Sustainable Energy Reviews. 2011;**15**(1):584-593

[8] García Prieto CV, Ramos FD, Estrada V, Villar MA, Diaz MS. Optimization of an integrated algae-based biorefinery for the production of biodiesel, astaxanthin and PHB. Energy. 2017;**139**:1159-1172

[9] Puyol D, Batstone DJ, Hülsen T, Astals S, Peces M, Krömer JO. Resource recovery from wastewater by biological technologies: Opportunities, challenges, and prospects. Frontiers in Microbiology. 2017;7:2106

[10] Wang S-K, Wang X, Tao H-H, Sun X-S, Tian Y-T. Heterotrophic culture of *Chlorella pyrenoidosa* using sucrose as the sole carbon source by co-culture with immobilized yeast. Bioresource Technology. 2018;**249**:425-430

[11] Acién Fernández FG, González-López CV, Fernández Sevilla JM, Molina Grima E. Conversion of CO2 into biomass by microalgae: How realistic a contribution may it be to significant CO2 removal? Applied Microbiology and Biotechnology. 2012;**96**(3):577-586

[12] Das AAK, Esfahani MMN, Velev OD, Pamme N, Paunov VN. Artificial leaf device for hydrogen generation from immobilised *C. reinhardtii* microalgae. Journal of Materials Chemistry A. 2015;**3**(41):20698-20707

[13] Zhang Y, Kong X, Wang Z, Sun Y, Zhu S, Li L, et al. Optimization of enzymatic hydrolysis for effective lipid extraction from microalgae *Scenedesmus* sp. Renewable Energy. 2018;**125**:1049-1057

[14] Halder P, Azad AK. Chapter 7 - recent trends and challenges of algal biofuel conversion technologies. In: Azad AK, Rasul M, editors. Advanced Biofuels. Cambridge: Woodhead Publishing; 2019. pp. 167-179

[15] Masojídek J, Torzillo G. Mass cultivation of freshwater microalgae. In: Jørgensen SE, Fath BD, editors. Encyclopedia of Ecology. Oxford: Academic Press; 2008. pp. 2226-2235

[16] Molina Grima E, Belarbi EH, Acién Fernández FG, Robles Medina A, Chisti Y. Recovery of microalgal biomass and metabolites: Process options and economics. Biotechnology Advances. 2003;**20**(7):491-515

[17] Ashokkumar V, Chen W-H, Ngamcharussrivichai C, Agila E, Ani FN. Potential of sustainable bioenergy production from *Synechocystis*

sp. cultivated in wastewater at large scale – A low cost biorefinery approach. Energy Conversion and Management. 2019;**186**:188-199

[18] Chisti Y. Biodiesel from microalgae. Biotechnology Advances. 2007;**25**(3):294-306

[19] Vicente G, Martínez M, Aracil J. Integrated biodiesel production: A comparison of different homogeneous catalysts systems. Bioresource Technology. 2004;**92**(3):297-305

[20] Ashokkumar V, Agila E, Sivakumar P, Salam Z, Rengasamy R, Ani FN. Optimization and characterization of biodiesel production from microalgae *Botryococcus* grown at semi-continuous system. Energy Conversion and Management. 2014;**88**:936-946

[21] George B, Pancha I, Desai C, Chokshi K, Paliwal C, Ghosh T, et al. Effects of different media composition, light intensity and photoperiod on morphology and physiology of freshwater microalgae *Ankistrodesmus falcatus* – A potential strain for bio-fuel production. Bioresource Technology. 2014;**171**:367-374

[22] Pancha I, Chokshi K, Maurya R, Trivedi K, Patidar SK, Ghosh A, et al. Salinity induced oxidative stress enhanced biofuel production potential of microalgae *Scenedesmus* sp. CCNM 1077. Bioresource Technology. 2015;**189**:341-348

[23] Damiani MC, Popovich CA, Constenla D, Leonardi PI. Lipid analysis in *Haematococcus pluvialis* to assess its potential use as a biodiesel feedstock. Bioresource Technology. 2010;**101**(11):3801-3807

[24] Nascimento IA, Marques SSI, Cabanelas ITD, Pereira SA, Druzian JI, de Souza CO, et al. Screening microalgae strains for biodiesel production: Lipid productivity and estimation of fuel quality based on fatty acids profiles as selective criteria. BioEnergy Research. 2013;**6**(1):1-13

[25] Mata TM, Martins AA, Caetano NS. Microalgae for biodiesel production and other applications: A review. Renewable and Sustainable Energy Reviews. 2010; **14**(1):217-232

[26] Talebi AF, Mohtashami SK, Tabatabaei M, Tohidfar M, Bagheri A, Zeinalabedini M, et al. Fatty acids profiling: A selective criterion for screening microalgae strains for biodiesel production. Algal Research. 2013;**2**(3):258-267

[27] Francisco EC, Jacob-Lopes E, Franco TT. Microalgae as feedstock for biodiesel production: Carbon dioxide sequestration, lipid production and biofuel quality. Journal of Chemical Technology and Biotechnology. 2010;**85**:395-403

[28] Mandotra SK, Kumar P, Suseela MR, Ramteke PW. Fresh water green microalga *Scenedesmus abundans*: A potential feedstock for high quality biodiesel production. Bioresource Technology. 2014;**156**:42-47

[29] Doan TTY, Sivaloganathan B, Obbard JP. Screening of marine microalgae for biodiesel feedstock. Biomass and Bioenergy. 2011;**35**(7):2534-2544

[30] Islam AM, Magnusson M, Brown JR, Ayoko AG, Nabi NM, Heimann K. Microalgal species selection for biodiesel production based on fuel properties derived from fatty acid profiles. Energies. 2013;**6**(11):5676-5702

[31] Chen X, Hu L, Xing R, Liu S, Yu H, Qin Y, et al. Ionic liquid-assisted subcritical water promotes the extraction of lipids from wet microalgae *Scenedesmus* sp. European Journal of Lipid Science and Technology. 2015;**117**(8):1192-1198

[32] Raja R, Hemaiswarya S, Kumar NA, Sridhar S, Rengasamy R. A perspective on the biotechnological potential of microalgae. Critical Reviews in Microbiology. 2008;**34**(2):77-88

[33] Zhu L, Nugroho YK, Shakeel SR, Li Z, Martinkauppi B, Hiltunen E. Using microalgae to produce liquid transportation biodiesel: What is next? Renewable and Sustainable Energy Reviews. 2017;**78**:391-400

[34] Venkata Mohan S, Rohit MV, Chiranjeevi P, Chandra R, Navaneeth B. Heterotrophic microalgae cultivation to synergize biodiesel production with waste remediation: Progress and perspectives. Bioresource Technology. 2015;**184**:169-178

[35] Baadhe RR, Potumarthi R, Gupta VK. Chapter 8 - lipase-catalyzed biodiesel production: Technical challenges. In: Gupta VK, Tuohy MG, Kubicek CP, Saddler J, Xu F, editors. Bioenergy Research: Advances and Applications. Amsterdam: Elsevier; 2014. pp. 119-129

[36] Ashokkumar V, Chen W-H, Kamyab H, Kumar G, Al-Muhtaseb AH, Ngamcharussrivichai C. Cultivation of microalgae *Chlorella* sp. in municipal sewage for biofuel production and utilization of biochar derived from residue for the conversion of hematite iron ore (Fe2O3) to iron (Fe) – Integrated algal biorefinery. Energy. 2019;**189**:116128

Chapter 12

Rapid Evaluation of Biomass Properties Used for Energy Purposes Using Near-Infrared Spectroscopy

Jetsada Posom, Kanvisit Maraphum and Arthit Phuphaphud

Abstract

The parameters corresponding to combustion and pyrolysis such as proximate parameter (emissions), calorific value, elemental component, pyrolysis characteristics (temperature), and thermal properties are necessary to the thermal conversion process and the trading of biomass. Traditionally, these parameters of wood chips, milled wood, and biomass pellets are determined with chemicals, time-consuming, and required technical experts, such as thermogravimetry, bomb calorimetry, dry oven, muffle furnace, and so on. The near-infrared (NIR) spectroscopy is a rapid, noncontact no-chemical measurement. For NIR spectroscopy, only 2–3 seconds are used for evaluation, and it could be used for online measurement. The application of NIR spectroscopy in the estimation of the biomass characteristics of wood chips, milled wood, and biomass pellets is described in this chapter.

Keywords: biomass properties, pyrolysis characteristics, proximate analysis, calorific value, near-infrared spectroscopy

1. Introduction

Nowadays, the utilization of alternative energy is increasing because energy produced from fossil fuels increase in greenhouse gasses which lead to global warming [1, 2]. Biomass produced from fast-growing tree and waste of agricultural activity is interesting because it can be used in the future [3]. Many countries have focused on the use of biomass fuel to be the resource of future energy. Therefore, utilization with high efficiency is a key [4]. To achieve efficiency of thermal conversion process, knowledge of biomass properties used for energy purposes is important. For example, the parameters that are important for combustion are calorific value (including gross calorific (GCV) or higher heating value (HHV) and net calorific value (NCV) or lower heating value (LHV)), proximate data or emission (including moisture content (MC), volatile matter (VM), fixed carbon (FC) and ash (A)), and elemental component (carbon (C), hydrogen (H), oxygen (O), nitrogen (N), and sulfur (S)), while the important parameter for pyrolysis process is pyrolysis characteristics (temperature) (including T_{onset}, T_{sh}, T_{peak}, T_{offset}, and DTG_{peak}), kinetic parameter, proximate parameter, and elemental component. On the other hand, parameter which is essential in designing of equipment for the drying and drying process is thermal properties (including thermal diffusivity, thermal conductivity, specific heat) [5]. These parameters are explained as follows:

1.1 Calorific value

Calorific value (CV) is the total energy released when the matter is burned. CV is divided into two types, i.e., gross calorific value (GCV) and net calorific value (NCV). CV is generally used to calculate, design, and operate thermal power plants [6]. It also can be used to interpret how biomass is used to achieve process efficiency [7, 8]. The GCV is "the amount of heat released from combustion of a certain amount of fuel and assuming its combustion product water had returned to liquid state at the end of a measurement in which it took the latent heat of vaporization of product water into account". NCV demonstrates the total of heat released under conditions where "the product water was still at vapor state and its latent heat of vaporization was not recovered" [9]. GCV is generally determined using bomb calorimetry.

Meanwhile, NCV is determined using a calculation method [10, 11] with theoretical function of the GCV by subtraction from moisture content (MC) as follows [12]:

$$NCV_{wb} = GCV_{wb} - 2.443 \times MC\% \quad (1)$$

where NCV_{wb} is the lower heating value (MJ/kg) and GCV_{wb} is the higher heating value (MJ/kg). GCV is expressed on a wet weight basis (MJ/kg). MC is percent moisture content expressed on a wet basis. Using the term NCV is better than GCV if it operates under actual operating conditions because the LHV represents a real energy content, which is defined as the efficiency of the thermal process [13].

1.2 Proximate analysis (emission)

Proximate data include four main parameters, i.e., MC, VM, FC, and Ash. MC is the amount of water in a substance which had a negative effect on the pyrolysis and combustion process, for example, a negative effect on both process and quality of bio-oil production where temperature dropped during pyrolysis and provided more water contents in bio-oil products. Measuring MC can be conducted by drying in an oven (ASTM E1756-08) [14]. VM is the vapor gas released after MC is released. Generally, VM has around 70–80% of dried biomass. VM content indicates whether the biomass was gasified easily or not and whether it was usable for boiler and gasifier design [15]. In fast pyrolysis and gasification process, VM was utilized to estimate the liquid fuel and gas fuel rate. VM was measured by muffle furnace, following the standard method of the ASTM E87282 (2006) and particular wood fuels [16]. FC is calculated by equation FC = 100-(MC% + VM% + A%) as-received base. FC is the residue contained after MC and VM are released and mostly consists of carbon content. Ash is the solid residue after biomass volatilizes and FC is combusted [17]. High ash content leads to low HV and negative factor of storage and transportation due to no energy content. Ash content is determined using the muffle furnace method following the ASTM E1755-01 [18]. Ash content leads to trouble in waste systems due to fouling and slagging.

Normally, according to standard method, the determination of proximate analysis requires a long time which leads to cost implications [17]. Thermogravimetry (TG) can be used to determine the proximate analysis also. TG was a direct method and convenient and required small sample to be used. The direct measurement of proximate data of biomass examined by TGA is illustrated in **Figure 1**. TGA is the method utilized to study the degradation behavior in biomass burned and degraded,

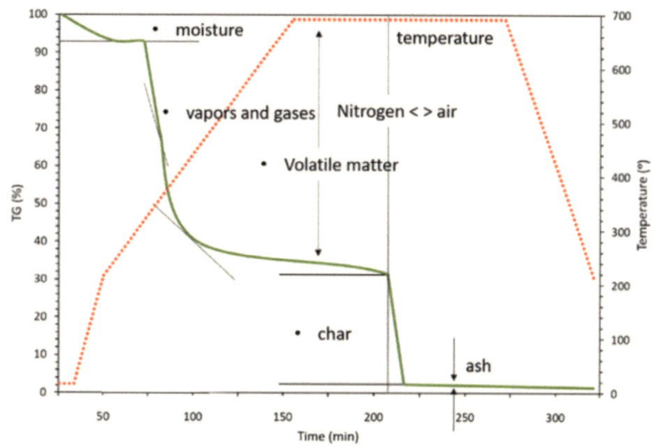

Figure 1.
Thermal behavior of biomass carried on by TGA [19].

recording weight loss as a time function versus temperature. MC, VM, FC, and A were estimated by direct measurement of weight changes from the TG chart [18].

1.3 Elemental compositions

The elemental components of biomass (i.e., C, H, N, O, and S) are defined as 100% = C + H + O + N + S + A. It is necessary in operating the overall thermal process, such as for calculation of heat content and balancing of the heat process [20]. For example, it is used to evaluate the total of flue gas (air) required to complete combustion. The burning of biomass containing more S and N leads to oxide emissions, i.e., SO_x, NO, and NO_x, released into the atmosphere, which leads to acid rain [17]. High C and H lead to increased HV, but high O content leads to decreased HV [21]. Elemental composition is determined using CHNS analyzer, which is carried on by burning a sample in a combustion chamber with pure O_2 and then measuring the gasses released (such as CO_2, N_2, SO_2, and H_2O) from combustion [20].

1.4 Pyrolysis characteristic

Pyrolysis is a thermal degradation process under the condition of absence oxygen [17], which demonstrates the behavior degradation of biomass. Pyrolysis characteristics, i.e., T_{onset}, T_{sh}, T_{peak}, T_{offset}, and DTG_{peak}, were defined as: "T_{onset} was the extrapolated onset temperature calculated from the partial peak that results from the decomposition of the hemicellulose component, T_{sh} was the temperature corresponding to the overall maximum of the hemicellulose decomposition rate, DTG_{peak} was the overall maximum of the cellulose decomposition rate (dm/dt at the highest peak, wt loss %/min), T_{peak} was the temperature corresponding to the overall maximum of the cellulose decomposition rate, and T_{offset} was the extrapolated offset temperature of the DTG_{peak} curves determined using thermogravimetric analysis (TGA)" [22]. The direct measurement of pyrolysis characteristics determined using DTG curve is shown in **Figure 2**. These parameters are applied to operate pyrolysis process [24]. T_{onset} is the start of the degradation of biomass, as it is needed to know when the biomass is degraded and to help to set the gas condensation time. T_{peak} is the maximum decomposition since the biomass that decomposed at temperature at T_{peak} can produce the highest gas production rate equaling to DTG_{peak}. The temperature of T_{offset} had no composition of biomass. If the process is heated above T_{offset}, it affects the cost of capital energy.

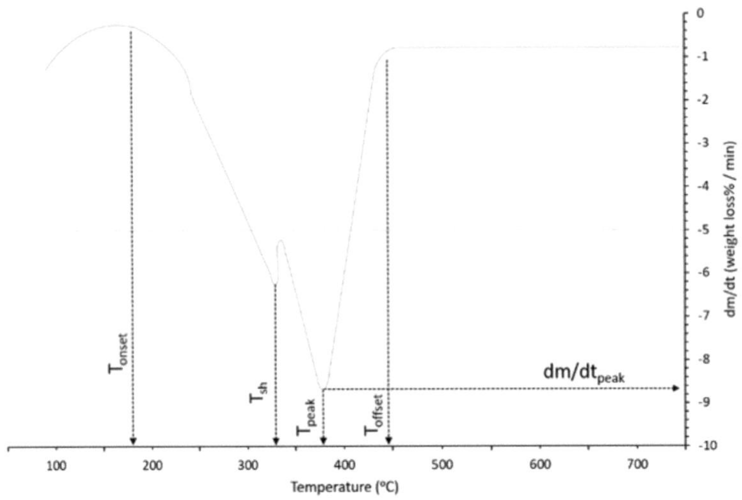

Figure 2.
DTG curve profile of milled bamboo [23].

In addition, many researchers have mentioned pyrolysis behavior as follows: Wannapeera et al. [25] and Lv et al. [26] mentioned that proportions of element compositions effect pyrolysis behavior. For example, high lignin shows slow degradation, while high hemicellulose and cellulose provide fast degradation [25, 26]. High cellulose provides high yields of bio-oil, high hemicellulose content provided high gas yields, and high lignin content provided high charcoal residue [27]. Yang et al. [28] reported that hemicellulose was degraded at 220–315°C, cellulose was decomposed at 315–400°C, and lignin was slowly degraded in a wide range between 160 and 900°C. Then, the pyrolysis behavior of biomass depends on the ratio of chemical content in biomass, such as hemicellulose, cellulose, and lignin. As such, understanding the degradation behavior is important in achieving efficiency of thermal conversion.

1.5 Kinetic parameters

Understanding of degradation behavior of biomass which explains the cracking mechanism required knowledge of kinetic behavior. Kinetic parameters, i.e., activation energy (Ea), pre-exponential factor (A), and reaction order (n) [29, 30], are employed to predict of the reaction behavior, which are usable for the optimization of the pyrolytic degradation process; they are examined using the information of biomass decomposition obtained from the thermogravimetric (TG) and derivative thermogravimetric (DTG) [29]. Kinetic parameters are utilized to design and achieve efficiency of processes [31, 32]. The minimum amount of energy needed to initiate chemical change is calculated using Ea, while the reaction rate is calculated using A and n [29].

2. Application of NIR spectroscopy on estimation of biomass properties

2.1 NIR spectroscopy for quantitative and qualitative analysis

In the 1800s, William Herschel discovered that the end of the visible band could provide heat [33]. In 1835, Ampere found that this was one of the electromagnetic waves which have its wavelength in light. It was called infrared radiation [34]. Until in 1960, much research was reported involving IR analysis and its application. The research showed that this was a new method which could be used in evaluating water content [35]. Meanwhile, Karl H. Norris succeeded in using the spectroscopic method, for example, visible and NIR for predicting the quality of agricultural products (i.e., grains). Since 1970, many publications have reported on the application of NIR spectroscopy with food and agricultural products. This was a wide use of NIR spectroscopy. Near infrared (NIR) ranged between 700 and 2500 nm (12,500–4000 cm^{-1}) and is the technique which studies the correlation between analyte and its corresponding NIR radian [36]. NIR radian is absorbed by the chemical bonds of C–H, N–H, S–H, C=O and O–H [36]. NIR technique had many advantages, such as being rapid, nondestructive, no-chemical use, environmentally friendly, and so on.

Despite NIR having many advantages, its weakness should be not overlooked. If the concentration of analyte is lower than 0.1% of the total weight of sample, the calibration equation is acceptable [37], because the peaks of analyte are overlapped and hidden from other substances. The basic NIR spectrometer contained three parts, i.e., light source, detector, and mathematical model [36]. The assumption of NIR protocol was that if the concentration of a sample changed, the absorption of the sample may differ. Basic NIR spectroscopy creates a mathematical equation that can provide the highest performance, which is indexed by accuracy of model. The accuracy of the calibration equation is indexed by statistical terms of coefficient of determination (R^2), standard error of prediction (SEP), residue prediction deviation (RPD), and bias. The application of NIR model is indicated by R^2 and RPD. If a model provided an R^2 of between 0.50 and 0.64, it could be used toward rough screening: 0.66 and 0.81, applied for screening and approximate calibrations; 0.83 and 0.90, applied for caution with most applications; and 0.92 and 0.96, applied for most applications [38] and, for RPD, between 0.0 and 2.3, not recommended, 2.4 and 3.0 applied for very rough screening, 3.1 and 4.9 applied for screening, 5.0 and 6.4 applied for quality control, 6.5 and 8.0 applied for process control; and RPD > 8.1 applied for any application [38]. Nicolaï et al. [39] and Zornoza et al. [40] indicted that $R^2 > 0.90$ and RPD > 3 was an excellent prediction; $0.81 < R^2 < 0.90$ and 2.5 < RPD < 3 was a good prediction, $0.66 < R^2 < 0.80$ and 2.0 < RPD < 2.5 was an approximate prediction, and $R^2 < 0.66$ and RPD < 2 was a poor prediction. For biomass, the physical forms included ground, chipped, and pelleted. **Figure 3** shows the different physical forms of biomass used for the thermal conversion process. The NIR spectroscopy used to estimate the biomass properties coupled with various physical forms is explained in the next topic.

a) b) c)

Figure 3.
(a) Ground biomass, (b) bamboo chips format, and (c) Leucaena *pellets.*

2.2 Proximate analysis and calorific value

The prediction of biomass properties used for energy purpose using NIR spectroscopy has been utilized by many researchers, such as cassava rhizome ground (MC, HHV, and LHV) [41], maize cob (GCV) [19], and recycled sawdust (GCV and ash) [42]. The measurement of MC was conducted by Shrestha and Sirisomboon [43], predicting the MC of bamboo wood chips using two different portable spectrometers, i.e., NIR-Gun (600–1100 nm) and MicroNIR (1150–2150 nm). **Figures 4** and **5** illustrate the scanning technique of two spectrometers through diffuse reflectance mode and the particle size of bamboo wood chips, respectively. The calibration equation was developed using PLS algorithm. The statistical terms of R^2, SECV, SEP, bias, and RPD were used to examine the performance of the NIR model, providing 0.924, 2.871% wb, 2.385% wb, −0.250% wb, and 3.656 for NIR-Gun. Meanwhile, a MicroNIR spectrometer gave 0.743, 4.349% wb, 4.499% wb, 0.026% wb, and 1.972, respectively. The results indicated that the model is suitable for screening and approximating the MC in bamboo chips. This technique benefits commerce in terms of setting prices and was usable in process control using the parameter of MC such as drying process, pelletization, and combustion. Portable spectrometers are convenient to use and suitable for use on field.

Prediction of MC and HHV of pellets fuel was investigated by Posom and Sirisomboon [44], who developed the NIR model to evaluate the MC and HHV of *Leucaena* pellets using an FT near-infrared spectrometer. **Figure 6a** illustrates the scanning technique for *Leucaena* pellets (8 mm diameter). Each sample was placed

a) b)

Figure 4.
Scanning of wood chips using portable NIR spectrometer (a) NIR-gun and (b) MicroNIR spectrometer.

Figure 5.
Particle size of bamboo wood chips.

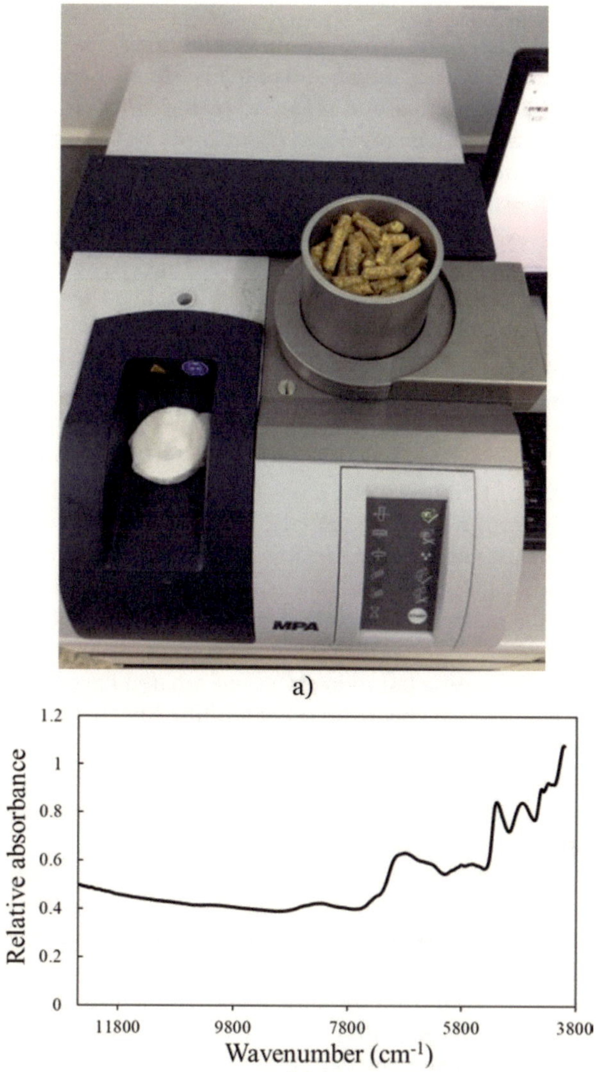

b)

Figure 6.
(a) Scanning of Leucaena *pellets in diffuse reflectance mode and (b)* Leucaena *pellets spectrum.*

in a quartz sampling cup (dimension with 87.0 mm diameter and 87.5 mm height) and scanned through the quartz window using a rotary mode. The scanning was done 64 times per sample and averaged to one spectrum. The spectrum of *Leucaena* pellets is illustrated in **Figure 6b**. In commercial setting where the MC and HHV are necessary, they are measured for quality assurance. The rapid method was used as an alternative method to the traditional MC and HHV due to complications including time, cost, and the requirement of a skilled technician. Based on quality assessment, NIR spectroscopy measured MC and HHV within 30 seconds.

For the MC model, the spectra pretreated by min-max normalization gave the affective calibration equation and wavenumber range of 7506–5446.3 and 4428–4242.9 cm^{-1}, providing the best result with a r^2 of 0.995, RMSECV of 0.187 (%wb), and RPD of 13.9. Meanwhile, the HHV model developed with the min-max normalization spectra, wavenumber range of 9403.8–7498.3, and 6102–5446.3 cm^{-1} also provided the optimal model, with r^2 of 0.964, RMSECV of 89.2 J/g, and RPD of 4.71. The MC and HHV models were excellent for any application.

Posom and Sirisomboon [45] developed the mathematical equation for the prediction of HHV, VM, FC, and ash of ground bamboo. Milled bamboo (particle size ≤3 mm (see **Figure 4a**)) was scanned using FT-NIR spectroscopy with diffuse reflectance mode. The absorbance was recorded by log 1/R unit, where R is the reflectance of milled sample. The purpose was to apply the NIR technique instead of bomb calorimetry and thermogravimetry (TG). The model was optimized using PLS regression. The result of HHV, VM, FC, and ash model provided r^2 of 0.92, 0.82, 0.85, and 0.51; RMSEP of 122 J g^{-1}, 1.15, 1.00, and 0.77%; and RPD of 3.66, 2.55, 2.62, and 1.44, respectively. The summary was that NIR spectroscopy was successful in estimating the HHV, VM, and FC; meanwhile the ash model was not recommended because of low RPD. It was recommended that the range of reference value should be as wide as possible. The spectra of the ground sample can give a high precision because of homogeneous sample. However, the use of ground sample was complicated in terms of sample preparation, which led to higher time and cost. Posom and Sirisomboon [46] updated the NIR model in prediction of HHV, LHV, MC, VM, FC, and ash of bamboo wood chips (see **Figure 4b**). This model replaced the previous model (ground bamboo model). The samples were collected for two seasons in 2017 and 2018. In real situations, wood chip preparation is easier than ground sample. In addition, even though the previous study on ground bamboo reported a good performance, in the prediction of ground sample, it was found to be inconvenient because of complicated sample preparation. NIR model were developed by directly scanning bamboo chips and created based on PLS regression. A big particle size led to lower spectrum repeatability, but this problem was solved using spectral preprocessing. The improvement of MC, ash, VM, and FC models (bamboo chips) could be applied for quality assurance. The HHV and LHV models can be utilized in most applications. A big particle size of wood chips can affect negatively in the calibration equation because the sample was inhomogeneous; however, this problem can be solved by spectral pretreatment technique. Therefore, the direct scanning in wood chips can eliminate the need process for grinding wood samples [46]. Moreover, a calibration equation can predict ash content in bamboo wood chips successfully. The reason why NIR spectroscopy can predict ash content is because: "the infrared bands for inorganic materials are broader, fewer in number and appear at lower wavenumbers than those observed in organic materials, and if an inorganic compound forms covalent bonds within an ion, it can produce a characteristic infrared spectrum" [47].

The heating value and proximate analysis (MC, VM, FC, and A model) model of biomass pellets was studied by Feng et al. [24], who constructed a 2D NIR hyperspectral image in the prediction of proximate analysis and illustrated a quality distribution map. The rapid measurements for properties of biomass pellets helped to monitor the feedstock pellet production. The models were constructed using an algorithm of PLSR and least-squares support-vector machine (LSSVM). The optimized calibration models constructed by successive projections algorithm (SPA)-LSSVM gave excellent efficiency in the prediction of MC, A, VM, and HV, with the R^2 of 0.94, 0.92, 0.94, and 0.90, respectively. The quality indexes of pellets can be visualized on predicting component maps by calculating prediction value to each pixel into the 2D hyperspectral images.

Nakawajana et al. [11] predicted the HHV, LHV, and ash of milled rice husk using FT-NIR spectroscopy. Illustrations of pure cellulose, lignin, and sample spectrum are displayed in **Figure 7**. Vibration band of cellulose (range 1) and lignin (range 2) was quite different, and some wavelengths of risk husk were quite similar to cellulose and lignin. The fact was that the rice husk contains lignocellulose. Range 1 was the vibration band of cellulose, and range 2 was the vibration band of lignin. Therefore, if we want to know which wavelength is the vibration band of analyte,

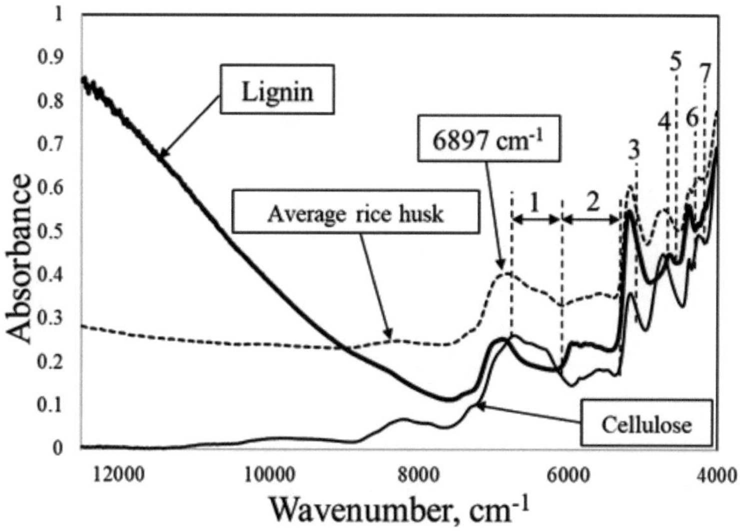

Figure 7.
Average rice husk, hemicellulose, cellulose, and lignin spectrum [11].

scanning of pure analyte with NIR radian should be taken to find obvious peaks in the absorption spectrum. We can assume that an obvious peak is the vibration band of analyte, which is suitable for model development. The calibration equation was validated by an unknown sample, which was obtained from another mill rice plant. The unknown sample was collected from different resources to set calibration. It ensured that the developed model could be used for future samples. HHV (J/g) was determined using bomb calorimeter, and LHV can be calculated using the knowledge of HHV and MC (%wet basis). LHV_{wb} was then determined as follows [7]:

$$LHV = HHV - 2433 \times MC \tag{2}$$

2443 is the latent heat of the vaporization of water at 25°C (J/g), and MC was the unit on a %wb determined from the air dry oven.

Ash content was determined using bomb calorimeter. It was the solid residue after the sample burned and was calculated as follows:

$$A\ (\%wt) = W_a/W_s \times 100 \tag{3}$$

W_a is the weight of the sample after burned, and W_s is the initial weight. The outliers were checked and removed using equation $\frac{(X_i - \overline{X})}{SD} \geq \pm 3$, where X_i is the reference value of sample i. X and SD are the mean and standard deviation of reference value, respectively. If any samples satisfied the criteria, they were cut from the data set. The result provides the relative standard error of prediction (RSEP) of 1.104, 1.159, and 5.975% for HHV, LHV, and A models. The results were recommended for screening as a quality assurance. The RSEP was the relative ratio between the absolute error of prediction and measured value; an RSEP of around 5% meant that every predictive value could provide the error of approximately 5% of reference value [48].

2.3 Elemental components

Posom and Sirisomboon [10] developed the NIR model to predict C, H, N, O, and S across FT-NIR spectroscopy. The scanning of samples was conducted with

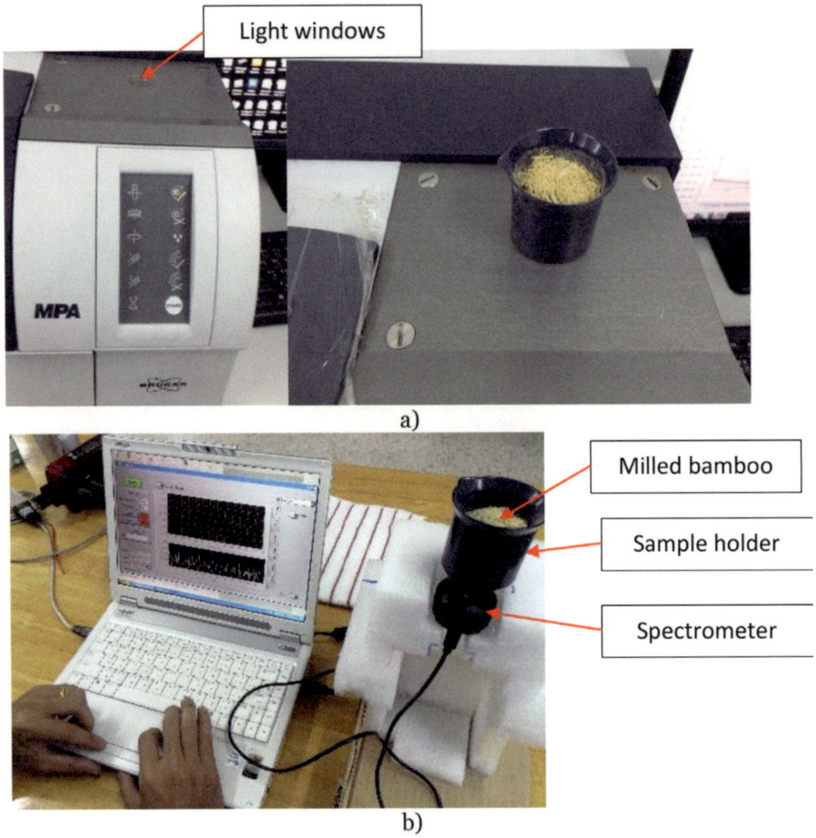

Figure 8.
Scanning of milled bamboo using (a) FT-NIR and (b) MicroNIR spectrometer.

diffuse reflectance mode, shown in **Figure 8a**. The ground sample was placed in a sample cup (43 mm diameter and 50 mm height). The bottom of the sample cup was made from quartz. The sample was scanned in diffuse reflectance mode 64 times and averaged to one spectrum. The sample at the bottom of the sample cup was obtained to be analyzed by traditional method. The model was established using PLS regression. The NIR models of C, H, N, S, and O provided r^2 and RPD of 0.803 and 2.31, 0.856 and 2.65, 0.973 and 6.6, 0.785 and 2.19, and 0.522 and 1.46, respectively. N and H model could be utilized for most applications. C and S models were fair. The O model was poor because the reference method was not precise. The reference method for O was calculated as O% = 100-C%-H%-N%-S%-Ash%, meaning that the predicted O included total error from another parameter. This result could be used to guide operation control into the thermal conversion process to predict flow rate of flue gas and air requirement in combustion and to predict flue gas component in combustion and pyrolysis. Posom and Saechua [49] predicted the elemental components (C, H, N, O, and S) of ground bamboo using low-cost spectrometer (MicroNIR spectrometer), which was a handheld spectrometer. **Figure 8b** displays the scanning method using a portable NIR spectrometer. It shows that only C (carbon) model was recommend for screening because the spectrum collected from the low-cost spectrometer was scanned once, while FT-NIR spectrometer scanned each sample for 64 times which was averaged.

2.4 Kinetic parameters

The activation energy (Ea) of ground bamboo was studied using online NIR spectroscopic method [48]. The online systems are shown in **Figure 9**, including NIR spectrometer (MicroNIR), computer, and conveyer belt. The main work was to evaluate the Ea of ground bamboo through real-time monitoring. The Ea value was determined using the Coats-Redfern method. The online prediction of Ea of the reaction order (n) at $n = 1$ and $n \neq 1$ was investigated. The model of online measurement was created using PLS algorithm. The model of Ea at $n = 1$ and Ea at $n \neq 1$ had r^2 of 0.781 and 0.714, respectively, and SEP of 5.249 and 6.858 kJ/mol, respectively. Both models were fair and could be applied to screening and monitoring. The vibration bands of lignocellulosic such as CH_2, hemicellulose, cellulose, and lignin influence prediction.

2.5 Pyrolysis characteristics

The identification of the pyrolysis characteristics of milled bamboo using NIR spectroscopy was conducted [23] using a FT-NIR spectroscopy, including T_{onset}, T_{sh}, T_{peak}, T_{offset}, and DTG_{peak}. The scanning method was performed using diffuse reflectance mode, which is shown in **Figure 8a**. "T_{onset} is the extrapolated onset temperature that is calculated from the partial peak resulting from the decomposition of the hemicellulose component; T_{sh} is the temperature corresponding to the overall maximum of the hemicellulose decomposition rate; DTG_{peak} is the overall maximum of the cellulose decomposition rate; T_{peak} is the temperature corresponding to the overall maximum of the cellulose decomposition rate; T_{offset} is the extrapolated offset temperature of the DTG_{peak} curves determined using thermogravimetric analysis (TGA)." The aim is to be used as an alternative for thermogravimetry method. The calibration equations were created by PLSR and validated using test set method. The results found that the models of T_{onset}, T_{sh}, T_{peak}, T_{offset}, and DTG_{peak} showed R^2 of 0.566, 0.845, 0.917, 0.973, and 0.671; RMSEP of 9.7 C, 4.36 C, 3.77 C, 2.66 C, and 0.428 wt loss %/min; and RPD of 1.52, 2.58, 3.48, 3.55, and 1.75, respectively. The vibration bands of water and hydrocarbon (CH_3, O-H stretch, Ar-OH, CH_2 and HC=CH, O-H and C-H, CH_2) influence the prediction.

2.6 Prediction of thermal properties

The use of NIR spectroscopy in the prediction of the thermal properties (thermal diffusivity, thermal conductivity, specific heat) of *Jatropha* kernels was studied by

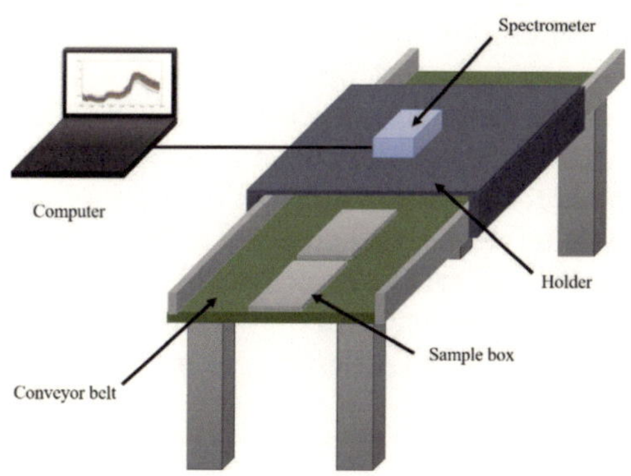

Figure 9.
Schematic diagram of the experimental design [48].

a) b)

Figure 10.
(a) Jatropha *kernels, (b) rotational diffuse reflectance scanning.*

Posom and Sirisomboon [50]. The samples were scanned using FT-NIR spectrometry across a wavenumber of 12,500–4000 cm^{-1}, with diffuse reflectance model. For the diffuse reflectance method, the scanning could confirm that there was not light transmitted through the sample. The thickness of the sample path should be enough. The calibration equations were created by PLS regression. **Figure 10a** and **b** shows *Jatropha* kernels and scanning of technique, respectively. The results of thermal diffusivity, thermal conductivity, and specific heat at 40 and 100°C had the R^2 of 0.5968, 0.7592, 0.7509, 0.4211, and 0.6396%, respectively; the RMSEP of $1.1 \times 10^{-6}\ m^2\ s^{-1}$, 0.0169 $W\ m^{-1}{}^{\circ}C^{-1}$, 0.0685 $W\ m^{-1}{}^{\circ}C^{-1}$, 5.88 $kJ\ kg^{-1}{}^{\circ}C^{-1}$, and 15.8 $kJ\ kg^{-1}{}^{\circ}C^{-1}$; the biases of $-2.52 \times 10^{-7}\ m^2\ s^{-1}$, $2.85 \times 10^{-3}\ W\ m^{-1}{}^{\circ}C^{-1}$, $2.52 \times 10^{-2}\ W\ m^{-1}{}^{\circ}C^{-1}$, 1.83 $kJ\ kg^{-1}{}^{\circ}C^{-1}$, and 4.69 $kJ\ kg^{-1}{}^{\circ}C^{-1}$; and RPD of 1.57, 2.04, 1.98, 1.28, and 1.64, respectively. NIR spectroscopy could be applied in the estimation of the thermal properties of *Jatropha* kernels, due to reduced times and costs. The use of NIR spectroscopy required only 2–3 minutes for measuring and can be applied for drying process control.

3. Caution in model development and application

Even through the NIR spectroscopy is rapid, environmental friendly, and low cost in long-term analysis, the application of NIR spectroscopy should be careful as follows: if the concentration of analyte has lower than 0.1% of the total concentration of the sample, the calibration equation was not acceptable [37] because the absorption peak of the analyte is overlapped by other peaks; the model could be updated monthly to reduce the impact of season; dark and white reference collection of NIR instrument should be done often during operation.

For model development, the caution is before model development, the repeatability and reproducibility of reference method could be firstly examined until the result is satisfied; the maximum R^2_{max} could be calculated before modeling; the calibration equation could be validated again using unknown sample in which that sample must be obtained from different areas and different seasons unless there are three seasons.

4. Conclusion

From previous study, many research demonstrated that this technology is suitable in predicting biomass quality; it is usable in commercial and process control.

NIR spectroscopy is rapid which has the ability in the evaluation of the proximate data. A good prediction was found in predicting HHV ($r^2 > 0.9$), MC ($r^2 > 0.9$), VM ($r^2 > 0.8$), and FC ($r^2 > 0.8$), which is suitable toward for any application [38]. The performance of the model will be good when the range of analyte developed is wide.

For evaluation of elemental component, a good performance was found in the prediction of C ($r^2 > 0.8$), H ($r^2 > 0.8$), and N ($r^2 > 0.8$), which could be utilized for most application [38]. A poor prediction is found in the prediction of O and S, which is not recommended. The O model is low performance due to its reference method which is not precise (calculation method), while the S model is low performance because its range is not narrow which seems that the S content in biomass is not different.

The pyrolysis characteristic model was a good accuracy for the prediction of T_{sh} ($r^2 = 0.845$), T_{peak} ($r^2 = 0.917$), and T_{offset} ($r^2 = 0.973$). This method is rapid (2–3 seconds per sample), which is usable in the process control of pyrolysis.

For the measurement of Ea with online NIR spectroscopic method [48], it could be used as an alternative technique to TGA and Coats-Redfern method. The model is fair and could be applied for screening, which is usable to process control. The vibration bands of CH_2, hemicellulose, cellulose, and lignin impact in prediction.

Conflict of interest

The authors declare no conflict of interest.

Author details

Jetsada Posom[1,2,3]*, Kanvisit Maraphum[1,2,3] and Arthit Phuphaphud[1,2,3]

1 Department of Agricultural Engineering, Khon Kaen University, Khon Kaen, Thailand

2 Applied Engineering for Important Crops of the North East Research Group, Khon Kaen University, Khon Kaen, Thailand

3 Bio-Sensing and Field Robotic Laboratory, Khon Kaen University, Khon Kaen, Thailand

*Address all correspondence to: jetspo@kku.ac.th

References

[1] Taner T. Economic analysis of a wind power plant: A case study for the Cappadocia region. Journal of Mechanical Science and Technology. 2018;**32**(3):1379-1389

[2] Taner T, Naqvi SAH, Ozkaymak M. Techno-economic analysis of a more efficient hydrogen generation system prototype: A case study of PEM electrolyzer with Cr-C coated SS304 bipolar plates. Fuel Cells. 2019;**19**(1):19-26

[3] Taner T. A flow channel with nafion membrane material design of PEM fuel cell. Journal of Thermal Engineering. 2019;**5**(5):456-468

[4] Taner T, Demirci OK. Energy and economic analysis of the wind turbine Plant's draft for the Aksaray City. Applied Ecology and Environmental Sciences. 2014;**2**(3):82-85

[5] Sirisomboon P, Posom J. Thermal properties of *Jatropha curcas* L. kernels. Biosystems Engineering. 2012;**113**:402-409

[6] Nhuchhen DR. Prediction of carbon, hydrogen, and oxygen compositions of raw and torrefied biomass using proximate analysis. Fuel. 2016;**180**:348-356

[7] Werther J, Saenger M, Hartge EU, Ogada T, Siagi Z. Combustion of agricultural residues. Progress in Energy and Combustion Science. 2000;**26**:1-27

[8] Sheng C, Azevedo JLT. Estimating the higher heating value of biomass fuels from basic analysis data. Biomass and Bioenergy. 2005;**28**:499-507

[9] Shi H, Mahinpey N, Aqsha A, Silbermann R.Characterization, thermochemical conversion studies, and heating value modeling of municipal solid waste. Waste Management. 2016;**48**:34-47

[10] Posom J, Sirisomboon P. Evaluation of lower heating value and elemental composition of bamboo using near infrared spectroscopy. Energy. 2017;**121**:147-158

[11] Nakawajana N, Posom J, Paeoui J. Prediction of higher heating value, lower heating value and ash content of rice husk using FT-NIR spectroscopy. Engineering Journal. 2018;**22**(5):45-56

[12] Komilis D, Kissas K, Symeonidis A. Effect of organic matter and moisture on the calorific value of solid wastes: An update of the Tanner diagram. Waste Management. 2014;**34**:249-255

[13] Cooper CD, Kim B, MacDonald J. Estimating the lower heating values of hazardous and solid wastes. Journal of the Air and Waste Management Association. 1999;**49**

[14] ASTM E1756-08 Standard Test Method for Determination of Total Solids in Biomass. USA: ASTM International; 2008

[15] Jameel H, Keshwani DR, Carter SF, Treasure TH. Thermochemical conversion of biomass to power and fuel. In: Cheng J, editor. Biomass to Renewable Energy Process. USA: CRC Press; 2010

[16] ASTM E872-82. Standard Test Method for Volatile Matter in the Analysis of Particulate Wood Fuels. USA: ASTM International; 2006

[17] Basu P. Biomass gasification and pyrolysis: Practical design and theory. In: Biomass Characteristics. USA: Academic Press is an Imprint of Elsevier; 2010

[18] ASTM E1755-1. Test Method for Ash in Biomass. USA: ASTM International; 2007

[19] Posom J, phuphanutada J, Lapcharoensuk R. Gross calorific and ash content assessment of recycled sawdust from mushroom cultivation using near infrared spectroscopy. MATEC Web of Conferences. 2018;**192**:03021

[20] Zhang K, Zhou L, Brady M, Xu F, Yu J, Wang D. Fast analysis of high heating value and elemental compositions of sorghum biomass using near-infrared spectroscopy. Energy. 2017;**118**:1353-1360

[21] Runge TM, Zhang C, Mueller J, Wipperfurth P. Economic and environmental impact of biomass types for bioenergy power plants. In: Environmental and Economic Research and Development Program. 2013. Available from: https://www.focusonenergy.com/sites/default/files/research/1010RungeFinalReportx.pdf

[22] El-Sayed SA, Mostafa ME. Pyrolysis characteristic and kinetic parameters determination of biomass fuel powders by differential thermal gravimetric analysis (TGA/DTG). Energy Conversion and Management. 2014;**85**:165-172

[23] Posom J, Saechua W, Sirisomboon P. Evaluation of pyrolysis characteristics of milled bamboo using near-infrared spectroscopy. Renewable Energy. 2017;**103**:653-665

[24] Feng X, Yu C, Shu Z, Liu X, Yan W, Zheng O, et al. Rapid and non-destructive measurement of biofuel pellet quality indices based on two-dimensional near infrared spectroscopic imaging. Fuel. 2018;**228**:197-205

[25] Wannapeera J, Worasuwannarak N, Pipatmanomai S. Product yields and characteristics of rice husk, rice straw and corncob during fast pyrolysis in a drop-tube/fixed-bed reactor. Songklanakarin Journal of Science and Technology. 2008;**30**(3):393-404

[26] Lv D, Xu M, Liu X, Zhan Z, Li Z, Yao H. Effect of cellulose, lignin, alkali and alkaline earth metallic species on biomass pyrolysis and gasification. Fuel Processing Technology. 2010;**91**:903-909

[27] Stefanidis SD, Kalogiannis KG, Iliopoulou EF, Michailof CM, Pilavachi PA, Lappas AA. A study of lignocellulosic biomass pyrolysis via the pyrolysis of cellulose, hemicellulose and lignin. Journal of Analytical and Applied Pyrolysis. 2014;**105**:143-150

[28] Yang H, Yan R, Chen H, Lee DH, Zheng C. Characteristics of hemicellulose, cellulose and lignin pyrolysis. Fuel. 2007;**86**:1781-1788

[29] Parthasarathy P, Narayanan KS, Arockiam L. Study on kinetic parameters of different biomass samples using thermo-gravimetric analysis. Biomass and Bioenergy. 2013;**58**:58-66

[30] Syed S, Qudaih R, Talab I, Janajreh I. Kinetics of pyrolysis and combustion of oil shale sample from thermogravimetric data. In: The 7th Jordanian International Engineering Conference JIMEC'7; 27-29 September 2010; Amman-Jordan; 2010

[31] Ceylan S, Topçu Y. Pyrolysis kinetics of hazelnut husk using thermogravimetric analysis. Bioresource Technology. 2014;**156**:182-188

[32] Lopez-Velazquez MA, Santes V, Balmaseda J, Torres-Garcia E. Pyrolysis of orange waste: A thermo-kinetic study. Journal of Analytical and Applied Pyrolysis. 2013;**99**:170-177

[33] Næs T, Isaksson T, Fearn T, Davies T. Multivariate Calibration and

Classification. Chichester: NIR Publication; 2004. 323 p

[34] Davies AMC. An Introduction to Near Infrared (NIR) Spectroscopy. 2017. Available from: https://www.impublications.com/content/introduction-near-infrared-NIR-spectroscopy

[35] Osborne BG, Fearn T. Near Infrared Spectroscopy in Food Analysis. Theory of Near Infrared Spectroscopy. New York, USA: Longman Scientific & Technical; 1986. pp. 36-40

[36] Osborne BG, Fearn T. Near Infrared Spectroscopy in Food Analysis. Theory of Near Infrared Spectroscopy, USA, 133. New York: Longman Scientific & Technical; 1986. pp. 36-40

[37] Burns DA, Ciurczak EW, editors. Handbook of Near-Infrared Analysis. New York, NY: Marcel Dekker Inc; 1992. p. 13

[38] Williams P. Near-Infrared Technology-Getting the Best Out of Light. Nanaimo, British Columbia, and Winnipeg, Manitoba, Canada: PDK Grain; 2007

[39] Nicolaï BM, Beullens K, Bobelyn E, Peirs A, Saeys W, Theron KI, et al. Nondestructive measurement of fruit and vegetable quality by means of NIR spectroscopy: A review. Postharvest Biology and Technology. 2007;**46**:99-118

[40] Zornoza R, Guerrero C, Mataix-Solera J, Scow KM, Arcenegui V, Mataix-Beneyto J. Near infrared spectroscopy for determination of various physical, chemical and biochemical properties in Mediterranean soils. Soil Biology and Biochemistry. 2008;**40**(7):1923-1930

[41] Nakawajana N, Posom J. Comparison of analytical ability of PLS and SVM algorithm in estimation of moisture content, higher heating value, and lower heating value of cassava rhizome ground using FTNIR spectroscopy. IOP Conference Series: Earth and Environmental Science. 2019;**301**:012032

[42] Posom J, Nakawajana N. Gross calorific value estimation for milled maize cob biomass using near infrared spectroscopy. MATEC Web of Conferences. 2018;**192**:03022

[43] Shrestha A, Panmanas Sirisomboon P. Rapid evaluation of moisture content in bamboo chips using diode array near infrared spectroscopy. MATEC Web of Conferences. 2018;**192**:03020

[44] Posom J, Shrestha A, Saechua W, Sirisomboon P. Rapid non-destructive evaluation of moisture content and higher heating value of *Leucaena leucocephala* pellets using near infrared spectroscopy. Energy. 2016;**107**:464-472

[45] Posom J, Sirisomboon P. Evaluation of the higher heating value, volatile matter, fixed carbon and ash content of ground bamboo using near infrared spectroscopy. Journal of Near Infrared Spectroscopy. 2017;**25**(5):301-310

[46] Sirisomboon P, Funke A, Posom J. Improvement of proximate data and calorific value assessment of bamboo through near infrared wood chips acquisition. Renewable Energy. 2020;**147**:1921-1931

[47] Stuart B, Molecules I. Infrared Spectroscopy: Fundamentals and Applications. The Atrium, Southern Gate, Chichester, West Sussex, England: John Wiley & Sons, Ltd; 2004

[48] Sirisomboon P, Posom J. On-line measurement of activation energy of ground bamboo using near infrared spectroscopy. Renewable Energy. 2019;**133**:480-488

[49] Posom J, Saechua W. Prediction of elemental components of ground

bamboo using micro-NIR spectrometer. In: Proceedings of the IOP Conference on Earth and Environmental Science. 2019

[50] Posom J, Sirisomboon P. Evaluation of the thermal properties of *Jatropha curcas* L. kernels using near-infrared spectroscopy. Biosystems Engineering. 2014;**125**:45-53

Chapter 13

Digestate: The Coproduct of Biofuel Production in a Circular Economy, and New Results for Cassava Peeling Residue Digestate

Sammy N. Aso

Abstract

Circular economic paradigm applies residue from one process as input material for another, fostering sustainable benefits for humanity. Anaerobic digestion (AD) is an attractive technology for biogas production in a circular economy. Digestate is the residual organic matter generated as coproduct of biogas. Because digestate is nutrient rich and largely stabilized, it has varied management options. Digestate is suitable for direct use as bio-fertilizer and is a good amendment material to improve soil physical properties. However, the quality, safety, and utility of digestate are dependent upon the characteristics of feedstock, digester process, pre- and post-digestion treatments. Digestates emanating from AD of animal manure, energy crops, food processing residues, and other feedstocks have been reported in published literature. On the other hand, there is dearth of reports on digestate emanating from AD process that utilized cassava peeling residue (CPR) as sole feedstock. This chapter presents relevant information on digestates including production, feedstock, quality and safety requirements, processing and treatment technologies, regulatory aspects, applications management options, cost implications, as well as challenges and opportunities. In addition, new results of nitrogen (N), phosphorus (P), and potassium (K) compositions of liquid fraction of CPR digestate are reported.

Keywords: anaerobic digestion, biofuel, biogas, cassava, cassava peeling residue, CPR, circular economy, digestate, management options, renewability, sustainability

1. Introduction

Linear economic model has been constructed on the premise of production, use, and disposal of used resources as wastes. However, there are serious limitations associated with the linear paradigm. These include nonrenewability, unsustainability, and environmental perturbations characterized by negative impacts on air, eco-diversity, soil, and water quality and safety. On the other hand, circular economic model maximizes the 3 (three)Rs of reduce, reuse, and recycle resources. In particular, circular economy applies residue from one process as input material for another process. This approach delivers sustainable benefits for humanity in terms of air, ecology, energy, environment, food, forest, housing, sanitation, soil and water quality, safety and security; as well as improvements in animal and human health, economic, social, and industrial developments.

On the predicate of biorefinery platform, biotechnological upgrading of biomass via biological, chemical, physical or some combinations of these would create bio-based energy, chemicals, and other beneficial metabolites and products within the domain of circular economic model. In this context, anaerobic digestion (AD) is an attractive technology as it would utilize organic resources in waste streams to generate biogas and digestate. However, the quality of digestate is dependent upon variables such as characteristics of feedstock, digester process, and treatment options. Digestates emanating from AD of animal manure, energy crops, agricultural residues, organic fraction of municipal solid wastes (OFMSW), and other feedstocks have been reported in published literature [1–3]. On the other hand, there is dearth of reports on nutrient properties of digestate generated from AD processes that utilized cassava peeling residue (CPR) as sole feedstock. This chapter presents relevant information on digestates in general, and new results of a technical experiment conducted to secure overview assessment of nitrogen (N), phosphorus (P) and potassium (K) compositions of liquid fraction of CPR digestate.

2. Anaerobic digestion (AD)

AD is a biochemical process that decomposes organic matter to generate flammable biogas and residual digestate. The process is achieved with the assistance of a suite of microorganisms in a near oxygen free environment. Biogas is basically composed of methane and carbon dioxide in the respective range of 40–75% and 25–40%. Other constituents are hydrogen, nitrogen, oxygen, hydrogen sulfide and other trace components ranging from 0.1 to 3% [4]. Successful AD operations are carried out within digester or reactor systems designed to supply nutrients required for metabolic activities of the microbes, as well as prevent conditions or elements that may become stressors or present inhibitory effects. AD digester operations and systems may be classified according to the following [5–7]:

- Optimal temperature regimen: psychrophilic (<20°C), mesophilic (30–38°C), and thermophilic (48–57°C);

- Total solid (TS) content: wet digestion (TS < 12%), semi-dry digestion (TS 12–20%), and dry digestion (TS > 20%);

- Feeding mode: batch, fed-batch, semi-continuous, and continuous;

- Process stage or step: single-stage (where all AD processes—hydrolysis, fermentation, acetogenesis, and methanogenesis are executed in one reactor), and multi-stage (where the processes are separated into two or more reactors);

- Fluid-dynamic mode: plug flow, completely stirred or mixed, and hybrid; as well as

- Digester design: anaerobic baffled reactor (ABR), anaerobic filter (AF), anaerobic dynamic membrane reactor (AnDMBR), anaerobic mixed biofilm reactor (AMBR), completely or continuous stirred-tank reactor (CSTR), covered lagoon, expanded granular sludge bed (EGSB), fixed dome, flexible balloon or tube, floating cover or drum, sequential batch anaerobic composting

(SEBAC), stirred anaerobic sequencing batch reactor (SASBR), up-flow anaerobic sludge bed (UASB) or up-flow multistage anaerobic reactor (UMAR).

Today there are millions of anaerobic digesters (domestic, medium, and large-scale versions) operating in the world and generating tremendous amount of biogas. In 2016 for instance, about 60.8 billion m^3 of biogas (1.31 EJ) was generated worldwide; most of it, 84%; in Europe (54%) and Asia (30%) [8]. The technical status of AD plants varies widely. Advanced state-of-the-art systems are prevalent in Europe and more low-tech installations in Africa, Asia and South America. However, irrespective of the level of sophistication, the two fundamental products of AD are biogas and digestate.

3. Digestate

Digestate is the residual organic matter generated as coproduct of biogas production. Digestate is suitable for direct use as bio-fertilizer, as raw material for production of bio-fertilizers, and as amendment material to improve soil physical

S/N	Feedstock	S/N	Feedstock
1	Agro-industrial residues	61	*Miscanthus sacchariflorus* (Maxim.) Hack silage
2	Animal manure	62	*Miscanthus sinesis giganteus* Silage
3	Barley straw	63	Molasses
4	Biodegradable plastics	64	Mozzarella Cheese Whey
5	Biodiesel wastewaters	65	Municipal solid waste
6	Biowastes	66	Municipal waste water
7	Blood industry residues	67	Oat silage
8	Buffalo farming wastewater	68	Olive oil mill wastewater
9	Buffalo manure	69	Olive Pomace, olive waste
10	Cacao	70	Orange peel waste
11	Cardboard	71	Organic fraction of municipal solid waste
12	Cattle/cow: manure/slurry	72	Paper
13	Cattle (beef) urine	73	Paper sludge
14	Cereal bran	74	Peach-juice pulp
15	Cereal-WPS	75	Peeled Cassava wash water
16	Cereals	76	Pharmaceutical industry sludge
17	Cheese Whey	77	*Phleum pratense* L. silage
18	Chicken manure	78	Pig urine
19	*Chroococcus* sp. (algal biomass)	79	Piggery wastewater
20	Coconut chips	80	Pig/swine effluent; manure; slurry
21	Coffee grounds	81	Plum stones
22	Corn	82	Potato chips production residues
23	Corn cob mix	83	Potato waste
24	Cornmeal	84	Potatoes

S/N	Feedstock	S/N	Feedstock
25	Corn residue	85	Poultry litter/manure/waste
26	Cover crops	86	Primary sludge
27	Crushed cassava juice	87	Pumpkin waste
28	Dairy manure	88	Rabbit manure
29	Distiller's waste	89	Rape residue
30	Dried blood of slaughterhouse waste	90	Restaurant food waste
31	Duck slaughterhouse sludge	91	Rice residues
32	Edible oil	92	Rye
33	Energetic crops	93	Sewage sludge
34	Energy maize	94	*Sida Hermaphrodita* Rusby silage
35	Fennel waste	95	Slaughterhouse waste
36	Fish by-product	96	Sludge from Slaughterhouse wastewater treatment plant
37	Food industry residues	97	Solid farmyard manure
38	Food waste	98	Sorghum silage
39	Fruits and distillery by-products	99	Source-separated organic household waste
40	Fruit Marc	100	Source-separated municipal solid waste
41	Garden wastes	101	Starch processing wastewater
42	Glycerin	102	Straws (cereal, pea)
43	Grape seeds	103	Sugar beet pulp
44	Grass (clover, Sudan); grass silage	104	Sugar sorghum (*S. saccharatum* L. Moench.) silage
45	Green waste	105	Sunflower residue, sunflower silage
46	Hemp	106	Tea leaves
47	Household kitchen waste	107	*Tetraselmis* sp. (algal biomass)
48	Household waste	108	Thin stillage (bioethanol by-product)
49	Human excreta	109	Triticale
50	Human urine	110	Triticale silage
51	Industrial and commercial wastes	111	Turkey manure
52	Jute Caddis	112	Vegetable waste
53	Kitchen waste	113	Vinasse
54	Landscape waste	114	Waste-activated sludge
55	Ley silage	115	Waste potato starch
56	Livestock waste	116	Wastewater
57	Maize stover	117	Wastewater sludge
58	*Medicago sativa* L. silage	118	Wheat
59	Milk (serum, whey)	119	Yeast production wastewater
60	Millet	120	*Zea mays* L. (corn, maize) silage

Source: Assembled from scientific literatures in the public domain, most of them cited in this present work.

Table 1.
Feedstocks used in digestate production and studies.

properties such as bulk density, hydraulic conductivity, and moisture retention capacity. Digestate is also attributed with improved sustainability and veterinary safety; reductions in odors, weed seeds, plant pathogens, food chain contamination risks and greenhouse gas emissions. The three basic types of digestate are: whole digestate, liquor (liquid fraction) digestate, and fiber (solid fraction) digestate. Whole digestate is the digestate as obtained leaving the digester at the end of AD process. It contains less than 15% dry matter. This whole digestate could be separated into liquid and solid fractions using appropriate technology and method. The liquid fraction constitutes up to 90% of the digestate by volume, contains 2–6% dry matter, particles <1.2 mm in size, and most of the soluble nitrogen and potassium, while the solid fraction retains most of the digestate phosphorus, and contains dry matter content > 15% [9, 10].

However, the quality, safety, and utility of digestate are dependent upon variables such as feedstock characteristics (pH, chemical composition, carbon-nitrogen ratio (C/N), particle size), digester process (temperature, inoculum, microbial community, hydraulic retention time (HRT)), as well as pre- and post-digestion treatments. Feedstock should possess balanced nutrients, including optimal C/N to satisfy physiological needs of the microorganisms. High or low C/N would disrupt biogasification and lead to reduced biogas output due to low buffer capacity (high C/N) or ammonia inhibition (low C/N). Generally, for biogas production, C/N of 20–30 is considered optimal. For food wastes, C/N of around 15 could be appropriate. Digestates within C/N range of 15–20 are regarded as safe for application to agricultural land without further treatment [11]. When sole feedstock lacks sufficient nutrients for adequate C/N, feedstocks with complimentary nutrients profile are co-digested to offset the limitations. **Table 1** highlights some feedstocks that have been used in AD operations and digestate studies.

4. Regulations, quality, and safety requirements

Perhaps the most important variable affecting the quality and safety of digestate is feedstock. Starting with a high-quality feedstock would virtually guarantee a safe and quality digestate. Source separation can be used to achieve high purity feedstock. The biological, chemical, and physical properties of digestate may be governed by regulations and quality assurance systems. The European Union (EU) and many European national governments have hygienic, quality and safety standards for digestate certification that consider feedstock source and other aspects such as digester process, treatment options, handling and storage requirements. The essential quality and safety requirements for digestate destined as biofertilizer must be achieved regardless of the initial raw material. Essential quality and safety parameters include nutrients content, dry matter and organic dry matter contents, homogeneity, pH, purity (free of inorganic impurities such as glass, metal, plastic, and stones), sanitized and safe for soil organisms and the environment with regards to biological status (pathogenic organisms) and chemical status (organic and inorganic contaminants/pollutants). Furthermore, the digestate should be free of odor, phytotoxicity and weed seeds; and be satisfactorily stabilized.

Quality assurance systems for digestate certification may comprise monitoring to ensure control; standardization to ensure repeatable performance; characterization label to identify product fitness; declaration to describe product constituents; application guidelines to ensure safe and proper use; and documentation to prove that the product received required treatments following approved protocols. **Table 2** presents established criteria and characteristics for the production and use of quality and safe digestates. In the EU, conformity with these criteria is enough to

Criteria	Process/parameter	Requirements
Hygiene	Pasteurization at 70°C	1 h
	Sterilization at 133°C	20 min
	Weed seeds and sprouting plant parts	≤2/L
	Odor	Free of annoying odors
Pathogens	*E. coli*	≤1000 CFU/g fresh matter
	Salmonella spp.	Absent in 25 g fresh matter
Heavy metals	Cadmium (Cd)	0.8–20 mg/kg DM
	Chromium (Cr)	75–1000 mg/kg DM
	Copper (Cu)	75–1000 mg/kg DM
	Lead (Pb)	80–900 mg/kg DM
	Mercury (Hg)	0.6–16 mg/kg DM
	Nickel (Ni)	30–300 mg/kg DM
	Zinc (Zn)	300–4000 mg/kg DM
Organic pollutants	Polycyclic aromatic hydrocarbons	3–6 mg/kg DM
	Dioxins and furans	20 ng TE/kg
	Chlorinated pesticides	0.5 mg/kg Product
	Polychlorinated biphenyls	0.2 mg/kg DM
	Absorbable organic halogens	500 mg/kg DM
	Linear alkylbenzene sulphonates	1300 mg/kg DM
	Nonylphenol and nonylphenolethoxylates	10 mg/kg DM
	DEPH: Di (2-ethylhexyl) phthalate	50 mg/kg DM
Inorganic pollutants	Non-stone impurities >2 mm (glass, metal, plastic, etc.)	0.5% m/m dry matter
	Stones > 5 mm	8% m/m dry matter
Stability	Volatile fatty acids	0.43 g COD/g VS
	Residual biogas potential	0.25 l/g VS
	Respiration rate	16 mg CO_2 g VS^{-1} day^{-1}
Declarations	Name of producer, type of product (whole, liquid, solid), mass of product, total nitrogen, ammonium nitrogen, total phosphorus, total potassium, soluble chloride, soluble sodium, dry matter, volatile solids, pH, bulk density, etc.	Relevant units where applicable (e.g., kg; kg/m^3; mg/(kg DM); mg/L; %;)
Additives and chemicals	Lime, iron chloride, iron oxide, bentonite, diatomaceous earth	
Feedstock sources	Agriculture (e.g., manure, harvesting by-products, silage, energy crops); animal by-products (e.g., manure, stomach intestine, raw milk); food industry (residues from food industry that contain food grade additives); food related shops (e.g., potatoes, dairy waste, bread, meat remnants, flowers, plants); forrest (e.g., bark, wood chips, sludge from the cellulosic industry); parks, gardens (e.g., leaves, grass); greenhouses (e.g., tops, peat products); households, kitchens, restaurants (e.g., fruit and vegetables residues, food, coffee and tea remainders, egg shells); etc.	

Source: [9, 12–16].

Table 2.
Quality and safety validation criteria for digestates.

ensure that digestate complies with European "End of Waste" criteria; and can be used without further waste management controls.

5. Treatment technology options

In the context of AD and digestate, we may distinguish between pre- and post-treatment processes. A pretreatment process refers to a processing operation applied upstream, before the digestate emerges from the digester. This could range from size reduction or thermochemical treatment of feedstock substrate; to process management (such as pH, temperature, and retention time control). On the other hand, a posttreatment process is that processing operation applied downstream of digestate harvest. This may also involve size reduction, other unit operations; composting, and end-product requirements that ensure the digestate sanitation. Post treatment may generate nutrient concentrates, liquid and solid fraction digestates conditioned to standardized biofertilizers, and final liquid effluent that could be discharged into a stream or sewage system. Benefits of posttreatment include enhanced marketability, reductions in handling, storage and transportation costs/requirements, and compliance with environmental regulations.

Depending on the starting feedstock and desired end product form of the digestate, similar technologies could be used for pre and post treatment processing. Applied technologies and methods may be classified as biological, chemical, or physical. The methods could also be used in combination. Biological treatment could be accomplished with the use of microorganisms and catalysts; chemical treatment with acids, alkalis and oxidants; and physical treatment by mechanical and thermal means. Physicochemical treatment combines physical and chemical techniques. Ammonia fiber explosion (AFEX), and supercritical CO_2 explosion are examples. The major classifications of treatment options and associated technologies are presented in **Table 3**.

Category/ method	Technology option	Example means/aids
Biological	Bacteria	*Clostridium* sp. strains LDC-8-c12, 5-8, CO6-72; *Rhodobacter sphaeroides* KD131; Thermosaccharolyticum strain M18
	Composting	Green waste, vine shoot pruning, wood chips
	Enzyme	*Carbohydrase*, *laccase*, lignin *peroxidase*
	Fungi	*Ceriporia lacerata*, *Ceriporiopsis subvermispora* (ATCC 96608), *Pleurotus ostreatus*
Chemical	Acids, organosolvs	Inorganic acids (hydrochloric, nitric, phosphoric, sulfuric); organic acids (fumaric, maleic). May be used in percolation, plug flow, shrinking-bed, batch, and countercurrent modes
	Alkalis	Ammonia, lime
	Ammonia recovery	Ion exchange; scrubbing, stripping, precipitation (struvite)
	Ionic liquids	1-Butyl-3-methylimidazolium hydrogen sulfate [bmim]HSO_4], 1-ethyl-3-methylimidazolium acetate (EMIM-OAc), 1-ethyl-3-methylimidazolium diethyl phosphate, 3-allyl-1-methyl-1H-imidazol3-ium chloride [Amim][Cl]
	Oxidants	Hydrogen peroxide, ozone

Category/ method	Technology option	Example means/aids
Physical	Mechanical	
	Dewatering:	Centrifuges, gravity tables, presses (belt, filter, rotary, screw)
	Disintegration/maceration (chipping, grinding, milling, shredding):	Ball mill, colloid mill, hammer mill, two-roll mill
	Extrusion:	Band, single screw, twin screw
	Homogenization:	High pressure homogenizers
	Lysis:	Lysis-centrifuges
	Membrane separation:	Electrodialysis, microfiltration, nanofiltration, pervaporation, reverse osmosis, ultrafiltration
	Sonication:	Ultrasound/sonoreactors (bath, flat plate, probe, tube)
	Irradiation	Electron beam, gamma ray
	Thermal	Drying/torrefaction, electric heating, evaporation, hot oil, hot water, hydrothermal, microwave, steam
Physicochemical	Expansion/explosion	Ammonia fiber expansion/explosion (AFEX), steam explosion, supercritical carbon dioxide (SC-CO_2) explosion

Source: [10, 15, 17–36].

Table 3.
Major categories of treatment and technology options for AD and digestate processing.

6. Applications management options for digestate

In the service of circular economy, there are many applications management options for digestate. These may include algae cultivation, energy production, bio-adsorbent production, building materials production, nutrients recovery/production, soil creation and other value-added commodities. Perhaps the two most widely recognized utilities of digestate are as land application for soil amendment and as biofertilizer.

6.1 Biofertilizer and soil amendment

Technological aids used in modern agriculture such as inorganic fertilizers and antibiotics have negative impacts on soil, water, and air quality and safety, and therefore pose health risks to humans and the ecosystem. Inorganic fertilizers for instance have caused environmental and soil quality degradation, eutrophication and heavy metals pollution. Similarly, field-spreading agricultural land with raw/untreated manures derived from medicated livestock contributes to dissemination of veterinary antibiotic residues and antibiotic-resistant pathogens. Lincomycin, monensin, and sulfamethazine antibiotics were reported to affect soil microbial community composition and respiration, denitrification and nitrogen transformations [37]. Applications of digestate for biofertilizer and soil amendment purposes could ameliorate some of these adverse effects.

Amendment propensity relates to capability to maintain soil fertility and humus balance. Dairy slurry digestate was found richer in humic substances than raw dairy slurry [38]. Researchers concluded that digestate enhanced soil biological stability, microbial biomass and enzymatic activities [39].

On the other hand, fertilizer properties relate to provision of nutrients necessary for good crop performance. Leaves of alfalfa plant fertilized with digestate had higher contents of N, P, and K in comparison to alfalfa fertilized with mineral fertilizers [40]. Digestate also produced higher yields of dent corn than the application of chemical fertilizers [38]; higher yield of potato (*Solanum tuberosum*) over the application of compost [41]; and 30% increase in yield over farm yard manure [42].

6.2 Nutrients recovery

Digestate is applied in recovery of nutrients, production of fertilizers and volatile fatty acids (VFAs). Livestock manure contains about 49 g N/kg TS and 6 g P/kg TS; energy crops, 17 g N/kg TS and 2.5 g P/kg TS; and agro-wastes, 27 g N/kg TS and 3 g P/kg TS [43]. Much of these nutrients remain in digestate after AD operation. For example, total N, P, and K values for digestates obtained from wet AD of agricultural wastes were reported respectively in the ranges 44–120, 8–42, and 28–95 g/kg DM [44]. These nutrients could be recovered/harvested with the technologies outlined in **Table 3**.

VFAs are important input organic acids used extensively in the bioenergy, food, chemical, cosmetic, pharmaceutical, textile, and other industries. Acetic acid (E 260), propionic acid (E 280) and butyric acid are examples; and are GRAS (generally regarded as safe) rated by the FDA. Acetic acid is used to defend against *Campylobacter*, *Escherichia coli*, *Listeria*, *Salmonella*, and other pathogens in beef, chicken, pork, turkey, carcasses, skin and hides [45]. Butyric acid is used in the textile industry to enhance heat and sunlight resistance of fibers. In the food industry, it is used as additive for flavor formulation and modification [46]. Similarly, propionic acid (E 280) is used as antibacterial and antifungal agent to decontaminate packaging films and coatings, and to protect meat and meat products such as sausages, bologna and ham. VFAs have been harvested from digestates generated from short-term dry AD of swine manure, generated from AD of food waste, and used in recovery of biological nitrogen and phosphorus from sewage sludge [47–49].

6.3 Energy production

Digestate can be deployed for energy generation. Recirculating digestate into the digester maximizes biogas production, at the same time minimizing methane emissions during digestate storage, transport, and use. Digestate was pyrolyzed (via the use of Pyroformer, quartz rotary kiln reactor, and thermo-catalytic reforming reactor) to produce biofuels: pyrolysis oil (biooil) and pyrolysis gas (syngas). The biooil generated by thermo-catalytic reforming process at 750°C had a higher heating value of 33.9 MJ/kg, and a total acid number of 4.9 mgKOH/g [50].

Algae have widespread applications and potentials in: biofuels, cosmetics, biofertilizer, infant formulas, nutritional supplements, livestock feeds, chemical and allied industries, and biodegradable packaging. Perhaps more importantly, digestate could be used for the cultivation and production of microalgae. In the context of biorefinery platform and circular economy, various compounds produced by microalgae and their applications have been reported [51, 52].

6.4 Other applications

Digestates have other utilities and management options. These include applications in aquaculture, gardening and horticulture, and the production of building materials and biochar.

6.4.1 Biochar

Biochar (charcoal) is the byproduct of thermal pyrolysis of carbonaceous biomass; and has carbon sink properties. Dairy waste and whole sugar beet digestate biochar were effective in eliminating heavy metals (Pb^{2+}, Cu^{2+}, Ni^{2+}, and Cd^{2+}) from aqueous solutions [53].

6.4.2 Gardening and horticulture

Due to its organic origin and physicochemical characteristics, digestate is useful in gardening and horticulture. It could be applied in soil creation or remediation, and has found applications in green houses, plant nurseries, and home gardening [54].

6.4.3 Building materials

A 50% substitution of wood with cattle manure digestate produced particleboard panels that met ANSI performance requirements [55]. USDA reported that medium-density fiberboard and wood/plastic composite engineered materials could be created using digestate solids without compromising mechanical or aesthetic values [56].

6.4.4 Aquaculture

Digestate is better than raw manure in fertilizing fish ponds. Firstly, digestate is hygienic because most of the bacteria, parasites and their eggs are destroyed in the AD process. Thus, pond sanitation is improved; minimizing fish diseases and the cost of veterinary services. Secondly, the digestate is largely stabilized and therefore does not consume and compete with fish for dissolved oxygen. Tilapia, Silver carp, Bighead carp, Silver barb and Mrigal fish species raised in pond fertilized with digestate matured faster and achieved higher net weight gain than counterparts raised in pond fertilized with chemical fertilizer or raw manure. By comparison, while chemical fertilizer increased net yield over raw manure by 27%, digestate increased net yield by 55% [57].

6.4.5 Bio-adsorbents and bedding

Digestates have been applied as bio-adsorbents to scavenge heavy metals from contaminated soils and water [58], and as chicken litter [54], and other livestock bedding [56, 59].

7. Cost implications

The big picture cost elements relevant to AD systems include land acquisition, site preparation/development, plant and machinery (including digester/reactor, pre and post treatment technologies), personnel, feedstock, environmental impact, other operating costs (electricity, logistics, regulations), and revenue from products (biogas and digestate). In the case of digestate, feedstock, treatment processes, and the logistics of storage, transport, handling and field application bear crucial concerns. Cost-effective digestate production process is presaged by efficient feedstock collection and sorting operations. A cost benefit analysis of municipal solid waste management system in Yangon, Myanmar, identified weak organizational structure and ineffective collection methods in the existing system that operated with just 32% waste collection efficiency. An alternative system with increased waste collection efficiency was then proposed. The new system required labor and vehicular productivity; using vehicles with container-hoist handling mechanism. The new system reduced operating and other costs associated with the old system by up to 42% [60]. It is noteworthy that consumer and public environmental behavior and cooperation on waste management could be modified by pecuniary and nonpecuniary information. In Surabaya city, Indonesia, researchers found that in the reference case in which the no information treatment was applied, mean WTP (willingness to pay) for marginal improvements in a waste collection and disposal program was estimated to be US$ 14.65. The researchers reported that pecuniary information increased WTP by 20.5%, whereas non-pecuniary information had a negative but statistically insignificant effect on WTP [61].

A situation where 50% of whole unprocessed digestate was applied on agricultural land near the generating biogas plant and the other 50% transported to a location 20 km away was studied. Cost for digestate utilization near the biogas plant was € 3.34 (US$ 3.73)/t, and that at a location 20 km away was € 5.47 (US$ 6.10)/t [62]. This study highlights the impact that location or site of digestate utilization could have on cost. Such distance related cost also applies to feedstock substrate. Generally, the farther the distance, the higher the cost.

Researchers performed specific cost analysis for six scenarios that involved direct land application of digestate as reference, and various treatment technology options that included screw press and decanter centrifuge separation, belt drying, evaporation concentration, purification by ultrafiltration and reverse osmosis, and nutrients recovery by ammonia stripping and precipitation. Result indicated that net specific costs ranged from € 1.94 (US$ 2.16)/m^3 of digestate for the reference scenario, to € 5.45 (US$ 6.08)/m^3 for stripping, to € 6.80 (US$ 7.58)/m^3 for belt dryer [62]. Similarly, the costs of AD were found to vary up to € 109 (US$ 122)/t of digestate from € 35 (US$ 39)/t for basic storage of digestate for aerobic conditioning, to € 70 (US$ 78)/t for digestate ready for direct land application, to € 79 (US$ 88)/t for on farm co-digestion [63].

Case studies were conducted for separation systems in three regions (Aachen, Borken, and Siegen) of Germany. The researchers determined that investment and variable costs were respectively € 23,000 (US$ 25,536) and € 0.47 (US$ 0.52)/m^3 for screw press; € 27,000 (US$ 29,977) and € 0.48 (US$ 0.53)/m^3 for screening drum press; and € 163,000 (US$ 180,970) and € 1.46 (US$ 1.62)/m^3 for decanter centrifuge. Further analysis revealed the unit cost of digestate disposal for screening drum press varied from € 4.1 (US$ 4.6)/m^3 in Aachen to € 4.8 (US$ 5.3)/m^3 in Borken, and Siegen [64].

The following were reported about AD in the UK. Least cost post treatment technology for digestate derived from a 10% solids content food waste was biological oxidation at £13.18 (US$ 16.97)/t of feedstock. At 20% solids content, least cost option was direct application of whole digestate to agricultural land at £8.76 (US$ 11.28)/t. The cost of treating 4000 t of slurry with a mechanical screen separator was £0.44 (US$ 0.57)/t per year, and treatment with decanting centrifuge cost £2.21 (US$ 2.85)/t per year. Furthermore, about £3.5M (US$ 4.5 M) would be required to construct a 1 (one) MWe AD plant utilizing farm wastes as feedstock [65–67].

In the continent of Africa, cost of establishing a 4 m^3 anaerobic digester was found to range from US$ 555 in Uganda to US$ 698 in Cameroun to US$ 979 in Rwanda [68]; while that of founding a family size floating drum plant was estimated at US$ 1667 [69].

Techno-economic analyses were performed for post treatment technologies used to recover nutrients from the digestates of five full scale farm AD systems. Results showed membrane technology had specific cost of € 6.97 (US$ 7.72)/m^3 of treated digestate. Drying was estimated at € 5.81 (US$ 6.44)/m^3, while stripping operated at € 5.44 (US$ 6.03)/m^3 [70]. In addition, the process economics of membrane-based nutrients extraction and fractionation from dairy manure digestate indicated cost of solid-liquid separation unit to be US$ 11,000; the microfiltration extraction unit cost US$ 30,000; the nanofiltration fractionation unit was priced at US$ 60,000; and the daily cost of operation (chemicals, energy and water) was approximately US$ 24 [71].

Finally, digestates are used as quilt for cattle bedding and poultry litter due to significant cost offsets to livestock farms. The cost of solid digestate as animal bedding (US$ 55 per dry ton) is cheaper than the cost of alternative wood-based replacement materials such as wood chips at US$ 65 per dry ton or sawdust and shavings at US$ 124 to US$ 248 per tonne [55, 59].

8. Challenges and opportunities

Digestates have good fertilizer qualities: nutrients, safety and other properties required for soil amendment and plants production. However, relative to mineral fertilizers, digestates are not well known in many countries. Therefore, their potential as mineral fertilizer alternative/substitute is limited. Perhaps, standardized quality assurance and control protocols, regulations, certifications, legal and other institutional management systems organized internationally could help demonstrate digestates' benefits, quality and safety, and thereby engender confidence in their utilization as sustainable fertilizer and soil amendment products. Reconciling and bringing such issues and their benefits to existence present challenges and opportunities. Presented in **Table 4** are some of these challenges and opportunities of the waste, AD and digestate system.

Issues	Challenges and opportunities
8.1. Concept of waste	Challenge: the conventional or customary status of looking at waste as a problem presents significant challenge. Opportunity: seeing waste as potential resource would help change perception and attitude, possibly stimulating salient management options. Opportunities may emerge in the areas of prevention, recovery, collection, sorting, reducing, reusing, and recycling. For developing countries these have implications for environmental hygiene and sanitation.
8.2. Biowaste	Challenge: because biodegradable waste could be a source of heavy metals and polluting organic compounds, it presents challenges to life generally, and to the environment. Opportunity: these challenges create opportunities to develop management options (e.g., biological treatments) to protect life, environment, and to benefit agriculture and ecosystem. Biowaste is reported to have potential to tackle climate change in the areas of nitrous oxide (NO_2) emissions mitigation, and sequestration capacity of agricultural soils [72].

Issues	Challenges and opportunities
8.3. E-waste	Challenge: problems and dangers of e-waste, heavy industry products and components; including electrical and electronic equipment, waste batteries, engine blocks, paint, etc. Opportunity: guidance/support for the informal (non or loosely regulated) establishments, to call attention to dangers and health risks that may be associated with used or discarded electronic devices/items (acids, other chemicals, radioactive materials, etc.).
8.4. Mineral waste	Challenge: mining of solid minerals do present health and environmental challenges. Opportunity: chances to implement safeguards for hazardous minerals and to divert safe wastes to beneficial applications. Examples are uses as substitute for backfill material in open pit mining, landfill, or as grit in construction materials. Production of concrete and brick for structural work (bridges, dams, launch pads, highways) are possibilities.
8.5. Source of feedstock	Challenge: the source of digestate feedstock and its treatment could present barriers. PAS 110 in the UK does not approve certification for digestate generated from mechanically biologically treated waste. Such digestates require proof of biodegradability test to be considered suitable for recycling; like land spreading. There is also the issue of digestate originating from co-digestion of industrial waste and household waste. In the Netherlands, the desire in AD electricity regime to maximize biogas production by mixing manure with other organic material conflicts with AD biofertilizer rules for spreading digestate from co-digested manure on farm land. Opportunity: some of these challenges are consumer-induced barriers and lack quantitative elements. Opportunities might lie in the sociocultural realm, such as modifying social and cultural attitudes and behaviors towards waste and its inherent heterogeneity.
8.6. Unrecovered organic matter	Challenge: AD is more adapted to easily putrescible carbohydrates (starch, sugar). Recalcitrant lignocellulosic components (lignin, etc.) remain undigested. Efficiency of organic matter conversion was quite low as >97% of lignin in maize stover was found undigested [73]. AD could thus lead to unrecovered organic matter still present in digestate Opportunity: prospects for advanced and innovative pretreatment technologies to fractionate, recover, purify and convert lignin or other recalcitrant organics to more digestible biopolymers. Alkaline treatment, gamma irradiation, membrane technologies, organosolv, steam explosion, wet oxidation, etc. may come to the rescue (**Table 3**).
8.7. Informal and low status	Challenge: AD and digestate are perceived to be in domain of informal waste management system and service; and therefore, relegated as only appropriate for the rural populace. Opportunity: integration of formal and informal systems. Training to abate misconceptions, lack of awareness, and raise public profile of digestate. These may purge image of biogas and digestate as products that are derived from wastes, and hence belong to poor/rural settings.
8.8. Legal barriers	Challenge: lack of binding global (and for developing countries, own country) coherent rules, laws, directives, regulations and policy frameworks. Opportunity: the formulation of these guidelines and laws on waste governance system. Implementing appropriate technologies and business models for waste management.

Issues	Challenges and opportunities
8.9. Data and waste reporting system	Challenge: lack of reliable data on waste management systems, design features, standard operating procedures (SOPs), etc. could limit exchange of ideas and retard progress. Opportunity: waste management value chain information is vital. Quantity, type, economic sector, source, and composition data could guide prioritization of strategies and enable trends forecast that deliver better outcomes. Global exchange of briefs would catalyze spread of best practices.
8.10. Standardization	Challenge: although digestate products have similar characteristics as commercial chemical fertilizers, they are not classified in any way, are poorly developed in most countries, and there is no overall guidance [20, 62, 70]. These barriers restrict utilization and trade. Opportunity: these challenges create opportunities to establish frameworks that enable digestate utilization through standardization, fair comparison, commerce development, and international trade.
8.11. Marketing	Challenge: regional nutrient availability, agricultural structure, season, feedstock and degree of upgrading have been reported to challenge and impact digestate prices and marketing [54]. Opportunity: upgraded products offer increased marketability due to their denser nutrients. Marketing to nutrient deficient regions, non-agricultural sectors and purposes represent prospects. Manufacturers of organic soils, particle- and fiber-boards, landscapers, and private customers all represent credible market outlets.
8.12. Cost barrier	Challenge: initial investment fund is a major issue. Cost of establishing a 4 m^3 AD digester in the continent of Africa ranges from US$ 555 to US$ 979 [68]; and the price for a family size floating drum reactor was reported at US$ 1667 [69]. In Sri Lanka, a family unit digester generating 6–10 m^3 of biogas per day cost Rs. 17,000 (US$ 5459); and described as difficult proposition for low-income families [74]. In the UK, a 1 MWe AD plant utilizing farm wastes as feedstock cost about £3.5M (US$ 4.5 M) to construct [67]. Also, costs associated with animal breeding and maintenance (veterinary care, feed, water, etc.) escalate operating costs, and constrain availability of manure for feedstock. Opportunity: easing cost barriers would require support with appropriate and necessary interventions (policies, credit facilities, subsidy schemes, preventive maintenance that promote solutions, prolong facilities productive lifespan, and minimize operating costs). Furthermore, transparency on proposals and bidding for new plants and projects could build confidence in the process.
8.13. Urban and rural dichotomy	Challenge: differences between metropolitan, urban, sub-urban, and rural areas can compromise AD projects. Segregation by infrastructure and income for example could affect waste collection and limit access to feedstock. Opportunity: prospects for rural development with public utilities, services, and infrastructure (roads, power, water, etc.) These would facilitate logistics for waste collection, AD processes, and digestate handling/evacuation.
8.14. Contamination of agricultural land	Challenge: most of the digestate produced in AD is used for soil amendment and as biofertilizer. There are risks of spreading animal pathogens, heavy metals, and other pollutants on soils due to the presence of these hazards in animal by-products used in AD. Sulfadiazine and oxytetracycline are antibiotics found in

Issues	Challenges and opportunities
	manure of medicated animals that affect soil quality. Twenty five percent of 70 digestate and compost samples assessed in Switzerland contained polycyclic aromatic hydrocarbons (PAHs) concentrations beyond the regulated threshold value of 4000 μg/kg_{dw} [75]. Opportunity: digestate is a sustainable fertilizer and soil improver; thus, necessary to assure its safety. The potential to contaminate soils with pollutants from digestate application beacons vigilance and chances to develop technical and monitoring strategies that sequester and purge the digestates of polluting hazards before their use.
8.15. Air pollution	Challenge: digestate has potential to emit substances and gasses that contaminate the air and influence global warming [11]. Challenges also exist due to lack of practical tools to monitor primary air pollutants [76]. Opportunity: advanced methods of digestate management and reutilization to minimize emissions of air pollutants (ammonia: NH_3, nitrous oxide: NO_2) and greenhouse gases (methane: CH_4, nitrogen dioxide: N_2O). Strategies may include processing (composting, curing, dewatering); alternative applications (in construction, aquaculture, regeneration activities); and storage. Development of software tools that enable quantitative monitoring of emissions from digestate soil applications on a routine basis is another prospect area.
8.16. Bad odors	Challenge: compared to raw manure slurry, digestate has fewer bad odors. However, this may not be true when compared to chemical fertilizer. There have been complaints of nuisance odors associated with land-spreading of digestate [77], and at landfills and composting plants [78]. Opportunity: this problem could be due to spreading practice and/or the spreading of unstable digestates. Application of good timing and spreading techniques (trailing-shoes, injection), and use of stabilized digestates (sufficient HRT, aerobic composting) would minimize odor issues.
8.17. Bad legacies	Challenge: there are challenges associated with bad reputation of AD systems and biogas plants around the world. A study in 2006 found that 60% of 600–700 domestic biogas plants in Ethiopia was not functioning [79]. During the 7 years period from 2009, more than 3600 biogas plants were installed in the Tigray region of Ethiopia; and a 2017 study reported that 58.1% of the installations was not operational [80]. The 21 biogas plants installed by Pakistan council for appropriate technology (PCAT) in the 1970s were reported to have failed to perform [81]. In 1986, a survey of the status of 25 biogas plants in Kenya found 36% to be alive, functional and maintained. Another 36% was described as dead, not functional, and not maintained. Unfinished projects accounted for 8%; while remaining plants were reported in disrepair, with varied patterns of being alive, dead, not functional, and not maintained [82]. The regional bioenergy program of the Latin American energy organization (OLADE), catalogs biogas technology projects in Latin American countries. Experience began in 1953 and by 1986 at least 22 countries including Bolivia, Colombia, Costa Rica, Dominican Republic, Ecuador, Grenada, Guatemala, Guyana, Haiti, Honduras, Nicaragua, Jamaica, and Peru had projects at varying levels of implementation. Out of the 3950 biodigesters inventoried, 60% was found operable and 40% was either shut down or functioning irregularly or completely abandoned [83]. Though China rebounded and emerged as a major reference on

Issues	Challenges and opportunities
	household digesters, about 50% of biogas tanks installed from 1958 into the 1970s were abandoned in the 1980s. By 1988 the seven million rural digesters existing in 1980 dropped to 4.7 million [84]. In 1986, a survey of biogas plants in Sri Lanka indicated that 61% was functional. By 1996 only 28.5% of completely surveyed 365 biogas systems was reported functional. At this point 16 units had been abandoned and the success rate for biogas systems implementation was reported as 32.9% [74]. In the Netherlands, for a period of over 30 years beginning in the 1970s, many AD projects using biomass were considerably delayed, suspended, abandoned and out rightly never realized. [85, 86]. These failures and circumstances taken together portrayed negative images and bad legacies for biogas plants. Opportunity: reasons adduced for failures included economic, social, technical, and policy components such as high investment and maintenance costs, urbanization and socio-cultural constraints, poor dissemination strategy, complicated permit regulations, shortage of feedstocks, lack of or inadequate training, poor digester design, etc. These reasons provide opportunities to create circumstances, provisions and tools that would promote and sustain biogas systems. Some examples are mobilization of local and external funds, more business-friendly policies and rules, appropriate and sustainable technologies, technical training, warranties for plant performance. Also, public dissemination of information and follow-up on successful programs could help.
8.18. Low diffusion rate	Challenge: in Latin America, the number of rural biogas plants installed yearly from mid-1985 to 1992 was less than 15% of that installed from 1982 to mid-1985. Challenges included technology adoption, technical manpower and materials of construction. However, non-technical reasons for biogas adoption failures accounted for up to 69%, 50% and 25%, respectively, in Cote d'Ivoire (Ivory Coast), Costa Rica and Tanzania [84]. Unstable institutional environment, lack of network and lobby activities, lack of initiatives between academia, research institutes, private sector entrepreneurs and stakeholders were cited nontechnical reasons. For the Netherlands, apart from technological problems; limited economic feasibility, fragmented support from the government, decreases in energy prices, and lack of financial support which made return on investment uncertain contributed to inadequate AD diffusion. Opportunity: cooperation between academia, government, industry and other stakeholders (farmers, energy sector, municipalities). Cooperative efforts that landed mutually beneficial outcomes should be highlighted, applauded and replicated. Well planned long-term, clear and supportive arrangements would facilitate continuity. Government policy that guide search for solutions, market formation and resources mobilization. Ease of technology adoption would also require reliable and sustainable infrastructure (technical assistance, manpower, cohesive farming approach with biogas and digestate, integration and dissemination of societal and cultural values and norms).
8.19. Inhibition of microalgae	Challenge: it has been shown that the green alga (*Raphidocelis subcapitata*) is sensitive to digestate, with ecotoxicity index; EC_{50} of 0.77% [87]. Similarly, *Scenedesmus bijuga*; and oil-rich *Chlorella* sp., including *C. minutissima* and *C. sorokiniana* were found sensitive to digestate. Also, the dark color of liquid digestate of algal biomass inhibited the growth of *Chroococcus* sp. Therefore, cultivation of algae for value added products recovery could be

Issues	Challenges and opportunities
	minimized in the presence of digestate. Opportunity: because algae are exploited for biofuels, and various other useful biotechnological metabolites production by valorization of digestate, the inhibitory effect of digestate on algae cultivation is of practical interest. Therefore, digestate pretreatment or at least its dilution before use [88], would aid good algal productivity.
8.20. Nomadic and free-range culture	Challenge: many developing nations have nomadic animal husbandry architecture and free-range culture. These make the gathering of animal manure as feedstock for digesters a major challenge. In Pakistan, for example, livestock farmers from time to time relocate to weather conditions more benign to their livestock. However, current digester designs used by rural populations such as the fixed dome and floating drum are sedentary and cannot be readily moved by the farmers with their livestock. Opportunity: perhaps this challenge creates opportunity for a mobile biogas system such as the portable biogas plant reported in the year 2016 [81].
8.21. Disparity between developed and developing countries	Challenge: the economic, political and technological mismatches and divides between industrialized and industrially developing countries are challenging local, regional and international waste management systems. Environmental and health dangers do not know or respect boundaries (local, regional, or international) by land, sea, air or space. Planet earth is perhaps at the cusp of the axiomatic global village and economy. Sooner than later, pollution and instability at one corner of the earth would reverberate and affect other parts (Plastics in the oceans? Heavy metals in food, aquatic and terrestrial biota? Ebola in America? Flood events in Zimbabwe, Mozambique, Puerto Rico and U.S. Virgin Islands? Wildfires in Australia, Brazil, Portugal and USA? Coronavirus (COVID–19) in Japan, Singapore, and USA?). Opportunity: cooperation and support are needed to enable developing nations to leapfrog and shorten the learning curve and development timescales. Developing nations need guidance and assistance to cope with technological demands and challenges, and eschew reinventing the wheel. Waste management offers an opportunity for cooperation among nations for the betterment of humanity and planet earth.

Table 4.
Challenges and opportunities of the waste, AD and digestate system.

9. Cassava peeling residue (CPR) digestate

N, P, and K are critical macro nutrients for crops production. N is considered the limiting nutrient in growth and yield [89]. P is required for energy transfer, signal transduction, photosynthesis, and macromolecular respiration [90]. K is responsible for metabolism of cell division, enzymatic reactions of amide formation, and amino acid activation during proteins biosynthesis and substrate phosphorylation [91]. To be a credible mineral fertilizer substitute, digestate must have the capacity to deliver the necessities and requirements of N, P, and K.

Table 1 presented a broad gamut of materials used in biogas and digestate creation. The table covered energy crops, agricultural byproducts, food processing residues, livestock effluents, organic fraction of municipal solid wastes, and pharmaceutical industry sludge. However, cassava peeling residue (CPR) was not

S/N	Nutrient	Value [mg/L]
1	Ammonia nitrogen	561
2	Ortho-phosphorus	20
3	Potassium	1066
4	Total Kjeldahl nitrogen	573
5	Total phosphorus	31

Table 5.
Nutrients values of liquid fraction of cassava peeling residue (CPR) digestate.

represented in the table. There is a published report on ammonium, potassium, total nitrogen, and total phosphorus contents of digestate generated from co-digestion of human urine, cow dung, and cassava effluent (a mixture of peeled cassava wash water and crushed cassava juice) [92]. CPR is a solid substrate abundantly generated during production of cassava root-based food systems such as gari and starch [93]. The present author is not aware of any report on nutrients value of digestate generated from the AD of CPR as sole feedstock. Therefore, a technical experiment was conducted to secure an overview assessment of N, P, and K compositions of liquid fraction of CPR digestate.

Some results of the research work on CPR as sole substrate for AD were reported earlier. These included proximate properties (e.g., moisture content, total solids, volatile solids), digester performance characteristics (methane content of biogas, pH, discharge effluent COD), feedstock materials, sampling procedures, analyses [94]. Presented in **Table 5** are results of nutrient values of liquid fraction of CPR digestate. **Table 5** results appear to be within the range of some published nutrients values for liquid digestates derived from other feedstocks such as algal biomass (*Chroococcus* sp.) [88], starch processing wastewater [95], source separated household waste [96], as well as liquid and solid manure slurries [97].

10. Conclusions

Cassava (*Manihot esculenta* Crantz) is perhaps third largest source of food energy for humans. Cassava supports the nutrition and subsistence of up to one billion persons in over 100 countries. Also, cassava is gluten free and could thus assuage medical complications for individuals with celiac disease. Cassava root processing byproduct such as CPR has organic matter content with applications in biogas and digestate production. This is a welcome development in views of biorefinery platform and the emergent circular economy. CPR digestate may be applied directly for agronomic uses or treated to generate products with varied applications and utilities. Treatment technologies may be biological, chemical, physical, or some combinations. Global benefits would include carbon sequestration, energy recovery, resource sustainability and recycling, waste reduction, profitability of AD process, biogas facilities, and agricultural systems in general. End effects of climate change mitigation, enhanced energy and food security, environmental and ecological protection, and sustainable development are good news for humanity and planet earth. These outcomes should motivate and provide consumers, farmers, regulators, managers, and other stakeholders in the emergent circular economy with insights to integrate and apply quality, safety, marketing, handling, storage, transportation, compliance with environmental regulations, and cost considerations and requirements strategies for digestate; into a renewable and sustainable energy production and waste management system.

Acknowledgements

All currency conversions to US$ were based on exchange rate taken at different times and days, during the period of last quarter of the year 2019, from the Foreign Exchange Converter Site: https://www1.oanda.com/currency/converter/

Conflict of interest

There is no conflict of interest (private or public) associated with this work.

Author details

Sammy N. Aso
Food Engineering Laboratory, Rivers State University, Port Harcourt, Nigeria

*Address all correspondence to: sammyasso@yahoo.com

References

[1] Gómez X, Cuetos MJ, García AI, Morán A. An evaluation of stability by thermogravimetric analysis of digestate obtained from different biowastes. Journal of Hazardous Materials. 2007; **149**(1):97-105. DOI: 10.1016/j.jhazmat.2007.03.049

[2] Tambone F, Orzi V, D'Imporzano G, Adani F. Solid and liquid fractionation of digestate: Mass balance, chemical characterization, and agronomic and environmental value. Bioresource Technology. 2017;**243**:1251-1256. DOI: 10.1016/j.biortech.2017.07.130

[3] Antoniou N, Monlau F, Sambusiti C, Ficara E, Barakat A, Zabaniotou A. Contribution to circular economy options of mixed agricultural wastes management: Coupling anaerobic digestion with gasification for enhanced energy and material recovery. Journal of Cleaner Production. 2019; **209**:505-514. DOI: 10.1016/j.jclepro.2018.10.055

[4] Salomon KR, Lora ES. Estimate of the electric energy generating potential for different sources of biogas in Brazil. Biomass and Bioenergy. 2009;**3**(9): 1101-1107. DOI: 10.1016/j.biombioe.2009.03.001

[5] Kusch S, Schäfer W, Kranert M. Dry digestion of organic residues. In: Kumar S, editor. Integrated Waste Management. Vol. 1. Croatia: IntechOpen; 2011. pp. 115-134. ISBN: 978-953-307-469-6. Available from: http://cdn.intechopen.com/pdfs/17433/InTech-Dry_digestion_of_organic_residues.pdf

[6] Aso SN, Teixeira AA, Achinewhu SC. Cassava residues could provide sustainable bioenergy for cassava producing nations. Chapter 13. In: Waisundara VY, editor. Cassava. Rijeka, Croatia: IntechOpen; 2018. pp. 219-240. DOI: 10.5772/intechopen.72166

[7] Liu H, Wang L, Zhang X, Fu B, Liu H, Li Y, et al. A viable approach for commercial VFAs production from sludge: Liquid fermentation in anaerobic dynamic membrane reactor. Journal of Hazardous Materials. 2019;**365**:912-920. DOI: 10.1016/j.jhazmat.2018.11.082

[8] WBA: World Bioenergy Association. WBA Global Bioenergy Statistics 2018. Stockholm, Sweden: World Bioenergy Association; 2018. Available from: https://worldbioenergy.org/uploads/181203%20WBA%20GBS%202018_hq.pdf

[9] BSI: British Standards Institution. Specification for whole digestate, separated liquor and separated fibre derived from the anaerobic digestion of source-segregated biodegradable materials. PAS 110:2010. British Standards Institution, London, UK. 2010. pp. 60. Available from: http://www.wrap.org.uk/sites/files/wrap/PAS110_vis_10.pdf

[10] Tampio E, Marttinen S, Rintala J. Liquid fertilizer products from anaerobic digestion of food waste: Mass, nutrient and energy balance of four digestate liquid treatment systems. Journal of Cleaner Production. 2016;**125**:22-32. DOI: 10.1016/j.jclepro.2016.03.127

[11] Zeshan S, Visvanathan C. Evaluation of anaerobic digestate for greenhouse gas emissions at various stages of its management. International Biodeterioration & Biodegradation. 2014;**95**(Part A):167-175. DOI: 10.1016/j.ibiod.2014.06.020

[12] EC: European Commission. Regulation (EC) No. 208/2006 of 7 February 2006 amending Annexes VI and VIII to regulation (EC) No. 1774/2002 of the European parliament and of the council as regards processing standards for biogas and composting plants and requirements for manure. Official Journal of European

Community. 2006;**L36**:25-31. Available from: https://publications.europa.eu/en/publication-detail/-/publication/eb5ea96c-1ee9-4654-931a-37a1f581b32e/language-en

[13] Petersson A. English summary of SPCR 120—Certification rules for digestate from biowaste by the quality assurance system of Swedish Waste Management. Swedish Gas Centre; 2008. Available from: http://www.fao.org/fileadmin/user_upload/nr/sustainability_pathways/docs/Certification%20rules%20for%20digestate%20from%20biowaste.pdf

[14] Siebert S. Quality requirements and quality assurance of digestion residuals in Germany. In: ECN/ORBIT Workshop: The future for Anaerobic Digestion of Organic Waste in Europe. Nüremberg, Germany. 2008. Available from: http://www.kompost.de/uploads/media/Quality_Requirements_of_digestion_residuals_in_Germany_text_01.pdf

[15] Al Seadi T, Lukehurst C. Quality management of digestate from biogas plants used as fertiliser. IEA Bioenergy. 2012:4-36. Available from: https://www.researchgate.net/profile/Teodorita_Seadi/publication/265227188_Quality_Management_of_Digestate_from_Biogas_Plants_Used_as_Fertiliser/links/54b62dec0cf26833efd35c9d.pdf

[16] Saveyn H, Eder P. End-of-waste criteria for biodegradable waste subjected to biological treatment (compost & digestate): Technical proposals. In: European Commission, Joint Research Centre, Institute for Prospective Technological Studies (IPTS). Seville: Spain; 2014. Available from: http://ftp.jrc.es/EURdoc/JRC87124.pdf

[17] Mottet A, Steyer JP, Déléris S, Vedrenne F, Chauzy J, Carrère H. Kinetics of thermophilic batch anaerobic digestion of thermal hydrolysed waste activated sludge. Biochemical Engineering Journal. 2009; **46**(2):169-175. DOI: 10.1016/j.bej.2009.05.003

[18] Alvira P, Tomás-Pejó E, Ballesteros M, Negro MJ. Pretreatment technologies for an efficient bioethanol production process based on enzymatic hydrolysis: A review. Bioresource Technology. 2010;**101**(13):4851-4861. DOI: 10.1016/j.biortech.2009.11.093

[19] Hjorth M, Christensen KV, Christensen ML, Sommer SG. Solid-liquid separation of animal slurry in theory and practice. A review. Agronomy and Sustainable Development. 2010;**30**:153-180. DOI: 10.1051/agro/2009010

[20] Mouat A, Barclay A, Mistry P, Webb J. Digestate market development in Scotland. Vol. OPR080-801. Stirling, UK: Zero Waste Scotland; 2010. Available from: http://www.wrap.org.uk/sites/files/wrap/Zero_Waste_Scotland_Digestate_Market_Development.pdf

[21] Moon HC, Song IS. Enzymatic hydrolysis of foodwaste and methane production using UASB bioreactor. International Journal of Green Energy. 2011;**8**(3):361-371. DOI: 10.1080/15435075.2011.557845

[22] Pilli S, Bhunia P, Yan S, LeBlanc RJ, Tyagi RD, Surampalli RY. Ultrasonic pretreatment of sludge: A review. Ultrasonics Sonochemistry. 2011;**18**: 1-18. DOI: 10.1016/j.ultsonch.2010.02.014

[23] van der Stelt MJC, Gerhauser H, Kiel JHA, Ptasinski KJ. Biomass upgrading by torrefaction for the production of biofuels: A review. Biomass and Bioenergy. 2011;**35**: 3748-3762. DOI: 10.1016/j.biombioe.2011.06.023

[24] Zhang Q, He J, Tian M, Mao Z, Tang L, Zhang J, et al. Enhancement of methane production from cassava

residues by biological pretreatment using a constructed microbial consortium. Bioresource Technology. 2011;**102**(19):8899-8906. DOI: 10.1016/j.biortech.2011.06.061

[25] Bustamante M, Alburquerque J, Restrepo A, de la Fuente C, Paredes C, Moral R, et al. Co-composting of the solid fraction of anaerobic digestates, to obtain added-value materials for use in agriculture. Biomass and Bioenergy. 2012;**43**:26-35. DOI: 10.1016/j.biombioe.2012.04.010

[26] Elliott A, Mahmood T. Comparison of mechanical pretreatment methods for the enhancement of anaerobic digestion of pulp and paper waste activated sludge. Water Science & Technology. 2012;**84**(6):497-505. DOI: 10.2175/106143012X13347678384602

[27] Liu X, Wang W, Gao X, Zhou Y, Shen R. Effect of thermal pretreatment on the physical and chemical properties of municipal biomass waste. Waste Management. 2012;**32**(2):249-255. DOI: 10.1016/j.wasman.2011.09.027

[28] WRAP: Waste and Resources Action Programme. Enhancement and Treatment of Digestates from Anaerobic Digestion. Banbury, Oxon, United Kingdom: WRAP; 2012. Available from: http://www.wrap.org.uk/sites/files/wrap/Digestates%20from%20Anaerobic%20Digestion%20A%20review%20of%20enhancement%20techniques%20and%20novel%20digestate%20products_0.pdf

[29] Zhao P, Shen Y, Ge S, Chen Z, Yoshikawa K. Clean solid biofuel production from high moisture content waste biomass employing hydrothermal treatment. Applied Energy. 2014;**131**:345-367. DOI: 10.1016/j.apenergy.2014.06.038

[30] Peleteiro S, Rivas S, Alonso JL, Santos V, Parajo JC. Utilization of ionic liquids in lignocellulose biorefineries as agents for separation, derivatization, fractionation, or pretreatment. Journal of Agricultural and Food Chemistry. 2015;**63**(37):8093-8102. DOI: 10.1021/acs.jafc.5b03461

[31] Carrere H, Antonopoulou G, Affes R, Passos F, Battimelli A, Lyberatos G, et al. Review of feedstock pretreatment strategies for improved anaerobic digestion: From lab-scale research to full-scale application. Bioresource Technology. 2016;**199**:386-397. DOI: 10.1016/j.biortech.2015.09.007

[32] Liguori R, Faraco V. Biological processes for advancing lignocellulosic waste biorefinery by advocating circular economy. Bioresource Technology. 2016;**215**:13-20. DOI: 10.1016/j.biortech.2016.04.054

[33] Zeng Y, Guardia AD, Dabert P. Improving composting as a post-treatment of anaerobic digestate. Bioresource Technology. 2016;**201**:293-303. DOI: 10.1016/j.biortech.2015.11.013

[34] Bhutto AW, Qureshi K, Harijan K, Abro R, Abbas T, Bazmi AA, et al. Insight into progress in pre-treatment of lignocellulosic biomass. Energy. 2017;**122**:724-745. DOI: 10.1016/j.energy.2017.01.005

[35] Duque A, Manzanares P, Ballesteros M. Extrusion as a pretreatment for lignocellulosic biomass: Fundamentals and applications. Renewable Energy. 2017;**114**:1427-1441. DOI: 10.1016/j.renene.2017.06.050

[36] Shi L, Xie S, Hu Z, Wu G, Morrison L, Croot P, et al. Nutrient recovery from pig manure digestate using electrodialysis reversal: Membrane fouling and feasibility of long-term operation. Journal of Membrane Science. 2019;**573**:560-569. DOI: 10.1016/j.memsci.2018.12.037

[37] D'Alessio M, Durso LM, Miller DN, Woodbury B, Ray C, Snow DD. Environmental fate and microbial effects of monensin, lincomycin, and sulfamethazine residues in soil. Environmental Pollution. 2019;**246**: 60-68. DOI: 10.1016/j.envpol.2018.11.093

[38] Tani M, Sakamoto N, Kishimoto T, Umetsu K. Utilization of anaerobically digested dairy slurry combined with other wastes following application to agricultural land. International Congress Series. 2006;**1293**:331-334. DOI: 10.1016/j.ics.2006.03.013

[39] Alburquerque JA, Fuente C, Campoy M, Carrasco L, Nájera I, Baixauli C, et al. Agricultural use of digestate for horticultural crop production and improvement of soil properties. European Journal of Agronomy. 2012;**43**:119-128. DOI: 10.1016/j.eja.2012.06.001

[40] Koszel M, Lorencowicz E. Agricultural use of biogas digestate as a replacement fertilizers. Agriculture and Agricultural Science Procedia. 2015;**7**: 119-124. DOI: 10.1016/j.aaspro.2015.12.004

[41] Garfí M, Ferrer-Martí L, Velo E, Ferrer I. Evaluating benefits of low-cost household digesters for rural Andean communities. Renewable and Sustainable Energy Reviews. 2012;**16**(1): 575-581. DOI: 10.1016/j.rser.2011.08.023

[42] Marchaim U. Biogas process for sustainable development. FAO Corporate Document Repository. M-09. 1992. ISBN: 92-5-103126-6. Available from: https://www.build-a-biogas-plant.com/PDF/BiogasSustainableDevolpment.pdf

[43] Giuliano A, Bolzonella D, Pavan P, Cavinato C, Cecchi F. Co-digestion of livestock effluents, energy crops and agro-waste: Feeding and process optimization in mesophilic and thermophilic conditions. Bioresource Technology. 2013;**128**:612-618. DOI: 10.1016/j.biortech.2012.11.002

[44] Monlau F, Sambusiti C, Ficara E, Aboulkas A, Barakat A, Carrère H. New opportunities for agricultural digestate valorization: Current situation and perspectives. Energy and Environmental Science. 2015;**8**(9):2600. DOI: 10.1039/c5ee01633a

[45] Carpenter CE, Smith JV, Broadbent JR. Efficacy of washing meat surfaces with 2% levulinic, acetic, or lactic acid for pathogen decontamination and residual growth inhibition. Meat Science. 2011;**88**(2): 256-260. DOI: 10.1016/j.meatsci.2010.12.032

[46] Li C, Wang Y, Xie G, Peng B, Zhang B, Chen W, et al. Complete genome sequence of *Clostridium butyricum* JKY6D1 isolated from the pit mud of a Chinese flavor liquor-making factory. Journal of Biotechnology. 2016; **220**:23-24. DOI: 10.1016/j.jbiotec.2016.01.003

[47] Huang WW, Huang WL, Yuan T, Zhao ZW, Cai W, Zhang ZY, et al. Volatile fatty acids (VFAs) production from swine manure through short-term dry anaerobic digestion and its separation from nitrogen and phosphorus resources in the digestate. Water Research. 2016;**90**:344-353. DOI: 10.1016/j.watres.2015.12.044

[48] Wang K, Yin J, Shen D, Na L. Anaerobic digestion of food waste for volatile fatty acids (VFAs) production with different types of inoculum: Effect of pH. Bioresource Technology. 2014; **161**(6):395-401. DOI: 10.1016/j.biortech.2014.03.088

[49] Liu H, Han P, Liu H, Zhou G, Fu B, Zheng Z. Full-scale production of VFAs from sewage sludge by anaerobic alkaline fermentation to improve biological nutrients removal in domestic

wastewater. Bioresource Technology. 2018;**260**:105-114. DOI: 10.1016/j.biortech.2018.03.105

[50] Neumann J, Binder S, Apfelbacher A, Gasson JR, Ramírez García P, Hornung A. Production and characterization of a new quality pyrolysis oil, char and syngas from digestate–Introducing the thermo-catalytic reforming process. Journal of Analytical and Applied Pyrolysis. 2015; **113**:137-142. DOI: 10.1016/j.jaap.2014.11.022

[51] Singh R, Parihar P, Singh M, Bajguz A, Kumar J, Singh S, et al. Uncovering potential applications of cyanobacteria and algal metabolites in biology, agriculture and medicine: Current status and future prospects. Frontiers in Microbiology. 2017;**8**:515. DOI: 10.3389/fmicb.2017.00515

[52] Koutra E, Economou CN, Tsafrakidou P, Kornaros M. Bio-based products from microalgae cultivated in digestates. Trends in Biotechnology. 2018;**36**(8):819-833. DOI: 10.1016/j.tibtech.2018.02.015

[53] Inyang M, Gao B, Yao Y, Xue Y, Zimmerman AR, Pullammanappallil P, et al. Removal of heavy metals from aqueous solution by biochars derived from anaerobically digested biomass. Bioresource Technology. 2012;**110**: 50-56. DOI: 10.1016/j.biortech.2012.01.072

[54] Dahlin J, Herbes C, Nelles M. Biogas digestate marketing: Qualitative insights into the supply side. Resources, Conservation and Recycling. 2015;**104**: 152-161. DOI: 10.1016/j.resconrec.2015.08.013

[55] Spelter H, Winandy J, Zauche T. Anaerobically digested bovine biofiber as a source of fiber for particleboard manufacturing: An economic analysis. BioResources. 2008;**3**(4):1256-1266. Available from: https://ojs.cnr.ncsu.edu/index.php/BioRes/article/viewFile/BioRes_03_4_1256_Spelter_WZ_ADBF_Particleboard/278

[56] USDA. Uses of solids and by-products of anaerobic digestion. Farm Energy. 2019;**3**:2019. Available from: https://farm-energy.extension.org/uses-of-solids-and-by-products-of-anaerobic-digestion/#Livestock_bedding

[57] Sophin P, Preston TR. Effect of processing pig manure in a biodigester as fertilizer input for ponds growing fish in polyculture. Livestock Research for Rural Development. 2001;**13**:60. Available from: https://www.lrrd.cipav.org.co/lrrd13/6/pich136.htm

[58] Garcia-Sánchez M, Garcia-Romera I, Cajthaml T, Tlustoš P, Száková J. Changes in soil microbial community functionality and structure in a metal-polluted site: The effect of digestate and fly ash applications. Journal of Environmental Management. 2015;**162**:63-73. DOI: 10.1016/j.jenvman.2015.07.042

[59] Alexander R. Digestate utilization in the U.S. Biocycle. 2012;**53**(1):56. Available from: https://www.biocycle.net/2012/01/12/digestate-utilization-in-the-u-s/

[60] Tin AM, Wise DL, Su WH, Reutergardh L, Lee SK. Cost-benefit analysis of the municipal solid waste collection system in Yangon, Myanmar. Resources, Conservation and Recycling. 1995;**14**(2):103-131. DOI: 10.1016/S0921-3449(95)80004-2

[61] Setiawan RP, Kaneko S, Kawata K. Impacts of pecuniary and non-pecuniary information on pro-environmental behavior: A household waste collection and disposal program in Surabaya city. Waste Management. 2019;**89**:322-335. DOI: 10.1016/j.wasman.2019.04.015

[62] Drosg B, Fuchs W, Al Seadi T, Madsen M, Linke B. Nutrient recovery by biogas digestate processing. IEA Bioenergy. 2015. Available from: https://

www.iea-biogas.net/files/daten-redaktion/download/Technical%20Brochures/NUTRIENT_RECOVERY_RZ_web2.pdf

[63] Hogg D. Eunomia Research & Consulting. Costs for municipal waste management in the EU. Final Report to Directorate General Environment, European Commission. 2002. Available from: http://ec.europa.eu/environment/waste/studies/pdf/eucostwaste.pdf

[64] Delzeit R, Kellner U. The impact of plant size and location on profitability of biogas plants in Germany under consideration of processing digestates. Biomass and Bioenergy. 2013;**52**:43-53. DOI: 10.1016/j.biombioe.2013.02.029

[65] Baddeley A, Ballinger A, Cessford I, Smith EM, Enviro A. Assessing the costs and benefits for production and beneficial application of anaerobic digestate to agricultural land in Wales. In: Project OMK007-203. WRAP: Waste and Resources Action Programme. Cardiff, UK; 2014. Available from: http://www.wrapcymru.org.uk/sites/files/wrap/Assessing%20the%20Costs%20and%20Benefits%20for%20Production%20and%20Beneficial%20Application%20of%20Anaerobic%20Digestate%20to%20Agricultural%20Land%20in%20Wales%202014.pdf

[66] Møller HB, Lund I, Sommer SG. Solid–liquid separation of livestock slurry: Efficiency and cost. Bioresource Technology. 2000;**74**(3):223-229. DOI: 10.1016/S0960-8524(00)00016-X

[67] Stiles WAV, Styles D, Chapman SP, Esteves S, Bywater A, Melville L, et al. Using microalgae in the circular economy to valorise anaerobic digestate: Challenges and opportunities. Bioresource Technology. 2018;**267**: 732-742. DOI: 10.1016/j.biortech.2018.07.100

[68] Mulinda C, Hu Q, Pan K. Dissemination and problems of African biogas technology. Energy and Power Engineering. 2013;**5**(08):506. Available from: http://www.build-a-biogas-plant.com/PDF/ProblemsAfricanBiogas2013.pdf

[69] Amigun B, Von Blottnitz H. Capacity-cost and location-cost analyses for biogas plants in Africa. Resources, Conservation and Recycling. 2010; **55**(1):63-73. DOI: 10.1016/j.resconrec.2010.07.004

[70] Bolzonella D, Fatone F, Gottardo M, Frison N. Nutrients recovery from anaerobic digestate of agro-waste: Techno-economic assessment of full scale applications. Journal of Environmental Management. 2018;**216**: 111-119. DOI: 10.1016/j.jenvman.2017.08.026

[71] Gerardo ML, Aljohani NHM, Oatley-Radcliffe DL, Lovitt RW. Moving towards sustainable resources: Recovery and fractionation of nutrients from dairy manure digestate using membranes. Water Research. 2015;**80**: 80-89. DOI: 10.1016/j.watres.2015.05.016

[72] Favoino E, Hogg D. Effects of composted organic waste on ecosystems—A specific angle: The potential contribution of biowaste to tackle Climate Change and references to the soil policy. Proceedings of the International Congress CODIS. 2008; **2008**:145-156. Available from: https://orgprints.org/13135/1/fuchs-etal-proceedings-codis-2008.pdf#page=151

[73] Hu Y, Shen F, Yuan H, Zou D, Pang Y, Liu Y, et al. Influence of recirculation of liquid fraction of the digestate (LFD) on maize stover anaerobic digestion. Biosystems Engineering. 2014;**127**:189-196. DOI: 10.1016/j.biosystemseng.2014.09.006

[74] de Alwis A. Biogas—A review of Sri Lanka's performance with a renewable energy technology. Energy for Sustainable Development. 2002;**6**(1): 30-37. DOI: 10.1016/S0973-0826(08)60296-3

[75] Brändli RC, Bucheli TD, Kupper T, Furrer R, Stahel WA, Stadelmann FX, et al. Organic pollutants in compost and digestate. Part 1. Polychlorinated biphenyls, polycyclic aromatic hydrocarbons and molecular markers. Journal of Environment Monitoring. 2007;**9**(5):456-464. DOI: 10.1039/B617101J

[76] Tiwary A, Williams ID, Pant DC, Kishore VVN. Assessment and mitigation of the environmental burdens to air from land applied food-based digestate. Environmental Pollution. 2015;**203**:262-270. DOI: 10.1016/j.envpol.2015.02.001

[77] Wallace P, Harris G, Frederickson J, Howell G, Tompkins D. Biofertiliser management: Best practice for agronomic benefit & odour control. In: Tompkins D, editor. Project OAV036-210. Cardiff: The Waste and Resources Action Program (WRAP); 2011. Available from: http://www.wrapcymru.org.uk/sites/files/wrap/Digestate%20odour%20management%20-%20Cymru.pdf

[78] Cheng Z, Sun Z, Zhu S, Lou Z, Zhu N, Feng L. The identification and health risk assessment of odor emissions from waste landfilling and composting. Science of The Total Environment. 2019;**649**:1038-1044. DOI: 10.1016/j.scitotenv.2018.08.230

[79] Eshete G, Sonder K, ter Heegde F. Report on the Feasibility Study of a National Programme for Domestic Biogas in Ethiopia. SNV Netherlands Development Organization: Addis Ababa, Ethiopia; 2006. Available from: http://www.bibalex.org/Search4Dev/files/338849/172350.pdf

[80] Berhe TG, Tesfahuney RG, Desta GA, Mekonnen LS. Biogas plant distribution for rural household sustainable energy supply in Africa. Energy and Policy Research. 2017;**4**(1):10-20. DOI: 10.1080/23815639.2017.1280432

[81] Mushtaq K, Zaidi AA, Askari SJ. Design and performance analysis of floating dome type portable biogas plant for domestic use in Pakistan. Sustainable Energy Technologies and Assessments. 2016;**14**:21-25. DOI: 10.1016/j.seta.2016.01.001

[82] Day DL, Chen TH, Anderson JC, Steinberg MP. Biogas plants for small farms in Kenya. Biomass. 1990;**21**(2):83-99. DOI: 10.1016/0144-4565(90)90051-K

[83] Caceres R, Chiliquinga B. Experiences with Rural Biodigesters in Latin America. Dordrecht: Springer; 1986. pp. 150-165. DOI: 10.1007/978-94-009-4313-1_21

[84] Ni J-Q, Nyns E-J. New concept for the evaluation of rural biogas management in developing countries. Energy Conversion and Management. 1996;**37**(10):1525-1534. DOI: 10.1016/0196-8904(95)00354-1

[85] Raven RPJM. Implementation of manure digestion and co-combustion in the Dutch electricity regime: A multi-level analysis of market implementation in the Netherlands. Energy Policy. 2004;**32**(1):29-39. DOI: 10.1016/S0301-4215(02)00248-3

[86] Negro SO, Hekkert MP, Smits RE. Explaining the failure of the Dutch innovation system for biomass digestion—A functional analysis. Energy Policy. 2007;**35**(2):925-938. DOI: 10.1016/j.enpol.2006.01.027

[87] Tigini V, Franchino M, Bona F, Varese GC. Is digestate safe? A study on its ecotoxicity and environmental risk on a pig manure. Science of the Total Environment. 2016;**551-552**:127-132. DOI: 10.1016/j.scitotenv.2016.02.004

[88] Prajapati SK, Kumar P, Malik A, Vijay VK. Bioconversion of algae to methane and subsequent utilization of digestate for algae cultivation: A closed loop bioenergy generation process.

Bioresource Technology. 2014;**158**: 174-180. DOI: 10.1016/j.biortech.2014.02.023

[89] Möller K, Habermeyer J, Zinkernagel V, Reents H-J. The impact and the interaction of nitrogen and *Phytophthora infestans* as yield-limiting and yield-reducing factors in organic potato (*Solanum tuberosum* L.) crops. Potato Research. 2006;**49**(4):281-301. DOI: 10.1007/s11540-007-9024-7

[90] Shenoy VV, Kalagudi GM. Enhancing plant phosphorus use efficiency for sustainable cropping. Biotechnology Advances. 2005;**23**(7-8): 501-513. DOI: 10.1016/j.biotechadv.2005.01.004

[91] Sobachkin AA. The physiological role of potassium in increasing the productivity of farm crops. In: Potassium Research and Agricultural Production. Proceedings of the 10th Congress of the International Potash Institute held in June 1974 in Budapest, Hungary. International Potash Institute. 1974. pp. 147-152. Available from: https://www.ipipotash.org/uploads/udocs/potassium_research_and_agricultural_production.pdf

[92] Edith KKN, Francis KY, Martin KK, Felix KK. Characterization of digestates from anaerobic co-digestion of manioc effluent, human urine and cow dung. Journal of Water Resource and Protection. 2019;**11**(06):777-788. DOI: 10.4236/jwarp.2019.116047

[93] Aso SN. Food engineering stratagem to protect the environment and improve the income opportunities of gari processors. Journal of Nigerian Environmental Society (JNES). 2004; **2**(1):31-36

[94] Aso SN, Pullammanappallil PC, Teixeira AA, Welt BA. Biogasification of cassava residue for on-site biofuel generation for food production with potential cost minimization, health and environmental safety dividends. Environmental Progress & Sustainable Energy. 2019. DOI: 10.1002/ep.13138

[95] Tan X, Chu H, Zhang Y, Yang L, Zhao F, Zhou X. *Chlorella pyrenoidosa* cultivation using anaerobic digested starch processing wastewater in an airlift circulation photobioreactor. Bioresource Technology. 2014;**170**: 538-548. DOI: 10.1016/j.biortech.2014.07.086

[96] Haraldsen TK, Andersen U, Krogstad T, Sørheim R. Liquid digestate from anaerobic treatment of source-separated household waste as fertiliser to barley. Waste Management & Research. 2017;**29**:1271-1276. DOI: 10.1177/0734242X11411975

[97] Pirelli T, Rossi A, Miller C. Sustainability of biogas and cassava-based ethanol value chains in Viet Nam: Results and recommendations from the implementation of the Global Bioenergy Partnership indicators. In: FAO Environment and Natural Resources Management Working Paper 69. Rome: FAO; 2018. Available from: http://www.fao.org/3/i9181en/I9181EN.pdf

Section 4

Hydroelectric Energy

Chapter 14

Formation and Transformation of Typical Pollutant from MSW by Hydrothermal Carbonization towards Biofuel Hydrochar Production

Wentao Jiao, Nana Peng and Zhengang Liu

Abstract

An unprecedented increase in municipal solid waste (MSW) is increasingly attractive in response to waste-to-energy. MSW pretreatment is an essential step due to the inherent properties of MSW. Hydrothermal carbonization (HTC) offers an efficient approach for converting MSW into carbonaceous hydrochars. In this chapter, the formation and transformation of heavy metals and polycyclic aromatic hydrocarbons (PAHs) during HTC of MSW were determined. The results indicated that HTC can homogenize the density and size of MSW and also increase carbon content. Moreover, the concentrations of heavy metals in the leachates of the hydrochars were lower than the United States Environmental Protection Agency (US EPA) maximum limits. Compared to MSW, the concentrations of Cr, Cd, Hg, and Zn in the hydrochars were low and the concentrations of Pb, As, Ni, and Cu were high. The concentrations of PAHs in the hydrochars increased with increasing temperature in the range of 1298.71–177698.20 μg/kg, which were much higher than that in MSW, except for H-160. The dominant PAH rings in MSW and the hydrochars were four-ring PAHs and three-ring PAHs, respectively. These findings suggest that 180°C is an appropriate hydrothermal temperature to reduce heavy metals and the toxicity PAHs of MSW.

Keywords: biomass, waste-to-energy, fuel quality, heavy metals, polycyclic aromatic hydrocarbons, hydrochar

1. Introduction

At present, municipal solid waste (MSW) generated in daily life is the major waste in the urban management, which consists of food waste, plastics, paper, garden waste, textiles, stone, glass, etc. [1]. An unprecedented increase in the amount of MSW is observed in all over the world, especially for the populous development countries, such as India and China due to the population growth, rapid urbanization, and industrialization [2, 3]. It is reported that about 1.70–1.90 billion metric tons per year of MSW is produced in the world [4]. In China, the MSW delivering quantity had grown to 0.20 billion metric tons in 2016 [5]. It is estimated that China will generate about 0.50 billion metric tons at 2025 and contribute nearly 25% to total world's MSW amount [6]. Therefore, the disposal of MSW has been and will continue to be an urgent and major issue. Landfill, incineration, and composting are commonly applied for the treatment of MSW [7–9]. Among those methods, landfill is the most common disposal method of MSW in China due to the cost-effective and low requirements of both separation and technique. However, the landfill suffers from many problems, such as serious secondary pollution from the leachate, the increasing cost of land acquirement due to the increase development of population and urbanization, and so on [10]. The incineration of MSW can achieve effectively energy recovery, reduce waste volume by 90% and waste weight by 70%, and decrease the toxicity [11]. Additionally, an increase in the combustible components of MSW such as plastics and papers improves the combustibility of MSW. On the other hand, the quick depletion of fossil fuels and worldwide environmental pollution promote the development of sustainable energy obtained from renewable resources like MSW. In view of those advantages, incineration as an economically feasible way for the effective energy recovery from MSW has been paid great attention in recent years. For example, in the last 10 years, an increase in the incineration treatment rate of MSW in China is observed in the range of 14.52–37.51% [5]. But the specific properties of MSW are the major drawbacks to employ it as an energy feedstock for incineration, such as the high moisture and oxygen contents, low carbon content, and the heterogeneity. To overcome these problems, pretreatment technology is need to homogenize raw MSW and improve its fuel property.

Hydrothermal carbonization (HTC), also known as hydrothermal upgrading, wet torrefaction, or coalification, is a wet and moderate temperature (180–350°C) process in an autogenous or elevated pressures reactor [12–15]. Water is a necessary component and plays an important role during HTC, which exists as a hot, expanded liquid or possibly supercritical state to participate in the reaction as reactants or contributes to the changes in the free energy of activation [13, 16]. A series of hydrolysis, decarboxylation, dehydration, and condensation reactions occur during HTC resulting in carbonaceous solid product known as the hydrochar, aqueous, and gases [17]. The distributions and characteristics of the products are strongly affected by the HTC conditions, and many studies present reaction temperature plays the most important role [14, 18]. Compared to other dry carbonization processes (such as gasification and pyrolysis), HTC do not need to spend extra energy on the drying process of raw biomass. On the other hand, HTC takes place in low temperature and has high conversion efficiency. For example, the solid products produced from HTC and pyrolysis were determined and compared based on the fuel qualities, and the results showed that the hydrochar obtained from HTC had higher energy density and thermal efficiency and lower pollutant emissions than that obtained from pyrolysis [19, 20], confirming that HTC can proceed with the same level of conversion efficiency as higher-temperature processes of dry carbonization [21, 22].

The first experiment about HTC was used in cellulose (which is known to pure material) to produce carbonaceous materials in 1913 [21, 23], and then early reviews were published in 1993 and 1994 [24, 25]. To date, the feedstocks supplied for HTC rang from pure materials to more complex biomass, such as agriculture waste [26–28], manure [29], alga [30], and so on [18, 31]. Recently, the acid or basic solution can also be applied into the feedwater in order to remove the metals and increase the hydrochar yield solution [26, 32, 33]. The existing literature covers a wide range of studies on HTC of biomass and has been proved that HTC as a novel thermal pretreatment process is a promising and viable method to homogenize the biomass, increase the carbon content, decrease moisture content, and increase grindability [21, 31]. Generally, the potential to use the hydrochar has concentrated on the alternative fuel for energy supply and additive agent for soil amendment.

With the application of the hydrochars, environmental issues should be considered, such as the pollution risk of heavy metals and the emissions of polycyclic aromatic hydrocarbons (PAHs).

Biomass contains various metals such as light metals and heavy metals during the metabolic demands for the growth process. However, those metals usually remain in the ash during the combustion because of their high melting point, leading to the slagging/fouling and the negative impact on the environment. Several investigations have been published about the removing of light metals during HTC of lignocellulosic biomass [20, 34]. The results show that light metals can be easily removed from raw biomass during HTC due to the water solubility properties [20]. In contrast to light metals, heavy metals exhibit the different chemical forms and is expected to result in the different remove rates. Additionally, considering the potential risk to the environment and health, heavy metals like chromium (Cr), cadmium (Cd), zinc (Zn), copper (Cu), nickel (Ni), arsenic (As), mercury (Hg), and lead (Pb) have been the worldwide concern [35]. However, the transformation of heavy metals during HTC process has not yet been studied in detail, especially for the MSW.

PAHs are known as persistent organic pollutants due to the long-range transport potential and the effect of mutagenicity, teratogenicity, and carcinogenicity [36]. Many studies provide the evidence that the environmental levels of PAHs have a significant inhibitory effect on human health and the growth of animals and plants [37, 38]. PAHs generated from the combustion or pyrolysis of solid fuels, such as MSW with the human industrialization, are widespread pollutants in the environment [36, 39]. Moreover, stricter standards on PAHs in recent years must be established to satisfy the social needs. Consequently, numerous researches on the decomposition and formation of PAHs have been reported in literature in order to control PAH emission from combustion or pyrolysis of coal [40], biomass [41], and so on [39, 42]. It is noteworthy that free PAHs are referred to the PAHs contained in the macromolecular structure of the feedstocks, which are easily able to emit into the environment [43]. Compared to PAHs generated from combustion or pyrolysis, limited investigations are known on free PAHs in coal [44], MSW ashes [45], and biochar obtained of biomass [46]. However, to our best knowledge, no study clearly has been done to investigate the free PAHs in MSW and corresponding hydrochars.

With the widespread application of the hydrochar, it is essential to investigate the formation and transformation of heavy metals and PAHs during HTC of MSW. Considering the significant effect of hydrothermal temperature, the concentrations of heavy metals and PAHs in MSW and corresponding hydrochars were determined in this chapter to evaluate the effect of HTC on the transformation of heavy metal and PAHs.

2. Preparation and characterization of biofuel hydrochar

Organic components of MSW were investigated in the present study, which was comprised of food waste (64.93 wt.%), plastics (15.07 wt.%), paper (12.94 wt.%), wood waste (1.48 wt.%), and textiles (3.11 wt.%) [10]. HTC of MSW was undertaken using a Morey-type reaction vessel that included a laboratory 50-mL Teflon reaction vessel and an SUS steel pressure vessel [47]. Briefly, the MSW and de-ionized water (a mass ratio of 1:3) was mixed using a stirrer and was fed to the vessel. Furthermore, the vessel was sealed and heated to the desired temperature using an oven. Experiments were conducted at different temperatures (160, 180, 200, 220, 240, and 260°C) for 10 h. After the reaction time was completed and the reactor had cooled down to room temperature, the solid product, called the hydrochar, was collected by vacuum filtration and dried at 50°C for 24 h before use. The hydrochar yield was calculated as the mass ratio of the hydrochar and the MSW. There are three replicates for each hydrochar. H-x represented the hydrochar produced at some hydrothermal temperature, where x was the centigrade temperature.

The inductively coupled plasma-optical emission spectroscopy (ICP-OES, USA) was used to determine the concentrations of heavy metals, except for Hg. About 0.1 g of MSW or the hydrochar was digested in Teflon reactor by the mixed acids (4 mL 65% HNO_3, 4 mL 30% H_2O_2, 2 mL 70% $HClO_4$, and 4 mL 48% HF). The reactor was sealed and operated at 170°C for 12 h in order to dissolve the sample. After the reactor was cooled to the ambient temperature, the liquid solution was transferred to FTFE beaker, which was placed on electric hot plate to remove the moisture. Furthermore, the residual was re-dissolved using 1:1 v/v ratio of HNO_3 to the desired concentration and filtered with the glass fiber filter for ICP-OES analysis.

The concentration of Hg was analyzed on the cold atomic fluorescence mercury meter after total dissolution. For each experiment, a sample of approximately 0.2 g and mixed acids (4 M HCl and 2 M HNO_3) were sent to the glass tube and heated to 95°C for 1.5 h. Furthermore, 10 mL 0.05% $K_2Cr_2O_7$ (diluted by 5% HNO_3) was added to the mixed solution after cooling down to ambient temperature and then diluted to the desired concentration. The solution was filtered and analyzed by the cold atomic fluorescence mercury meter.

The toxicity characteristic leaching procedure (TCLP) was carried out according to the ASTM method 1311 in order to elucidate the potential mobility of heavy metals in the hydrochars [48]. 0.5 g of dried hydrochar was weighed and dissolved in 10 mL of TCLP extraction fluid. After rotated for 18 h, the solution was filtered through glass fiber (0.6–0.8 μm) and acidified with 1 M HNO_3 to pH 2 for ICP-OES analysis.

To determine the concentrations of PAHs, 0.5 g of the sample was Soxhlet extracted in dichloromethane (DCM) as the extraction solvent for 24 h. The extract was concentrated by a rotary evaporator at temperature around 38°C. A total of 15 mL of n-hexane was added and reconcentrated to 1 mL for solvent exchange [49]. The silica column was used for cleanup, which filled with silica gel particles (100–200 mesh), anhydrous sodium sulfate, and aluminum oxide from the bottom to the top. The extract was transferred to the silica column and eluted with n-hexane and a mixed solvent (n-hexane and DCM 7:3 (v/v), 70 mL). The fraction in the mixed solvent was solvent-exchange to n-hexane, concentrated to nearly dryness with a gentle blow of N_2, and diluted to 1 mL for analysis. The fraction was analyzed on a gas chromatograph-mass spectrometer (GC-MS, GC model Agilent 6890 and MS model 5793, USA) equipped with a DB-5MS column (30 m × 0.25 mm i.d., 0.25 μm film thickness). The GC oven was set to an initial temperature of 50°C, held in 2 min, and heated to 300°C at the rate of 6°C/min with final hold time at 300°C of 5 min.

An external standard calibration was used for the quantification of 16 priority PAHs. The examined 16 priority PAHs were shown in **Table 1**. Toxic equivalent factor (TEF) is used to take BaP as a toxic benchmark, and the corresponding TEF of individual PAHs can be found elsewhere [50]. The surrogate standards including naphthalene-d8, perylene-d12, acenaphthene-d10, and phenanthrene-d10 were applied for determining the recovery rate of PAHs. The recoveries for surrogate standards were in the range of 71.45–124.06%.

The surface morphology of the hydrochars was investigated using a field emission scanning electron microscopy (FESEM) model JSM-7500F. The sample dried at 105°C for 24 h was kept onto a carbon tape, coated with Au, and then placed under high vacuum condition. For each sample, at least five different sites were taken to increase the accuracy.

Ring number	Individual PAHs	Abbreviations	TEF
2	Napthalene	Nap	0.001
3	Acenaphthylene	Acy	0.001
	Acenapthene	Ace	0.001
	Fluorene	Flu	0.001
	Phenanthrene	Phe	0.001
	Anthracene	Ant	0.010
4	Fluoranthene	Fla	0.001
	Pyrene	Pyr	0.001
	Benz[a]anthracene	BaA	0.100
	Chrysen	Chr	0.010
5	Benzo[b]fluoranthene	BbF	0.100
	Benzo[k]fluoranthene	BkF	0.100
	Benzo[a]pyrene	BaP	1.000
	Dibenz[ah]anthracene	DaA	1.000
6	Indeno[1,2,3-c,d]pyrene	IcP	0.100
	Benzo[ghi]perylene	BgP	0.010

Table 1.
The 16 priority PAHs and corresponding TEF value.

3. Characteristics of MSW and biofuel hydrochars

3.1 Physicochemical properties of biofuel hydrochar

The yields of the hydrochars obtained from MSW during HTC are summarized in **Figure 1**. As expected, the yields of hydrochars in the range of 37.68–70.37% gradually decreased with increasing hydrothermal temperature from 160 to 260°C. A significant decrease of hydrochar yield was observed with the increase from 160 to 200°C. The hydrothermal temperature had no obvious effect on the hydrochar yield when the temperature was higher than 200°C. The quickest decrease of hydrochar yield was observed again at hydrothermal temperature from 240 to 260°C. The highest decrease of hydrochar yield at low temperature can be linked with easy decomposition of the protein, carbohydrate, and lipid in MSW. Furthermore, the decomposition of the char effectively caused the obvious decrease of hydrochar yield when the temperature was above 240°C.

Table 2 showed the ultimate analysis, the ratios of H/C and O/C, and HHV of MSW and corresponding hydrochars. The carbon content and oxygen content of MSW were 42.13 and 49.10%, respectively. As expected, compared to MSW, the carbon contents of the hydrochars were higher than that of MSW, except for H-160, while the contents of hydrogen, sulfur, nitrogen, and oxygen were roughly lower. It was noteworthy that compared to MSW, the H/C and O/C ratios of the hydrochars were observed to be low, indicating that the hydrochars had higher heating values and reduced energy loss during the combustion, which was consistent with the previous studies [51, 52]. This may be explained by the decrease of low-energy H—C and O—C bonds and the increase of high-energy C—C bonds [51]. The higher heating values (HHV) of MSW and corresponding hydrochars were investigated, and the results showed that the HHV of the hydrochars (in the range of 16.06–31.76 MJ/kg) almost increased with increasing

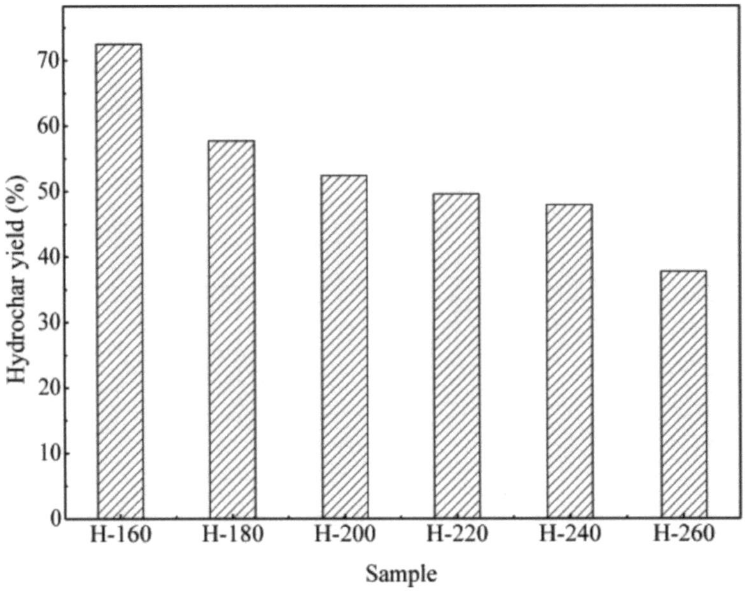

Figure 1.
The yields of the hydrochars obtained from MSW during HTC.

		MSW	H-160	H-180	H-200	H-220	H-240	H-260
Ultimate analysis (db, %)	C	42.13	41.92	49.3	65.15	73.88	67.16	75.71
	H	6.93	5.73	5.64	5.98	5.96	5.84	6.06
	S	0.29	0.28	0.26	0.26	0.20	0.25	0.17
	N	1.55	0.94	1.02	1.21	1.13	1.27	1.23
	O[a]	49.10	51.13	43.78	27.40	18.83	25.48	18.06
H/C		1.97	1.64	1.37	1.10	0.97	1.04	0.96
O/C		0.87	0.91	0.67	0.31	0.19	0.28	0.17
HHV (MJ/kg)		17.73	16.06	20.27	26.91	30.79	27.64	31.76

[a]*By difference.*

Table 2.
Ultimate analysis, the ratios of H/C and O/C, and HHV of MSW and corresponding hydrochars.

temperature, which were higher than that of MSW (17.73 MJ/kg), except for H-160. It is reported that the HHV of the solid fuel must be higher than 20 MJ/kg in order to ensure autothermal combustion [53]. In the present study, the HHV of the hydrochars exceeded 20 MJ/kg, except for H-160. The above results demonstrate that the hydrochar obtained from MSW above 160°C can be considered as a promising solid fuel.

3.2 The surface morphology of biofuel hydrochar

Figure 2 showed the appearances of MSW and the hydrochars obtained from HTC of MSW. It should be noticed that the individual components of MSW and H-160 were still observed. When the hydrothermal temperature was higher than 180°C, the hydrochars were homogenized and grinded into particles, indicating that HTC was an effective method for homogenizing the MSW.

The effect of hydrothermal temperature on the morphology characteristic of the MSW was shown in **Figure 3**. It was found that the hydrochar had a rough surface.

Figure 2.
The appearances of MSW and corresponding hydrochars.

Figure 3.
SEM of the hydrochars obtained from MSW.

With the increasing hydrothermal temperature from 160 to 240°C, the microspheres appeared, and the number of those microspheres increased. The formation of microsphere number increased due to the decomposition or carbonization of MSW. Protein, cellulose, and other components in MSW firstly decompose into small fragments and then are melted and carbonized with the rise of hydrothermal temperature, resulting in the formation of carbon microspheres [54]. When the temperature was higher, however, the microspheres were damaged.

4. The transformation of heavy metals during HTC of MSW

Table 3 presented the mobility of heavy metals in the hydrochars using the TCLP. The concentrations of Cu in the TCLP leachates of all the hydrochars were highest than other heavy metals, except for Ni in H-180, while the concentrations of Hg and As were not detected at most temperature. The United States Environmental Protection Agency maximum thresholds of determined heavy metal were also illustrated in **Table 3**. No stipulated standards for heavy metals usually are calculated as the drinking water standard multiplied by 100 [55], such as Ni. As shown in **Table 3**, the concentrations of heavy metals in TCLP leachates of the hydrochars were all lower than the US EPA permissible limits, showing that the hydrochars can be considered as non-hazardous materials.

The concentrations of heavy metals in MSW and corresponding hydrochars, such as Cr, Cd, Pb and so on, were determined in **Figure 4**. The results showed that for all determined heavy metals in MSW, Cr had a maximum concentration at a value of 93.29 μg/g and the lowest concentration of heavy metals was Hg (0.15 μg/g). The concentrations of Hg in the hydrochars were still the minimum, while the highest concentration among the heavy metals was Cr at temperature from 160 to 220°C, and As and Zn were the significant dominant metals at 240 and 260°C, respectively. Compared to MSW, the concentrations of Cr, Cd, and Hg in the hydrochars were low, and the other metals exhibited a different trend, which was ascribed to differences in the transformation of heavy metals during HTC. In addition, H-260 had the lowest concentrations of Cr, Pb, and Hg, followed by H-180. As for other heavy metals, except for Cu, the lowest concentrations were observed at H-180. This result indicated that the HTC at 180 and 260°C were a promising method to reduce the content of heavy metals in the hydrochar obtained from MSW.

Heavy metals cannot be destroyed or formed and be just transferred from biomass to waste during HTC. In order to determine the transformation of heavy

Metal	Sample							
	Cr	Cd	Pb	Hg	As	Zn	Ni	Cu
H-160	0.21	0.07	0.81	—	0.31	2.93	0.66	0.22
H-180	0.35	0.24	0.06	—	0.09	1.30	1.46	0.13
H-200	0.30	0.25	0.24	—	—	1.70	0.50	0.40
H-220	0.41	0.35	0.35	—	—	3.49	1.99	0.69
H-240	0.31	0.34	0.06	—	—	3.44	1.63	0.45
H-260	0.12	0.32	0.24	—	—	1.37	0.29	0.64
EPA limit	5.00	1.00	5.00	0.20	5.00	25.00	2.00	100.00

—, not detected.

Table 3.
The concentrations (μg/g) of heavy metals in TCLP leachates of the hydrochars.

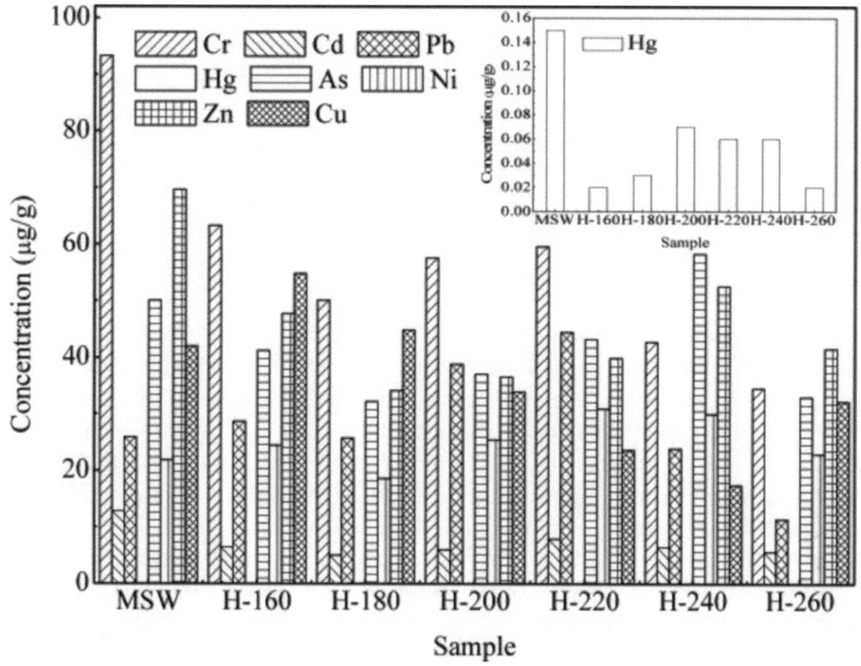

Figure 4.
The concentrations (μg/g) of metals in MSW and corresponding hydrochars.

Metal	Sample					
	H-160	**H-180**	**H-200**	**H-220**	**H-240**	**H-260**
Cr	49.14	30.94	31.74	31.94	21.94	13.96
Cd	36.43	22.46	24.33	30.55	24.06	16.46
Pb	80.33	57.32	78.65	85.05	44.30	16.63
Hg	10.45	9.65	24.54	21.49	18.42	6.52
As	59.67	37.02	38.79	42.78	55.77	24.75
Ni	81.13	49.16	60.89	70.20	65.84	39.61
Zn	49.81	28.23	27.51	28.43	36.16	22.48
Cu	94.57	61.65	42.33	27.90	19.90	28.82

Table 4.
The retention rates (%) of heavy metals in the hydrochars at different hydrothermal temperatures.

metals during HTC, the retention rate was determined. The retention rate of individual metal is defined as the percentages of that metal content in the hydrochar over that in the raw material [20]. **Table 4** illustrated the retention rates of heavy metals in the hydrochars at different temperatures. It is observed that Cu exhibited the highest retention rates in H-160 and H-180 at the percentage of 94.57 and 61.65%, respectively. The highest retention rates in H-200 and H-240 were for Pb. As for H-240 and H-260, significant retention rates of Ni were observed. The lowest retention rates in the samples were for Hg at most temperatures, indicating that compared to other determined heavy metals, Hg in MSW was easily transferred and removed during HTC due to the easy volatility of Hg. With respect to the effect of hydrothermal temperature, a quick decrease of retention rate of Cu were observed at temperature from 160 to 240°C and then slightly increased. As for other heavy metals, the decreases in retention rates were accompanied by an increase in temperature from 160 to 180°C. When the temperature was in the range of 180–220°C, the retention rates increased, while the decreased trend with further increasing temperature was observed except for Hg from 200 to 260°C and Zn and As from 240 to 260°C. This can be explained by the inherent properties and specific chemical forms of heavy metals in MSW.

5. The formation and transformation of PAHs during HTC of MSW

Figure 5 summarized the concentrations of total free PAHs in MSW and corresponding hydrochars. In general, the concentration of PAHs in MSW was 2181.16 μg/kg. The concentrations of PAHs in the hydrochars were in the range of 1298.71–177698.20 μg/kg, which were much higher than that in MSW, especially at high temperature, except for H-160. The experimental results showed that the free PAHs in the hydrochars are not only obtained from MSW, but there also other ways to produce PAHs.

Considering the effect of hydrothermal temperature, the concentrations of PAHs increased with the increase of temperature from 160 to 240°C and conversely decreased when the temperature continued to increase. The competitive decomposition and formation of PAHs during HTC are responsible for the variation of PAH concentrations in the samples. At temperature from 160 to 240°C, the formation of PAHs was the dominant reaction, which promoted the increase of PAHs. The dehydration, decarboxylation, and condensation reactions occurred during the HTC process and became more intensive with the increasing hydrothermal temperature, resulting in the higher degree of coalification and the formation of amorphous hydrochar [56]. The amorphous hydrochar, containing aliphatics, aromatic compounds, and so on, has been proved to cause the increased PAH concentrations [46]. Furthermore, the free radicals like •OH produced from the decomposition of water or the coalification of MSW promote the formation of PAHs [50]. Additionally, the formations of PAHs are endothermic reactions indicating that the formation of PAHs can be promoted by the increasing temperature. In summary, the formation of PAHs was the dominant reaction at hydrothermal temperature from 160 to 240°C. Conversely, the concentrations of PAHs reduced with further increasing temperature due to the breakdown of structure of the hydrochar. As evidenced in **Figure 4**, the amount of microspheres increased with the temperature from 160 to 240°C, indicating that more produced and higher observed. However, the microspheres are damaged at 260°C, showing that the porous structure collapsed. It is reported that the porous structure of the hydrochars played an important role in absorbing the sugars, furfurals, and so on due to the abundant potential active sites, resulting in increased PAH concentrations [57]. As a result, the porous structure of H-260 is damaged leading to the decrease of PAH concentrations. It is interesting to note that the concentrations of metals also have the effect of the formation of PAHs. For example, Zn promoted the formation of PAHs, and Cu was not favorable for the formation of PAHs, which was consistent with the result in the present study [58]. In detail, the highest concentration of PAHs was obtained at 240°C, which the lowest concentration of Cu and the highest concentration of Zn were observed in H-240. It can be seen that PAHs can be transferred and formed during HTC, which is different from heavy metals, and the transformation and the formation of PAHs are affected by the structure and metal concentration.

The concentrations of 16 individual PAHs in MSW and corresponding hydrochars were illustrated in **Table 5**. The results showed that the concentration of BaA in MSW was 938.01 μg/kg at a value of 43.01% of the total PAHs, followed by Phe

(436.41 μg/kg) and Fla (199.40 μg/kg). Meanwhile, Nap, Flu, and Phe in the hydrochars, except for H-260, accounted for the majority of total PAHs. As for H-260, Flu, Phe, and Chr were the dominant individual PAHs at values of 25.50, 27.79, and 18.59%, respectively. It was thought that the molecular sizes of dominant PAHs in the hydrochars were lower than that in MSW.

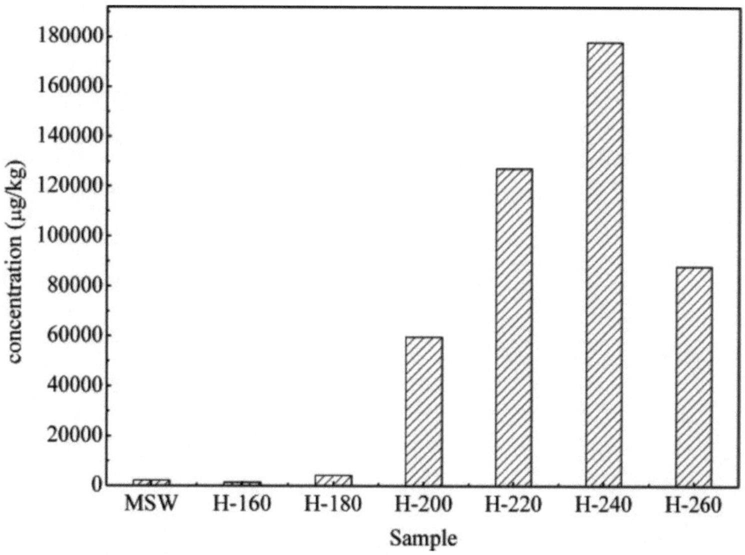

Figure 5.
The concentrations (μg/kg) of total free PAHs in MSW and corresponding hydrochars.

PAHs	Sample						
	MSW	H-160	H-180	H-200	H-220	H-240	H-260
Nap	101.91	319.13	740.34	11561.17	38824.40	46665.05	9504.29
Acy	2.27	1.75	2.45	3.50	43.27	63.24	49.33
Ace	10.36	10.96	30.17	245.37	618.05	815.80	533.56
Flu	118.65	580.92	1373.72	25429.68	38284.80	45850.40	22438.64
Phe	436.41	272.21	1605.30	14813.56	31264.00	45048.85	24455.65
Ant	28.45	11.85	32.62	340.73	1404.53	2646.07	4125.61
Fla	199.40	30.09	116.01	322.66	1024.09	2082.74	1709.28
Pyr	156.49	21.09	83.71	187.06	377.09	794.50	647.95
BaA	938.01	3.40	13.10	634.05	1653.12	3809.17	2133.97
Chr	83.95	16.68	164.94	5834.41	12674.34	25374.60	16356.79
BbF	36.04	2.48	2.93	96.96	453.23	2086.83	2414.22
BkF	19.19	1.14	6.13	67.24	247.10	613.68	984.55
BaP	9.97	2.06	2.06	49.12	114.45	364.21	669.98
IcP	16.07	1.82	2.14	6.63	11.85	150.01	183.71
DaA	10.51	1.39	3.59	55.84	183.85	717.15	975.35
BgP	13.48	1.74	3.09	22.88	155.25	615.88	799.38

Table 5.
The concentration (μg/kg) of 16 individual PAHs in MSW and corresponding hydrochars.

In order to investigate the change of molecular size during HTC process, the 16 individual PAHs are classified by ring number according to **Table 1**. The result was illustrated in **Figure 6**. As can be seen from **Figure 6a**, in MSW, the two-, three-, four-, five- and six-ring PAHs comprised about 4.67, 27.33, 63.17, 3.47, and 1.35% of total PAHs, respectively, indicating that four-ring PAHs were the most prevalent in MSW. As for the hydrochars, the concentrations of three-ring PAHs exhibited a maximum in the range of 53.14–72.79% of the hydrochars obtained at temperature from 160 to 260°C. Additionally, the concentrations of two-, three-, and four-ring PAHs in the hydrochars increased with an increase of hydrothermal temperature, showed a maximum at 240°C, and decreased with further increasing temperature. However, an increase in the concentrations of five- and six-ring PAHs in the hydrochars was accompanied by an increase in the temperature. **Figure 6b** showed the effect of hydrothermal temperature on the molecular weight of PAHs in MSW and

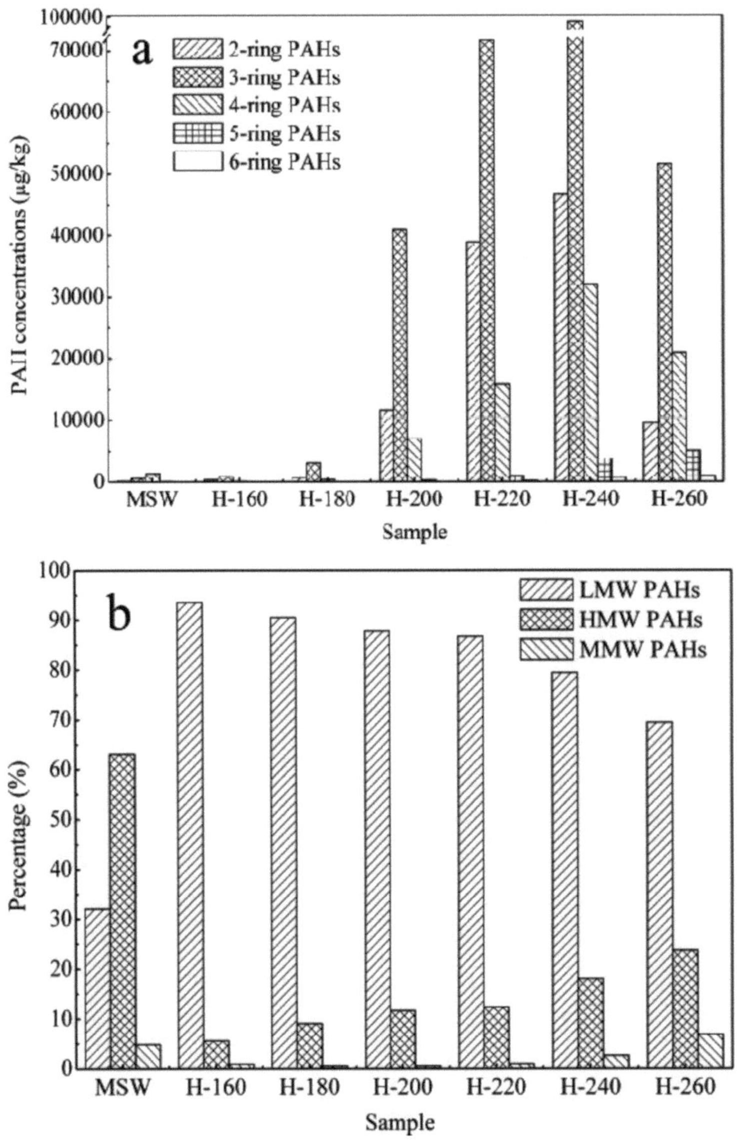

Figure 6.
The ring number (a) and percentages of LMW, MMW, and HMW PAHs (b) in MSW and corresponding hydrochars.

corresponding hydrochars. In detail, PAHs contained low molecular weight PAHs (LMW PAHs, two-and three-ring PAHs), middle molecular weight PAHs (MMW PAHs, four-ring PAHs), and high molecular weight PAHs (HMW PAHs, five- and six-ring PAHs) [39]. Considering the MSW, significant percentage of MMW PAHs was observed, at a value of 63.14% of total PAHs, while the percentages of LMW and HMW PAHs were 32.04% and 4.82%, respectively. As for the hydrochars, the LMW PAHs had maximum percentages of 69.45–93.59% of total PAHs, which reduced with an increase of hydrothermal temperature. Compared to LMW PAHs, however, the MMW PAHs and HMW PAHs (in the range of 5.57–23.69% and 0.47–6.85%, respectively) exhibited a different trend with hydrothermal temperature. The above results confirmed that the dominant PAHs in MSW were higher than that in the hydrochar.

The higher ring number and molecular weight of PAHs are expected to have higher toxicity [58]. In order to analyze the toxicity of PAHs in MSW and corresponding hydrochars, the TEQ value of 16 priority PAHs is used according to the following equation [50]:

$$TEQ = \Sigma(TEF_i \times C_i) \tag{1}$$

where TEF values of the 16 individual PAHs are shown in **Table 1**. C represents the concentration of 16 individual PAHs in the hydrochars obtained at different hydrothermal temperatures. i denotes each 16 individual PAHs.

According to Eq. (1), the TEQ values of MSW and corresponding hydrochars are shown in **Figure 7**. Considering the effect of the hydrothermal temperature (160–260°C), significant increase in TEQ values of the hydrochars from 5.87 μg TEQ/kg to 2489.13 μg TEQ/kg occurred. The TEQ value of MSW (123.7 μg TEQ/kg) was at least eight times higher than that of H-160 and H-180 and lower than that of other hydrochars. It was interesting to note that the total PAHs of MSW was lower and the TEQ was higher than that of H-180. It is indicating that, considering the concentration and toxicity of PAHs, 160 and 180°C are the suitable hydrothermal temperatures to reduce the concentrations and toxicity of PAHs in MSW.

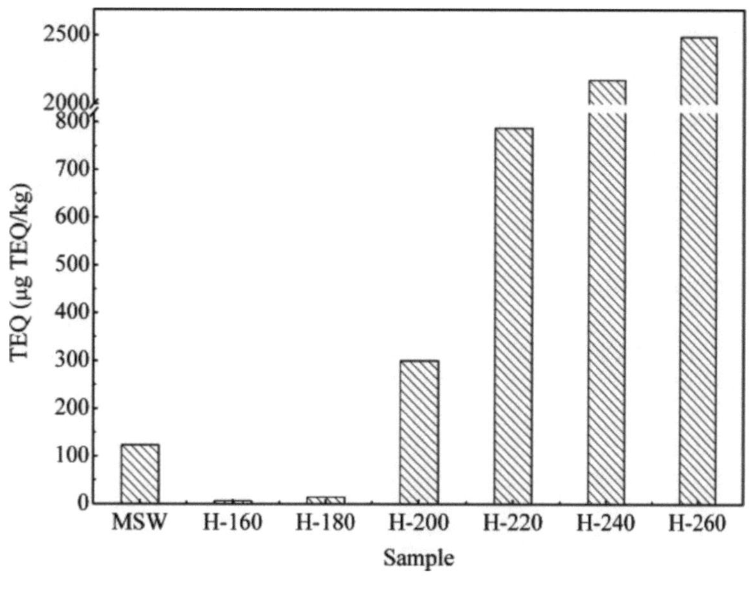

Figure 7.
The TEQ values (μg TEQ/kg) of MSW and corresponding hydrochars.

6. Challenges

Commercialization of HTC of MSW for biofuel production should overcome the following major challenges:

- HTC requires a large heat input for its carbonization reactions and for maintenance of its moderately high reaction temperature. This heat requirement greatly reduces energy conversion efficiency and greatly increases its capital cost.

- The pretreatment and feeding of wet MSW, which is fibrous and widely varying in composition, is another major challenge.

- Separation of product may require the addition of energy consumption, and this can greatly increase the system's cost and reduce its overall energy efficiency.

Further research is required to address this important challenge. A final problem that might inhibit commercialization of HTC of MSW is the corrosion of the reactor wall.

7. Conclusions

In this chapter, the formation and transformation of heavy metals and free PAHs during HTC of MSW were analyzed through determining the concentrations of heavy metals and free PAHs in MSW and corresponding hydrochars. The results showed that with the increasing hydrothermal temperature, the hydrochar yield decreased in the range of 37.68–70.37%. But HTC can homogenize the MSW and increase carbon content and HHV, especially at temperature above 160°C. As for the heavy metal, the concentrations of heavy metals in TCLP leachates of the hydrochars were all lower than the US EPA permissible limits. Compared to MSW, the concentrations of Cr, Cd, Hg, and Zn were low, while Pb, As, Ni, and Cu in the hydrochars exhibited a different trend. The PAH concentrations in MSW was 2181.16 μg/kg, which were much lower than that in corresponding hydrochars, except for H-160. The TEQ values were in the order H-260 > H-240 > H-220 > H-20 0 > MSW > H-180 > H-160. Considering the fuel properties and the contentions of heavy metal and PAHs, the hydrochar obtained at 180°C is a promising alternative solid fuel with high fuel property and environmentally friendly characteristic.

Acknowledgements

The authors are grateful for the financial support from Zhengang Liu from the "100 Talents" Program of the Chinese Academy of Sciences.

Conflict of interests

The authors declare no competing financial interest.

Author details

Wentao Jiao[1,2], Nana Peng[1,3] and Zhengang Liu[1,2*]

1 Research Center for Eco-Environmental Sciences, Chinese Academy of Sciences, Beijing, China

2 University of Chinese Academy of Sciences, Beijing, China

3 College of Environmental Science and Engineering, Beijing Forestry University, Beijing, China

*Address all correspondence to: zgliu@rcees.ac.cn

References

[1] Tozlu A, Özahi E, Abuşoğlu A. Waste to energy technologies for municipal solid waste management in Gaziantep. Renewable and Sustainable Energy Reviews. 2016;**54**:809-815. DOI: 10.1016/j.rser.2015.10.097

[2] Kulkarni B. Environmental sustainability assessment of land disposal of municipal solid waste generated in Indian cities—A review. Environmental Development. 2020;**33**:100490. DOI: 10.1016/j.envdev.2019.100490

[3] Zheng L, Song J, Li C, Gao Y, Geng P, Qu B, et al. Preferential policies promote municipal solid waste (MSW) to energy in China: Current status and prospects. Renewable and Sustainable Energy Reviews. 2014;**36**:135-148. DOI: 10.1016/j.rser.2014.04.049

[4] Chen P, Xie Q, Addy M, Zhou W, Liu Y, Wang Y, et al. Utilization of municipal solid and liquid wastes for bioenergy and bioproducts production. Bioresource Technology. 2016;**215**:163-172. DOI: 10.1016/j.apenergy.2016.06.148

[5] National Bureau of Statistics of the people's Republic of China. Available from: http://www.stats.gov.cn/ztjc/ztsj/hjtjzl/2014/201609/t20160918_1400854.html

[6] Hoornweg D, Bhada-Tata P, Kennedy C. Environment waste production must peak this century. Nature. 2013;**502**:615-617. DOI: 10.1038/502615a

[7] Gupta D, Mahajani SM, Garg A. Effect of hydrothermal carbonization as pretreatment on enrgy recovery from food and paper wastes. Bioresource Technology. 2019;**285**:121329. DOI: 10.1016/j.biortech.2019.121329

[8] Li Y, Zhao X, Li Y, Li X. Waste incineration industry and development policies in China. Waste Management. 2015;**46**:234-241. DOI: 10.1016/j.wasman.2015.08.008

[9] Maqhuzu A, Yoshikawa K, Takahashi F. Stochastic economic analysis of coal-alternative fuel production from municipal solid wastes employing hydrothermal carbonization in Zimbabwe. Sciences of The Total Environment. 2019;**23**:135337. DOI: 10.1016/j.scitotenv.2019.135337

[10] Peng N, Liu Z, Liu T, Gai C. Emissions of polycyclic aromatic hydrocarbons (PAHs) during hydrothermally treated municipal solid waste combustion for energy generation. Applied Energy. 2016;**184**:396-403. DOI: 10.1016/j.apenergy.2016.10.028

[11] Fonts I, Gea G, Azuara M, Ábrego J, Arauzo J. Sewage sludge pyrolysis for liquid production: A review. Renewable and Sustainable Energy Reviews. 2012;**16**:2781-2805. DOI: 10.1016/j.rser.2012.02.070

[12] Yang W, Wang H, Zhang M, Zhu J, Zhou J, Wu S. Fuel properties and combustion kinetics of hydrochar prepared by hydrothermal carbonization of bamboo. Bioresource Technology. 2016;**205**:199-204. DOI: 10.1016/j.biortech.2016.01.068

[13] Basso D, Patuzzi F, Castello D, Baratieri M, Rada EC, Weiss-Hortala E, et al. Agro-industrial waste to solid biofuel through hydrothermal carbonization. Waste Management. 2016;**47**(Part A):114-121. DOI: 10.1016/j.wasman.2015.05.013

[14] Kambo HS, Dutta A. Comparative evaluation of torrefaction and hydrothermal carbonization of lignocellulosic biomass for the production of solid biofuel. Energy Conversion and Management.

2015;**105**:746-755. DOI: 10.1016/j.enconman.2015.08.031

[15] Tekin K, Karagöz S, Bektaş S. A review of hydrothermal biomass processing. Renewable and Sustainable Energy Reviews. 2014;**40**:673-687. DOI: 10.1016/j.rser.2014.07.216

[16] Akiya N, Savage PE. Roles of water for chemical reactions in high-temperature water. ChemInform. 2002;**33**:293-293. DOI: 10.1002/chin.200243293

[17] Benavente V, Calabuig E, Fullana A. Upgrading of moist agro-industrial wastes by hydrothermal carbonization. Journal of Analytical and Applied Pyrolysis. 2015;**113**:89-98. DOI: 10.1016/j.jaap.2014.11.004

[18] Mäkelä M, Benavente V, Fullana A. Hydrothermal carbonization of lignocellulosic biomass: Effect of process conditions on hydrochar properties. Applied Energy. 2015;**155**:576-584. DOI: 10.1016/j.apenergy.2015.06.022

[19] Xiao L, Shi Z, Xu F, Sun R. Hydrothermal carbonization of lignocellulosic biomass. Bioresource Technology. 2012;**118**:619-623. DOI: 10.1016/j.biortech.2012.05.060

[20] Liu Z, Balasubramanian R. Upgrading of waste biomass by hydrothermal carbonization HTC and low temperature pyrolysis LTP: A comparative evaluation. Applied Energy. 2014;**114**:857-864. DOI: 10.1016/j.apenergy.2013.06.027

[21] Berge ND, Ro KS, Mao J, FloraJoseph RV, Chappell MA, Bae S. Hydrothermal carbonization of municipal waste streams. Environmental Science & Technology. 2011;**455**:5696-5703. DOI: 10.1021/es2004528

[22] Funke A, Ziegler F. Hydrothermal carbonization of biomass: A summary and discussion of chemical mechanisms for process engineering. Biofuels, Bioproducts and Biorefining. 2010;**4**:160-177. DOI: 10.1002/bbb.198

[23] Titirici MM, Thomas A, Antonietti M. Back in the black: Hydrothermal carbonization of plant material as an efficient chemical process to treat the CO2 problem? New Journal of Chemistry. 2007;**31**:787-789. DOI: 10.1039/B616045J

[24] Bobleter O. Hydrothermal degradation of polymers derived from plants. Progress in Polymer Science. 1994;**19**:797-841. DOI: 10.1016/0079-6700(94)90033-7

[25] Hoekman SK, Broch A, Robbins C. Hydrothermal carbonization HTC of lignocellulosic biomass. Energy & Fuels. 2011;**25**:1802-1810. DOI: 10.1021/ef101745n

[26] Reza MT, Rottler E, Herklotz L, Wirth B. Hydrothermal carbonization HTC of wheat straw: Influence of feedwater pH prepared by acetic acid and potassium hydroxide. Bioresource Technology. 2015;**182**:336-344. DOI: 10.1016/j.biortech.2015.02.024

[27] Huff MD, Kumar S, Lee JW. Comparative analysis of pinewood, peanut shell, and bamboo biomass derived biochars produced via hydrothermal conversion and pyrolysis. Journal of Environmental Management. 2014;**146**:303-308. DOI: 10.1016/j.jenvman.2014.07.016

[28] Sabio E, Álvarez-Murillo A, Román S, Ledesma B. Conversion of tomato-peel waste into solid fuel by hydrothermal carbonization: Influence of the processing variables. Waste Management. 2016;**47**(Part A):122-132. DOI: 10.1016/j.wasman.2015.04.016

[29] Dai L, Tan F, Wu B, He M, Wang W, Tang X, et al. Immobilization of phosphorus in cow manure during

hydrothermal carbonization. Journal of Environmental Management. 2015;**157**:49-53. DOI: 10.1016/j.jenvman.2015.04.009

[30] Lu Y, Levine RB, Savage PE. Fatty acids for nutraceuticals and biofuels from hydrothermal carbonization of microalgae. Industrial & Engineering Chemistry Research. 2015;**54**:4066-4071. DOI: 10.1021/ie503448u

[31] Burguete P, Corma A, Hitzl M, Modrego R, Ponce E, Renz M. Fuel and chemicals from wet lignocellulosic biomass waste streams by hydrothermal carbonization. Green Chemistry. 2016;**18**:1051-1060. DOI: 10.1039/C5GC02296G

[32] Zuo X, Liu Z, Chen M. Effect of H_2O_2 concentrations on copper removal using the modified hydrothermal biochar. Bioresource Technology. 2016;**207**:262-267. DOI: 10.1016/j.biortech.2016.02.032

[33] Lynam JG, Coronella CJ, Yan W, Reza MT, Vasquez VR. Acetic acid and lithium chloride effects on hydrothermal carbonization of lignocellulosic biomass. Bioresource Technology. 2011;**102**:6192-6199. DOI: 10.1016/j.biortech.2011.02.035

[34] Reza MT, Lynam JG, Uddin MH, Coronella CJ. Hydrothermal carbonization: Fate of inorganics. Biomass and Bioenergy. 2013;**49**:86-94. DOI: 10.1016/j.biombioe.2012.12.004

[35] Peng N, Li Y, Liu T, Lang Q, Gai C, Liu Z. Polycyclic aromatic hydrocarbons and toxic heavy metals in municipal solid waste and corresponding hydrochars. Energy & Fuels. 2017;**31**:1665-1671. DOI: 10.1021/acs.energyfuels.6b02964

[36] Košnář Z, Mercl F, Perná I, Tlustoš P. Investigation of polycyclic aromatic hydrocarbon content in fly ash and bottom ash of biomass incineration plants in relation to the operating temperature and unburned carbon content. Science of the Total Environment. 2016;**563-564**:53-61. DOI: 10.1016/j.scitotenv.2016.04.059

[37] Perera FP, Tang D, Wang S, Vishnevetsky J, Zhang B, Diaz D, et al. Prenatalpolycyclicaromatichydrocarbon PAH exposure and child behavior at age 6-7 years. Environmental Health Perspectives. 2012;**120**:921-926. DOI: 10.1289/ehp.1104315

[38] Alves CA, Vicente AMP, Gomes J, Nunes T, Duarte M, Bandowe BAM. Polycyclic aromatic hydrocarbons PAHs and their derivatives oxygenated-PAHs, nitrated-PAHs and azaarenes in size-fractionated particles emitted in an urban road tunnel. Atmospheric Research. 2016;**180**:128-137. DOI: 10.1016/j.atmosres.2016.05.013

[39] Peng N, Li Y, Liu Z, Liu T, Gai C. Emission, distribution and toxicity of polycyclic aromatic hydrocarbons PAHs during municipal solid waste MSW and coal co-combustion. Science of the Total Environment. 2016;**565**:1201-1207. DOI: 10.1016/j.scitotenv.2016.05.188

[40] Chen Y, Zhi G, Feng Y, Fu J, Feng J, Sheng G, et al. Measurements of emission factors for primary carbonaceous particles from residential raw-coal combustion in China. Geophysical Research Letters. 2006;**33**:1-4. DOI: 10.1029/2006gl026966

[41] Rajput P, Sarin MM, Sharma D, Singh D. Atmospheric polycyclic aromatic hydrocarbons and isomer ratios as tracers of biomass burning emissions in Northern India. Environmental Science and Pollution Research International. 2014;**21**:5724-5729. DOI: 10.1007/s11356-014-2496-5

[42] Zhou H, Wu C, Onwudili JA, Meng A, Zhang Y, Williams PT. Influence of process conditions on the formation

of 2-4 ring polycyclic aromatic hydrocarbons from the pyrolysis of polyvinyl chloride. Fuel Processing Technology. 2016;**144**:299-304. DOI: 10.1016/j.fuproc.2016.01.013

[43] Dong J, Li F, Xie K. Study on the source of polycyclic aromatic hydrocarbons PAHs during coal pyrolysis by PY-GC-MS. Journal of Hazardous Materials. 2012;**243**:80-85. DOI: 10.1016/j.jhazmat.2012.09.073

[44] Verma SK, Masto RE, Gautam S, Choudhury DP, Ram LC, Maiti SK, et al. Investigations on PAHs and trace elements in coal and its combustion residues from a power plant. Fuel. 2015;**162**:138-147. DOI: 10.1016/j.fuel.2015.09.005

[45] Shen C, Tang X, Yao J, Shi D, Fang J, Khan MI, et al. Levels and patterns of polycyclic aromatic hydrocarbons and polychlorinated biphenyls in municipal waste incinerator bottom ash in Zhejiang province. China. Journal of Hazardous Materials. 2010;**179**:197-202. DOI: 10.1016/j.jhazmat.2010.02.079

[46] Keiluweit M, Kleber M, Sparrow MA, Simoneit BRT, Prahl FG. Solvent-extractable polycyclic aromatic hydrocarbons in biochar: Influence of pyrolysis temperature and feedstock. Environmental Science & Technology. 2012;**46**:9333-9341. DOI: 10.1021/es302125k

[47] Etoh J, Kawagoe T, Shimaoka T, Watanabe K. Hydrothermal treatment of MSWI bottom ash forming acid-resistant material. Waste Management. 2009;**29**:1048-1057. DOI: 10.1016/j.wasman.2008.08.002

[48] US EPA. Tests methods for evaluating solid waste, Physical Chemical Methods. SW-846, Method 1311. Washington, DC: US Environmental Protection Agency; 1992. Available from: http://wwwepagov/epaoswer/hazwaste/test/pdfs/1311pdf

[49] Liu S, Tao S, Liu W, Liu Y, Dou H, Zhao J, et al. Atmospheric polycyclic aromatic hydrocarbons in North China: A winter-time study. Environmental Science & Technology. 2007;**41**:8256-8261. DOI: 10.1021/es0716249

[50] Liu S, Wang C, Zhang S, Liang J, Chen F, Zhao K. Formation and distribution of polycyclic aromatic hydrocarbons PAHs derived from coal seam combustion: A case study of the Ulanqab lignite from Inner Mongolia, northern China. International Journal of Coal Geology. 2012;**90-91**:126-134. DOI: 10.1016/j.coal.2011.11.005

[51] Liu Z, Quek A, Hoekman KS, Balasubramanian R. Production of solid biochar fuel from waste biomass by hydrothermal carbonization. Fuel. 2013;**103**:943-949. DOI: 10.1016/j.fuel.2012.07.069

[52] Kim D, Lee K, Park KY. Hydrothermal carbonization of anaerobically digested sludge for solid fuel production and energy recovery. Fuel. 2014;**130**:120-125. DOI: 10.1016/j.fuel.2014.04.030

[53] Liu Z, Balasubramanian R. Hydrothermal carbonization of waste biomass for energy generation. Procedia Environmental Sciences. 2012;**16**:159-166. DOI: 10.1016/j.proenv.2012.10.022

[54] Zhao L, Zhang FS, Chen M, Liu Z, Wu DBJ. Typical pollutants in bottom ashes from a typical medical waste incinerator. Journal of Hazardous Materials. 2010;**173**:181-185. DOI: 10.1016/j.jhazmat.2009.08.066

[55] Skodras G, Grammelis P, Prokopidou M, Kakaras E, Sakellaropoulos G. Chemical, leaching and toxicity characteristics of CFB combustion residues. Fuel.

2009;**88**:1201-1209. DOI: 10.1016/j.fuel.2007.06.009

[56] Keiluweit M, Nico PS, Johnson MG, Kleber M. Dynamic molecular structure of plant biomass-derived black carbon biochar. Environmental Science & Technology. 2010;**44**:1247-1253. DOI: 10.1021/es9031419

[57] Reza MT, Uddin MH, Lynam JG, Coronella CJ. Engineered pellets from dry torrefied and HTC biochar blends. Biomass and Bioenergy. 2014;**63**:229-238. DOI: 10.1016/j.biombioe.2014.01.038

[58] Chen Y, Zhao R, Xue J, Li J. Generation and distribution of PAHs in the process of medical waste incineration. Waste Management. 2013;**33**:1165-1173. DOI: 10.1016/j.wasman.2013.01.011

Chapter 15

Hydro Energy Potential for Electricity Generating on Selected Regions in Turkey

Ibrahim Yuksel, Omer Yuksek and Hasan Arman

Abstract

The aim of this chapter is to provide that hydroelectric energy production in Sakarya and Eastern Black Sea Basins which are selected as pilot region in Turkey. Hydroelectric energy has an important place in energy sector is generated from 26 main Basins in Turkey. Sakarya and Eastern Black Sea Basins are very important water resource in Turkey. In this study, hydro energy potential for electricity generating has been investigated in Turkey and hydro energy potential determined for the selected small streams in the Sakarya and Eastern Black Sea Basins by using hydro energy calculating methods.

Keywords: hydropower, energy potential, electricity generation, Sakarya Basin, Eastern Black Sea Basin, Turkey

1. Introduction

The Energy Market Regulatory Agency (EMRA) has license fee exemption for renewable energy investors and the Turkish Electricity Trading Company, TETAS, can provide buying guarantees to renewable energy, further supporting inward investment.

In Turkey, present and future installed electricity capacity by energy sources is given in the **Table 1**. According to the projection, the share of hydro capacity will be

Energy source	2010	2012	2014	2016	2018	2020
Lignite	8280	8280	8280	8280	8280	8280
Hard coal	560	560	560	560	560	560
Imported coal	1840	3470	4680	4690	4690	4690
Natural gas	14,840	17,560	18,400	18,400	18,400	18,400
Geothermal	90	90	90	90	90	90
Fuel oil	1900	1900	1900	1900	1900	1900
Other	29,382	33,920	35,970	35,980	35,980	35,980
Hydropower	16,393	19,880	19,880	21,094	21,094	21,112
Total (MW)	45,255	53,320	55,370	56,594	56,594	56,612

Table 1.
Present and future installed electricity capacity by energy sources (MW) in Turkey.

Renewable energy sources	2000	2005	2010	2015	2020
Primary energy supply					
Hydropower (ktoe)	2656	4067	4903	7060	9419
Geothermal, solar and wind (ktoe)	978	1683	2896	4242	6397
Biomass and waste (ktoe)	6457	5325	4416	4001	3925
Renewable energy production (ktoe)	10,091	11,074	12,215	15,303	19,741
Share of total domestic production (%)	38	48	33	29	30
Share of TPES (%)	12	12	10	9	9
Generation					
Hydropower (GWh)	30,879	47,287	57,009	82,095	109,524
Geothermal, solar and wind (GWh)	109	490	5274	7020	8766
Renewable energy generation (GWh)	30,988	47,777	62,283	89,115	118,290
Share of total generation (%)	25	29	26	25	25
Total final consumption					
Geothermal, solar and wind (ktoe)	910	1385	2145	3341	5346
Biomass and waste (ktoe)	6457	5325	4416	4001	3925
Renewable total consumption (ktoe)	7367	6710	6561	7342	9271
Share of total final consumption (%)	12	10	7	6	6

Table 2.
Renewable energy supply in Turkey.

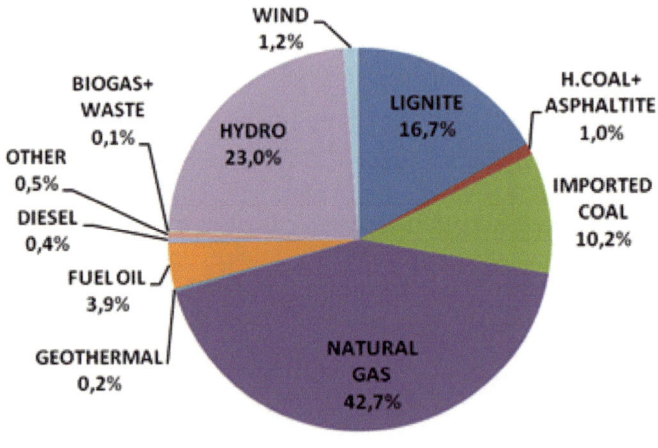

Figure 1.
Share of sources in electricity generation in 2020.

highest with 37.4% and share of natural gas will be 32.6%. Renewable energy supply for at the present and in the future in Turkey is given in the **Table 2** [1–7] and the hare of sources in electricity generation in 2020 is given ın the **Figure 1**.

2. Hydropower capacity and hydropower development in Turkey

In the last decade a lot of dams and small hydropower have been planned to build in Turkey in Eastern Black Sea and Sakarya Basins also. Most of them have been

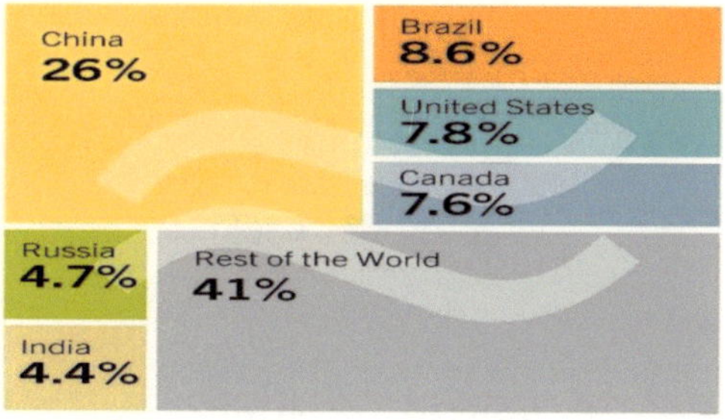

Figure 2.
Hydropower global capacity, shares of top six countries, 2013.

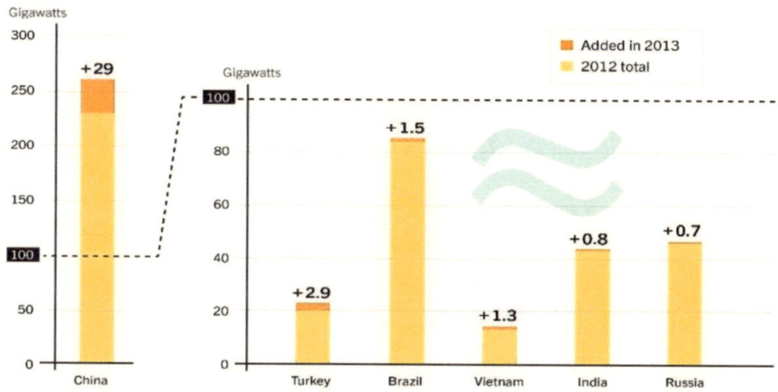

Figure 3.
Hydropower capacity and additions, top six countries for capacity added, 2013.

already completed (for example Deriner Dam 669.60 MV in Artvin) and others under construction (for example Yusufeli Dam 558 MV in Yusufeli) in the eastern Black Sea region. Similarly, Adasu Small Hydropower (10 MV) has been completed and Akçay Dam (12 MV) under construction in Pamukova in Sakarya Basin.

On the other hand, in the last two decades a lot of dams and small hydropower (SHP) either completed or under construction in Turkey. State Hydraulic Works (DSI) provided equipment for the 140 MW Kigi project in Turkey, including three 46.6 MW turbines with pre-turbine gate valves, three generators of the same capacity, switchgear, and all other equipment for the turbine hall [8].

Figures 2 and **3** show hydropower global capacity, shares of top six countries and hydropower capacity and additions, top six countries for capacity added, 2013 respectively [8, 9].

3. Calculation method of hydroelectric potential

There are different methods for calculating hydroelectric energy. Since the amount of energy depends mostly on the water drop and water discharge values it is decided to use the method which is mentioned from some literatures [10, 11]. The method and it equations as follows:

$$N = \gamma * H * Q \tag{1}$$

The equation, $N = \gamma^*H^*Q$, $N = 9.81^*H^*Q^*\,\eta$ ($\eta = 0.85$–0.92) can be written for N and η is 0.85 the N is calculated as:

$$N = 8 * H * Q \tag{2}$$

and then E is calculated as:

$$E = N * 24 * 365 \tag{3}$$

In these equations:

N = Power (tm/s), (1 tm/s= 9.81 kW = 13.3 HP) and it goes to (kW, MV, GW)
γ = Specific weight (t/m^3)
H = Water drop (head of water) (m)
Q =Water discharge (m^3/s)
E = Energy value of water resource (kWh, MWh, GWh)

4. Hydro energy potential for electricity generating on Sakarya and Eastern Black Sea Basins

4.1 Water resources and hydro energy capacity in Sakarya and Eastern Black Sea Basins

Out of 26 main basins, Sakarya and Eastern Black Sea Basins are very important about hydro energy potential in Turkey.

In 2016, Sakarya and Eastern Black Sea Basins have 63.358 km^2 and 22.845 km^2 area and 5.16 km^3 and 16.46 km^3 annual average flow respectively.

In this study, some streams which are observed their water discharge and the others hydraulics and hydrological characteristic were selected in the Sakarya and Eastern Black Sea Basins [12, 13].

When considered the alternative energy resources, hydroelectric generation is not generally achieved in Turkey, especially in Sakarya and Eastern Black Sea Basins. However, with some laws and regulations enacted, in many activities and opportunities such as construction, production and distribution related to investments carried out exclusively by the Turkish Government, but private sectors have not used enough these opportunities yet. Therefore, by using these facilities well, it will be possible for the private sector to close the energy gap in Turkey or to minimize it by evaluating the river resources in the region [14]. In this study, some small streams have been selected and their hydro energy potential determined for Sakarya and Eastern Black Sea Basins in Turkey.

4.2 Hydroelectric values for selected small streams in the Sakarya and Eastern Black Sea Basins

There are three types of potential which are gross (theoretical) potential, technical potential and economical potential [12, 15, 16].

Economical potential in generally depends on some geological and technical conditions such as geographical structure of basin, technical team and equipment facilities etc. To calculate of hydropower potential for selected streams in the Sakarya and Eastern Black Sea Basins the equations 1, 2 and 3 are used.

Hydroelectric Power (N) and hydroelectric energy (E) have been calculated by using this method. Hydroelectric power and hydroelectric energy values for selected streams in the Sakarya and Eastern Black Sea Basins are given in the **Tables 3** and **4** respectively.

Stream and station name	Water discharge (Q), m^3/s	Water drop (head of water) (H), m	Hydroelectric power (kW)	Hydroelectric energy (GWh)
Sakarya River - Dogancay	121	41	4217	364
Kocasu - Rustumkoy	17.9	198	3013	260
Sakarya River - Aktas	7.19	837	5115	442
Aladağ Stream - Karakoy	13.3	505	5709	493
Mudurnu Stream - Dokurcun	7.75	286	1884	163
Total			19,938	1722

Table 3.
Hydroelectric power and hydroelectric values for selected streams in the Sakarya Basin.

Name of selected small stream	Average water discharge (m^3/s)	Average water drop (altitude) (m)	Hydroelectric power (kW)	Hydroelectric energy (GWh)
Iskefiye Stream	1.3	535	5.564	48.74
Sana Stream	1.5	455	5.460	47.83
Orta Stream	1.05	285	2.394	20.97
Horyan Stream	1.12	230	2.061	18.05
Canakci Stream	2.36	405	7.646	66.98
Yanbolu Stream	4.90	60	2.352	20.60
Kalyan Stream	2.45	135	2.646	23.18
Fol Stream	2.91	155	3.608	31.61
Kucuk Stream	2.34	110	2.059	18.04
Durana Stream	1.71	90	1.231	10.79
Kadiralak Stream	2.90	135	3.132	27.44
Manahoz Stream	4.98	65	2.590	22.68
Baltaci Stream.	5.72	75	3.432	30.06
Altıntas Stream	4.11	95	3.124	27.36
Acisu Stream	1.56	155	1.934	16.95
Akhisar Stream	1.60	105	1.344	11.77
Solakli Stream	2.13	105	1.789	15.67
Holo Stream	2.65	95	2.014	17.64
Askaroz Stream	6.26	105	5.258	46.06
Kokasor Stream	4.03	105	3.385	29.65
Pilahoz Stream	2.50	95	1.900	16.64
Potomya Stream	0.84	145	974	8.54
Hako Stream	0.83	105	697	6.11
Hongra Stream.	0.36	105	302	2.65
Buyuk Stream	1.56	105	1.310	11.48
Vanazit Stream.	2.56	70	1.434	12.56

Name of selected small stream	Average water discharge (m^3/s)	Average water drop (altitude) (m)	Hydroelectric power (kW)	Hydroelectric energy (GWh)
Zekere Stream	1.85	105	1.554	13.61
Kizilev Stream	6.34	105	5.326	46.65
Koyunham Stream	1.47	105	1.235	10.82
Baltama Stream	1.76	105	1.478	12.95
Tokmadin Stream	1.28	245	2.509	21.98
Bal Stream	0.67	155	831	7.28
Catalcam Stream	0.89	125	890	7.80
Kirikli Stream	0.80	95	6.08	5.33
Yagli Stream	13.01	55	5.724	50.15
Total			89.798	786.61

Table 4.
Hydroelectric power and hydroelectric values for selected streams in the Eastern Black Sea Basin (some data are from [17]).

5. Conclusion

In this study shows that Turkey has abundant hydropower energy sources and great renewable energy potential and is keen to reduce its dependence on fossil fuels by increasing its use of renewable energy resources. The hydropower sector is a good example of the increasing interest in generation of electricity with renewable resources. It is expected that the Government will call for tenders for hydropower licenses.

It is well known that the environmental impacts of hydro plants are minimal compared with alternative resources. They make use of our renewable "green energy" resource, without causing pollution. It is not enough for governments to support the development of renewable energy technologies. They must also support their commercial application in the country.

On the other hand, hydroelectric facilities are very valuable resources in Sakarya and Eastern Black Sea Basins, which has a great potential due to meteorological, topographic, hydraulics and hydrological conditions. Using this potential, the economic situation of people most of whom are unemployed and poor and live rural areas will be significantly improved.

Since small hydropower plants are river type hydropower plants, small hydropower potential should be used in the country especially in the Sakarya and Eastern Black Sea Basins. Because river type power plants are superior than the other plants, this type plants have not only cost and operational features but also good environmental effects.

List of symbols and abbreviations

E	energy value (kgm)
N	power (kgm/s)
γ	specific weight (kg/m^3)
Q	water discharge (m^3/s)
H	head of water drop (m)
η	efficiency factor of turbine (%)

S	statistic
V	variance
t_i	the number of ties of length (m).
Z	the statistic S is then standardized
d	difference between ranks
n	number of data
R	rank value
S	Sen's slope
x	denotes the variable
i and j	indices
b	median S_k
TWh	terra watt hour
GWh	giga watt hour
MWh	mega watt hour
kW	kilo watt

Author details

Ibrahim Yuksel[1*], Omer Yuksek[2] and Hasan Arman[3]

1 Engineering and Architecture Faculty, Department of Civil Engineering, Istanbul Rumeli University, Istanbul, Turkey

2 Engineering Faculty, Civil Engineering Department, Karadeniz Technical University, Trabzon, Turkey

3 College of Science, Geology Department, United Arab Emirates University, Al-Ain, UAE

*Address all correspondence to: yukseli2000@yahoo.com

References

[1] MENR, Ministry of Energy and Natural Resources. Energy Statistics of Turkey in 2010. Available from: http://www.enerji.gov.tr [Accessed: 22 October 2016]

[2] IEA, International Energy Agency. Energy Policies of IEA Countries: Turkey 2005 Review. Paris: OECD/IEA; 2005

[3] Yuksel I. Renewable energy status of electricity generation and future prospect hydropower in Turkey. Journal of Renewable Energy. 2013;**50**(1):1037-1043

[4] TEIAS, Turkish Electricity Transmission Company. Turkish electricity statistics in 2010. Ankara, Turkey: TEIAS; 2010. Available from: www.teias.gov.tr

[5] EIE, Electrical Power Resources Survey and Development Administration. Hydroelectric Power Plant Projects Carried Out by EIE, 2010. Ankara, Turkey; 2011

[6] WECTNC, World Energy Council Turkish National Committee. Energy Situation of Turkey in 2010. Ankara, Turkey: WECTNC; 2010

[7] Kaygusuz K. Wind energy status in renewable electrical energy production in Turkey. Renewable and Sustainable Energy Reviews. 2010;**14**:2104-2112

[8] By David Appleyard with additional reporting by Bethany Duarte; Hydro Industry FAQS and HRW-Hydro Review Worldwide, "Reaching for Turkey's Hydropower summit". Available from: http://www.hydroworld.com

[9] REN21, Renewable Energy Policy Network for the 21st Century. Renewables Global Status Report, Paris. 2014. Available from: http://www.ren21.net/

[10] Akdoğar M. Energy resources and study of Eastern Black Sea hydropower potential balance [M.Sc. Thesis]. Tarabzon, Turkey: Karadeniz Technical University, Graduate School of Natural and Applied Sciences; 2006

[11] Agıralioglu N, Erkek C. Water Resources Engineering. Istanbul, Turkey: Beta Publisher (in Turkish); 1993

[12] Yuksel I, Demirel İH. Determınatıon of hydroelectric potential in Sakarya in Turkey by using trend analysis. Fresenius Environmental Bulletin. 2018;**27**(11):7257-7264

[13] EIE, Electrical Power Resources Survey and Development Administration. Between 1935 and 2005 Monthly Water Flow Averages. Ankara, Turkey; 2008

[14] Yuksel I. Investigation of Hydraulic and Hydrological Properties of River for Designing of Small and Medium Fall Hydroelectric Power Plants in Sakarya Basin. Sakarya, Turkey: Sakarya University, Scientific Research Coordinator. Final Report; 2007

[15] ESHA, European Small Hydropower Association. 2005. Small hydropower for developing countries. Available from: http://www.esha.org/ [Accessed: 06 March 2016]

[16] Yuksel I. Development of hydropower: A case study in developing countries. Energy Sources, Part B: Economics, Planning, and Policy. 2007;**2**(2):113-121

[17] Serencam U. The analysis of the hydropower potential of small streams in the Eastern Black Sea Region [M.Sc. Thesis]. Sakarya, Turkey: Graduate School of Natural and Applied Sciences, Sakarya University; 2007

Section 5

Geothermal Energy

Chapter 16

An Overview of Geothermal Energy Production in Germencik, Turkey

Kaan Yamanturk and Cihan Dogruoz

Abstract

As it is known, the utilization and production of renewable energy resources are very important in recent years. Due to its geological structural formations, Turkey has a serious geothermal energy potential as a renewable energy resource comparing with the other countries. West side of Turkey has also a critical role to use the geothermal energy resources. In these fields, geothermal is mostly used in electricity generation, greenhouse heating and locational requirements. The components while producing the geothermal water from wells such as heating pumps, re-injection pipes and other equipment are also significant. In this study, coefficient of performance (COP) utilizing in heat pumps has been investigated and the new approach to find out the parameter has been identified. Based on COP equation, the formula of COP has been re-coded on Dev C++ compiler by using C++ computer language in order to focus on the importance of computer aided applications in geothermal energy sector. There are no more studies showing the COP with C++ codes in literature. On the other hand, Germencik region, in the west side of Turkey, has been evaluated and the production processes by Guris Construction and Engineering Co. Inc. have been explained in the study. Moreover, the potential of Turkey has also been mentioned in this study. The aim of the study is to examine the Germencik region geothermal energy potential and to improve the coefficient of performance by using C++ in heat pumps. The result of this study shows us the Germencik region has an important potential and the computer aided technologies can also be adapted easily into the processes while producing geothermal energy.

Keywords: geothermal energy, renewable energy, coefficient of performance, computational aided, heat pump

1. Introduction

The basis of European Union policy on renewable energy was made in 1997 when the share of renewable energy was 6% of gross internal energy consumption in European Union (EU). The share of renewable energy could reach between 55% and 75% of gross final energy consumption in EU in 2050. This means, the utilization and requirement of renewable energy resources are indispensable [1, 2]. The utilization of renewable energy has come to a share of 17% in gross final energy consumption in 2015 in EU and projections reveal that it will exceed the target of 20% in 2021. The production of renewable electricity in EU increased almost 75% since 2005 to 927 TWh in 2015 and it seems to be likely to lead to 1210 TWh renewable electricity in 2021 [3].

Utilization	Capacity
Total Electricity Production	1527 MWe
Geothermal District Heating	116,000 Residences Equivalence (1033 MWt)
Greenhouse Heating	4,3 Million m^2 (820 MWt)
Heating of Thermal Facilities	46,400 Residences Equivalence (420 MWt)
Het Energy of Thermal Water Use in Hotels, Spas and Time Share Facilities	450 Geothermal Spa (1205 MWt) (20 Million guests/annual)
Agricultural Drying	1,5 MWt
Geothermal Cooling	0,1 MWe (0,35 MWt)
Heat Pumps; GSHP	120 MWt; 8,5 MWt
Total Heat Use	3488 MWt (336,000 Residences Equivalence)
Carbondioxide Production	400,000 tonnes/year

Table 1.
Geothermal utilization capacities in Turkey [10].

As one of the renewable energy resource, Turkey has a serious geothermal energy potential comparing with the other countries. West side of Turkey has also a critical role to use the geothermal energy resources. In these fields, geothermal is mostly used in electricity generation, greenhouse heating and locational requirements. The components while producing the geothermal water from wells such as heating pumps, re-pipes and other equipment are also significant As a result of growth in industry and world's population, the demand of energy has increased and fossil fuel resources have been reducing in the World. Furthermore, fossil fuels are not preferred by most people due to its environmental effects. For that reason, there is an urgent need to produce and utilize sustainable and environmentally clean energy sources such as geothermal energy.

As it is known, geothermal energy as a renewable energy resource can play an important role in human life. Geothermal Power Plants work with over % 90 capacity factor, therefore it acts like a "Baseload power plant" [4]. There are many studies including geothermal energy aspects in literature [5, 6]. In Turkey, there is a great potential of geothermal energy. The main utilizes of geothermal energy in Turkey are electricity generation, domestic hot water supply and indoor space heating, greenhouse heating, heat pumps, CO2 and dry –ice production processes. According to Bejan et al., conservation of mass and energy principles have been utilized by exergy method together with the second law of thermodynamics [7]. Quantity and quality of heat losses and location of energy degradation can be indicated by exergy analysis [8]. Moreover, concept of exergy analysis for evaluation of geothermal power plants have been used by Bodvarsson and Eggers in 1972. Exergy of saturated water for sink conditions of different temperatures have been tabulated by them [9].

In this study, some of the potential geothermal energy resources of Turkey have been examined and the computer applications of geothermal energy in heat pumps have been carried out using coefficient of performance parameter. **Table 1** shows the geothermal utilization capacities in Turkey according to Mertoglu et al. [10]. Furthermore, the renewable energy projection of Turkey has also been investigated in the study.

2. The potential of geothermal energy in Turkey

In Turkey, Geothermal and hydro energy resources had high share of 22.10% in all electricity production in 2018. In 2005, renewable law provided supporting mechanism for buying electricity from geothermal [11]. By Pfister in 1995, Bursa Kukurtlu thermal water was studied and the result of age range was confirmed independently based on the results of geothermal modeling predicting residence times of the order of 10^4 to 10^5 years [12]. Lake sediments can be deposited within waters which comprised sources from geothermally heated groundwater such as Ulubey formation in Usak basin, Turkey [13]. Another example of geothermal heated energy resource is Denizli basin situated in western Turkey with fissure ridge travertines [14]. Sıcak Çermik located in Sivas city is also known as geothermal region 25 km west of the city center where hot springs permeate Upper Miocene-Lower Pliocene sediments. Piper and others have studied the fissure fill travertine from the geothermal field in central Turkey with several objectives [15]. A thermodynamic model and energy and exergy balance equations have been determined by Ganjehsarabi et al. in Dora II geothermal power plant which is located in Aydın city, Turkey [16].

There are several investigations in Denizli-Kızıldere field based on geothermal brine. One of these studies include the silica removal from simulated solutions of geothermal brine in Kızıldere, Denizli [17]. Moreover, **Figure 1** shows 500 m and 1000 m depth temperature distribution maps of geothermal energy in Turkey by Basel et al. [18, 19].

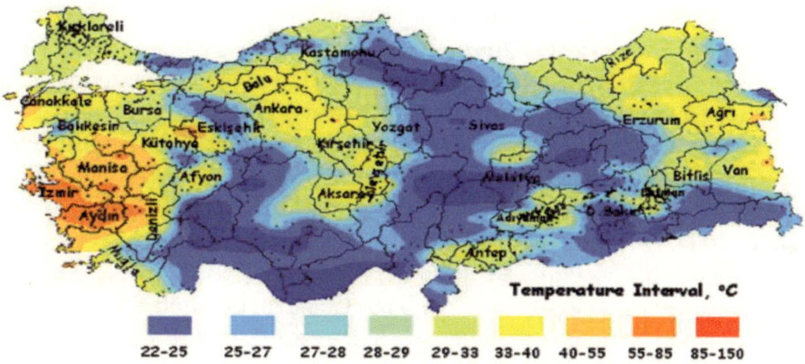

(a)

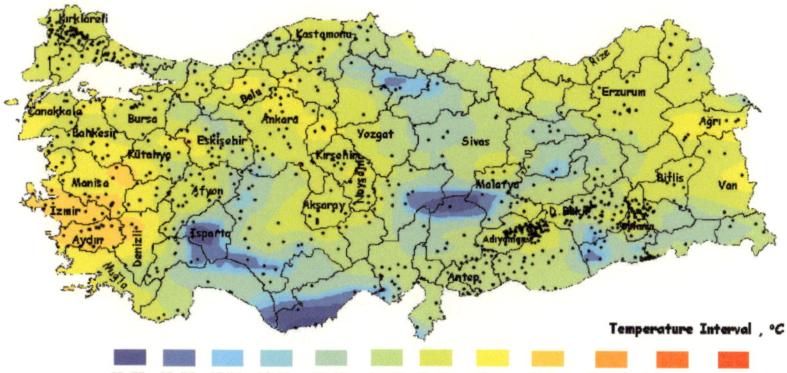

(b)

Figure 1.
Depth temperature distribution map of 500 m (a) and 1000 m (b) of geothermal energy resources in Turkey [18, 19].

Kamara geothermal area has also potential which is located West side of Denizli basin. Brogi and his friends have studied travertine deposition and hydrothermal fluids circulation in Kamara location [20]. Marmara sea has also potential of geothermal energy resources in Turkey. According to Pfister et al., the geothermal area in Marmara sea can be carried out by local zones of vertical hydraulic permeability because of active transtensional faulting of the crust [21].

3. Geothermal energy in Guris Germencik Omerbeyli field

The Germencik geothermal field is located at the western side of the tectonic valley known as Buyuk menderes Graben in SW Turkey. The reservoir is water dominated with high reservoir temperature up to 276°C. The geothermal reservoir is fault controlled hydrothermal geothermal system. The first geothermal well (OB-1) was drilled at 1982 by Mineral Research and Exploration General Directorate (MTA) and discovered high temperature reservoir in the region.

In 2003, Guris Construction and Engineering Co. Inc. engineers started to examine the field studies and carried out good results in-situ. After ending field working in 2005, already drilled geothermal production wells have been evaluated for generating electricity in Germencik field. Nowadays, 106 wells drilled at license area. There are totally 75 geothermal wells belonging to Guris Construction and Engineering Co. Inc., including 39 production and 36 re-injection wells which are actively used to generate the electricity in Germencik field. Each wells are varying from 865 m to 3500 m in depth and the casing diameters are changing from 7 inch to 20 inches. In Germencik field, each well life is approximately between 30 and 40 years. Re-injection fluid temperature varies from 85–105°C in the field. According to the calculations, 1 MWe production includes 50 tones geothermal fluid in wells. Each production well, wellhead pressure is changing from 18 atm to 25 atm in the field. Distances between the production and re-injection wells are around 500 meters in Germencik. Moreover, there is a North/South and East/West slip fault zone in Germencik region. The Guris Construction and Engineering Co. Inc. area in Germencik field is 3530 hectare license area and the geothermal power plant installed capacity is 209,9 MW in this area as shown in **Table 2**.

Power plant name	CDO	Plant type	MWe
Galip Hoca	April 2, 2009	Double Flash	47,4
Efe-2	October 1, 2014	ORC (Binary)	22,5
Efe-3	March 6, 2015	ORC (Binary)	22,5
Efe-4	July 3, 2015	ORC (Binary)	22,5
Efe-1	August 26, 2015	Double Flash	47,4
Efe-6	August 20, 2017	ORC (Binary)	22,6
Efe-7	October 12, 2018	ORC (Binary)	25

Table 2.
Guris Germencik power plants [by Guris construction and engineering Co. Inc].

4. Heat pumps and computational applications of COP

There is a heat source and a cold sink for heat pumps while utilizing the geothermal energy resources. The refrigerant flows through the pipes under the ground while the system is working. The heat from the water raises the temperature of fluid flowing in the pipes which then returns to the power plant. Based on the system, the fluid gets vaporized to turn a turbine or advances directly into cooling or heating household by condenser.

In heat pumps, the energy performance is translated by the (COP) coefficient of performance which is expressed by the formula;

$$\mathrm{COP} = \mathrm{T_C} / (\mathrm{T_H} - \mathrm{T_C}) \tag{1}$$

Where T_H is the temperature of hot source in Kelvin and T_C is the temperature of cold source in Kelvin. The coefficient of performance (COP) of a heat engine can be clarified as the ratio between the production of amount of heat energy and energy consumption for fulfilling the energy transfer. The equation of finding COP can also be determined by using C++ coding application utilizing C++ compiler based on the source coding as shown below;

```
#include < iostream>
using namespace std;
int main()
{
float Tc,Th,COP;
cout < <"Temperature of cold source in Kelvin: ";
cin> > Tc;
cout < <"Temperature of hot source in Kelvin: ";
cin> > Th;
COP = (Tc)/(Th-Tc);
cout < <"Coefficient of Performance: "<<COP;
}
```

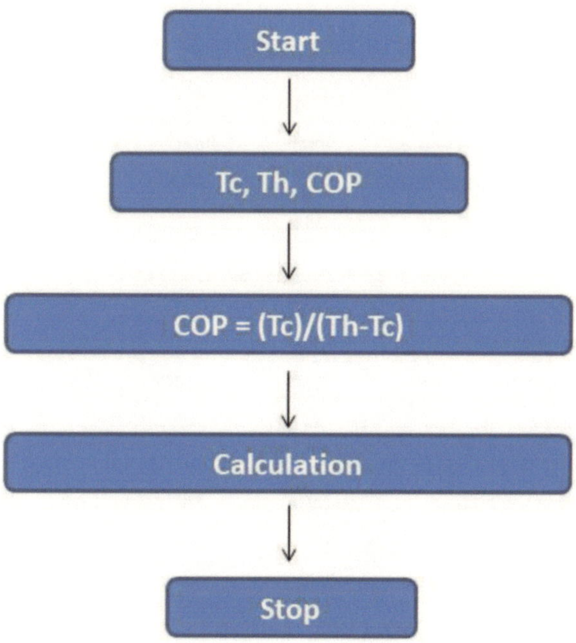

Figure 2.
Algorithm of C++ application of COP equation.

The C++ coding is the open source codes which can be used in available compilers such as Dev C++. **Figure 2** shows the algorithm of COP equation which can be used to find out the coefficient of performance. In this study, the application of C++ is created based on the formula of coefficient of performance (COP) by using Dev C++ compiler. The main purpose of this application is to reveal that the computer aided applications are making the processes easier and they are also used in geothermal energy resources software programs.

5. Renewable energy projections in Turkey

In Turkey, renewable energy supply is dominated by hydropower and biomass, however scarcity of supply and environmental effects have led to a decline in biomass use mainly for residential heating. Renewable energy supply in total declined from 1990 to 2004 because of a decrease in biomass supply.

While investigating the projection of geothermal, solar and wind energies, the number increases from 6397 ktoe to 10,526 ktoe as primary energy supply between 2020 and 2030. Moreover, it becomes from 8766 GWh to 11,686 GWh between 2020 and 2030. As total final consumption, the number increases from 5346 ktoe to 9513 ktoe between 2020 and 2030. It means, geothermal energy utilization is increasing every year in all over the World.

In recent years, the composition of renewable energy supply has varied and wind power is beginning to dominate market share. On the other hand, geothermal energy supply is also increasing due to its utilization in market. **Table 3** shows renewable energy supply and **Table 4** shows renewable energy projections for future generations in Turkey, respectively [22, 23].

Renewable energy resources	1990	1995	2000	2002	2005
Primary Energy Supply					
Hydropower (ktoe)	1991	3057	2656	2897	4067
Geothermal, solar and wind (ktoe)	461	654	978	1142	1683
Biomass and waste (ktoe)	7208	7068	6457	5974	5325
Renewable energy production (ktoe)	9660	10,779	10,091	10,013	11,074
Share of total domestic production (%)	38	40	38	40	48
Share of TPES (%)	18	17	12	13	12
Generation					
Hydropower (GWh)	23,148	35,541	30,879	33,684	47,287
Geothermal, solar and wind (GWh)	80	86	109	153	490
Renewable energy generation (GWh)	23,228	35,627	30,988	33,837	47,777
Share of total generation (%)	40	41	25	26	29
Total Final Consumption					
Geothermal, solar and wind (ktoe)	392	580	910	1048	1385
Biomass and waste (ktoe)	7208	7068	6457	5974	5325
Renewable total consumption (ktoe)	7600	7648	7367	7022	6710
Share of total final consumption (%)	18	15	12	12	10

Table 3.
Renewable energy supply in Turkey [22, 23].

Renewable energy resources	2010	2015	2020	2025	2030
Primary Energy Supply					
Hydropower (ktoe)	4903	7060	9419	11,214	14,214
Geothermal, solar and wind (ktoe)	2896	4242	6397	8426	10,526
Biomass and waste (ktoe)	4416	4001	3925	3365	5665
Renewable energy production (ktoe)	12,215	15,303	19,741	21,342	24,343
Share of total domestic production (%)	33	29	30	28	26
Share of TPES (%)	10	9	9	8	7
Generation					
Hydropower (GWh)	57,009	82,095	109,524	129,876	150,876
Geothermal, solar and wind (GWh)	5274	7020	8766	9786	11,686
Renewable energy generation (GWh)	62,283	89,115	118,290	135,678	165,678
Share of total generation (%)	26	25	25	24	24
Total Final Consumption					
Geothermal, solar and wind (ktoe)	2145	3341	5346	7413	9513
Biomass and waste (ktoe)	4416	4001	3925	3246	3646
Renewable total consumption (ktoe)	6561	7342	9271	10,786	11,786
Share of total final consumption (%)	7	6	6	5	5

Table 4.
Renewable energy projections in Turkey [22, 23].

6. Conclusions

Geothermal energy resource as renewable energy is increasing at all over the world in recent years. As it is known, geothermal energy is clean, renewable and sustainable energy resource and also can be utilized everywhere in the world. The projection of renewable energy also shows the importance of geothermal utilization in the future (**Table 4**).

Some components utilizing to produce the energy resource from underground are very significant such as heat pumps. Heat pumps are significant in geothermal energy production due to the heat transformation. The heat is used by geothermal energy from the earth to deliver heat and power. For that reason, coefficient of performance in heat pumps is also important. The computer aided applications in geothermal energy sector are commonly utilized. Different softwares are adapted into the systems. In this study, coefficient of performance (COP) equation has been investigated and the equation of COP has been modified by using C++ computer language coding. One of the main purpose is to reveal the capability of modifying the equations from numerical to computerized applications.

The potential of geothermal energy resources in Turkey have also been studied in this study. The results show that there is a great potential of geothermal energy sources especially west side of Turkey. The Germencik region was evaluated and the status of geothermal energy in the field constructed and operated by Guris Construction and Engineering Co. Inc. has been investigated in this study. Monitoring of the wells is showing high performance in a very good environmentally and healthy conditions. All operations carried by Guris Construction and Engineering Co. Inc. are in high standards which resulted in elimination of any environmentally harmful emissions and wastes. Geothermal power plants with high quality engineering perform with high capacity factor, including coefficient of performance of heat pumps, can act like a base load power plant. Geothermal power plants of Guris Construction and Engineering Co. Inc. have the highest capacity factor proves the study. In conclusion, where the resource is suitable, Geothermal Energy Power Plants with high capacity factors are necessary to be replaced with the fossil typed Thermal Power Plants.

Acknowledgements

The study was supported by Guris Construction and Engineering Co. Inc., Turkey. The support is greatly acknowledged.

Author details

Kaan Yamanturk[1] and Cihan Dogruoz[2*]

1 Guris Construction and Engineering Co. Inc., Golbasi, Ankara, 06830, Turkey

2 Mining Engineering Department, Dumlupinar University, Kutahya, 43100, Turkey

*Address all correspondence to: cihan.dogruoz@dpu.edu.tr

References

[1] COM (97) 599 Final, Communication from the Commission: Energy for the Future: Renewable Sources of EnergyeWhite Paper for a Community Strategy and Action Plan, European Commission, 1997.

[2] COM(2011) 885 final, Energy Roadmap 2050 Communication from the Commission to the European Parliament, the Council, the European Economic and Social Committee and the Committee of the Regions, 2011.

[3] Scarlat, N., Dallemand, J.F. and Fahl, F., 2018, "Biogas: Developments and perspectives in Europe", Renewable Energy, 457-472

[4] IRENA (International Renewable Energy Agency), 2017, "Renewable Cost Database", http:// costing.irena.org/ irena-costing.aspx.

[5] Hettiarachchi HDM, Golubovic M, Worek WM, Ikegami Y., 2007, "Optimum designcriteria for a n Organic Rankine cycle using low-temperature geothermal heat sources". Energy;32(9):1698e706.

[6] Shi Hua, Michaelides Efstathios E., 2007, "Binary dual-flashing geothermal power Plants". International Journal of Energy Research;13:127e35.

[7] Bejan A, Tsatsaronis G, Moran M., 1996, "Thermal design and optimization". New York: Wiley;

[8] Rosen MA., 2002, "Does industry embrace exergy?" International Journal of Exergy;2 4:221e3.

[9] Bodvarsson G and Eggers DE., 1972, "The exergy of thermal power". Geothermics;1:93e5.

[10] Mertoglu, O., Simsek, S., Basarir, N., 2020, "Geothermal Energy Use-Projection, Country Update for Turkey", Proceedings World Geothermal Congress 2020.

[11] TSI, Turkish Statistical Institute, 2020, "National Inventory Report for submission under the United Nations Framework Convention on Climate Change", Turkish Greenhouse Gas Inventory 1990-2018

[12] Pfister, M., 1995, "Geothermische Untersuchungen in der Region Marmara". Diss. ETH-Z Nr. 11054, 135.

[13] Maddy, D., Veldkamp, A., Demir, T., Aytaç, A.S., Schoorl, J.M., Scaife, R., Boomer, R., Stemerdink, C., van der Schriek, T., Aksay, S., Lievens, C.,2020, "Early Pleistocene River Terraces of the Gediz River, Turkey: The role of faulting, fracturing, volcanism and travertines in their genesis", Geomorphology, pp.107-102

[14] Westaway, R., 1990, "Block rotation in western Turkey. 1. Observational evidence". Journal of Geophysical Research 95, 19857-19884.

[15] Piper, J.D.A., Mesci, L., Gursoy, H., Tatar, O., Davies, C.J., 2007, Palaeomagnetic and rock magnetic properties of travertine: Its potential as a recorder of geomagnetic palaeosecular variation, environmental change and earthquake activity in the Sıcak C̦ ermik geothermal field, Turkey, Physics of the Earth and Planetary Interiors 161 (2007) 50-73

[16] Ganjehsarabi, H., Gungor, A., Dincer, İ., 2012, "Exergetic performance analysis of Dora II geothermal power plant in Turkey", Energy 46 (2012) 101e108

[17] Badruk, M. And Matsunaga, I., 2001, "Experimental results of silica removal from simulated solutions of geothermal brine of Kizildere field, Turkey", Geothermics 30 (2001) 561-570

[18] Basel, D.K., Serpen, U., Satman, A., 2010, "Turkey's Geothermal Energy

Potential: Updated Results", Thirty-Fifth Workshop on Geothermal Reservoir Engineering Stanford University, Stanford, California,

[19] Serpen, U., N. Aksoy, T. Onur and E.D. Korkmaz, 2009, "Geothermal Energy in Turkey: 2008 Update" Geothermics, V. 38, pp. 227_237.

[20] Brogi, A., Alcicek, M.C., Yalciner, C.C., Capezzuoli, E., Liotta, D., Meccheri, M., Rimondi, V., Ruggieri, G., Gandin, A., Boschi, C., Buyuksarac, A., Alcicek, H., Bulbul, A., Baykara, M.O., Shen, C.C., 2016, "Hydrothermal fluids circulation and travertine deposition in an active tectonic setting: Insights from the Kamara geothermal area (western Anatolia, Turkey)", Tectonophysics 680 (2016) 211-232

[21] Pfister, M., Rybach, L., Simsek, S., 1998, "Geothermal reconnaissance of the Marmara Sea region (NW Turkey): surface heat flow density in an area of active continental extension", Tectonophysics 291, 77-89

[22] Ministry of Energy and Natural Resources (MENR). Energy Statistics in Turkey. available from http://www.enerji.gov.tr (accessed date 26 May 2007).

[23] International Energy Agency (IEA). Energy Policies of IEA Countries: Turkey,2005 Review, OECD/IEA, Paris, 2005.

Section 6

Other Applications of Renewable Energy System

Chapter 17

Design of Three-Term Controller Using a PIC18F452 Microcontroller

Mostefa Ghassoul

Abstract

Microcontrollers are used in almost any applications that come across one's mind, from small control applications such as home appliances to aerospace. Microcontroller-based controllers are cost-effective and flexible to modify the design to meet the requirement for any control of any industrial plant. Microcontrollers do not require external hardware interface, memories, counter/timers, and ADCs, because they are all integrated inside the chip. Those controllers could be programmed online and do not require any backup memories except for big applications. This chapter presents the implementation of the three-term PID controller using a Microchip PIC18F452 microcontroller. To read data into the controller, a 10-bit integrated ADC is used; and to read data out of the machine, an external 12-bit serial DAC is used. Before programming the PIC microcontroller, the task to be tested could be off-line using a software simulator to make sure that it is working according. When that is the case, it could be then fired into the controller on-line in a mater seconds. Not only that, if the user decides to use different algorithm, he only programs the controller again online.

Keywords: PIC microcontroller, PID, timer, Digital to analog converter, serial digital to converter, liquid crystal display

1. Introduction

The PID (Proportional Integral Differential) algorithm is the most popular feedback controller used within the process industries. It has been successfully used for over 50 years. It is a robust easily understood algorithm that can provide excellent control performance despite the varied dynamic characteristics of process plant. It is designed to generate an output that causes some corrective effort to be applied to a process so as to drive a measurable process variable towards a desired value, known as the set point. The concept is based (as shown in **Figure 1**) on the re-input of the system own output according to certain laws (hence the name "feedback"). It is desired for the system output to follow the set point. All feedback controllers determine their output by observing the difference, called error, between the set point and the actual process variable measurement. The PID looks at (a) the current value of the error, (b) the integral of the error over a recent time interval, and (c) the current derivative of the error signal to determine not only how much of a correction to apply, but for how long. Each of those three quantities are multiplied by a (tuning constant) and added together. Thus the PID output is a weighted sum. Depending on the application one may want a faster convergence speed or a lower overshoot. By adjusting the weighting constants, K_p, K_i, and K_d, the PID is set to give the most desired performance.

As a result of enormous development in microcomputer technology, analog controllers have been replaced by digital controllers either in small or large industry. It is now a common practice to implement PID controllers in its digital version, which means that they operate in discrete time domain and deal with analog signals quantized in a limited number of levels. The trend toward digital rather than analog control is mainly due to: (1) versatility where programs can be easily modified or completely changed, (2) sophistication where advanced control laws could be implemented, (3) cost effectiveness where microcontrollers are available at very low costs compared to PLCs, industrial computers, RTUs or DCS. A typical digital feedback control system is shown in **Figure 2**. In digital feedback systems, the controller input and output are digital (sampled) rather than continuous signals. Thus, the continuous signal from the measurement device (sensor/transmitter) is sampled and converted periodically to a digital signal by an analog-to-digital converter (ADC). A digital control algorithm is then used to calculate the controller output as a digital signal. Because most final

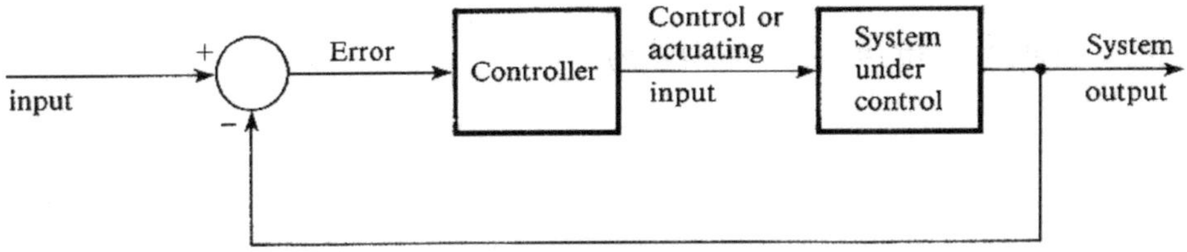

Figure 1.
Typical closed loop control system.

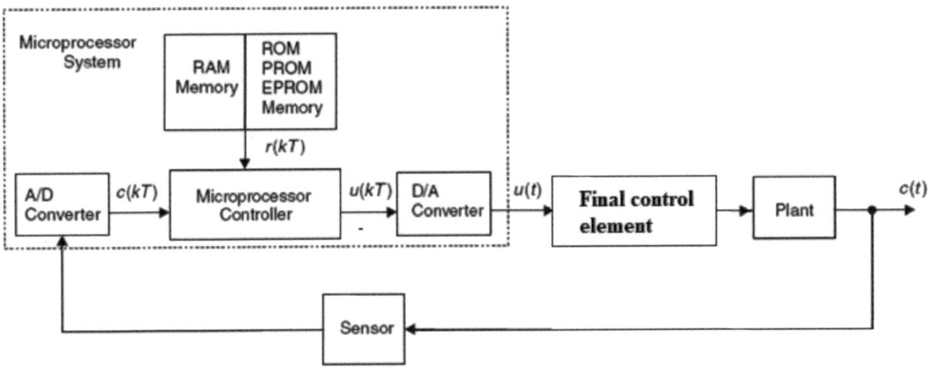

Figure 2.
Digital closed loop based on a microcontroller.

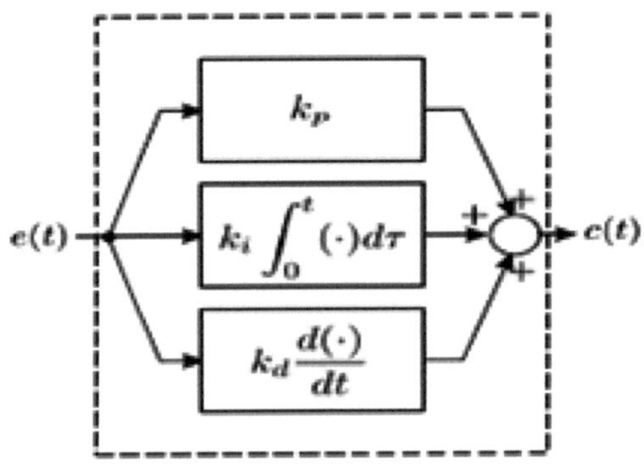

Figure 3.
The three-term PID controller.

control elements are analog devices, the digital output signal is usually converted to a corresponding analog signal by a digital-to-analog converter (DAC).

In feedback control, the objective is to reduce the error signal to zero where

$$e(t) = y_{sp}(t) - y_m(t) \tag{1}$$

where $e(t)$ = error signal, $y_{sp}(t)$ = set point and $y_m(t)$ = measured value of the controlled variable. For the PID controller, the three terms (proportional, integral, and derivative) are combined to minimize the error as much as possible. The most common combination of these three terms is in parallel as shown in **Figure 3**. The PID equation [1, 2] is given by:

$$P(t) = \overline{P} + K_C\left[e(t) + \frac{1}{\tau_I}\int_0^t e(t)dt + \tau_D\frac{de(t)}{dt}\right] \tag{2}$$

where $P(t)$ = controller output; $\overline{P}$ = bias (steady-state) value; K_C = controller gain; τ_I = integral time; τ_D = derivative time.

A straightforward way of deriving a digital version of the parallel form of the PID controller is to replace the integral and derivative terms by finite difference approximations,

$$\int_0^t e(t)dt \approx \sum_{j=1}^{k} e_j T_S \tag{3}$$

$$\frac{de}{dt} \approx \frac{e_k - e_{k-1}}{T_S} \tag{4}$$

where T_S = the sampling time (the time between successive measurements of the controlled variable); and e_k = error at the kth sampling instant for k = 1, 2, 3, ...

There are two alternative forms of the digital PID control equation, the position form and the velocity form. Substituting Eqs. (3) and (4) into (2) gives the *position form*:

$$P_k = \overline{P} + K_C\left[e_k + \frac{T_S}{\tau_I}\sum_{j=1}^{k} e_j + \frac{\tau_D}{T_S}(e_k - e_{k-1})\right] \tag{5}$$

where P_k is the controller output at the Kth sampling instant. Eq. (5) is referred to as the position form of the PID control algorithm because the actual value of the controller output is calculated [3, 4].

In the *velocity form*, the change in controller output is calculated. The velocity form can be derived by writing the position form of Eq. (5) for the $(k - 1)$ sampling instant:

$$P_{k-1} = \overline{P} + K_C\left[e_{k-1} + \frac{T_S}{\tau_I}\sum_{j=1}^{k-1} e_j + \frac{\tau_D}{T_S}(e_{k-1} - e_{k-2})\right] \tag{6}$$

Note that the summation still begins at j = 1 because it is assumed that the process is at the desired steady state for $j \leq 0$, and thus e_j = 0 for $j \leq 0$. Subtracting Eq. (6) from (5) gives the velocity form of the digital PID algorithm:

$$\Delta P_k = P_k - P_{k-1} = K_C\left[(e_k - e_{k-1}) + \frac{T_S}{\tau_I}e_k + \frac{\tau_D}{T_S}(e_k - 2e_{k-1} + e_{k-2})\right] \tag{7}$$

$$P_k = P_{k-1} + K_C\left[(e_k - e_{k-1}) + \frac{T_S}{\tau_I}e_k + \frac{\tau_D}{T_S}(e_k - 2e_{k-1} + e_{k-2})\right] \tag{8}$$

In this study, velocity form is chosen because of the following advantages:

1. It does not need initialization. The position form requires the initial value of the controller output $\overline{P}$, which is not normally known in practice. For example, an operator keeps the control loop in the manual mode until a desired steady state operation has been reached. At this point the error is zero and the position of the control valve would correspond to the $\overline{P}$ value. Therefore, if the operator would like to transfer the control from manual to automatic, he or she should enter in the position control algorithm the value of $\overline{P}$ which is not normally known. This difficulty can be bypassed with the velocity form of the control algorithms, which do not need initialization.

2. It is protected against integral windup. The integral mode of a controller causes its output to continue changing as long as there is a nonzero error. Often the errors cannot be eliminated quickly enough and given enough time they produce larger and larger values for the integral term, which in turn keeps increasing the control action until it is "saturated" (e.g., the valve completely opens or closes). This condition is called *integral windup.* Then, even if the error returns to zero, the control action will remain saturated. The position form with its continuous summation of errors will produce integral windup and special attention will be required. The velocity form, on the other hand, is protected from integral windup for the following reason: The control action changes continuously until it becomes saturated. But then as soon as the error changes sign, the control action can return within the control range in one sampling period.

3. It protects the process against computer failure. With the velocity algorithm one can send out a signal which is used to drive an integrating amplifier or a stepper motor. These devices will retain the last calculated position of the control valve (or other final control element) in case the computer fails, thus avoiding total loss of control of the process.

2. PIC18F452 background

As mentioned earlier, the implementation is based on a Microchip PIC18F452 microcontroller, where the controller plays the role of the brain of the control system [5]. The right choice of the microcontroller is essential, as it will be the core of the final design. The PIC18F452 from Microchip has been chosen for the following advantages:

1. Speed: with its maximum internal clock rate of 20 MHz and its 16-bit-wide instruction bus, the CPU can execute most of its instructions at a single machine cycle of four clocks which is equivalent to a 0.2 μs.

2. Math support: unlike classical microprocessors, the controller in hand has got a hardware multiplier and divider for multiple-bytes, fixed-point numbers and for floating-point numbers so multiplication is carried out in a single instruction.

3. Flexible timer resources: four independent timers modules support timing measurements and output interval control with a timing resolution as fine as 0.1 μs. Those timers could be used to produce up to three pulse width modulations which could be used for electrical motor control.

4. Free software tools: Microchip's Development Package MPLAB® (consisting of assembler, simulator, and user interface) as well as all manuals and application notes are available at no cost from their Web site (www.microchip.com).

5. Development tool versatility: it supports in-circuit debugger which permits the loading and execution of a user program as well as the use of breakpoints, memory/ register modification, and single stepping.

6. Build-in ADCs: it has analogue-to-digital converters with 10 bits resolution.

7. Built-in serial peripheral interface: it has a variety of serial bus interfaces like USART, I^2C & SPI.

8. C programmable: it could be programmed using C language with the use of a variety of built in C libraries developed by microchip.

The PIC18F452 microcontroller is a 40 or 44-pin depending on the package, where in the 40 pins configuration, a dual inline package is used; whereas in the 44 pins configuration, either thin quad flat package or dual flat no leads package is used. Its design is based on Harvard technology where the program and data have different buses. This type of microcontrollers is very cheap, small in size, and could be customized. It could be easily programmed on-line using either assembly language, BASIC or C language. In fact, it is ideal for small application such as the one in hand. The controller has a 24 kbytes of flash memory and 2048 bytes of SDRAM. It also has a 8 × 10 bits analog to digital channels. It also has 5 bidirectional digital ports with 33 inputs/outputs, configured as follows: 3 × 8 digital I/O ports (PORTB, PORTC and PORTD), one six digital I/O port (PORTA) and one three digital I/O port (PORTE). Unfortunately, one of the drawbacks of microcontrollers, it is very seldom to find one with a digital to analog converter. Luckily, they are few manufacturers around including microchip, which make serial DACs which could be programmed through Serial Port Interface (SPI) using only three wires. The PIC18F452 has four timer/counters which could be programmed either as 8 or 16 bit timers/counters. It also has two ports which could be configured either as capture, compare or pulse width modulation (PWM). It has two serial peripheral interfaces: (SPI) and an inter-integrated circuit (I^2C). An asynchronous port (USART) is also provided. For the microcontroller to output analogue data, an MCP4921 device is used. The device is a 12-bit buffered single voltage output Digital-to-Analog Converter (DAC). The device operates from a single 2.7 V to 5.5 V supply with an SPI compatible Serial Peripheral Interface. The user can configure the full-scale range of the device to be VREF or 2*VREF by setting the gain selection option bit (gain of 1 of 2). The user can shut down the device by setting the Configuration Register bit. In Shutdown mode, most of the internal circuits are turned off for power savings, and the output amplifier is configured to present a known high resistance output load (500 kΩ, typical). The device includes double-buffered registers, allowing synchronous updates of the DAC output using the LDAC pin. The device also incorporates a Power-on Reset (POR) circuit to ensure reliable powerup. The device utilizes a resistive string architecture, with its inherent advantages of low Differential Non-Linearity (DNL) error and fast settling time. The device is specified over

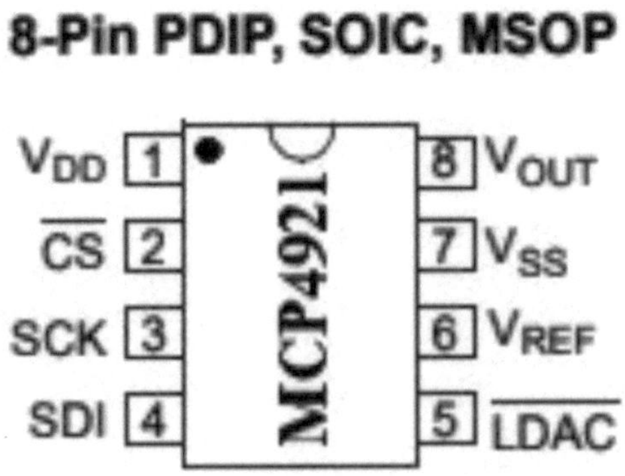

Figure 4.
MCP4921 pin configuration.

W-x	W-x	W-x	W-0	W-x	W-x	W-x	W-x	W-x	W-x	W-x	W-x	W-x	W-x	W-x	W-x
0	BUF	$\overline{GA}$	$\overline{SHDN}$	D11	D10	D9	D8	D7	D6	D5	D4	D3	D2	D1	D0
bit 15															bit 0

Figure 5.
Write command register for MCP4921 (12-bit DAC).

the extended temperature range (+125°C). It provides high accuracy and low noise performance for consumer and industrial applications where calibration or compensation of signals (such as temperature, pressure and humidity) is required. The MCP4921 device is available in the PDIP, SOIC, MSOP and DFN packages. **Figure 4** shows the chip pin configuration. The MCP4921 device is designed to interface directly with the Serial Peripheral Interface (SPI) port, which is available on the PIC18F452 microcontroller and supports Mode 0,0 and Mode 1,1. Commands and data are sent to the device via the SDI pin, with data being clocked-in on the rising edge of SCK. The communication is unidirectional; this means the data cannot be read out of the MCP4921. The CS (chip select active low) pin must be held low for the duration of a write command. The write command consists of 16 bits and is used to configure the DAC's control and data latches. Register shown in **Figure 5**, details the write command which is loaded into the input register that is used to configure and load the DAC register [6].

The write command is initiated by driving the CS pin low, followed by clocking the four Configuration bits and the 12 data bits into the SDI pin on the rising edge of SCK. The CS pin is then raised, causing the data to be latched into the DAC's input register. The MCP4921 utilizes a double-buffered latch structure to allow the analog output to be synchronized with the LDAC pin, if desired. By bringing the LDAC pin down to a low state, the content stored in the DAC's input register is transferred into the DAC's output register (VOUT), and VOUT is updated. The write to the MCP4921 device is 16-bit words. Any clocks past the 16th clock will be ignored. The Most Significant 4 bits are Configuration bits. The remaining 12 bits are data bits. No data can be transferred into the device with CS high. This transfer will only occur if 16 clocks have been transferred into the device. If the rising edge of CS occurs prior to that, shifting of data into the input register will be aborted. The most four significant bits are defined as follows:

bit 15 0 = Write to DAC register
1 = Ignore this command

bit 14 BUF: VREF Input Buffer Control bit
1 = Buffered
0 = Unbuffered

bit 13 $\overline{GA}$: Output Gain Selection bit
1 = 1x (VOUT = VREF * D/4096)
0 = 2x (VOUT = 2 * VREF * D/4096)

bit 12 $\overline{SHDN}$: Output Shutdown Control bit
1= Active mode operation. VOUT is available.
0 = Shutdown the device. Analog output is not available.
VOUT pin is connected to 500 kΩ (typical).

3. Liquid crystal display (LCD)

This module is designed to display the value of the temperature detected by the temperature sensor and to guide the user in changing the parameters of the controller. The LCD is a 16 × 2 alphanumeric display with the built-in Hitachi 44780 controller and LED backlighting. It works with an 8-bit data bus, which means it will require a total of 11 data lines. Three control lines (connected to port E) plus the 8 lines for the data bus (connected to port D) [7].

4. System design

The system is design around a stand-alone PIC18F452 controller, where the measured variable (MV) is read through channel0 (pin 2). The MV is subtracted from the set point automatically by the controller. The error is treated by the PIC PID and produces a digital control variable. This control variable is outputted through PIC serial data output pin (SDO pin 24) together with serial clock pin (SCK pin 18) to synchronize the conversion process. For the conversion to take place, the serial DAC chip select (CS) has to be pulled low. The CS is connected to pin RC0. The positive reference voltage is connected to +5 V (pin 6) and the negative reference voltage (pin 7) is tied to zero volt. The analog output is read through pin8 (Vout). This voltage is small to drive an electric motor. This voltage is pulled up to +12 V through the non-inverting operational amplifier (LM358). The Darlington transistor 2SD1409 is used to bust the current. The motor is connected to the emitter follower so that the driving current is sufficient enough to drive the motor. Needless to say that the diode 1N4148 is used to protect the Darlington transistor against any spike due to the change of current. **Figure 6** shows the schematic of the system. The LCD is used to display the measured temperature. To manipulate the setting of different parameters, six push buttons are used as follows:

Six push buttons were used in the project to allow the user to change the setting and the controller parameters. Their functions are as follows:

1. Reset: To reset the microcontroller.

2. Stop: interrupt the program to allow the user to change the controller settings

3. Run: To run the program

4. Mode: To allow the user to change between setting modes.

5. Increment: To increment the controller variables by 1 or 0.1.

6. Decrement: To decrement the controller variables by 1 or 0.1.

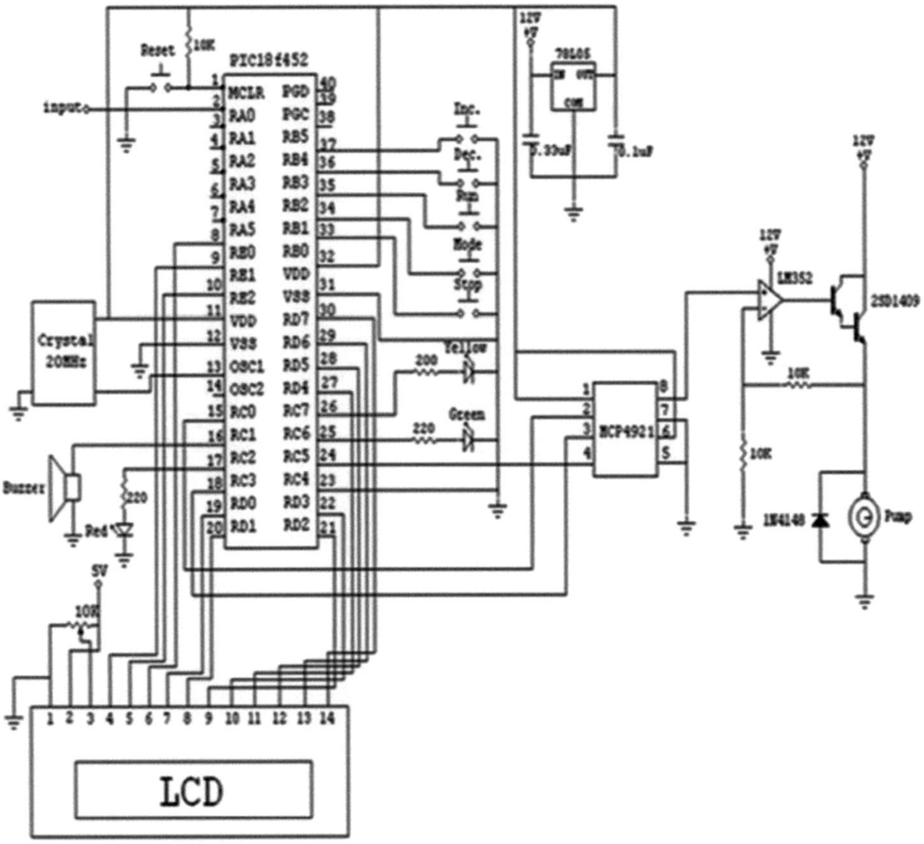

Figure 6.
System schematic circuit showing all the connection to the microcontroller, as well as the liquid crystal display and the final control element.

These switches are connected to PIC PORTB to allow the user to use the internal build-in pull up resistors to prevent floating instead of using external pull-down resistors. The reset has got a separate button connected to MCLR pin. A buzzer is used as an alarm to indicate that the temperature is more than what the user specifies. Three LEDs were used to show the user the status of the microcontroller program. The three colors green, yellow, and orange were used as follows:

1. Green: means that the PID controller is working properly.

2. Yellow: means that the program is interrupted by (STOP) push button.

3. Red: means that the alarm is triggered.

5. Software design

To implement the control program, three major routines are used; the main routine along with the timer and external interrupts. The program starts with the main routine which contains all the configurations of the external pins whether outputs or inputs. It also contains the configurations of timer and external interrupts, so when one of these interrupts is triggered, the microcontroller will stop its current execution and perform another action. The trigger will be caused by either an overflow in timer register or a change on an external pin (RB0/INT0).

Because the time is a crucial element in digital control, the PID algorithm is controlled through a timer interrupt. This choice allows the user the ability to calculate the sampling time accurately. On the other hand, an external interrupt (INT0) is used to interrupt the program in order to allow the user a chance to modify the controller parameters. In the following we discuss in some details about the functions of each routine.

5.1 Main routine

5.1.1 Routine function

This routine, as mentioned earlier, is dedicated to configure the direction of external pins as well as interrupt sources. It also allows the user to choose the measured variable (temperature, flow, level or others). The flow chart of this routine is shown in **Figure 7**.

5.1.2 External ports configuration

First PORTA (pin RA0) is configured as an analog input channel0 and PORTB as input digital port which is connected to the push button switches; while all other pins are configured as outputs.

5.1.3 LCD configuration

The configuration of the LCD was performed by separate software from Microchip called Application Maestro [8]. With the aid of this software, a configuration code was produced after modifying the module parameters. It was then incorporated into the project. Once incorporated, the LCD is configured and ready to work. One feature of using Application Maestro is its ability to use the prewritten code that this software provides to initialize or to write to the LCD.

5.1.4 Timer0 configuration

Timer0 can operate as a timer or as a counter. In Timer mode, the Timer0 module will increment with every instruction cycle (without prescaler). It is configured by setting a special function register called T0CON (timer0 control byte). This register is a readable and writable register that controls all the aspects of Timer0, including the prescale selection. In the design in hand, T0CON register is set to 0x85 (0b10000101) as shown below [9, 10].

TMR0ON	T08BIT	T0CS	T0SE	PSA	T0PS2	T0PS1	T0PS0
1	0	0	0	0	1	0	1
bit 7							bit 0

This value will configure the timer0 as follows:

●Bit7	TMR0ON	= 1	: Timer0 is enabled	
●Bit6	T08BIT	= 0	: Timer0 is configured as a 16-bit timer	
●Bit5	T0CS	= 0	: Internal instruction cycle clock	
●Bit4	T0SE	= 0	: This bit is used only with external clock	
●Bit3	PSA	= 0	: Timer0 prescaler is assigned	
●Bit2	T0PS2	= 1	: Bit2: T0PS2 =1:	} 1:64 prescaler value
●Bit1	T0PS1	= 0	: Bit1: T0PS1 =0:	
●Bit0	T0PS0	= 1	: Bit0: T0PS0 =1:	

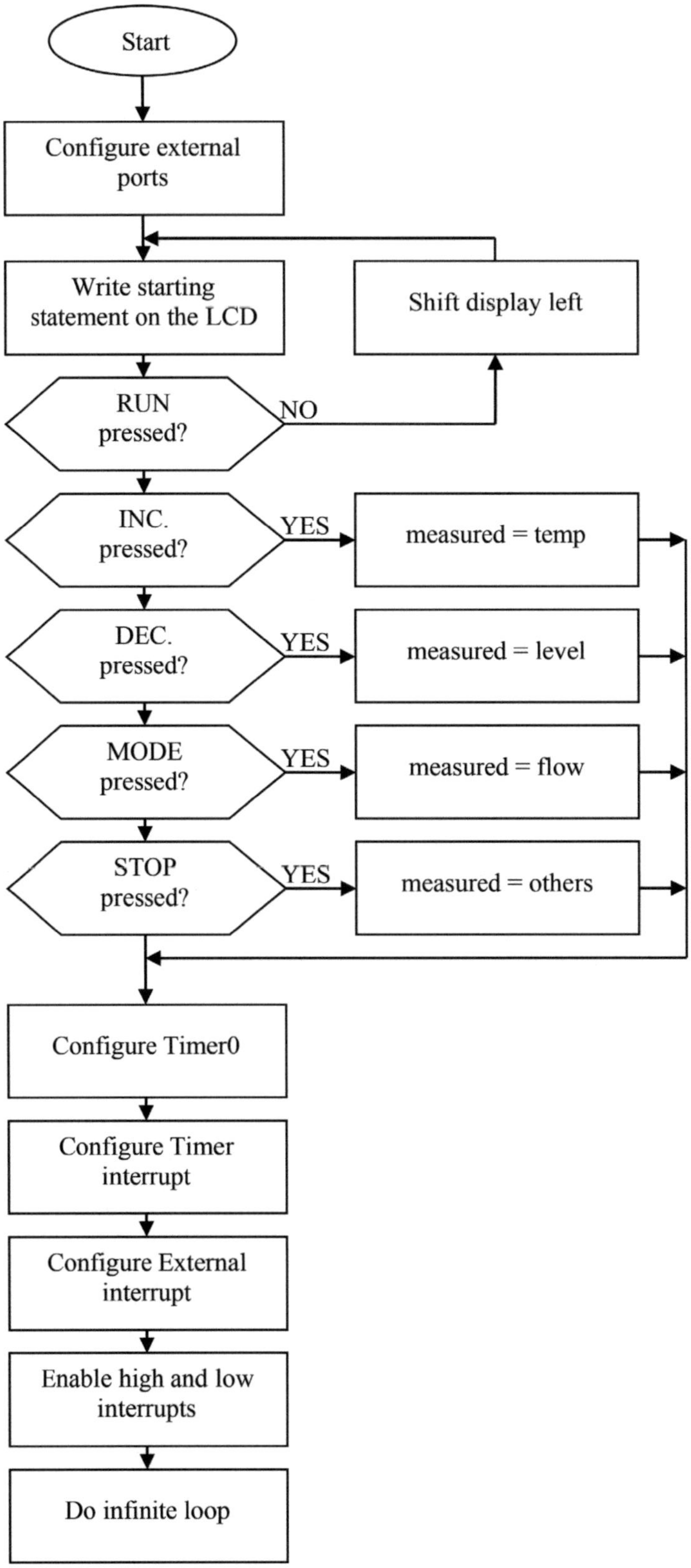

Figure 7.
Main routine.

5.1.5 Interrupt configuration

There are ten registers which are used to control internal and external interrupt operations to accommodate a variety of interrupts [11]. In the project in hand, only two interrupts are required INT0 and timer0 interrupt. To do so, only three control registers are required. These registers are INTCON, INTCON2, and RCON. INTCON register contains various enable bits as well as several interrupt flags. RCON is the Reset Control register which contains flag bits that allow differentiation between the sources RESET. Timer0 interrupt is enabled by setting TMR0IE bit (<5>) while external interrupt is enabled by setting INT0IE (INTCON<4>). Note that the interrupt flags are reset before enabling the interrupt in order to avoid unwanted interruptions.

To start the interrupt, the global interrupt bit GIE/GIEH (INTCON<7>) must be set. If set, it enables all unmasked interrupts, so if more than one interrupt source is used (as in our case) the Interrupt Priority Enable bit IPEN (RCON<7>) must be set and the interrupt sources should be specified either as high or low priority interrupt. The interrupt priority bit TMR0IP (INTCON2<2>) is used to specify the interrupt priority for Timer0. This bit is reset so timer0 interrupt is set to low priority. On the other hand, no need to specify the priority of the external interrupt (INT0), because it is already set to high priority by default.

After configuring the interrupts, the program will enter an infinite loop until one of the interrupt sources is triggered.

5.2 Timer interrupt routine

5.2.1 Routine function

The main purpose of this routine is to calculate the controller output and send it to the DAC serially through the synchronous SPI module [12]. **Figure 8** shows the routine function.

5.2.2 Timer reloading

Because of the importance of time in calculating the timed controller output, timer0 is used as an accurate hardware timer. The source clock of the timer is the crystal oscillator which is fed to the clock pin of Timer0 internally. The clock used is a 20 MHz derived from a stable crystal oscillator. This frequency is automatically divided by 4 because the controller machine cycle is 4 clocks to give a 5 MHz which is fed to the timer. The timer is exactly clocked every 0.2 μs and takes 13107.2 μs (16-bit mode) to count from zero to zero again. However, by loading the timer with a suitable value, a smaller time interval could be obtained. For example, by loading the timer with the value 4095 (0xFFF), the overflow would occur after 12288.2 μs. Alternatively, the time period can be extended by using a prescaler as was done in the main routine. If a divide by 64 prescaler is selected, timer0 only overflows after 838.848 ms. This is obtained as follows:

$$\frac{5 \text{ MHz}}{64} = 78,125 \text{ Hz}$$

$$(78,125 \text{ Hz})^{-1} = 12.8 \text{ μs}$$

$$12.8 \text{ μs} \times 65,535 = 838.848 \text{ ms}$$

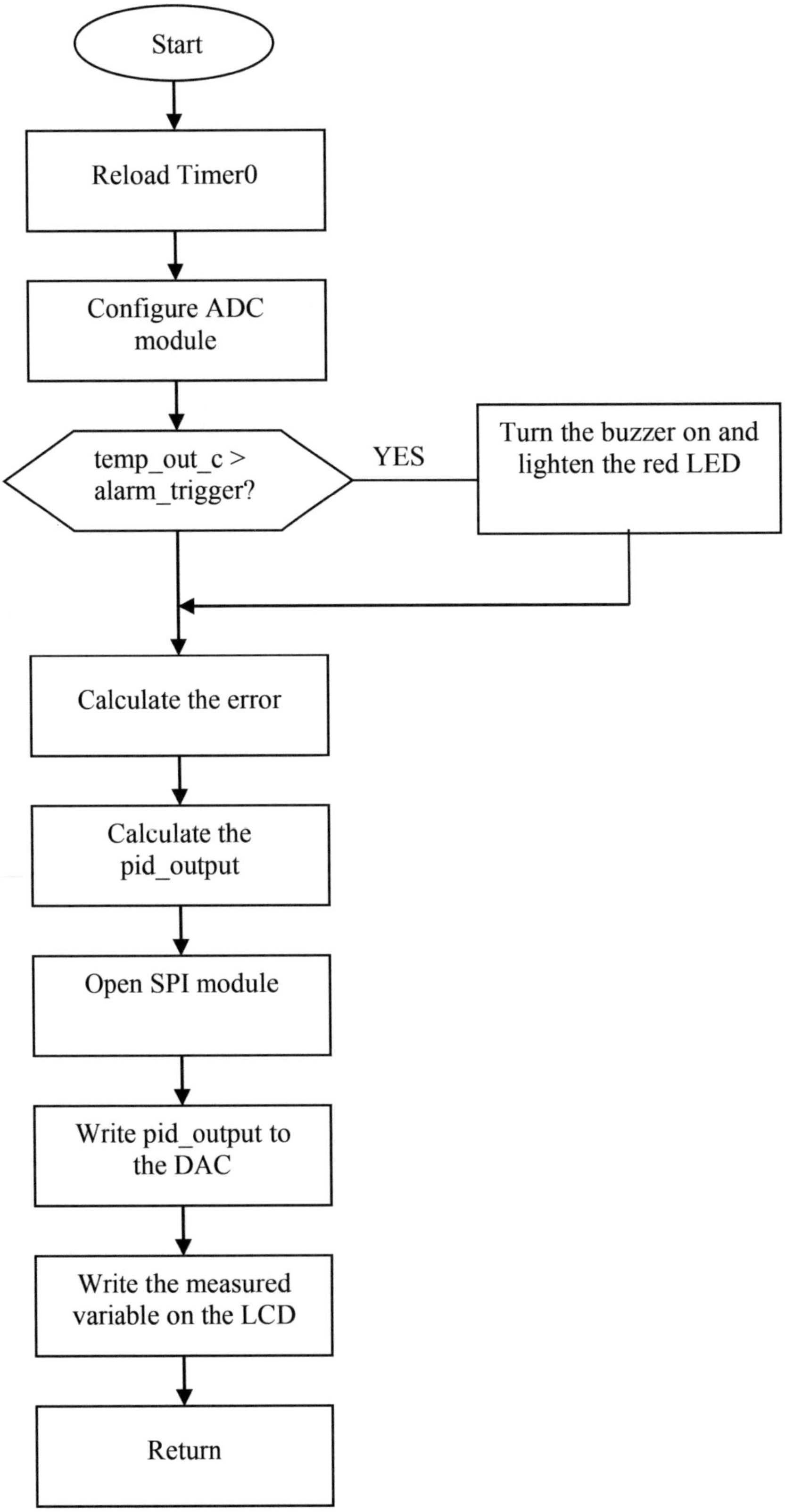

Figure 8.
Timer interrupt routine.

This time period is less than one second, while a one second sampling time is required for the design in hand. To obtain a one second sampling time, the timer should count 78,125 pulses.

Because timer0 register is only 16 bit wide, it is only limited to count up to 65,535 pulses. The interruption is trigged several times to obtain one second timing, after which the controller computes the control action and sends it to the DAC. By using MPLAP simulator, it was found that 5362 cycles are required to calculate the controller output and send it to the DAC besides 51 extra cycles needed to reload the timer with time constant. If the interruption is required to repeat itself five times before calculating the controller output, one needs 5362 + 51 × 5 = 5617 cycles (1.1234 ms). Thus, in order to get exactly one second sampling time, the timer register (TMR0) has to be reloaded with a value that interrupts the program every 998.8766 ms (1 s–1.1234 ms). The following shows how this value is obtained:

$$\text{No.of cycles for } 1\text{ s} = 5 \times 10^6 \text{ cycles}$$

$$\text{Therefore no.of cycles between interrupts} = 5 \times 10^6 - 5617 = 4,997,383 \text{ cycles}$$

By using a timer with 64 prescaler:

$$\text{No. of counts} = \frac{4,997,383}{64} = 78,084.10938$$

When we repeat the interrupt for 5 times:

$$\text{No. of counts} = \frac{78,084.10938}{5} = 15,616.82188$$

But because the timer counts in ascending order (from 0x0000 to 0xFFFF):

$$\text{Reload value} = 65,535 - 15616.82188 = 49918.17812$$

```
repeat=5*sampling_time;
term_1=((repeat)*51.0+5362.0);
term_2=(5000000.0*sampling_time-term_1)/64.0;
term_3=term_2/repeat;
cycle=65535-term_3;
```

However, the timer register accepts only integer numbers, thus the final value that should be added to the timer register is 49918. Because we omitted the numbers after the decimal point, our error will be ± 1 count which is equal to 64 cycles. Therefore, our error in calculating the sampling time will be:

$$\text{Timer error} = 64 \times 0.2\ \mu\text{s} = 12.8\ \mu\text{s}$$

This calculation is for getting 1 s sampling time. To expand the calculation in order to enable the user to change the sampling time, one defines two integer variables (repeat and cycle). The first variable repeat is to determine how many times we need to repeat the interrupt, while the second one cycle is the final value that should be added to the timer register. The following pseudo code shows the general formula used to reload the timer register.

5.2.3 Analog to digital converter module

The ADC module normally operates at 10-bits resolution, giving output digital values 0–1024 [13]. It needs a reference voltage to set the maximum and minimum values for the input conversion. This reference can be provided internally as V_{dd} and V_{ss} (supply values) or externally through V_{ref}^{+} and V_{ref}^{-} pins. To configure this module, OpenADC function from Microchip C library is used. This function performs a bitwise AND operation ("&") between its arguments which are defined in the file adc.h. The parameters of this function along with their meaning of each argument are discussed below [1]

```
OpenADC(ADC_FOSC_32 & ADC_RIGHT_JUST & ADC_8ANA_0REF,
ADC_CH0 & ADC_INT_OFF);
```

- ADC_FOSC_32: FOSC/32.

A clock divider to allow the minimum specified conversion time (about 20 μs). A 32 prescaler was chosen because the clock source is 20 MHz

- ADC_RIGHT_JUST: Right justified.

Because the ADRES register pair (where the converted values are loaded) is 16-bit wide. But the ADC is only 10bit wide. The ADC module could either be configured as right or left justified. In this project, right justified is chosen as shown in **Figure** 9. This sets the 6 most significant bits of register ADRES to zeros.

- ADC_8ANA_0REF: $V_{REF+} = V_{DD}$, $V_{REF-} = V_{SS}$

The supply values are chosen as the voltage references to the ADC.

- ADC_CH0: Channel0 (AN0) is selected

- ADC_INT_OFF: Interrupts of ADC interrupts are disabled.

- Once the A/D conversion is completed, the result is stored in an integer variable called result. After reading the analog value by the ADC module, the result will be compared with the variable alarm-trigger which was previously specified. If the result is greater than this value, the microcontroller triggers the buzzer and lights the red LED.

5.2.4 Controller calculation

Due to the limitation in the microcontroller's memory, the PID equation is divided into three terms (term_1, term_2, and term_3) and after calculating each term separately, they are added together along with the previous output to give the controller

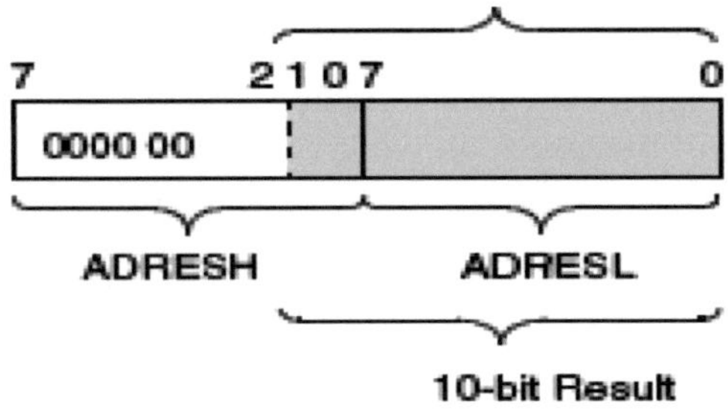

Figure 9.
Choosing right justified for data input.

output which will be sent to the DAC. The following code shows how to calculate the controller output

```
term_1 = KC*(error-last_error);
term_2 = KC*(sampling_time/TI)*error;
term_3 = KC*(TD/sampling_time)*(error-(2*last_error)+prev_error);
pid_output_v = term_1 + term_2 + term_3+pid_output_v;
```

5.2.5 SPI module

To send the control variable to the final control element, the serial DAC, which is interfaced to the Serial Peripheral Interface (SPI) port, is used. The SPI is initiated using Microchip C library called OpenSPI. This function also performs a bitwise AND operation between its arguments which are defined in the file SPI.h according to the following formula.

- SPI_FOSC_16: Master mode and the clock = FOSC/16
- MODE_00: Mode 0,0 (change takes place on the rising edge)
- SMPEND: Input data sample at end of data out

```
OpenSPI(SPI_FOSC_16, MODE_00, SMPEND);
```

After configuring the module, it is time to write a command to the DAC in order to convert it into analog signal. The write command is initiated by driving the CS pin low, followed by clocking the four configuration bits and the 12 data bits into the SDI pin on the rising edge of SCK. The CS pin is then raised, causing the data to be latched into the DAC's input registers and when the LDAC pin is pulled down through RC1, the values held in the DAC's input registers are transferred into the DAC's output registers to provide the analog signal. It is important to mention here that we wrote the write command in two steps (as shown in the following code) because the SPI module send only 8 bit at a time.

```
LATCbits.LATC0 = 0; // Chip Select is set
WriteSPI(pid_output_16_high); //4MSB as command + 4LSB as data
WriteSPI(pid_output_16_low); // 8 bit data
LATCbits.LATC0 = 1; // Chip Select is reset
LATCbits.LATC1=0; // enable LDAC for data output to DAC
LATCbits.LATC1=1; // disable LDAC
CloseSPI();
```

5.2.6 Writing on the LCD

To write characters to the LCD, required prewritten functions are provided by Application Maestro. Some of These functions are listed in following table:

XLCDInit()	It is used to initialize the LCD module according to the Application Maestro options
XLCDPut(data)	It sends the clocking signal and data to be displayed to the LCD

XLCDL1home()	Points to the first address location of line one of the LCD
XLCDL2home()	Points to the first address location of line two of the LCD
XLCDClear()	Clears the DDRAM content of the LCD and points to the 00 address location
XLCDPutRomString(addr)	Displays String in Program memory
XLCDPutRamString(addr)	Displays String in Data memory
XLCDCommand(Command)	It sends clocking signal and instructions to the LCD

For numbers to be displayed, they are first converted into strings (characters) before being sent to the LCD, since the latter only accepts strings. To do so a C function called sprintf is called upon. This function saves the number in an array after converting it into string. The subroutine to do so is shown below [4]:

```
sprintf (buf,"%d",temp_set);
XLCDPutRamString(buf);
```

5.3 External interrupt routine

5.3.1 Routine function

The main function of this routine is to allow the user to change the controller parameters. The routine is initiated by pressing the push button (STOP) which is connected to the external interrupt pin (RB0/INT0). Once initiated, the user is able to change all the parameters of the controller (K_C, τ_I, τ_D, sampling time, alarm trigger and sensitivity) by using three push buttons (MODE, INCREMENT and DECREMENT [11].

To determine which action the microcontroller should take if any push button is pressed, we defined two integer variables (present_button and present_mode) to be used as statuses. That is, each bit of them has specific meaning as described below:

- present_button

Np	___	___	___	___	___	___	___	___	___	___	___	___	dec	inc	Mod
bit15															bit0

- bit 15 **Np:** set if there are no push buttons pressed
- bit 14-3 Unimplemented
- bit 2 **dec:** set if the DECREMENT push button pressed
- bit 1 **inc:** set if the INCREMENT push button pressed
- bit 0 **mod:** set if the MODE push button pressed

- present_mode

___	Srt	Spt	KC	TI	TD	Stm	Sen	Alm	Tun	Dp	Tp	Kp	___	___	___
bit15															bit0

- bit 15 Unimplemented
- bit 14 **Srt:** set in the starting mode

- bit 13 **Spt:** set in the set point mode
- bit 12 **KC:** set in the controller gain mode
- bit 11 **TI:** set in the integral time mode
- bit 10 **TD:** set in the derivative time mode
- bit 9 **Stm:** set in the sampling time mode
- bit 8 **Sen:** set in the sensitivity mode
- bit 7 **Alm:** set in the alarm mode
- bit 6 **Tun:** set in the tuning mode
- bit 5 **Dp:** set in the process delay mode
- bit 4 **Tp:** set in the process time constant mode
- bit 3 **Kp:** set in the process gain mode
- bit 2-0 Unimplemented

Initially, before pressing any push button, present_button variable is loaded with 0x8000 (no push button pressed), and present_mode with 0x2000 (starting mode).Then if any push button is pressed, the corresponding bit of that push button will be set, giving a specific value of present_button which indicates the push button that was pressed by the user. So by performing a bitwise OR operation between the two variables (present_button and present_mode) we will come up with a number indicates the push button pressed and the present mode and based on that number we can decide the proper action to be taken by the microcontroller. The following code shows how to perform the OR operation after checking which of the push buttons was pressed. Beside changing the controller variables, this routine has another feature, it gives the user preliminary values of the controller parameters after entering the process variables. The result is derived based on Cohen-Koon tuning method. However, this feature is impractical if the sampling time is big [14].

```
if (mode_pin==0){
   Delay10KTCYx(70);
  present_button=mode_pushed;
}
else if (inc_pin==0){
        Delay10KTCYx(70);
  present_button=inc_pushed;
   }
else if (dec_pin==0){
        Delay10KTCYx(70);
present_button=dec_pushed;
   }
else if(end_pin==0)
     return;
action= present_mode | present_button;
```

6. Testing and verification

To test the system, a first order system given by the equation below was used. To run the control action, the system was converted into a difference equation given by Eq. (10).

$$G_p = \frac{10}{5S+1} \tag{9}$$

The process transfer function is first order, thus the discrete transfer function obtained using Zero-Order Hold will be:

$$HG(z) = \frac{az^{-1}}{1 - bz^{-1}} \tag{10}$$

where:

$$a = k_p\left[1 - \exp\left(\frac{-T_s}{\tau_p}\right)\right]$$
$$b = \exp\left(\frac{-T_s}{\tau_p}\right)$$

If $k_p = 10$, $\tau_p = 5$, and $T_s = 1$, The discrete transfer function will be:

$$HG(z) = \frac{1.813z^{-1}}{1 - 0.8187z^{-1}} \tag{11}$$
$$\Rightarrow (1 - 0.8187z^{-1})Y(z) = 1.813z^{-1}C(z)$$
$$Y(z) - 0.8187z^{-1}Y(z) = 1.813z^{-1}C(z)$$
$$y_n - 0.8187y_{n-1} = 1.813c_{n-1}$$

Therefore the difference equation of the output is:

$$y_n = 0.8187y_{n-1} + 1.813c_{n-1} \tag{12}$$

After getting the difference equation, the control scheme was tested and the output of **Figure 10** was obtained with the parameters set to: $K_c = 0.2, T_i = 4.0$ and $T_d = 0.0$. The parameters were then changed to: $K_c = 0.1, T_i = 3.0$ and $T_d = 0.2$. The response is shown in **Figure 11**.

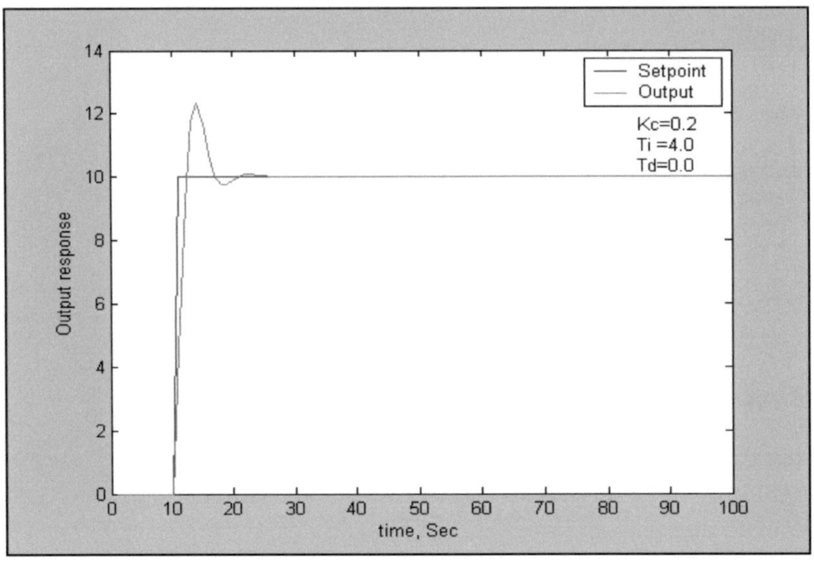

Figure 10.
Controller response with $K_c = 0.2, T_i = 4.0$ and $T_d = 0$.

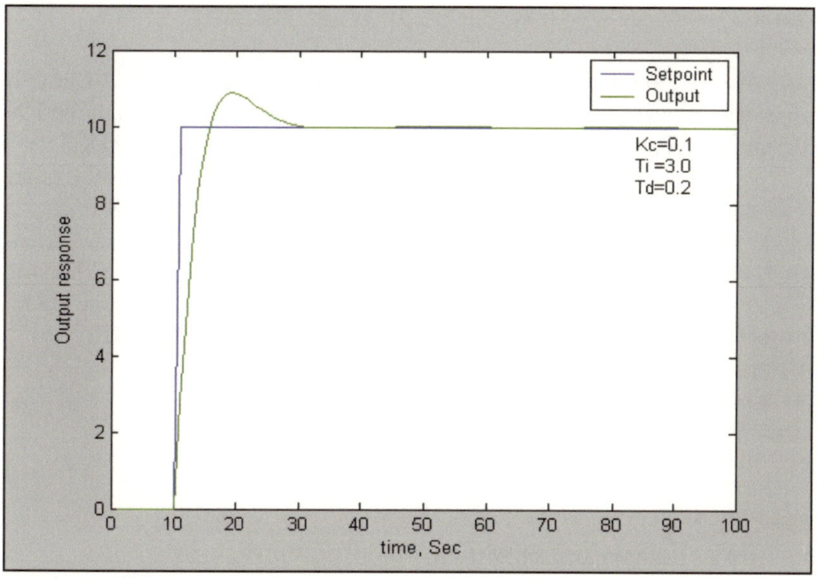

Figure 11.
Controller response with $K_c = 0.1, T_i = 3.0$ and $T_d = 0.2$.

7. Conclusion

By referring to the previous graphs, it could be concluded that the response tracks the set point as expected. In addition, the increase in controller gain (K_c) does speed up the response but at the expense of the overshoot. Based on these results, it could be concluded that the three-term controller is working according to plan. Because of the flexibility of the microcontroller and its programming, any control scheme could be developed and implemented in the manner as described in this chapter. Not only that, the scheme could be transferred to several high range microcontrollers from the same company such as 16 or 32 bits with the use of the benefits those types of controllers offer.

Author details

Mostefa Ghassoul
Chemical Engineering, University of Bahrain, Bahrain

*Address all correspondence to: mghassoul@uob.edu.bh

References

[1] Ogunnaike BA, Ray WH. Process Dynamics, Modeling, and Control. 1st ed. USA: Oxford University Press; 1994

[2] Stephanopoulos G. Chemical Process Control: An Introduction to Theory and Practice. USA: Prentice Hall; 1983

[3] Seborg DE, Edgar TF, Mellichamp DA, Doyle FJ. Process Dynamics and Control. 4th ed. New York, USA: John Wiley and Son; 2017. pp. 115-117

[4] Phillips CL, Nagle HT, Chakrabortty A. Digital Control System Analysis and Design. UK: Pearson; 2015. pp. 279-335

[5] Microchip PIC18FXX2 Data Sheet "High-Performance, Enhanced Flash Microcontrollers" Microchip 2006 (DS39564C)

[6] Microchip MCP4921/4922 datasheet "12-Bit DAC with SPI™ Interface" Microchip 2007

[7] HD44780 LCD starter guide. 2001

[8] Microchip application maestro software user's guide. 2003

[9] Microchip PIC18FXX2 Data Sheet "High-Performance, Enhanced Flash Microcontrollers: Timers" Microchip 2006 (DS39564C). pp. 103-115

[10] Bates M. PIC Microcontrollers: An Introduction to Microelectronics. Newnes; 2011

[11] Microchip PIC18FXX2 Data Sheet "High-Performance, Enhanced Flash Microcontrollers: External interrupts" Microchip 2006 (DS39564C). pp. 73-85

[12] Microchip PIC18FXX2 Data Sheet "High-Performance, Enhanced Flash Microcontrollers: Timer0 interrupt" Microchip 2006 (DS39564C). p. 85

[13] Microchip PIC18FXX2 Data Sheet "High-Performance, Enhanced Flash Microcontrollers: 10 bit ADC" Microchip 2006 (DS39564C). Chandler, Arizona USA; pp. 181-188

[14] Deitel PJ, Deitel HM. C How to Program. 6th ed. New Jersey USA: Pearson Prentice Hall; 2010

Chapter 18

Lessons Learnt from Some Natural Energy Sources

Dan Serbanescu

Abstract

The chapter presents in a systematic manner the lessons learnt from the natural energy systems (NES) and their specific features. The conclusions are based on the evaluation of the risk impact on environment and for the improvement of the risk evaluation methodologies of such systems. A specific feature of the NES is the interdependence between them and society/mankind and the environment. Risk analyses for such systems have specific features underlined while compared with the features of the artificial (man-made) energy systems (MMES). Previous works illustrated in detail the NES versus MMES differences. This chapter presents the main aspects of such a review, when applied to a specific NES, the natural nuclear fission reactor in Oklo, Gabon (NES_Oklo). NES_Oklo operated about two billion years ago for about two hundreds millions of years. The lessons drawn from studying how this reactor was built, operated and self-decommissioned are of high importance for nuclear energy and not only. There are conclusions drawn from the study of Oklo reactor, which seem to shake some taboo issues in Physics, like for instance the light speed limit and other fundamental aspects of Quantum Mechanics, which have also important philosophical implications.

Keywords: natural energy system (NES), man made energy systems (MMES), risk, nuclear reactor, quantum mechanics, philosophy of science, speed of light, ecology, energy systems life cycle (ES_LC), knowledge process (KP), topological structure (K(i)), theory of a given topological structure (Th(K(i)), Backward engineering)

1. Introduction

Two and a half billion years ago a natural fission reactor operated on the Earth (Oklo). The discovery of this natural energy source created a series of theories and had implications yet to be evaluated both on the man-made artifacts of similar type and on some fundamentals considered so far as improbable to be challenged in quantum physics, biology, ecology, nuclear reactor theory. It also has an impact on knowledge management, on the epistemology and ethics. Aspects of the implications for mankind and the lessons learnt so far on the actions to build a sustainable civilization are presented in this chapter.

In 1972 the international community involved in the research, design and operation of MMES of fission type reactors was surprised and challenged by a discovery of the remains of an ancient natural fission reactor, in Oklo (Gabon). It was a NES type reactor (NES_Oklo).

However the discovery was predicted long time before by PK Kuroda [1]. The reactor in Gabon operated, intermittently, two and a half billion years ago for about two hundreds millions year and had an approximate power of 100 kW. It operated with uranium ore (using the isotope U235) and water [2–5].

As the reactor physics classic results show, this would not be possible, provided the concentration of U235 (considered as a constant for the whole universe) being presently 0.71% was not higher (around 3.3%) by the time the reactor started operating. And this is not all. The reactor had to have a concentrated amount of U235 in a place forming a geometry and a configuration of cooling (with cooling water) of a very specific precise type. Apparently cyanobacteria concentrated the uranium and the water from the underground, pushed by the geological moves by that time (Africa and South America were splitting apart) created actually the reactor core, as called in the nuclear engineering. Even more than that, the type of soil assured the retention of the radioactive elements resulted from fission, which actually did not migrate further than the site.

All those aspects were very troubling for the nuclear community. In addition the calculations for the MMES reactors were seriously challenged when they were used to describe NES_Oklo.

Findings did not stop here, as series of other theories were developed, as for instance:

- Theories related to how the oxygen formation (taking place exactly by that time) were related to the activity of the geyser nuclear reactor splitting water vapors, as water got overheated, to the atmosphere.

- As for the biology the time of NES_Oklo operation is also coincident with the appearance of eukaryotes, living beings having cells with nucleus in a membrane, to which we also belong.

- As a top of troubling discoveries, the site evaluations challenged some fundamentals of quantum mechanics and relativity, related to the alpha constant and the speed of light.

- Not to mention the fact that new theories and observations started to assume that, may be even the Earth core is a nuclear fission reactor and may be Oklo was not the only surface reactor.

- More than that evidence on existence of fission reactors is found also in our neighboring planets (Mars), all taking place at a certain time of evolution of energy chains of the universe, of the solar system and of our Earth. Operation of such NES reactors appears to give serious inputs on how an ecological type of such source of energy might be designed by mankind. All those aspects are really of high interest and researches are going on.

- A troubling set of correlations and coincidences illustrate for this particular case how various phenomena with their lifecycles, their appearance, and development are connected to each other and how Mother Nature gives us lessons on how to manage complicated lifecycles of high energies without damaging it.

There is a vast literature on Oklo reactor, of which the references are representative in our view. The references could be started with the works of PK Kuroda, who predicted the first the possibility of the existence of a natural fission reactor on Earth.

2. Evaluation method

This chapter will focus only on the lessons learnt so far. However, there are more than only natural sciences implications, but also on the manner we acquire knowledge, on how we build models and interact with their reality and how we related to their lifecycles.

Therefore the chapter will not address the details of the researches on Oklo, but rather the lessons learnt to the humanity for such a discovery. The approach adopted in the presentation of Oklo lessons in this chapter is also based on some author's researches on the philosophy of science and models proposed to consider, model and interact with the energy sources, by describing their creation/emergence, their lifecycle and their interaction with mankind and its knowledge.

For this endeavor, a systematic approach was adopted and presented previously [6–9]. Based on this approach the NES and MMES are evaluated in their interaction and development/transformation from one to another in a systematic manner, which is based on some assumptions, as follows:

1. Energy sources create systems, which might be considered Complex Systems (CS) [6] These systems are composed of elements and connectors between them defined as categories, in the mathematical sense [6].

2. For the ES considered as CS, defined by NES and MMES, because they have a behavior of topological nature and for their models, a topological description is possible, as they

    - are described by invariants, that preserve their nature after transformations,

    - create complex networks fractal like structures and

    - their emergence/transformation from one phase/state/form/source to another takes place step by step [10].

The KP of a given ES for a given NES cannot be predicted in detail, but in its general features. The proposed approach considers that the KP generates a topological structure (K(i)) based on a set of relationships between the objects modeled and it is developed in accordance with a certain Theory (Th(K(i))). The topological structure resulting from the KP is in isomorphism with the topological structure describing the emergence rules of the NES from one state to another. The method is based on three principles [10]:

Principle 1: The topological structure K(i) is described by the notion of category considered as:

- reflecting a hierarchical "matrioshka" type of structure

- being a general description of cybernetic description of objects and models as "black-boxes" for each level of construction and for each object.

- being described by objects, morphisms, and identity morphisms

Principle 2: KP is performed in iterations on the categories for each object and each level up to the moment of reaching a critical status due to number and type of paradoxes that result at each step.

- The set of invariants (syzygies) is continuously optimized from diverse points of view (using tools from different sciences) and based on the existing results on them a final set of minimal syzygies for a given model —using a given scientific tool—is reached (Hilbert's syzygy theorem).

- The process of reaching a status for a set of syzygies is therefore predictable and has an end. However the end state described by the resultant set of syzygies in that KP phase may not correspond to the real object. Therefore, a new iteration using another type of methods—analogy from another science that the previous iteration—is used for a new iteration.

- The KP with these new tools will lead to another set of syzygies and have a status of paradoxes in comparison with the real object that will require a new iteration etc.

- An example of NES group is presented in this paragraph. NES are assumed to consist of the following levels of energy sources (NES):

    - Subquantic (SQ)

    - Quantic (Q)

    - Electromagnetic (EM)

    - Molecular (MO)

    - Molecular and life (MOL)

    - Conscious planetary life (CPL)

    - Stellar and universe not alive (SUNA)

    - Stellar and universe life (SUA)

    - Conscious stellar and universe (CSU)

Principle 3: KP is asymptotically stable and complete. However the resultant final structure of this process, which is a CAS, may not be known by its detailed phenomenological characteristics, nor predicted, but rather known for its dominant syzygies.

3. The invariants are called syzygies and they are in the format described by formulas (1) and (2).

$$\text{Gen}\,[\text{NES}] = [\text{EnTh EnI Sy Em NlnCx Fr}] \tag{1}$$

$$\text{Syzygy}\,[\text{NES}] = \text{f}\,(\text{Gen}[\text{NES}]) \tag{2}$$

There are some specific generators (in the sense of syzygy theory) for a K(i) structure built for NES:

- Exergy (Ex) of a NES (defined as the maximum useful work possible during a process that brings the system into equilibrium with a heat reservoir), as a measure of the efficiency of an energy conversion process. This generator has some specific characteristics:

- It is conserved only when all processes of the system and the environment are reversible
- It is destroyed whenever an irreversible process occurs.

- Entropy in a thermodynamic (EnTh) interpretation as a measure of disorder
- Information entropy (EnI) (as a measure of knowledge limits themselves)
- Synergy (Sy) as a measure of a resultant set of features for a NES appearing from the existence and interaction of all systems and subsystems, leading to a set of characteristics for the whole NES than exist in the sum of its parts
- Emergence (Em) from one level to another (in the example for NES presented from SQ to CSU) a process in which larger entities, patterns, and regularities arise through interactions among smaller or simpler entities that themselves do not exhibit such properties and evolve to new levels.
- Nonlinearity (even for simple systems) and/or complexity (NlnCx) of NES as sources of chaotic structure and behavior
- Features of CAS—fractals type of structure (Fr) of NES and K(i) knowledge topological structures built for them.

4. The physical meaning of the dominating syzygies, defining the phase change of ES (NES and MMES) is that they are a triadic set of characteristics of the state of the ES/syzygies and are [10, 11]:

   - Energy (E)
   - Mass (m)
   - Entropy (ψ)

These are optimal descriptors of each ES state and are described by the formulas (3)–(5)

$$E^{(k)} = E_0^{(k)} + \sum_{j=1}^{8} E_j^{(k)} * i_j^{(k)} \tag{3}$$

$$m^{(k)} = m_0^{(k)} + \sum_{j=1}^{8} m_j^{(k)} * i_j^{(k)} \tag{4}$$

$$\psi^{(k)} = \psi_0^{(k)} + \sum_{j=1}^{8} \psi_j^{(k)} * i_j^{(k)} \tag{5}$$

where

- E_0, m_0, ψ_0 –and $E_1{}^{(k)} {*} i_1{}^{(k)}$; $m_1{}^{(k)} {*} i_1{}^{(k)}$ $\psi_1{}^{(k)} {*} i_1{}^{(k)}$ (Noted for the states 0 and 1) define the term called real energy/mass/entropy; examples of energy in such states are the energies perceived at Earth level by a human observer (including such as NES_Oklo), defining the Real Reality.

- indexes 2 and 3 the simple complex part (for the states 2 and 3); examples of states of this type are the paranormal phenomena, energies, information channels perceived by a human observer becoming part of the observed object, defining the Intuition Reality of the second level Realm (cosmic) and

- the rest of components are the hyper-complex part (for the states 4–8); examples are states of paradoxical situations coming from other realities and totally unexplainable for a human observer, but managing them by enantiotropy feedback chain (entropy of states of the triadic ES) and they are our connection to the Universe Realm and diverse realities (Universes) (formula (6))

- The entropy has the following dominant syzygies for each state, as follows [11]:

    - Thermodynamic entropy, for the states 0 and 1 for the real states

    - Shannon entropy for the states 2 and 3, for the simple complex states

    - Enantiotropy for the states 4–8

The triadic set of syzygies defined the set of Realities (as in formula (6))

$$R^{(k)} = R_0^{(k)} + \sum_{j=1}^{8} R_j^{(k)} * i_j^{(k)} \tag{6}$$

5. ES and their models define topological algebraic spaces, which might be represented as polyhedral type, describing their states and illustrating the optimal cases.

6. The description of emergence/transformation of one source in another or of passage from one phase to another is based on the method presented in [6, 10].

7. ES and their models exist in two types of interconnections, with:

    - Other natural phenomena

    - At a given level of civilization

For instance NES_Oklo appeared 2.5 billion years ago, while the "Reactor designer" had at its disposal:

- A certain geological configuration

- A certain status of living beings

- A certain status of interface with cosmos

- No existing civilization

- Environment as we know being under construction

However, the interpretation we make of this source is done at a certain level of our civilization (in its very early beginnings, judging by the criteria of what kind of energy we could harness) [6]. We are far away by several centuries before being able to harness the energy of our sun, which is quite a primitive phase. On the other side, our KP is based on an extremely advanced tool (the interdisciplinary and trans disciplinary one) which may push us to advance much faster than we may envisage now. However, the stronger the forces we harness, the higher the risk to get to the finish of civilization by self-destruction.

We are at a crossroad of the evolution and lessons from NES like Oklo are extremely useful, as they show us how to harness better high energy with high risk sources [6].

8. In our present knowledge the KP assumes for the ES cases a set of assumptions generated by the paradigms, creating paradoxes, as for instance [6]:

    - **Paradigm 1-ES as a CS:** A modeling system has to be built in order to represent Risk Analyses for ES (RES) as a complex system, too. RES is converging to a stable unique real state. However the KP results, including those RES are limited by our present knowledge, as described by the real Earth level mentioned above.

    - **Paradigm 2:** ES model involves knowledge of the risks associated to a certain source of energy. However, usually we actually are not aware of the real risks and we know very little about the interconnections of lifecycle dangers for interfering processes (energy level, emergence correlated with civilization one or with geological one etc.)

    - **Paradigm 3:** Details of ES and their lessons learnt. We design ES (MMES) for which Nature already indicated the optimal solutions. However, due to our reduced technical and scientific level at a certain moment we cannot understand the lessons from the beginning, but step by step.

    - **Paradigm 4:** Understanding the ES risks (RES) and defining them is a difficult task as we design first of a kind MMES and as we are not aware of all the aspects of the lifecycle. The MMES are challenged inevitably by serious events, which apparently test the design continuously.

    - **Paradigm 5:** ES risk analyses results are seen as inputs to decision making risk calculation results are used for decisions. However we are facing decisions under high uncertainties and the use of lateral thinking is decisive.

    - **Paradigm 6:** In the ES risk analyses results there are limits and biases specific to the level of knowledge of that issue, but also there are "hidden" biases due to the level of KP in the whole civilization at that moment. Inter and trans disciplinarity is not just a desired option, but a mandatory one to minimize such biases.

    - **Paradigm 7:** RES results evaluation for further iterations in the.KP is an iterative process and the Principle 3 mentioned above applies. The result could be a better understanding by the use of diverse tools, as for instance the information one can get by "backward engineering" from natural examples.

3. Lessons learnt from NES_Oklo

NES_Oklo sends to us messages. By diverse evaluations one could mention so far messages as the following:

1. The issue of the meaning of risk analyses for ES is very important, as the lessons learnt from NES_Oklo show. NES_Oklo was a combined non-live living organisms operation to produce energy. This is a high important topic for the future MMES to be designed by assuming the use of Artificial Intelligence, may be also natural and living organisms, etc. The evolution of our civilization and/or possible future interactions at cosmic level require a clear strategy on how to proceed if combined (natural, artificial, living non-living, etc.) energy sources production is to be evaluated and designed.

2. NES_Oklo teaches us on the absolute importance of intrinsic safety (the reactor operated, got decommissioned without being of any harm to its environment, but on the contrary, being part of the evolution "plan").

3. NES_Oklo has the following features of importance for future evolutive MMES to be designed, built and operated by the mankind:

    a. The limits of NES_Oklo were very well defined for all its lifecycle phases

        i. During operation

            1. Geometry stability of the core assured by the rocks configuration (the concrete part of any MMES)

            2. Climate was stable in the parameters of the period

            3. Interface with living organism was designed to be not only harmless, but also useful for both sides (cyanobacteria were prosperous for several millions of years).

        ii. During decommissioning

            1. There was no migration beyond the site of the heavy radioactive solid waste.

            2. The aerosols were actually part of the plan to rebuild the Earth atmosphere and generate new living beings—eukaryotes.

            3. Apparently the design assumed how to better decommission it at the end of the lifecycle. Thinking of decommissioning from the research phase is a mandatory requirement for a well-designed MMES.

    b. There is a fractal like design of the whole NES_Oklo reactor, as for instance the manner the following reactor functions were assured, as reflection at lower levels of the same principles:

        i. Fuel load (uranium 235) to the reactor core, assured by cyanobacteria, as an intrinsic self-regulated process, in mirror with the operation of the whole reactor.

ii. Diffusion of small distances in the specific rock of the site (several meters for more than 2 billion years [12]).

iii. Radioactive radio-sols were part of the creation of new living organisms; therefore the containment was the whole atmosphere, without damaging it, but helping it.

iv. There was an intrinsic safety assured by delayed neutrons, preventing transformation of the reactor into a bomb

c. The validity of reactor physics codes used for MMES was highly challenged. Although it seems so far that they could reproduce the reactor core design, there are yet issues to be clarified.

d. NES_Oklo has a direct impact on the lifecycle preparation of existing and future MMES, as follows:

i. Review the type of best plant control—centralized versus decentralized

ii. Review of the safety analyses models for all the lifecycles and especially for decommissioning

iii. Review existing researches on the future man machine interface for new reactors, role of artificial intelligence and the role of KP and generations to operate the plants

iv. Set the goal of maximum simplification of MMES, counting to the highest extent possible on passive features and intrinsic safety protection.

v. Review the manner various phenomena are modeled for the reactor in coupled computer codes and either use higher computing capacities or simplify them

vi. Design MMES as part of regional/global energy sources systems, integrated in the environment, based on ecological principles.

e. Several aspects from fundamental quantum mechanics and theory of relativity are yet to be reviewed, as the NES_Oklo measurements are challenging some of them

i. How constant is the alpha constant and the role of the amazing number 137 in the architecture of the universe

ii. It appears that some constants are not so constant (for instance speed of light). If so the impact is very high on many aspects already considered confirmed and taboo to be challenged. An epistemic revolution is to be generated in Physics on the way to change the existing paradigms.

iii. There is an amazing set of coincidences to have a reactor core designed (geological, biological, cosmic, etc.). If the rare coincidence might be more or less accepted, the troubling finding that the NES_Oklo is not the only one of this type leads to the debate about anaphatic and kataphatic approaches to the understanding of the *Designer* of the world.

iv. The NES_Oklo operated from the design to decommissioning phase as a cybernetic machine understandable with high level cybernetics considering all the three levels from formulas (3)–(6) —real, simple complex and hyper-complex. The hyper-cybernetics, governed by the feedback control via the enantiotropy (entropy of the optimal ES states) is a very possible answer to previous questions. High level cybernetics—the cybernetics of CS states is indicated as describing such systems.

f. NES_Oklo raises a series of philosophical debates, too:

i. The evolution of life on Earth, the meaning of life and the role of randomness (if any) in its emergence and evolution.

ii. The future of our civilization and how to use better the lessons so that to avoid destroying ourselves by the time we harness more and more powerful energy sources.

iii. Why and how was it possible at a certain moment in time to have NES_Oklo? How to explain strange coincidences of NES_Oklo with eukaryotes, Earth terraforming and conditions for us to appear in the evolution (or what?) chain.

iv. How to understand/manage messages for which we do not have yet the capability to understand, as they are from the category of hyper complex reality?

4. Conclusions

NES_Oklo had so far a significant impact on nuclear physics and nuclear engineering. However, its impact is yet to be completed, as new investigations and interdisciplinary works discover unexpected facts of the lessons transmitted by Oklo to us.

NES_Oklo is an example of how to build and operate an optimal, environmental friendly, for all lifecycle phases, nuclear fission reactor.

Summarizing, its lessons are related to:

- Improvement of the design strategies for new MMES
- Lessons on how to solve the waste management problem
- The high advantages of using combined live-non alive elements in the fuel cycle
- Foster the fundamental research in quantum mechanics, as the lessons are that, we are not yet understanding even basic aspects (as for instance the role of various universal constants)
- Review the models we build for the Physics and ES and improve the KP for those aspects by using systematic approaches

Author details

Dan Serbanescu[1,2]

1 Division of Logic and Models in Science and Technology of the Romanian Committee for Science and Technology, Romanian Academy, Bucharest, Romania

2 Romanian National Nuclear Electricity Company Nuclearelectrica SA, Romania

*Address all correspondence to: dan.serbanescu1953@yahoo.com; dserbanescu@nuclearelectrica.ro

References

[1] Kuroda PK. The Origin of the Chemical Elements and the Oklo Phenomenon. Springer; 1982. ISBN: 978-3-642-68667-2

[2] Gil L. IAEA Office of Public Information and Communication, Meet Oklo, the Earth's Two-Billion-Year-Old Only Known Natural Nuclear Reactor. 2018. Available from: https://www.iaea.org/newscenter/news/meet-oklo-the-earths-two-billion-year-old-only-known-natural-nuclear-reactor

[3] https://en.wikipedia.org/wiki/Oklo

[4] Meshik AP. The Workings of an Ancient Nuclear Reactor. Scientific American on January 26, 2009, originally appeared in the October 2005. Available from: https://www.scientificamerican.com/article/ancient-nuclear-reactor/

[5] Lederman LM. Symmetry and the Beautiful Universe. Prometheus Books; 2004. ISBN: 9781591022428

[6] Serbanescu D. Selected Topics in Risk Analyses for Some Energy Systems. Lambert; 2015

[7] Spiridon LV, Serbanescu D, Sticlaru G. O Privire Asupra Unor Lectii de Cunoastere Date de Cavalcada Modelelor in f Zica. IYL; 2015

[8] Serbanescu D. Energetica si Fizica Nucleara Descoperiri, Accidente, Lectii ale Naturii, Comitetul Roman de Istoria si Filozofia Ştiinţei şi Tehnicii (CRIFST) Curs de Initiere in Istoria si Filozofia Stiintei Seria a IXoa. 2015

[9] Serbanescu D. On Some Natural Energy Systems and Lessons Learnt from Their Analysis [Despre Unele Sisteme Energetice Naturale si Invatamintele Studierii]. ISBN: 9783668669192; ISBN (Book): 9783668669208

[10] Şerbănescu D. An integrated perspective on knowledge and existence. In: Noema XVI; 2017. pp. 185-216

[11] Serbanescu D. On Some Unifying Approaches in Physics and Mind Sciences [Despre Unele Abordări Integratoare ale Fizicii și Științelor Minții—(A Main Text in Romanian with Extended Presentations in English of the Models and Results)]. DOI: 10.13240/RG.2.2.29959

[12] Yucca Mountain Project (YMP). Oklo: Natural Nuclear Reactors. Office of Civilian Radioactive Waste Management (OCRWM); 2004. Fact sheets. DOE. YMP-0010. Archived from the original on 2004

Chapter 19

Research and Development of the New Progressive Construction Press Machines

Ľubomír Šooš

Abstract

Throughout the world in the last two decades, solid noble biofuels produced on the basis of wood and agricultural biomass have been widely used. Many European Union countries have enacted legislation for solid noble biofuels oriented primarily to energy carriers manufactured from wood raw materials. The production of noble biofuels is a fitting direction for the recovery of biomass and other energy wastes. Fuel in the twenty-first century must, in addition to environmental and economic criteria, fulfil the criterion of high comfort and safety in its combustion. Modern energy carriers must have uniform size, density, moisture and shape. One of the most progressive means of waste recovery is the compaction of particulate matter into different shapes and sizes. The technologies of transforming biomass into biofuels with the required properties are compacting, briquetting and pelleting. I am working on a research of compaction technologies and the development of compacting machine design at the Institute of Production Systems, Environmental Technology and Quality Management of the Faculty of Mechanical Engineering of STU in Bratislava for more than 26 years. In its scientific research activity, the definition of equations describing the dynamic process of compaction, experimental tests of the compaction process, design and testing of new press machine were undertaken. Over the stated period, we have designed and developed several machines that are today in serial production. Chief among these is the BL 55–280 briquetting press, the DZ 240 disintegrating machine, the KUDO 2009 mobile unit for treatment of wood waste, the TR 200 solid material separator, the PLG 150 round pelleting press, the ZBL-2-860 double-screw briquetting press or the RCM 650 ring compacting machine. We have submitted over 53 patents and industrial models. The aim of this chapter is to describe the new original and progressive designs of these press machines.

Keywords: research, press technology, original design, machine, patent, fuels

1. Design of the new press machines

Research into compaction technologies and the development of compacting machine construction have been undertaken by the Institute of Production Systems, Environmental Technology and Quality Management of the Mechanical Engineering Faculty at the Slovak University of Technology in Bratislava (ÚSETM SjF STU) for more than 18 years. In its scientific research activity, the definition of equations

 IntechOpen

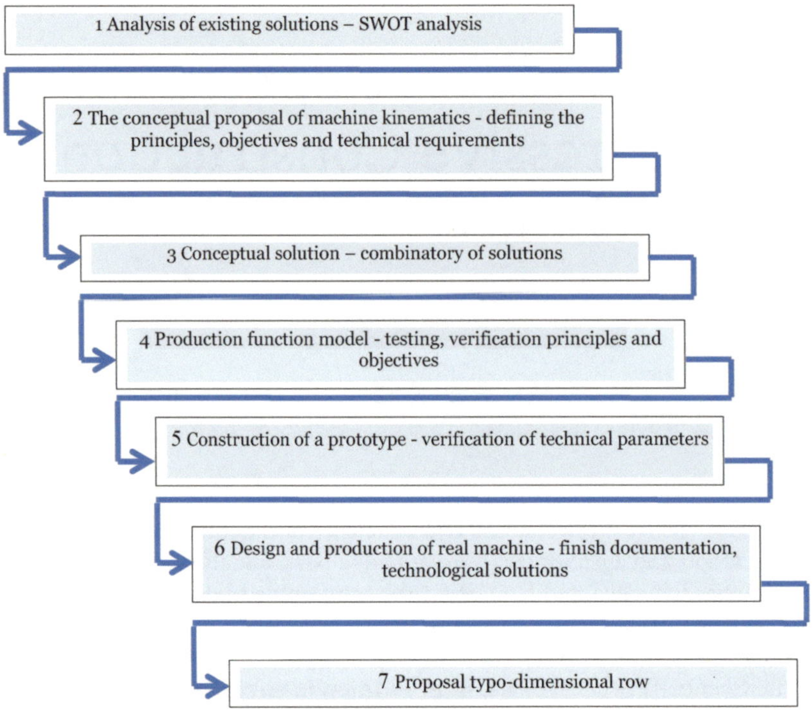

Figure 1.
Project management and coordination design of a new machine [1].

describing the dynamic process of compaction, experimental tests of the compaction process, and the analysis of existing legislation in the area of solid enriched biofuels were undertaken. Over the stated period, the ÚSETM SjF STU has designed and developed several machines that are today in serial production. Chief among these is the BL 55–280 briquetting press, the DZ 240 disintegrating machine, the KUDO 2009 mobile unit making use of wood waste, the TR 200 solid material separator, the PLG 150 round pelleting press, the ZBL-2-860 double-screw briquetting press and others. In this area, members of the institute have submitted over 55 patents and industrial models. The workplace has carried out some 64 national and 48 international projects in this area. Over 50 pelleting and briquetting production lines have been designed and subsequently implemented here, of which projects over 10 were carried out abroad.

When designing a new machine, we used a standard process customary in our department (**Figure 1**).

2. Research of compacting machines

The production of enriched biofuels is a fitting direction for the recovery of biomass and other energy wastes. Fuel in the twenty-first century must, in addition to the environmental and economic criteria, fulfil the criterion of high comfort and safety in its combustion. Modern energy carriers must have uniform size, density, moisture and shape. One of the most progressive means of waste recovery is the compaction of particulate matter into moulds with different shapes. The press technologies of transforming biomass into biofuels with the required properties are compacting, briquetting and pelleting. A common mark of these technologies is the pressing of raw materials under a certain pressure so that the mouldings achieve the required strength and shape. Compaction has several advantages:

- Higher calorific value when incinerating high-quality mouldings compared to combustion of the same material in an uncompressed state.
- A more uniform and longer process of burning of the mouldings than the material in an uncompressed state.
- Sufficient strength of mouldings guarantees high biofuel comfort during transport.
- High density of mouldings minimalizes costs for transport and storage.
- The high temperature and pressure during press process of the biofuel prevent the onset of biodegrading processes.
- The processing of otherwise hard-to-use waste into the form of enriched biofuel and others.

Immense progress in the usage of compacting technologies for the production of solid enriched biofuels has been recorded throughout the world. For instance, in Germany, Denmark and Austria, the demand for such biofuels exceeds domestic production. On the other hand, a certain limit in the energy usage of solid enriched biofuels from biomass has been reached at present, since many times their price is not commercially comparable with using fossil fuels [2]. The price of enriched biofuels is increased directly by high transport and in particular energy costs for their production. A reduction of such expenses is only possible by the application of new advanced progressive technologies and new constructions of the machinery necessary for production. These machines must be more economically effective with long working part lifetimes. Among the dominant problems for the producers of solid enriched biofuels are:

- Varying, often low-quality, mouldings (density, abrasion, strength) due to insufficient familiarity with the process.
- Excessive wearing of active parts of the compacting machines and devices.
- Inexact mathematic models, by which the compacting machines are designed.
- Mould shapes are not optimal, as mouldings are now produced as cuboid, e-angled (briquettes) or cylindrical (briquettes and pellets). These shapes are not optimal from the viewpoint of automatic supply in the combustion process.
- A lack of pure wooden biomass forces producers to search for new materials suitable for the production of enriched biofuels. These include agricultural raw materials or fitting industrial or separated municipal waste.

Compaction machines often are not constructed or designed for these materials, and therefore overloading, breakdowns and shortened operational life occur, as well as higher costs per unit of production and under- or over-dimensioning since the machines do not have optimised drive power. These all lead to increased energy demand which forces up the prices for such biofuels.

At present, it is necessary to develop and introduce into practice the highly effective press machines with extended lifetimes of their functioning parts and devices, optimised on the basis of precise mathematical models derived from experiments, and machinery with low energy and operation expenses per unit of compacted material. By such increases in effectiveness and reductions of production costs, it will be possible to produce biofuel at low prices that will make its use more widely applicable and its cost competitive with fossil fuels. This is a tool for achieving one of the primary goals of the European Union—increasing the implementation of renewable energy resources.

The contemporaneity of the project's intentions confirms the ever-increasing interest in producing energy carriers, increases in the number of users of biomass furnace for producing heat and also the continuing growth of production and consumption of solid enriched biofuels in Europe and around the world, not to mention in Slovakia.

3. Development briquetting machine

At the present time, we know three basic principles for operating briquetting presses: mechanical, hydraulic and screw (**Figure 2**).

Briquettes are suitable for kitchen stoves, brick ovens, level and central heating and fireplaces. Burning time of briquettes is from 30 to 60 min, depending on their shape and size. The moulds produced are generally in the form of cylinders, blocks or n-angles (**Figure 3**). In conformity with DIN 51731, the characteristic size is greater than 30 millimetres. Larger pressed fuels catch fire less easily and burn worse, and their combustion tends to be less ideal. Briquettes with larger diameters are also made with holes.

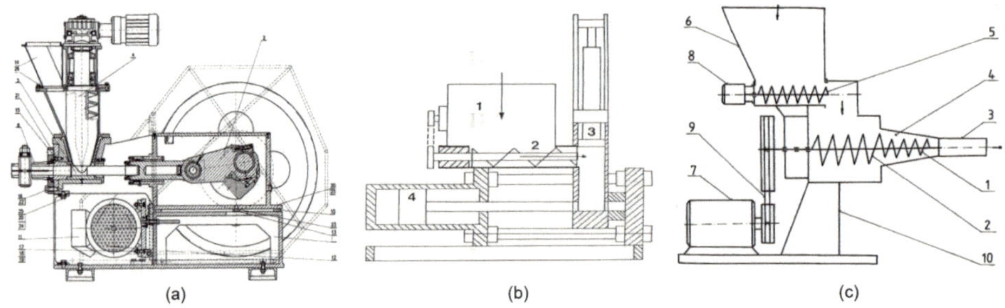

Figure 2.
Different principles of briquetting presses. (a) Mechanical, (b) hydraulic and (c) screw.

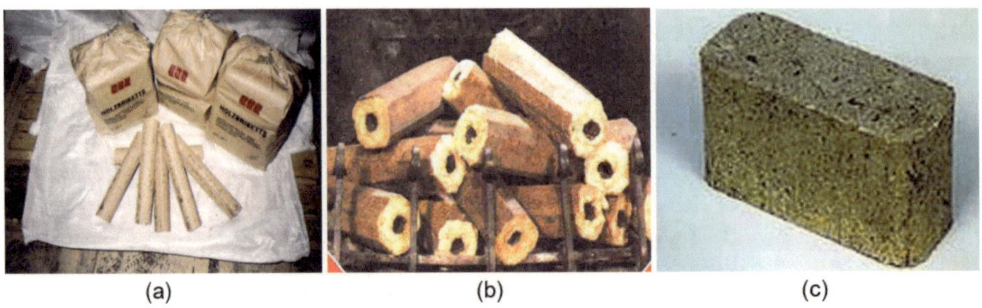

Figure 3.
Actual shapes of briquettes. (a) Cylindrical briquettes, (b) octagonal briquettes with holes and (c) block briquettes.

DOI: http://dx.doi.org/10.5772/intechopen.91405

3.1 Development of the screw briquetting press

The highest quality and strength of briquettes are achieved on screw briquette presses. But with this method, there is greater wear on the working equipment—the screw press. Screw briquetting presses (**Figure 2c**) clearly produce the best-quality briquettes (**Figure 3b**). The advantage of this sort of briquetting is that, as opposed to the other technologies, the moulds are produced continuously. The use of screw presses has a whole range of positive attributes. The briquettes are accordingly compact, with high densities, and can be produced in a number of shapes and sizes. The machinery is not noisy and do not create vibration comparable to other mechanical presses.

Despite the stated and undeniable benefits of this technology and the desirability of its moulds on the mark, such machines are almost non-existent. The primary reason is the very low service life of the tool and the lifetime of its axial bearings. The greater the expansion of screw briquetting presses, the shorter the lifespan of the work machine. The average life of screw and tools for such a press machine produced by today's manufacturers is from 50 to 120 hours, depending on the pressed material. This low durability and the consequent high cost for repair of screw or purchasing new tools outweigh the undisputed advantages of briquettes produced by this technology. Current designs of screw briquetting presses have several structural drawbacks. Among them are:

- High wear of the end of the screw
- Low axial bearing durability
- Heating of the rotating screw during start-up and cooling during operation
- Exerting high torque through the small diameter of the screwed shaft

Research into screw briquetting presses has also been intensively addressed. We have had successful cooperation with the Austrian company, Pini Kay. The result of the research is apart from other things a mathematical model for the design and optimization of the geometry of screw presses (**Figure 4a**), the design of a suitable abrasion-resistant material or the design of a rapidly replaceable screw head (**Figure 4b**).

The result of this research has been a solution for removing faults in the existing constructions of screw presses, not removing the reasons. With regard to the above, there emerged the original idea of designing a totally new conception of the construction and kinematics of the screw compactor. In order to eliminate the high axial load of bearings, we designed the original design *(UV 6045)* [4]—the

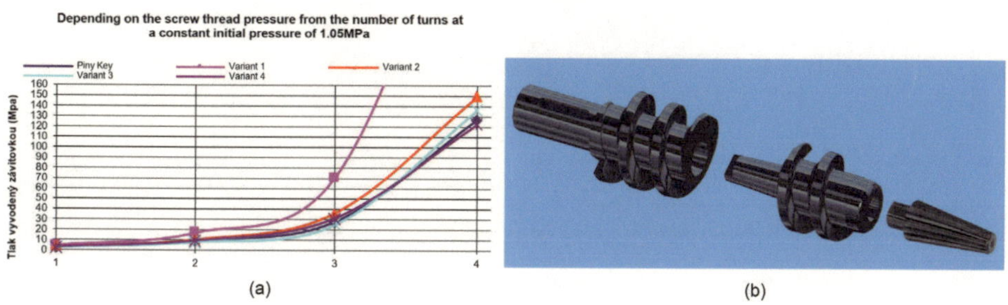

Figure 4.
Results of research on the screw press. (a) The course of pressures and (b) replaceable end (SK 288206) [3].

backward arrangement of the screws (**Figure 5**). A real outcome of this research is the prototype of a double-chamber screw press with a backscrew arrangement (**Figure 6**).

3.2 Development of the modular conception of mechanical briquetting machine

Constructional development mechanical briquetting press can be divided into development of an overall modular conception of a briquetting device and research into original design nodes. The basic requirement in the design of the modular conception was to achieve maximum universality of the individual construction elements and, with respect to the number of pressing chambers, the range of manufactured production offered. Research into original design nodes is focused on a significant expansion of the basic functions and technical parameters of the compaction machine as well as the possibilities of compacting waste with increased moisture and the production of briquettes and pellets on a single compacting machine.

3.2.1 Briquetting mechanical press BZ 50-300

The BZ 50–300 briquetting press is a mechanical briquetting press with an open pressing chamber (**Figure 7**) [5]. The press is driven by an electric motor, a pulley and a flywheel. The transformation of rotational movement to rectilinear is performed by the crank mechanism. The cylindrical piston tool performs direct rectilinear motion. The advantage of the press is its simple and reliable construction.

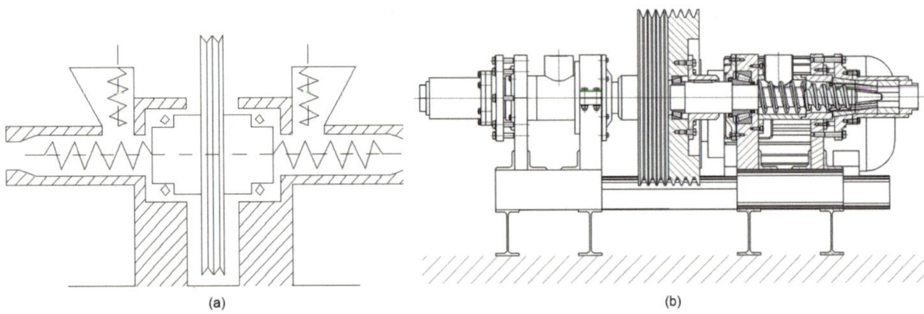

Figure 5.
Bicameral press of pressures. (a) Schema of press and (b) real design.

Figure 6.
Results of research on the screw press. (a) Prototype and (b) press in the production line.

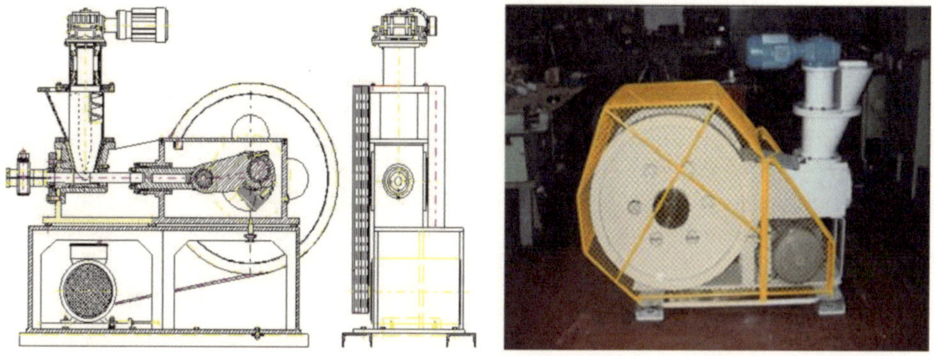

(a) (b)

Figure 7.
Briquetting mechanical press BZ 50-300. (a) Press construction and (b) actual look.

The diameter of the briquettes produced is from 50 to 60 mm, the total press input power is 17.5 kW, and the hourly press performance is 250–350 kg of briquettes. Up to the present, more than 80 units have been produced, not only for the Slovakia market but also for export to the Czech Republic, Estonia, Hungary and Austria. The press is manufactured by Konštrukta Industry Trenčín, and at this time its production is also being prepared at the Vural Žilina company.

3.2.2 Dual-chamber briquetting press BZ 2-50-600

In cooperation with the Vural Žilina company, work on a dual-chamber briquetting press for the BZ 2-55-600 (**Figure 8**) is in preparation [6]. The advantage of the new resolution is lower machine input per unit of output. The hourly output is doubled while the press input power is less than twice. Accordingly, costs for the press are reduced because up to 83% of the pieces are not dependent on the diameters of the briquettes produced nor on the number of pressing chambers.

The individual parts as well as the entire machine are verified for strength according to the required diameter of the product and the type of material processed. In the design and dimensioning of the construction of the dual-chamber press, we needed to know, just like with the single-chamber press, the course of

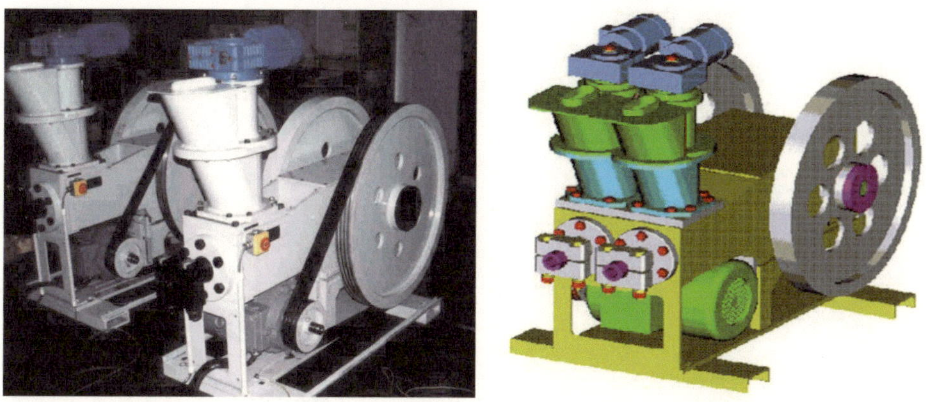

(a) (b)

Figure 8.
Briquetting press. (a) Two single-chamber presses BL 55-300 and (b) dual-chamber press BL 2-50-600.

power during lifting and its maximum value so that we would be able to correctly dimension the amount of drive, the flywheels, the balance of the crank mechanism and the strength of the individual components. This strength changes quite dramatically in dependence on the type and current state of the material to be compressed. From the standpoint of dimensioning and of the production itself, the new crankshaft (**Figure 9a**) was critical. A completely new part of the press was the construction design and the dimensioning of the crankcase, for example, the crankshaft (**Figure 9b**).

3.2.3 Unification of the parts of the briquetting press

The final goal of this task was to rework the existing construction of the briquetting press to a modular state with varying dimensions of the briquettes or pellets produced, in single- or dual-chamber versions. According to the requirements of the client, we are able to design a machine tailor-made from the viewpoint of the type of processed material and the diameter of the produced mouldings and from the standpoint of the required output of mouldings hourly production.

For design optimization "Modullis" application software was created [7]. The software also includes an analytic part containing mathematical models for dimensioning and control of the individual parts of the press. The combinatorics of the design allows for a rapid choice of the suitable parts according to the selected press version (single-chamber or dual-chamber) and the diameters of the produced briquettes (∅ 50, 55, 60, 65, 70 mm) or pellets (∅ 6, 8, 10, 12, 16, 20, 24 mm). The advantage of the designed module make-up is that the parts are made separately from the final assembly, which greatly reduces the manufacturing costs. The optimization result is the choice of a final set of parts for assembly and guaranteed machine output for compression of the required material.

In **Table 1** all variants of the briquetting press (briquette diameters ∅50–70 mm, single-chamber or dual-chamber versions) are shown; for each variant a list of the optional parts is written. The idea of this sketch can be transformed into a computer program on the basis of briquette diameter requirement. The number of pressing chambers determines how many new parts will have to be used for the refitting of the existing briquetting press whose construction forms our starting point, as well as the hourly output of the selected briquetting press. The parts of the single-chamber press are identical, for easy fitting into the single-chamber press. The universality of usage of the individual parts is clear from **Table 1** and the graphs in **Figure 10**.

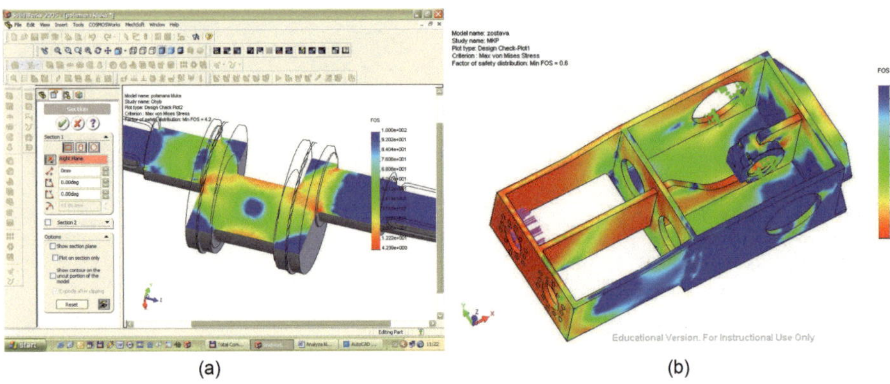

Figure 9.
Simulation of burdening of press parts [6]. (a) Control of crankshaft for bending and (b) pressure course.

Dual-chamber briquetting press		
Subgroup	**New**	**Adapted from BL 50-250**
Frame	x	
Main drive (electric motor)	x	
Flywheel		X
Pulley		X
Crusader		x
Pressing chamber		x
Screw feeder		x
Spring collet		x
Fan belts		x
Feeding socket		x
Crank box	x	
Crank shaft	x	
Crank shaft roller bearing		x
Crank shaft sleeve bearing	x	

Table 1.
Unification nodes.

3.3 Development of original parts of compacting machine

Part of the development of compacting machines is the research of the original construction nodes focusing on the significant expansion of the basic functions and the raising of the technical parameters of produced machines. Here the "breathing hood", the "pelleting insert" or the "folded piston" of the machine belong primarily.

3.3.1 Breathing briquetting hood

All of the compaction technologies analysed up to the present allow compaction of only materials with relative moisture lower than 18%. At the same time, drying is demanding in both an energy and an investment sense. This problem can be solved in two ways.

The first method is the replacement of the technology of drying by a new, less energy-demanding, technology. This refers to centrifuging, pressure, drainage and chemical technologies and their mutual combinations.

The second method is managing the technological compacting process with a higher value of relative moisture than the limit (18%). For this purpose we developed in our workplace a so-called breathing press hood [8], which allows for the draining of water from the hood of the press (**Figure 11**). In this way it becomes possible to achieve quality mouldings with input raw materials with 22% relative moisture.

On the stated hood, we briquetted wood waste with input moisture of 22%, with the mouldings showing sufficient quality. Relative moisture w_r is defined by the ratio of the weight of the water and the total weight of the pressed raw material. Due to the high pressure in the pressing process, the water is in a liquid state. After the exiting of the briquette from the pressing chamber, the pressure falls, and the water changes into vapour and escapes from the briquette, while tearing it. The essence of the idea is to allow the escape of the vapour directly from the pressing

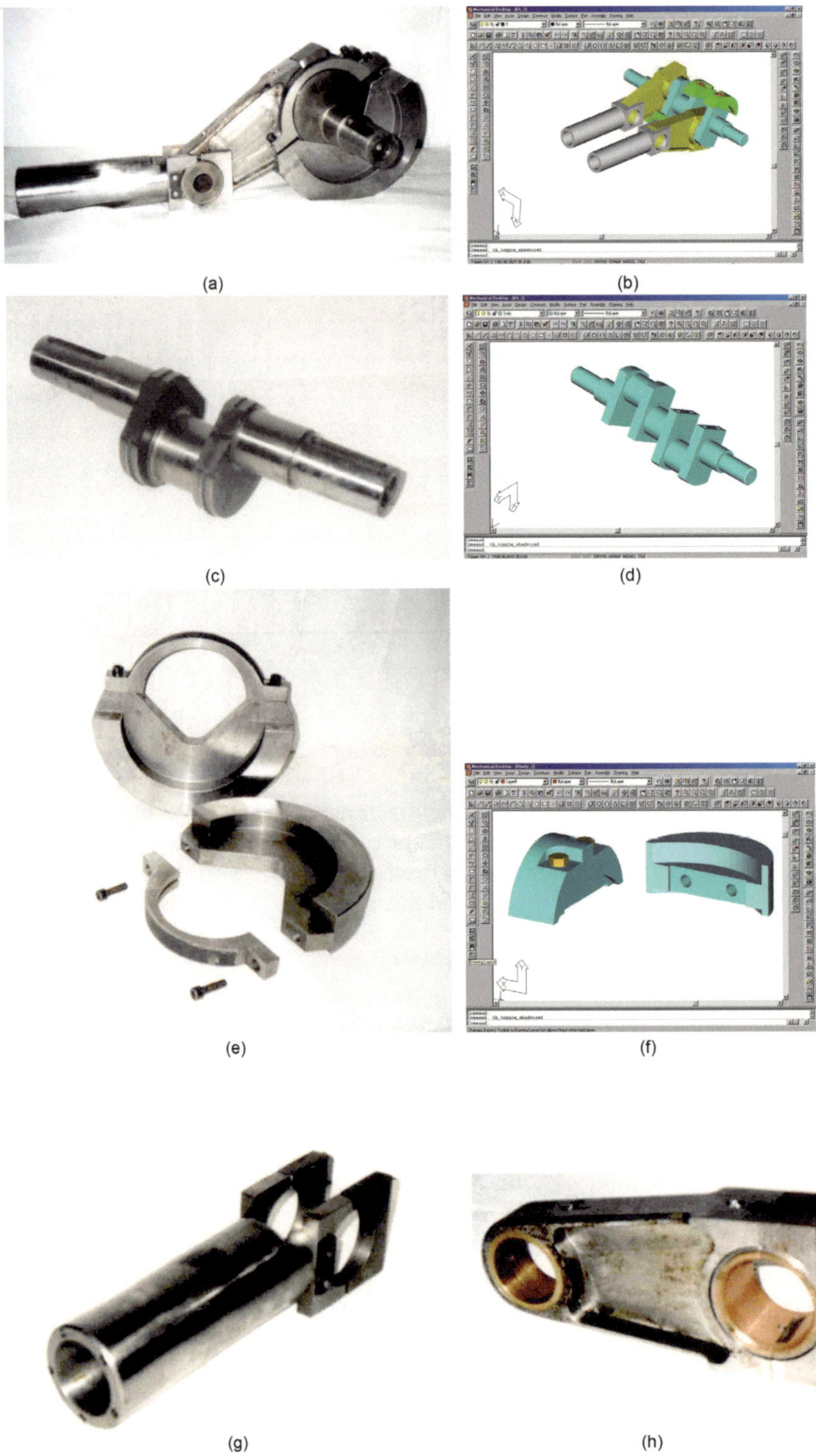

Figure 10.
Selection of parts for modular conception of a press. (a) Crank gear of single-chamber briquetting press BZ 50–250, (b) crank gear of dual-chamber briquetting press 2BZ 50–500. (c) Crank shaft of single-chamber briquetting press BZ 50–250, (d) crank shaft of dual-chamber briquetting press 2BZ 50–500, (e) counterweight of single-chamber briquetting press BZ 50–250, (f) counterweight of dual-chamber briquetting press 2BZ 50–500, (g) crusader – usable in both versions and (h) connecting rod.

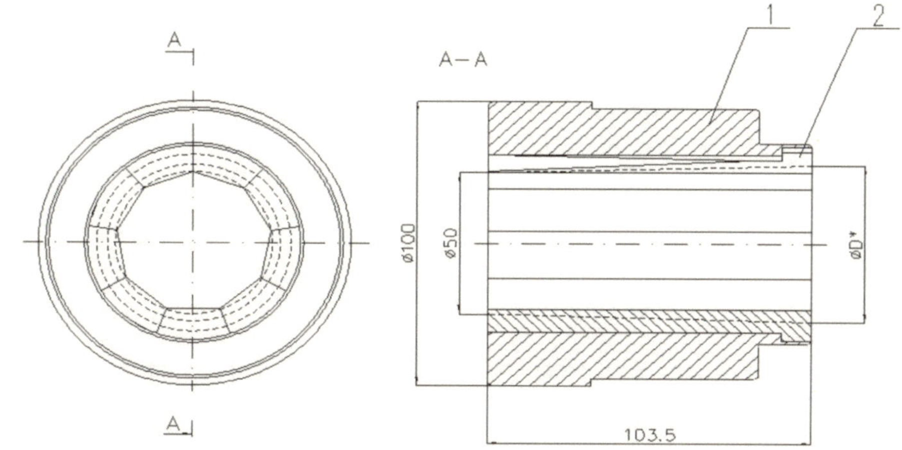

Figure 11.
Sketch of a breathing hood, 1 – hood body, 2– pressing segments, [8].

chamber during the reverse movement of the piston and a drop in pressure. Adjustments to the piston or drilling holes in the hood were not successful because their rapid shutdown resulted. The centre point of the proposed hood is that the formed vapour can escape through slots between the individual press segments. The slots close completely during the reverse movement and the drop in pressure.

3.3.2 Pelleting hood

The original result of the research into the construction of a briquetting press is the multi-technological construction of a press hood for a compacting machine [9]. Our attempt is to also produce pellets through a simple change of the compacting machine's press hood. The principle of the change is clear from **Figure 12**. In **Figure 12a**, we can see the briquetting hood of a crank briquetting press, which standardly produces briquettes with the required diameter. By adding (inserting) a pelleting insert (**Figure 12b**) to the briquetting hood, on that same press, we can make pellets with the required diameter (**Figure 12c**). In the two-dimensional range, we have designs of inserts that allow producing pellets with diameters of 8–20 mm. An example of a four-holed insert with a pellet diameter of 20 millimetres is seen in **Figure 12**.

Verification functioning tests were performed on straw and wood at the OPS company in Lehota pod Vtáčnikom. The positive result expanded the usage of the already existing briquetting press in producing economical pellets.

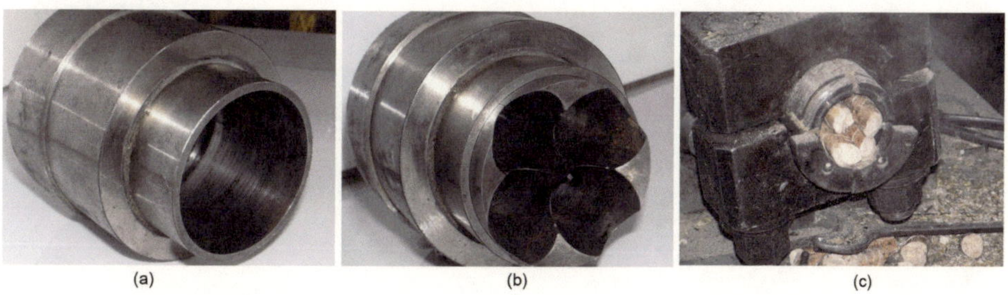

Figure 12.
Briquetting press pelleting Hood [9]. (a) Briquetting hood, (b) hood with pelleting insert and (c) production of pellets.

3.3.3 Folded press piston conception

A very quick wearing part of a briquetting press is the pressing piston, its front part in particular (**Figure 13a**). This problem can be solved through the use of quality materials and their thermal chemical treatment [10]. Another approach to the effective resolution of this problem is the construction design of a folded piston (**Figure 13b**).

4. Development of pelleting machine

The common feature of all press technologies, which include pelleting, is the compression of materials under high pressure [11].

A screw extruder is known, in which the screw feeder is also the tool of this press (**Figure 14a**). The disadvantage of this machine is the need for a cooling system and its low hourly rate of output during the moulding production. Another one is a horizontal pelletising machine with cylindrical rollers and a cylindrical die (**Figure 14b**). The disadvantage of this machine is the uneven supply of materials under both rollers. A horizontal pelletising machine with cylindrical die and compression rotor is also similarly designed. The disadvantage of this design is the increased friction between the tool and die and the resulting faster rotor wear. The construction of a horizontal pelletising machine with gears is also known (**Figure 14c**). Such a pelleting machine comprises a pair of hollow cylinders with gearing around their perimeter. There are holes drilled in the gearing through which the material is extruded into the interior of the cylinders. Cutters located in the interior of the cylinders cut the compressed pellets. The disadvantage of this machine is the small number of teeth on the cylinders, as a result of which it achieves a low hourly rate of output. There is also a vertical pelleting machine with conical rollers and a flat die (**Figure 14d**). The disadvantage of the machine is the high and uneven wear on the rollers.

The percent share of pellets on the refined fuels market is increasing dramatically. Currently, pellets are made in a cylindrical shape, with diameters of 6–25 millimetres and lengths of pellets equal to two to five times the diameter (**Figure 15**).

Greater sized pellets are used only for combustion in large furnaces and cement-making. The greatest advantage of pellets is that, due to the size and homogeneity of the fuel, a fully automated combustion process is possible. The disadvantages of pellets are the high ratio of the fuel's surface and its volume. With this the burning of volatile materials even at low temperatures (above 200°C), the very rapid

Figure 13.
Press piston. (a) Wearing of a press instrument and (b) stacked piston.

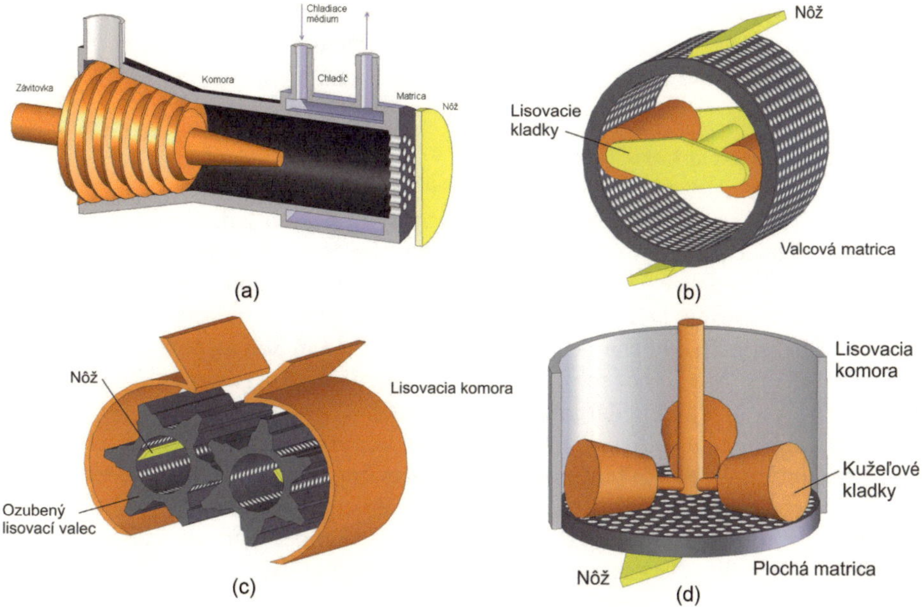

Figure 14.
Pelleting machine. (a) Horizontal screw, (b) cylindrical rollers and cylindrical die, (c) horizontal with gears and (d) vertical with conical rollers and a flat die.

Figure 15.
Pellets.

burning process (10–20 min) and also the high wear of the functioning parts (die, roller.) of the pelleting mills are related. Another disadvantage is the high demand associated with the production technology. The material must be of high quality, homogeneous, able to disintegrate into very small particles and have optimal humidity. The high investment costs associated with the manufacture of the fuel and with its usage are also disadvantageous. The input power for a unit of volume output "$P_{j,W}$" [W/(kg h^{-1})] is lower for pellet presses than for briquettes. This has to do with the fact that the ratio of the area of the press openings to the total area of the die on which the press pressure acts is less than 1. Only in special furnaces is it energy effective to recover pellets. The very high demands on investments also apply to the user of the fuel.

4.1 Pellet mill with spherical tool

As a consequence of the persistent problems with pelleting methods and with the design of pellet mills, there is an opening for the design of a pellet mill which would offer reduced energy costs and reduced wear of machine components [12]. The result of this effort is the pellet mill described further in the submitted invention. The first principle of the pellet mill was described in Slovak patent SK 286877 [13], titled as the "Method of pressing pellets of particulate organic and/or inorganic materials or raw mix and pellet press" (**Figure 16**).

Pellets compression is realised when the raw material with a required fraction and of maximum water content of 18% continues in to the open-work pressing chamber in the shape of the spatial effect of the V-ring.

The raw material is pulled in the work pressing chamber from the rotating screw and/or rotating balls. The rotating circular parts touch at the bottom. Open-work pressing chamber partially encloses a rotating spherical shape. After that the raw material is injected from the pressing chamber into the extrusion orifices or a whole die, from which mouldings emerge. Pellet mill consists of a hollow wedge-shaped body (1) with circular recesses for the wedge-shaped surface, wherein the wedge surface has a first and a second flange (2, 3). In the first flange (2) the first drive

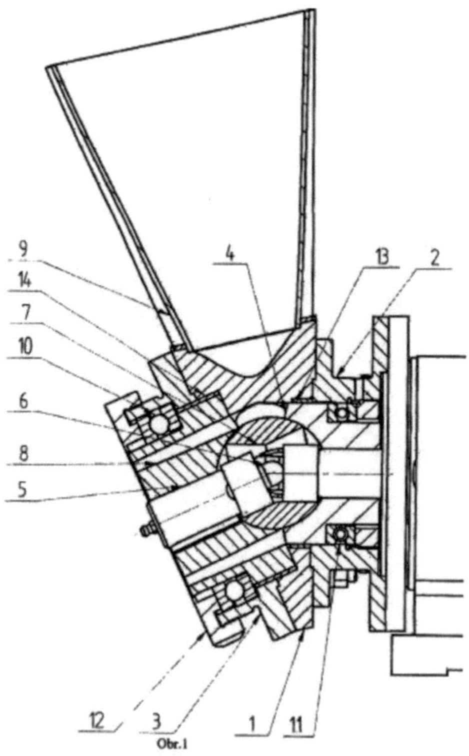

Figure 16.
Cut of the pelletizer PLG2010 [13]. 1 – Wedge body, 2 – first flange, 3 – second flange, 4 – first drive disk, 5 – second drive disk, 6 – the kinematics of the drive mechanism, 7 – ball tool, 8 – die, 9 – hopper, 10 – axial bearing, 11 – axial bearing, 12 – die, 13 – sliding sleeve, 14 – sliding sleeve.

plate is concentrically and rotatable mounted (4), which extends its face into a solid circle of the wedge surface of the wedge hollow body (1). The inner faces of the first drive plate (4) and the second disc (5) have a spherical recess which is fitted with a ball (7), the radius of which is equal to the radius of the spherical recesses. The second roll (5) comprises a concentric circle on the through-holes (8) by its width; the axial deflection of the first drive plate (5), together with a hollow wedge-shaped body (1) and the ball (7), is working in the pressing chamber of the spatial shape of the V-ring.

The main advantage of this idea is that press tool has got spherical shapes and contact between tool and die is theoretical in the point. Due to the small contact area (point), we obtained a small force-requiring pressing power, and we do not need the big motor power. For the construction of the mill, it is characteristic that the die is powered and its drive comes from the drive body. The spherical tool is hollow, and the kinetics of the drive mechanism for the die passes through the spherical tool. This can be achieved by coupling the die with a drive through the cardan joint or pivot joint. An important parameter of the mill is an axial shift of the drive's face from the die's face. In this design, the peripheral speed of the spherical tool during each revolution is not constant. This results in the uneven wear of the working tools, reducing the output efficiency of the mill and increasing the cost of the mill's repair due to the need of replacing the damaged parts.

To achieve a synchronised generating pressing space, it was considered appropriate to use the principle of cardan. The concept of the machine consists of two cylinders with axial and rotary roller with independent various axes of rotation, between which one ball was placed. These three members create the press chamber by the relative motion, in which the material is entrained and compressed on the principle of continuous changes in the geometry of the press jacket and thus its volume (**Figure 17**).

The first functional model was prepared at our institute in 2010 [13]. This machine was used to verify the proposed principles. The device has a ball diameter of 71.6 mm, a motor of 1 kW, an output per hour of 40–50 kg/h and a number of holes in a die of 21 x Ø 7 mm and belongs to the group of small pelleting presses. The production of individual parts of the pellet press is not technologically and financially very demanding, which can be an affordable solution of the pellet press with a high degree of financial recovery of biomass waste. The accuracy of the proposed structure was verified by testing the functional and technical parameters such as verification of pull material into the pressing chamber, the ability of pelleting, the performance verification and the validation of the quality of compression mouldings. The tests included pressed materials such as wood waste, MDF waste, straw, cecina, peat, black alluvial areas, wastewater treatment plant waste - sludge, rape, California earthworm humus, rat poisoning, cocoa waste. This principle was patented by the authors in 2006 and subsequently the basis for modification of the design and production of the pellet press prototype was created.

There was a development of a new prototype machine launched in 2011. The design of the machine has been subjected to a comprehensive analysis of the deficiencies of the machine. The analysis concerned the construction of the machine but also investigated the tribology machine and damage to various parts of the former variants of the machine. The current version of the ball pelleting mill is shown in **Figure 18**.

The current version of the machine has some fundamental changes from previous variants, which are its main advantages. The concept of the machine allows easy and rapid exchange of press tools, depending on the type of the pressed material. The tests of various materials using different matrices can be tried.

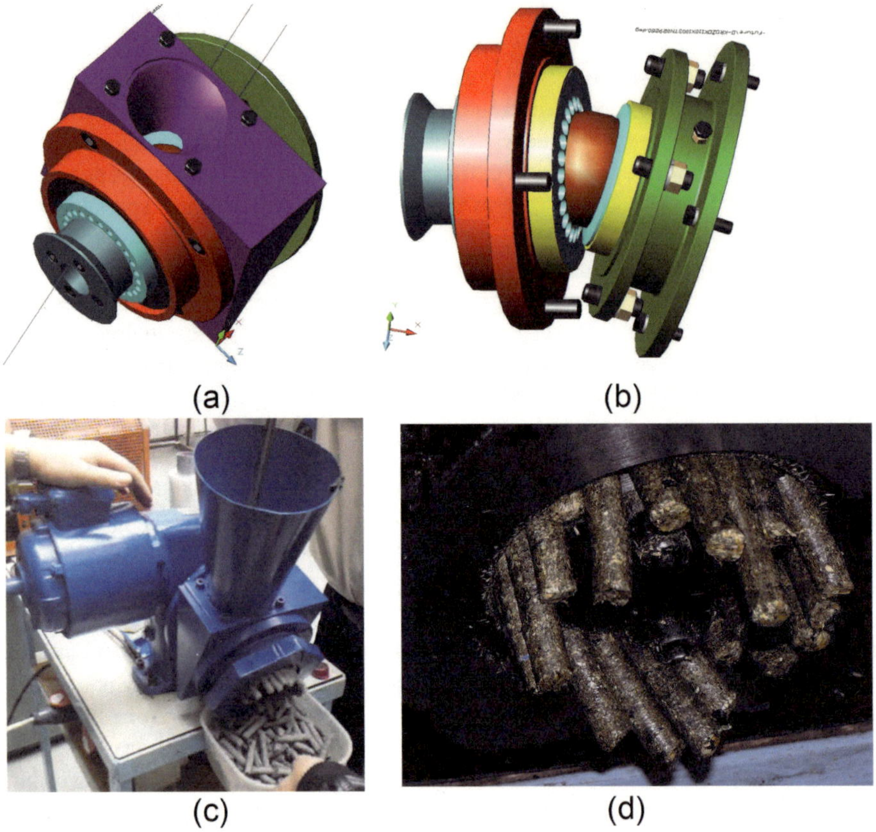

Figure 17.
The first concept of the pressing principle. (a) View of the 3D model, (b) view of the pressing chamber, (c) functional model and (d) view of the die.

It is also possible to change the length of the pressing channel by means of calibration matrices that can be added or removed. We can say that after the application is the modularity of the machine at a higher level which provides a simpler implementation of experiments. The final drive solution is modular machine tools on the same basic platform. Drive instruments may be by means of universal joints and friction bonds formed between tools using tools or gear using punch with projections. Each of these solutions helps to better ensure the transfer of torque from the stamp to the die.

There are ongoing long-term pilot plant verification test pelleting various types of materials. In the picture we can see pelleting paper sludge [14]. During the tests, the machine demonstrates compliance with the required operating parameters, quality and process reliability. Today, we are prepared to introduce its mobile technology lines for pellet horse dung. This line was started to be produced with the company IMC Slovakia, Banská Bystrica [1].

In the 2013, we prepared absolutely a new design of the modular pellet mills [14]. A pellet mill according to the invention will now be described in more detail in the sections with the accompanying drawings, where **Figure 19a** shows the mill mechanism with a freely placed spherical tool in a side cross-section. **Figure 19b** shows the mill mechanism with a hemispherical tool in a side cross-section, forming one block with the punch. **Figure 19c** is a side cross-section of the mill mechanism with a self-driven spherical tool.

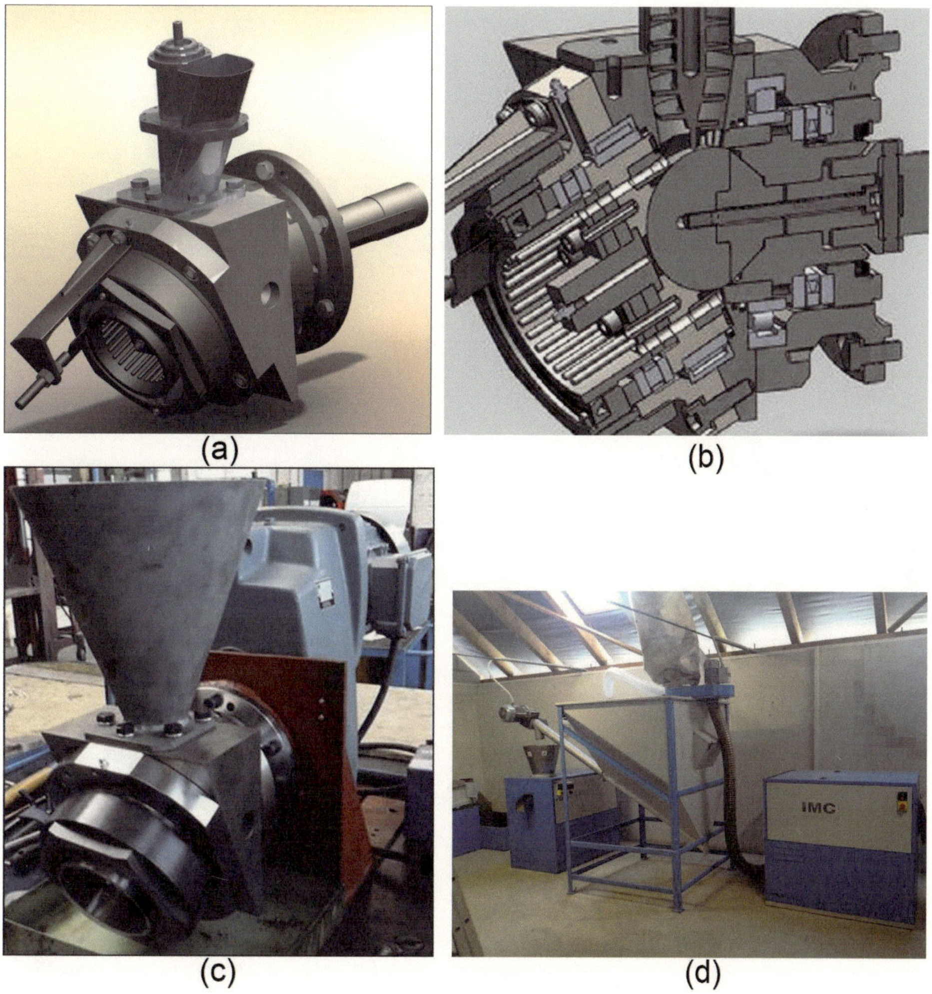

Figure 18.
Innovative spherical pellet machine (V4). (a) 3D model, (b) cut of the chamber, (c) prototype, variant V4 and (d) pellet in IMC line.

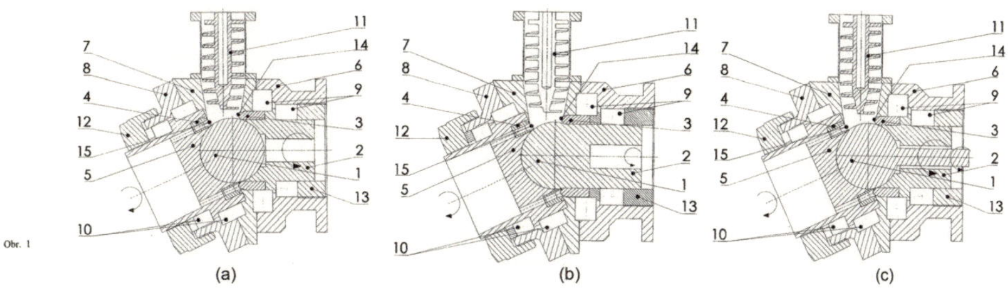

Figure 19.
Innovative spherical pellet machine [14]. (a) with loose ball, (b) mill with a spherical member associated with punch and (c) with a self-driven balls.

5. Compacting machine for produce new shape biofuels

The concept of an ideal briquette arose from a comparison of the properties, advantages and disadvantages of the briquettes of individual technologies (**Figure 20**). It is an attempt to create a briquette that would contain the advantages

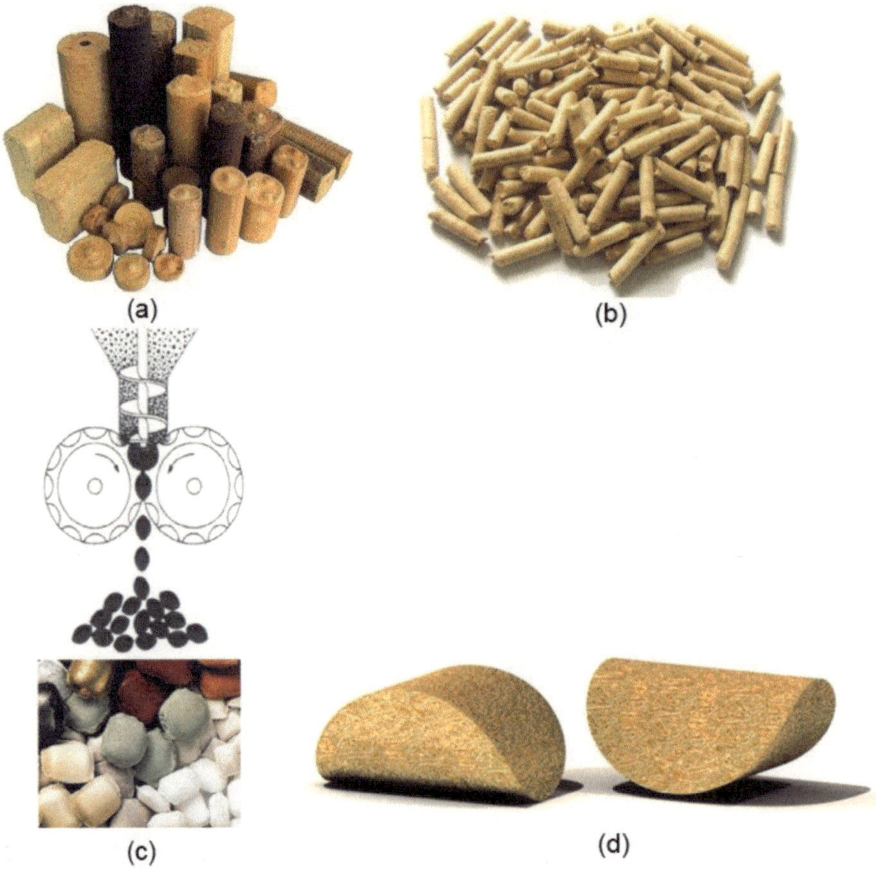

Figure 20.
Scheme of idealised granule [4]. (a) Briquetting, (b) pelleting, (c) compacting and (d) briquettes were valuated by: briquette burning process, automatization of combustion process, machine wear, and briquette storage and transit.

of pelleting, briquetting and compacting while at the same time eliminating the disadvantages of the individual technologies. A new briquette shape would be produced on the principle of compacting, which means being pressed between two rollers. In contrast with compacting, the new technology would have to allow for the briquette to be for a certain time under pressure, by which cooling would occur, as is common in pelleting and briquetting. Therefore, there arose the need to design a shape and, following that, a compactor machine that would manage to make newly shaped briquettes.

On the basis of the input requirements and later theoretic analysis, we created the almost-round briquette, which is designed as the intersection of two half cylinders each rotated through 90° and is suitable for automatic feeding. With such a briquette, it is possible to regulate the ratio of the surface to its volume. The edges at the intersection of the two cylinders allow optimal lighting of the briquette [15].

5.1 Kinematic

After the design of a suitable shape for the briquette, we continued with a combination of potential solutions and a design of a kinematic compactor that would be capable of producing such briquettes. The result of this combination was a proposal for a completely new principle "ring compaction machine". In 2012, we received a granted patent SK 287505 [16] for the stated principle. The invention belongs to the area of compaction of bulk organic and inorganic materials (**Figure 21**).

The essence of the new patented solution is a large ring (1) which has in its interior circumference a great number of slots (3) in semi-cylindrical shape whose axes are parallel with the axis of rotation of the ring. The compression instrument is also a disc (4), which on its outside circumference has a cylinder slot (5), which fits into the slots of the ring (1) and is turned 90° from the slots of the ring (1). So both the ring and the compression disc have independent movement. The hold-off stage under the pressure at which the moulds are cooled is ensured by the calibration struts (6). In stage A, there takes place the filling of the press chambers that appear during rotation, in stage B compression, in stage C partial expansion and finally in stage D gradual cooling of the briquette under pressure and its calibration. The effect of cooling of the binder brings the binder (lignin) from the plastic to the firm state, by which the briquette achieves the required rigidity [17].

5.2 Design and production of functional model

A further task arising out of the work on the overall project was designing the compacting machine itself. At the beginning, an analysis of kinematics of motions and the combination of possible structural solutions was carried out. In order to achieve the required briquette shape and kinematics of motions, we designed a functional model of the compacting machine as shown in **Figure 22**.

Design of force elements was based on a theoretical analysis of the forces calculated during the densification process. On the basis of such forces, calculations were made for the required input of the main drive, input of the drive of the compacting disc and the drive of the feeding screw. The final task in this part of the project was to prepare the manufacturing sketches and the final manufacture of the functional model (**Figure 22**). In this part of the task, two basic types of testing were

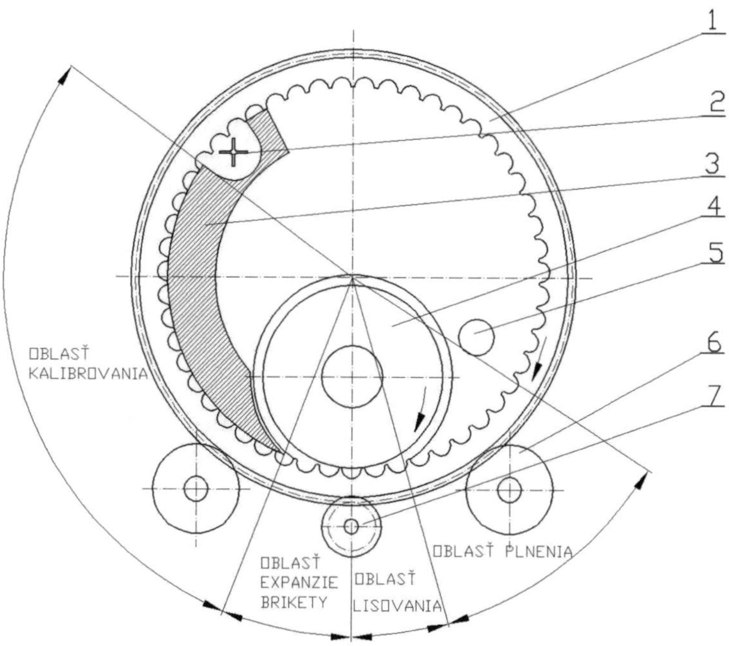

Figure 21.
Principle arrangement of annual compacting machine with internal pressing disc [16].

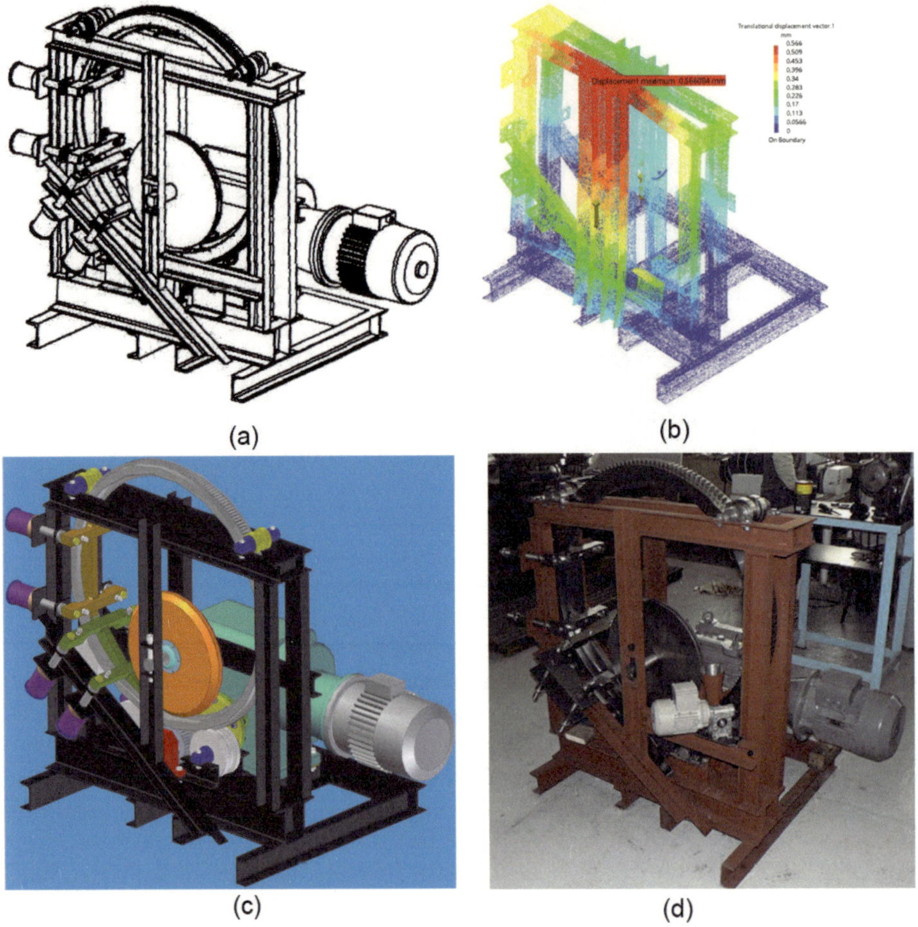

Figure 22.
Design of the ring compacting press. (a) Axonometric view, (b) FEM analysis, (c) 3D model and (d) function model.

performed on the functional model. The objective of the first set of measurements was to determine the required density of a briquette on the functional model. The second very important task in this experiment was to determine the maximum pressing force for the set density of the briquette.

6. Inverted kinematics

The subject of the solved project is the development and optimization of a new progressive, patent-protected construction of the inverted kinematics of a compaction screw briquetting press. There is the justified assumption that the solved construction would remove the stated faults.

The essence of the proposed solution of the compacting machine with "inverted kinematics" is that the rotational part of the machine is a body in which the screw is firmly mounted and in the centre of the rotational drum is a fixed non-rotating mandrel. So the rotational action is not performed by the press but a pressing chamber whose important part is the screw [18]. And on the other hand, the compacting instrument which in this case we call the mandrel is firmly held in the machine's frame and performs no movement.

The variable construction of the press with a rotational pressing chamber and a stationary instrument according to UV 7380 (**Figure 23**) is made up of a cylindrical or conical pressing chamber (1) with a pivotal bearing in its axial axle; on the inside the covering of which is a firmly held open orthogonal or clinogonnal helix (2) with a progressive or constant incline. In the axis of the cylindrical or conical pressing chamber (1), a cylindrical or conical stationary instrument (4) is firmly held in the frame (3). The pressing chamber (1) the drive mechanism (5) is connected with the moving unit (6) either directly or through a gearbox.

A compaction machine with inverted kinematics is made up of three main constructional sections—the pressing chamber, the rotating screw firmly connected with the rotating chamber, and the fixed pressing instrument, the mandrel. Each of the three main construction elements can be in different versions. In the case of the pressing chamber, this could be a cylindrical (**Figure 23a**) or conical shape (**Figure 23c**) with a diminishing cone diameter in the flow direction of the compacted material. The screw is prepared either with progressive inclination (**Figure 23a**) or with constant inclination (**Figure 23b** and **c**), and the pressing instrument is manufactured like the pressing chamber, either in conical or cylindrical shape (**Figure 23a** and **c**) or with conical shape (**Figure 23b**). The difference is that in the case of the conical version, the cone's diameter will increase in the direction of the material flow.

The proposed possibilities allow the creation of eight basic solution combinations for the compacting inverted kinematic machine.

The inverted kinematic compacting machine brings along a number of advantages [19]:

- Low energy demand due to the rotating cover. For rotations on a larger diameter, the lower torque is sufficient for producing pressing power equally great to in the case of a conventional apparatus.
- The simpler constructional solution for cooling or heating the pressing machine thanks to its rigid unmoving mounting.
- The construction eliminates the wearing of the front parts of the screws more than with the classical design. In the proposed solution, the tip of the stationary instrument is not moulded, does not turn and is located in the phase of the process which could be called calibrating/compacting. The tip just defines the space and the material simply bypasses it.
- The bearings on which the rotary compression chamber is mounted are not axially loaded because the force acts in a critically small diameter.

In line with industrial model SK 7380, it is possible to power the inverted kinematic compacting machine in a number of ways:

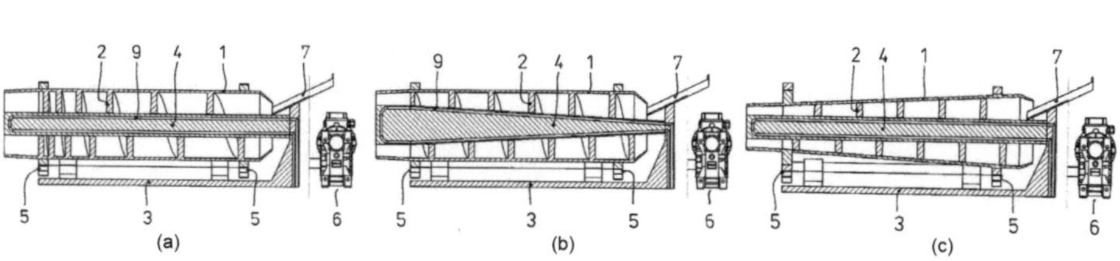

Figure 23.
Possible combinations of the proposed patent solution [18]. (a) Cylindrical body, progressive ascension of the screw, cylindrical mandrel, (b) cylindrical body, constant ascension of the screw, conical mandrel and (c) conical body, constant pitch of the screw, cylindrical mandrel.

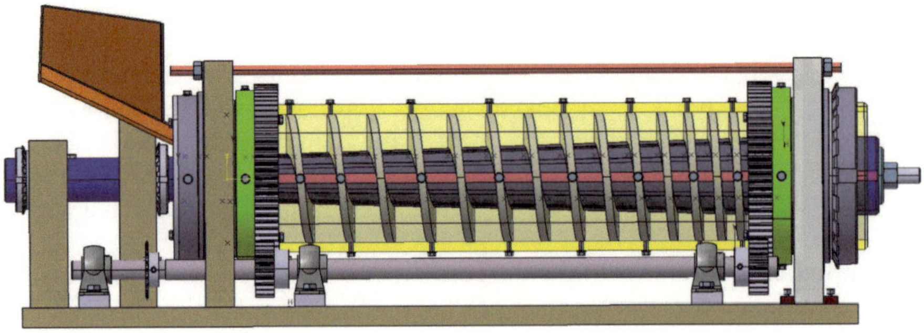

Figure 24.
New original design of the screw briquetting press.

- The machine drive by means of a gearbox can be worked out in two methods. The first method, where the axle of the drive shaft is identical or parallel to the axis of the drive shaft, is called direct drive through the gearbox. The second method, where the axle of the drive shaft is perpendicular to the drive shaft axle, is referred to as cranked drive through the gearbox.

- Machine drive through chain or belt drive. In this method, the torque contribution is very advantageous because the connection element, be it belt or chain, takes on the role of a safety element.

The various combinations of the make-up of the briquetting press and its drive methods offer sufficiently varied opportunities for research, development and selection of the optimal solution. In **Figure 24**, a 3D model is shown [20].

7. Design of the briquetting and pelleting lines

Until now we have carried out projects on 51 lines for briquetting wood waste. Within this number there are nine installations in Slovakia, two in Estonia and one each in the Czech Republic and Hungary [21]. Those in Slovakia include lines for the LandR company in Pezinok, IMEKO Malacky, Excellent Bratislava, Kéri Trnava, NORBA Prievidza, Defektospol Údavské, Obecný podnik Lehota pod Vtáčnikom, Colorspol Novoť, Ekomix-Natúrprodukt Vrbové and JUGA Lučenec (**Figure 25**).

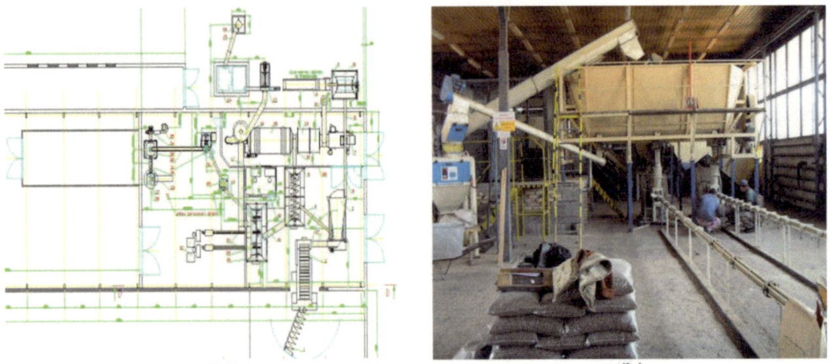

(a) (b)

Figure 25.
Combined line for briquette and pellet production for JUGA. (a) Layout of machines and (b) real view of the line.

In Estonia we designed two lines in Tabasalu for the KRK Moigu company, in the Czech Republic for the Dřevoterm firm in Náchod and in Hungary for the GUEM company in Salgotarjan.

The implemented projects show the positive contribution brought about by the original resolutions for compacting machines.

Author details

Ľubomír Šooš
Institute of Production Systems, Environmental Technology and Quality Management, Faculty of Mechanical Engineering of STU in Bratislava, Bratislava, Slovakia

*Address all correspondence to: lubomir.soos@stuba.sk

References

[1] Šooš Ľ, Križan P, Beniak J, Matúš M. New designs of compacting machines and shapes of pressing bio-biofuels. International Science Index. 2015; **17**(10):550-555. ISSN 1307-6892

[2] Križan P, Beniak J, Šooš Ľ, Kolláth Ľ, Matúš M. Experimental research of mechanical properties and parameters of waste raw materials based wood-plastic composites. In: American Advanced Materials Congress: Proceedings and Abstracts Book. Miami, USA: VBRI Press AB; 2016

[3] Ondruška J, Šooš Ľ, Križan P. Závitovka strojových zariadení: Patentový spis SK 288206. Banská Bystrica: Úrad priemyselného vlastníctva SR; 2014. p. 7

[4] Šooš Ľ, Ondruška J, Matúš M. Sústava konštrukčných uzlov na elimináciu axiálnych síl v strojových zariadeniach: Číslo úžitkového vzoru: 6045 SK, Dátum nadobudnutia: 2.3. Banská Bystrica: Úrad priemyselného vlastníctva SR; 2012. p. 11

[5] Šooš Ľ. a kol.: Výskum, vývoj, výkresová dokumentácia briketovacieho lisu BZ-50-250. Konštukta Industry. Výkresová dokumentácia a správa 244 strán; 1996

[6] Šooš Ľ, Gros P. Dual-chamber briquetting press. In Zeszyty Naukowe Politechniki Bialostockiej. 2002;**9**: 447-453 ISSN 0860-9292

[7] Riegel T. Dvojkomorový briquetting press [Diplomová práca] Katedra výrobnej techniky SjF STU v Bratislave; 2003. p. 62

[8] Šooš Ľ. Skladaná hubica briketovacieho lisu: Číslo patentu: SK 286889. Dátum udelenia: 4.5; 2009

[9] Šooš Ľ. Multitechnologická hubica briketovacieho lisu : Číslo patentu: SK 286878. Dátum udelenia: 4.5; 2009

[10] Šooš Ľ, Križan P. Variabilný piest zhutňovacieho stroja: Patent č. 288168 SK. Banská Bystrica: Úrad priemyselného vlastníctva SR; 2014. p. 11

[11] Šooš Ľ. Using off non-traditional raw for produced off bio-pellets and briquetts. In: Časopis. CZ Pro-Energy magazín. (ČR)-č. 2007;**2**:44-48. ISSN 1802-4599

[12] Grman M. Nová konštrukcia peletovacieho stroja [Diplomová práca]. Katedra výrobnej techniky SjF STU v Bratislave; 2001. p. 68

[13] Šooš Ľ, Grman M. Spôsob lisovania peliet zo sypkej organickej a/alebo anorganickej suroviny alebo surovinovej zmesi a lis na pelety: Číslo patentu: SK 286877, Dátum udelenia; 2009

[14] Šooš Ľ, Ondruška J, Biath P, Matúš M, Zegzulka J. Lis na pelety: patentový spis č. 288360. Banská Bystrica Úrad priemyselného vlastníctva SR; 2016. p. 11.

[15] Šooš Ľ, Bábics J, Beniak J, Križan P, Kováč P, Matúš M. Design and testing functional model compacting machine for produce new shape biofuels. In: IOP Conference Series: Materials Science and Engineering. Vol. 501, no.1; 2019. p. 012008

[16] Šooš Ľ. Kontinuálny spôsob lisovania biomasy do optimálnych výliskov a zhutňovací stroj: Číslo patentu: SK 287505, Dátum nadobudnutia; 2010

[17] Matúš M. Nizkoenergetický zhutňovací stroj. Písomná práca k dizertačnej skúške. ÚSETM SjF STU v Bratislave; 2008

[18] Šooš Ľ, Štefanka M, Ondruška J, Križan P, Matúš M. Zhutňovací stroj s rotujúcim bubnom a nehybným tŕňom.

Banská Bystrica Úrad priemyselného vlastníctva SR; 2019. p. 9

[19] Šooš Ľ, Beniak J, Križan P, Kováč P, Ondruška I, Matúš M. Inverted kinematics compacting machine. In: Energy reliability. REMOO [elektronický zdroj]: Book of Abstracts of the 9th International Conference and Workshop. 1st vyd. Erlangen: Verlag e. K; 2019. p. 1. ISBN: 978-3-9820758-1-5

[20] Štefanka M. Návrh funkčného modelu s obrátenou kinematikou. Diplomová práca SjF STU; 2019. p. 87

[21] Šooš Ľ, Matúš M, Beniak J, Križan P. Development of the compaction machine for the production of new shapes of pressed biofuels. In: TSME-IcoME: Abstract Book of the 8th TSME International Conference on Mechanical Engineering. Bangkok, Thailand. 1st vyd. Bangkok: King Mongkut's University of Technology North Bangkok; 2017. p. 55

Chapter 20

Performance, Gaseous and Particle Emissions from a Residential Pellet Stove

Md. Obaidullah and Jacques De Ruyck

Abstract

The objectives of this chapter are to present experimental results on performance, gaseous and particle emissions obtained from a modern bottom feed pellet stove of 2.5 kW output in part load heat and 5 kW output in nominal heat. Two experiments in part load and four experiments in nominal load output were conducted in a stove manufacturing plant in the southern part of Belgium. The particle emissions measurements are mass concentrations of PM_1 and $PM_{2.5}$, number concentrations and their particle size distributions. Particle emission measurements were conducted continuously from a partial flow dilution tunnel using an Electrical Low Pressure Impactor Plus (ELPI+). The CO emissions were analyzed continuously from the flue gas by a Siemens Ultramat 6 gas analyzer, CO_2 and O_2 concentrations were measured continuously using a Horiba PG-250 gas analyzer. A performance analysis in terms of combustion efficiency together with different losses of the pellet stove is also discussed. The results show that PM_1 and $PM_{2.5}$ concentrations obtained from the combustion phase of the nominal load experiments varied from 43.3 to 276 mg/Nm^3 and 66 to 36 mg/Nm^3 respectively, while the particle number concentrations varied from 1.4×10^7 to 8.8×10^7 particles/cm^3. The CO emissions obtained from the main combustion phase of the nominal load heat varied output from 50 to 145 mg/Nm^3.

Keywords: gaseous emissions, particulate matter, mass concentrations, number concentrations, size distributions, performance analysis

1. Introduction

Small scale combustion appliances are mainly used for the purpose of residential heating. Several studies show that combustion of biomass fuels in small scale heating appliances is a common source of both particulate matter (PM) and gaseous emissions such as fine particles, polycyclic aromatic hydrocarbons (PAH), volatile organic compounds (VOC) and carbon monoxide, carbon dioxide, nitrogen oxide, sulfur oxide, etc. [1–8]. PM is a dynamic mixture of particles in the flue gas released directly from the combustion devices. The aerodynamic diameter is generally used to indicate the particle size since the particles have different shapes and densities. It is defined as the diameter of a spherical particle with a mass density of 1000 kg/m^3 that has the same inertial properties in the flue gases [9–11]. Several particles size fractions of PM are defined in the literature: $PM_{0.1}$ (nano particles: <0.1 μm), PM_1 (ultrafine particles: <1 μm), $PM_{2.5}$ (fine particles: <2.5 μm), PM_{10} (coarse particles: <10 μm) [9, 10, 12].

In comparison to liquid and gaseous fuels, the emissions of particulate matter from biomass combustion are high [9, 13–17]. Majority of the particles is less than 1 μm (micrometer) in size and emitted straight away to the ambient air from the combustion appliances [18, 19]. Numerous studies have demonstrated that increased particle emissions in the ambient air correlate with severe health effects in the exposed population, including respiratory and cardiovascular illnesses as well as increased mortality [15, 20, 21]. Further, it has been mentioned that in the case of combustion related fine particle fraction is more dangerous to human health and environmental effects [2, 15, 20–22].

Using wood pellets as biomass fuel is gradually increasing due to their high energy density, easy transportability and the lower amount of gas emissions from its production and transportation comparing to oil, coal and natural gas [23]. Nowadays, combustion of wood pellets in small scale heating appliances is efficient and produces significantly lower emissions than the old wood log combustion appliances [4].

Most of the heating appliances in the market claim an optimized combustion with low emissions of gaseous pollutants at nominal operational load. However, operation at full load is only required for a short peak winter period [4]. For the rest of the year, the combustion appliances may work at lower operational loads as far as continuous operation is considered. The emissions of these pollutants are significantly different if the heating appliance is operating at lower loads i.e. part load. Carbon monoxide (CO) emissions from residential pellet heating devices mainly report during stationary operation, however, a considerable part of un-burnt fuels is emitted during the startup and burnout phases [24, 25].

This chapter presents the experimental results regarding particle and gaseous emissions from a modern bottom feed pellet stove operated with nominal load (5 kW) and part load (2.5 kW) heat output. The particle emission measurements include mass concentrations of PM_1 and $PM_{2.5}$, number concentrations and their particle size distributions measured continuously using a partial flow dilution tunnel together with an Electrical Low Pressure Impactor Plus (ELPI+) with a flow rate 10 lpm and cut-off size of the 14 stages from 6 nm to 10 μm.

2. Materials and methods

This section briefly discusses the previous work on particle emissions, experimental setup, fuel characteristics and combustion appliance related to the emissions measurements.

2.1 Literature review

There is lack of information regarding the characterization of particulate emissions from small scale biomass combustion. Several studies on particle emission from biomass stoves were carried out in EU countries. For example, Boman et al. [26] investigated six types of different pellet fuels in three different commercial pellet burners (10–15 kW) and observed that fine particles (<1 μm) contain a significant part of the total PM emissions.

Sippula et al. [27] investigated the effect of wood pellet combustion on the particle emissions from a top feed pellet stove with a heat output of 8 kW using an ELPI. Their results show that particle number emissions varied from 1.3×10^7 to 4.4×10^7 particles/cm^3 and the PM_1 varied from 69 to 343 mg/Nm^3. Gaegauf et al. [28] investigated particle emissions by using an SMPS on a pellet boiler with a capacity of 17 kW. They observed that the major part of the particle emissions were in the range between 30 and 300 nm.

Bari et al. [29] studied particle mass and number emissions from a pellet stove with a nominal output of 5 kW. The measurements were conducted from the stack using a Berner Low Pressure impactor (BLPI) and a Scanning Mobility Particle Sizer (SMPS) for mass concentrations and number size distributions respectively. The results show that the PM_{10} concentrations were between 31 and 201 mg/Nm^3, while number concentrations varied between 1.5×10^7 and 5.4×10^7 particles/cm^3. They observed that the particle mass size distributions were unimodal with maximum concentrations in the fine fraction.

Bäfver et al. [30] experimentally studied particle and CO emissions from modern and old type residential stoves of various heat capacity fired with wood logs and wood pellets. Measurements were performed using a Dekati Low Pressure Impactor (DLPI) for mass size distribution while an ELPI was used for number size distributions. Modern pellets stoves showed lower mass concentration of particles as well as lower CO concentrations than the old type wood stoves. They found that in all cases, the particle mass emissions were dominated by fine particles and there was only small fraction of coarse particles.

Qie et al. [31] studied particle emissions in a small scale pellet boiler (50 kW) using a Dust Trak-II Handheld Aerosol Monitor from 100 nm to 10 μm. Three types of biomass pellets, i.e. wood pellets, Miscanthus pellets and straw pellets were combusted. PM_{10} concentrations of wood pellets, Miscanthus pellets and straw pellets were 72.7, 100 and 150 mg/Nm^3, respectively. PM concentration results show that wood pellets are better than Miscanthus and straw pellets.

Johansson et al. [32] investigated particle emissions from a domestic pellet stove with a capacity of 6 kW. The stove was fired with wood pellets. Particle characterizations were done with an Electrical Low Pressure Impactor (ELPI). PM_{10} mass concentration was 47 mg/Nm^3, number concentrations varied between 1.8×10^7 and 8.7×10^7 particles/cm^3.

The above review briefly illustrates that a number of studies on particulate matter concentrations related to the small scale heating appliances at nominal load operations [19, 27, 33, 34]. Particle emissions from residential heating devices are documented mainly for stationary operation, however a considerable part of un-burnt fuels are emitted during the startup and burnout phases. PM emission characteristics at each phase of pellet stove operations are therefore important to be able to reduce the annual emissions from residential pellet combustion.

2.2 Experimental setup

The measurements were conducted according to the European standard EN 14785 for residential space heating appliances fired by wood pellets [35]. Two experiments (A and B) in part load and four experiments (C, D, E and F) in nominal load heat output were conducted for the emissions measurements from a bottom feed modern pellet stove. The stove was operated in different fan speeds, which regulate air flow into the combustion chamber. Experiments A and B were operated with low speed fan at 900 rpm, C and D with medium speed fan at 1250 rpm, E and F with high speed fan at 1400 rpm. Fan speed settings of each experiment are presented in **Table 1**. The wood pellets are transported through two screws from the pellet storage hopper to the burner cup. The rotation of screw-1 connected to the pellet storage hopper was 1.6/6 sec for the part load measurements, while 3.2/6 sec for the nominal load experiments. Screw-2 connected to the burner cup was operated at 2 rpm for all the experiments. The heat output of the stove was modified by controlling the rotation of screw-1, which controls the fuel supply.

Experiments	Stove load	Fan speed (rpm)
A	Part Load	Low, 900
B	Part load	Low, 900
C	Nominal load	Medium, 1250
D	Nominal Load	Medium, 1250
E	Nominal load	High, 1400
F	Nominal Load	High, 1400

Table 1.
Fan speed settings for the different experiments [7, 17].

2.3 Fuel characteristics

The elemental composition, moisture content and lower heating value (LHV) of the commercial pellets used in the combustion experiments are presented in **Table 2**. The pellets are made from soft wood, certified by DINPlus standard and available in the European market.

2.4 Combustion appliance

The combustion appliance used in the experiments was a bottom feed pellet stove with a nominal heat output of 5 kW. The pellet stove was setup on a balance to monitor the fuel consumption. The pellet stove is equipped with an internal pellet storage, where the pellets are supplied through two screws into the burner cup. The combustion takes place in the burner cup. A step motor is used to supply the pellet into the combustion chamber. The combustion air consisting of primary and secondary air is supplied through the holes under the grid of the burner cup. The air supply is fan assisted and depends upon the selected thermal output. A short cleaning period is set to occur every 30 min in the stove. During cleaning, the fuel supply decreases and the air supply increases for 1 min, removing the ash gathered on the grid. The front side of the stove is covered with a high temperature transparent glass window. The top of the combustion chamber is equipped with the baffle plate made

Parameter	Commercial pellets	DINplus
Length (mm)	<45	<45
Diameter (mm)	6.06 ± 0.1	6 ± 0.5
Durability (%)	98.9	>97.7
Fine content (%)	0.13	<1
Volumetric mass (kg/m^3)	675	>650
LHV (MJ/kg)	18.7	>16.9
Moisture (%)	8.6	10
Ash (wt %)	0.3	0.7
C (wt %)	49.1	—
H (wt %)	5.8	—
O (wt %)	44.8	—

Table 2.
Chemical properties of the pellet used in the combustion experiments [7, 17].

of vermiculite materials and the sides of refractory ceramic bricks made of calcium silicate. The flue gases are drawn out by an exhaust fan.

2.5 Gaseous emission measurements

A partial flow from the stack at about 2 m height from the pellet stove was withdrawn through an externally insulated steel probe of 12 mm diameter. The opening of the probe was positioned towards the flow of the stack. The CO emissions were analyzed continuously from the flue gas by a Siemens Ultramat 6 gas analyzer, CO_2 and O_2 concentrations were measured continuously using a Horiba PG-250 gas analyzer. Both gas analyzers cannot withstand the hot and humid flue gases for direct analysis. Before the analyzers, the flue gas samples passed through the chiller to remove moisture and to cool down the gas. The gas analyzers were calibrated with an appropriate gas mixture, before and after each combustion experiment. The measurement principles of the gas analyzers were galvanic analyzer for O_2 and non-dispersive infra-red for CO, CO_2. The analyzers have the measurement error of ±2% full scale in linearity and ± 0.5% full scale in repeatability. Temperature of the indoor air and flue gas in the stack were measured by the K-type thermocouples.

2.6 Particle emission measurements

Particle emissions were measured continuously from a partial flow dilution tunnel using an ELPI+ with a flow rate 10 lpm and cut-off size of the 14 stages from 6 nm to 10 μm. Sample particles entering the ELPI+ are first charged in the charger. The charged particles collected in each impactor stage produce an electrical current which is recorded by the respective electrometer channel. This current is proportional to particle numbers via mathematical algorithms [36]. Aluminum foils, greased with a mixture of acetone and Apiezon-L were placed on each of the impactor stages to prevent particle bouncing effects. The flue gases were diluted through a two steps partial flow dilution tunnel with pre-filtered dilution air before reaching the ELPI+. The dilution tunnel consists of a porous tube diluter (PRD), an ejector diluter (ED) and an air heater. The first stage dilution air injected (17.5 lpm) into the PRD was heated to match the raw sample temperature to reduce the risk of condensation. The second stage dilution air injected (49 lpm) into the ED, which was operated at ambient temperature to further dilute the sample and to reduce the sample temperature to the ambient condition. Dilution air is taken from outside the building to simulate the field conditions.

The CO_2 and O_2 concentration from the undiluted flue gas were analyzed continuously by a Horiba PG250 gas analyzer. CO_2 concentration was also measured continuously from the diluted sample by a Vaisala Carbocap analyzer to calculate the dilution ratio (DR). The details of the DR measurement were presented other works [4, 5, 37, 38].

3. Results and discussions

A total of six combustion experiments on gaseous and particle emissions from a bottom feed pellet stove were conducted. As the objectives of this chapter were to evaluate the emissions from different combustion phases of each experiment, the emission results from each experiment are presented as the startup, the combustion and the burnout phases.

3.1 CO emissions

The CO emissions obtained from the startup, combustion, burnout phase of all the experiments are presented in **Figures 1–3** respectively. The error bars present the uncertainty of the measurements. The CO emission values presented here are normalized with 13% dry oxygen content. It is clearly observed from all the experiments that CO emissions in the burnout phase were significantly higher than that in the startup phase followed by the combustion phase. The air excess (λ) in the burnout phase for all the experiments was quite higher than that in the other two phases. High excess air in the burnout phase cools the combustion chamber, resulting in high CO emissions.

It can be seen from **Figure 1** that the CO emissions obtained from the startup phase of the part load heat output varied from 1710 to 2370 mg/Nm^3 for experiments A to B, while in the nominal load heat output varied from 908 to 2294 mg/Nm^3 for the experiments C to F. The duration of startup phase is about 20 min for all the experiments. The stove operated in the medium speed fan gives lower CO emission in the startup phase. However, CO emissions in the startup phase for the experiments E and F were higher than the measurements C to D. At the startup phase, the combustion temperature was not high enough to provide sufficient burnout condition. This might be a reason for increasing CO emissions during the startup phase.

Figure 2 shows the comparison of CO emissions obtained from the main combustion phase of all the experiments with the required limit value of the NBN EN 14785 standard and the literature [39]. CO emission in Y axis is presented in logarithm scale. The CO emissions from the medium and high speed fan stove operated with nominal load output satisfied the required limit value of the NBN EN 13229 standard and lower than other work [39]. CO emissions obtained from the low speed fan operated with part load heat output did not meet the required limit value of the standard.

It can also be seen from **Figure 2** that the CO emissions obtained from the main combustion phase of the part load heat output varied from 1215 to 1450 mg/Nm^3 for experiments A and B, while in the nominal load heat output varied from 50 to 145 mg/Nm^3 for the experiments C to F. The duration of the main combustion phase varied from 3 h 50 min to 5 h 45 min for all the measurements. The lower CO emissions obtained from the stove operated with high speed fan than the medium speed fan followed by the low speed fan. Higher CO emission in the part

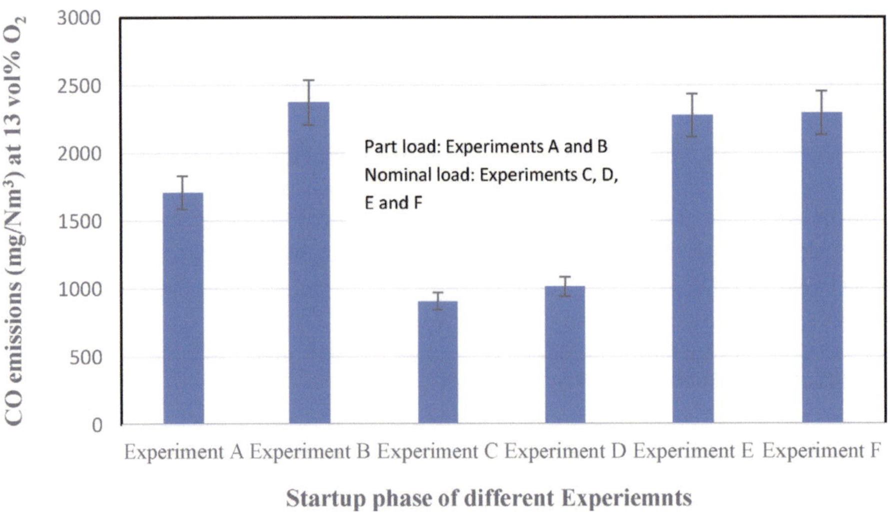

Figure 1.
CO emissions obtained from the startup phase of all the experiments [7].

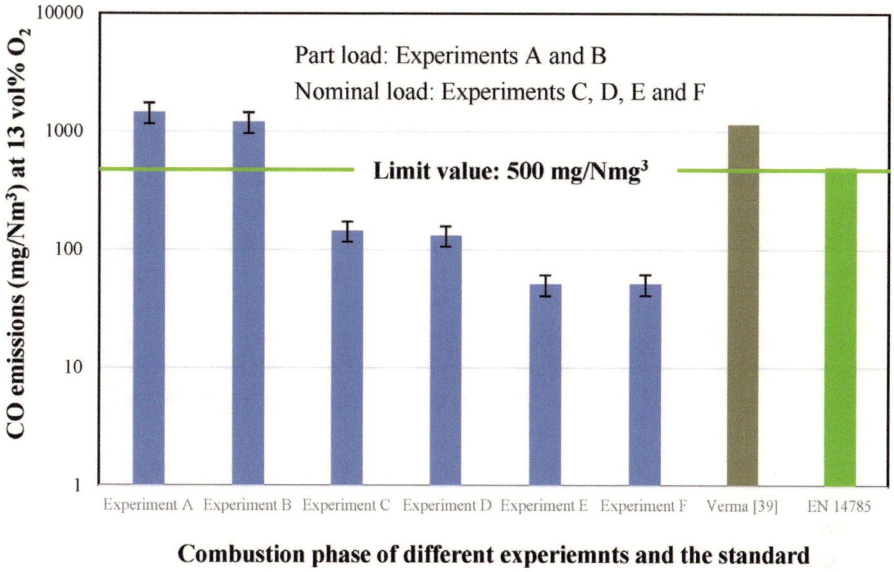

Figure 2.
Comparison of CO emissions obtained from the combustion phase of all experiments with the standard [7].

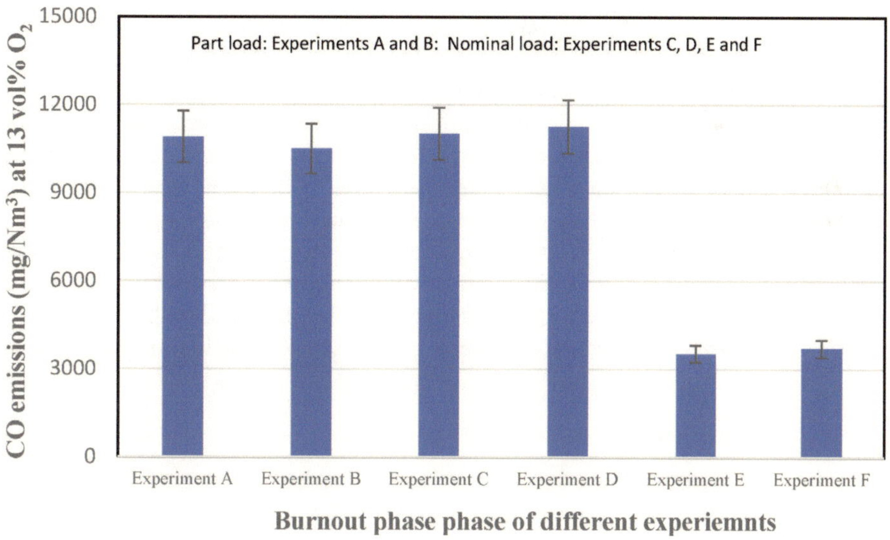

Figure 3.
CO emissions obtained from the burnout phase of all experiments [7].

load experiments was probably due to the higher air excess factor (about $\lambda = 4.35$) obtained in low speed fan, which gives lower combustion temperature, leading to high CO emissions. On the other hand, a correctly matched air excess factor (about $\lambda = 2.5$) for the medium and high speed fan operated experiments created favorable combustion conditions, leading to less CO emissions. Besides, the average flue gas temperature was much lower in the low speed fan (64°C) operated experiments than the medium speed (85°C) and high speed fan (101°C).

It can be seen from **Figure 3** that experiments E and F had the lower CO emissions in the burnout phase because of the different configuration of the fan speed from other measurements A, B, C and D. Also, the fan speeds for the experiments E and F were higher than the other experiments. This means that sufficient amounts of combustion air were supplied to burn the combustible gases; as a result, CO emissions were lower. Experiments conducted in the high speed fan gives lower CO emission in the burnout phase.

The total CO emissions showing in **Figure 4** are relatively higher in part load combustion experiments compared to nominal load output. This was due to the lower combustion temperatures caused by high air excess at the part load combustion experiments. The total CO emissions obtained in the part load experiments can be compared with values found in other work. For example, Schmidl et al. [40] investigated gaseous emissions from a 3 kW pellet stove in part load power output. The CO emissions in their study were 751 mg/Nm^3 which is quite lower than that in our study. The total CO emissions obtained from the nominal load output experiments are higher than other studies. For example, the CO emissions results of Bäfver et al. [30] investigated from a pellet stove with 5 kW capacity range between 140 and 405 mg/Nm^3.

It is clearly observed that CO emissions in the burnout phase from all the experiments were significantly higher than that in the startup phase followed by the combustion phase. For example, CO emissions in the burnout phase for experiments C to D were about 12 fold higher than in the startup phase and 75 fold higher than in the combustion phase. The air excess (λ) in the burnout phase for all the experiments was quite higher than that in the other two phases. High excess air in the burnout phase cools the combustion chamber, resulting in high CO emissions. It can be mentioned from the experimental results that the impact of higher CO emissions in the startup and burnout phase has influence on the total CO emissions.

Several studies [34, 41, 42] mention that higher combustion temperature, better turbulence mixing fuel with necessary oxygen and sufficient residence time can play a major role in combustion optimization and consequently emissions reduction. The CO emissions from a combustion device might represent the incomplete combustion caused by low combustion temperature, insufficient oxygen, short residence time, poor fuel and air mixing or a combination of these factors [34, 43].

CO emissions from small scale biomass combustion appliances can be reduced using flue gas cleaning technologies such as catalytic combustors, which consist of a metal wire mesh covered with catalytic material, platinum and palladium. The catalytic combustors are attached to a steel frame which can be inserted compactly

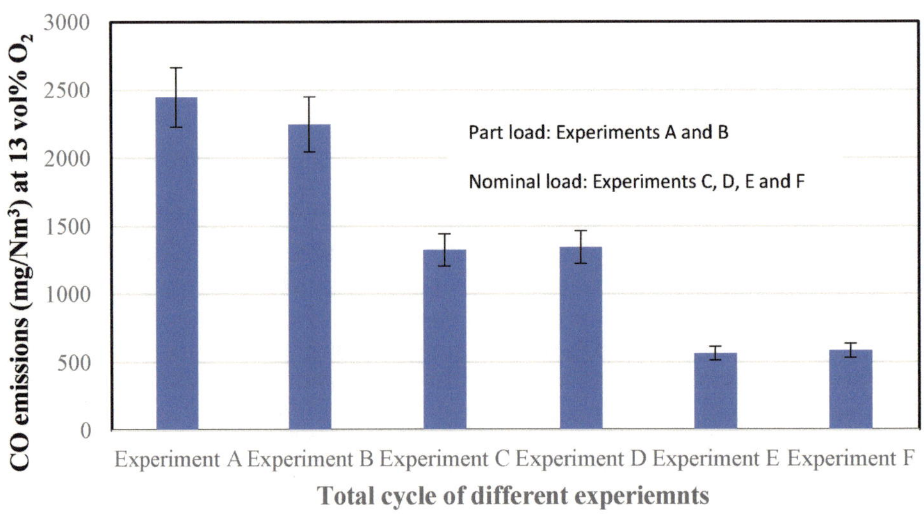

Figure 4.
Total CO emissions obtained from all the experiments [7].

inside the stack through an opening. Smoke gases pass through the catalytic element and ignite at a much lower temperature around 250°C. The result is that harmful substances are more completely burned. The fuel produces more heat through an extended clean burn. Hukkanen et al. [44] investigated reduction of gaseous emissions using a catalytic combustor from a 15 kW stove. Their results show that reductions of CO reached about 25% during the whole combustion cycle. Such gas cleaning systems are however quite expensive for small scale applications.

3.2 Particle mass concentrations

Figures 5 and **6** present the comparison of particle mass concentrations of PM_1 and $PM_{2.5}$ obtained from the startup, combustion, burnout phases and the total cycle of all the combustion experiments. **Figure 5** shows PM_1 concentrations obtained from the startup phase of the part load heat output. Experiments A to B varied from 26.1 to 38.4 mg/Nm^3 for the startup phase, 20.4 to 29.8 mg/Nm^3 for the combustion phase, 9.3 to 10.2 mg/Nm^3 for the burnout phase and 20.5 to 27.6 mg/Nm^3 for the total cycle, while in the nominal load heat output experiments C to F varied from 22 to 106 mg/Nm^3 for the startup phase, 43.3 to 276 mg/Nm^3 for the combustion phase, 4.7 to 12 mg/Nm^3 for the burnout phase and 44 to 236 mg/Nm^3 for the total cycle.

The PM_1 results obtained from the combustion phase in nominal load output of experiments C and D are significantly higher than those from the other experiments. In the part load heat output measurements, the startup phase of experiment A gave the highest PM_1 emissions. The variation of particle mass concentrations among all the experiments is due to the configuration of the burner, which operated with the different fan speeds.

The particle mass emissions reported by Sippula et al. [27], Johansson et al. [18] were also measured in a partial flow dilution tunnel. It should be mentioned that some conditions in the sampling line such as the dilution ratio, temperature and the measurement equipments were different. Therefore, it is difficult to formulate a direct comparison of the particle emissions results. Our particle mass concentration results for the PM_1 fraction obtained from the combustion phase in nominal load heat output can be compared with results from another study. For example, Sippula et al. [27] investigated particle mass emissions from a top

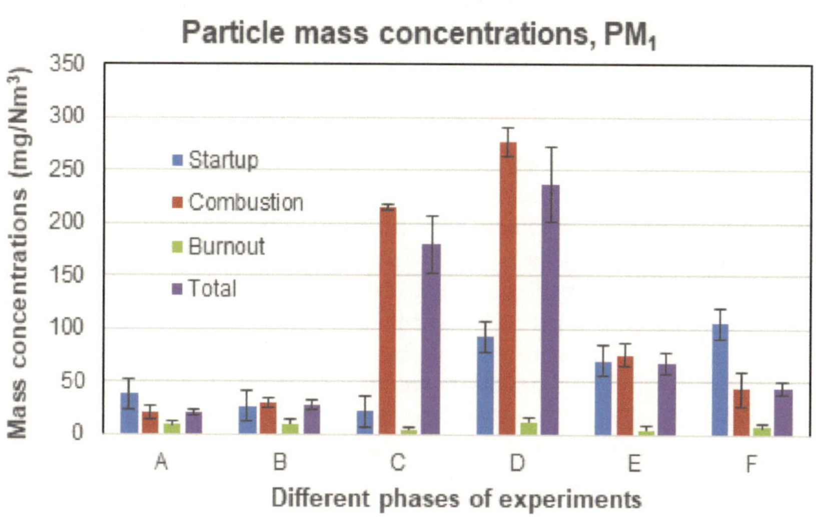

Figure 5.
Comparison of PM_1 concentrations obtained from all the combustion experiments [17].

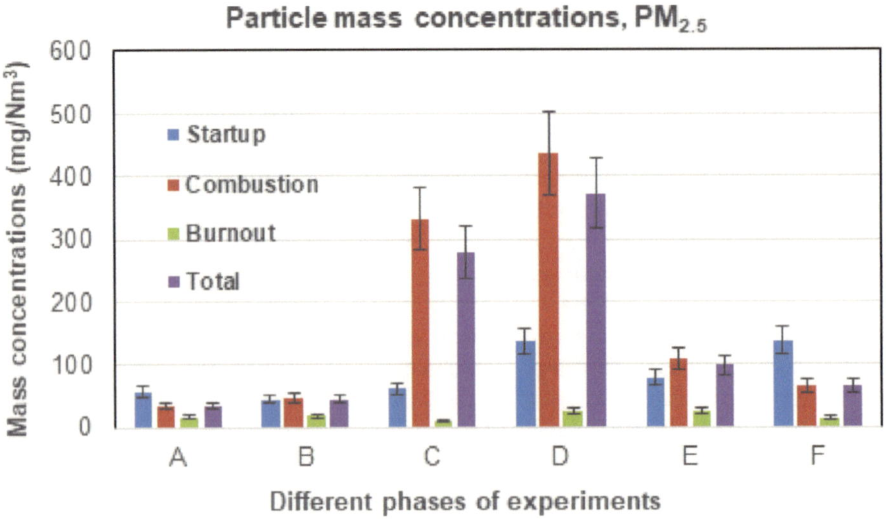

Figure 6.
Comparison of $PM_{2.5}$ concentrations obtained from all the combustion experiments [17].

feed pellet stove with a heat capacity of 8 kW using filter samples. Average mass concentrations of PM_1 in their study varied between 69 and 343 mg/Nm^3 which are slightly higher values than in our study.

Figure 6 shows the comparison of $PM_{2.5}$ concentrations obtained from different experiments at nominal load and part load heat output. In part load experiments, $PM_{2.5}$ concentrations varied from 45.3 to 57.8 mg/Nm^3 for the startup phase, 34.4 to 48 mg/Nm^3 for the combustion phase, 17 to 18 mg/Nm^3 for the burnout phase and 34.2 to 45 mg/Nm^3 for the total cycle, while in the nominal load experiments, varied from 61.5 to 138 mg/Nm^3 for the startup phase, 66 to 436 mg/Nm^3 for the combustion phase, 10.3 to 26.3 mg/Nm^3 for the burnout phase and 65.7 to 373 mg/Nm^3 for the total cycle.

It is observed that the particle mass fractions of PM_1 and $PM_{2.5}$ concentrations from the experiments operated with medium fan speed are about 4-folds higher than the high speed fan experiments. This might be due to the lower air excess factor $\lambda = 2.8$ for the medium speed fan experiments, while $\lambda = 2.4$ for the high speed fan experiments, which gives higher combustion temperature and creates favorable combustion condition.

In nominal load experiments C, D and E, particle mass fractions of PM_1 and $PM_{2.5}$ obtained from the combustion phase are significantly higher than in the other two phases of startup and burnout. The PM_1 and $PM_{2.5}$ concentration levels can also vary from one experiment to another, which are typical for the biomass combustion. The PM_1 concentrations of all the nominal load experiments accounted for 61, 68 and 50% of the $PM_{2.5}$ concentrations for the startup, combustion and burnout phase respectively, while 62, 61 and 55% for the part load experiments, which are lower than the values found in literature [26]. This analysis from both nominal and part load het output experiments clearly shows that more than 50% of particle mass concentrations is the fine fractions of PM_1.

It is reported in the literature that the fine particle mass fractions (PM_1 and $PM_{2.5}$) are generally formed from easily volatile inorganic species (K, S, Na and Cl) and heavy metal elements (Zn and Pb) that have vaporized during combustion, which later saturate and form fine particles by nucleation. The nucleated particles grow further by coagulation, agglomeration, condensation and surface reactions. In the gas phase, these species undergo reactions resulting in the formation of alkaline metal sulphates, chlorides and carbonates as well as heavy metal oxides. Organic species represent the fraction of fine particle emissions. These particles are mainly due to incomplete combustion and to the condensation of the unburned hydrocarbon during the cooling phase of the flue gas [9, 10, 12, 27, 45].

3.3 Particle mass size distributions

Figures 7–9 present the particle mass size distribution graphs obtained from the startup, combustion and burnout phases of all the experiments respectively. The abscissa represents the particle aerodynamic diameter against the ordinate which shows the ratio of particle mass concentration dM to the logarithm of the channel width dlog(Dp), where Dp is the aerodynamic diameter.

Figure 7 shows that maximum particle concentrations obtained for all the experiments in the fine mode are at the particle size of 320 nm. Particle size between 10 and 80 nm contains very small amounts of mass and are probably soot particles, therefore these particles are not seen in the mass size distribution graphs. **Figure 8** shows that experiments A, B, E and F had quite similar mass size distributions, with maximum particle concentrations in the fine mode at the particle size of 320 nm, while experiments C and D had the peak particle emissions at the particle size of 750 nm.

In **Figure 9**, it can be seen that particle mass size distributions of all the experiments (except experiments C and D) from the burnout phase are quite similar with a maximum particle concentration at the particle size of 320 nm. The mass size distributions graphs showed that all the experiments had a fraction of particles appearing in the coarse mode.

3.4 Particle number concentrations

Figure 10 shows the comparison of particle number concentrations obtained from the startup, combustion, burnout and total cycle of all the experiments. Particle number concentrations obtained from the part load experiments A and B varied from 9.5×10^6 to 1.2×10^7 particles/cm^3 for the main combustion phase and 1.0×10^7 to 1.3×10^7 particles/cm^3 for the total cycle, while in the nominal load heat output experiments C, D, E and F varied from 1.4×10^7 to 8.8×10^7 particles/cm^3 for the combustion phase and 1.4×10^7 to 7.6×10^7 particles/cm^3 for the total cycle. Particle number emissions from the startup and burnout phase have the less impact on the total particle number emissions.

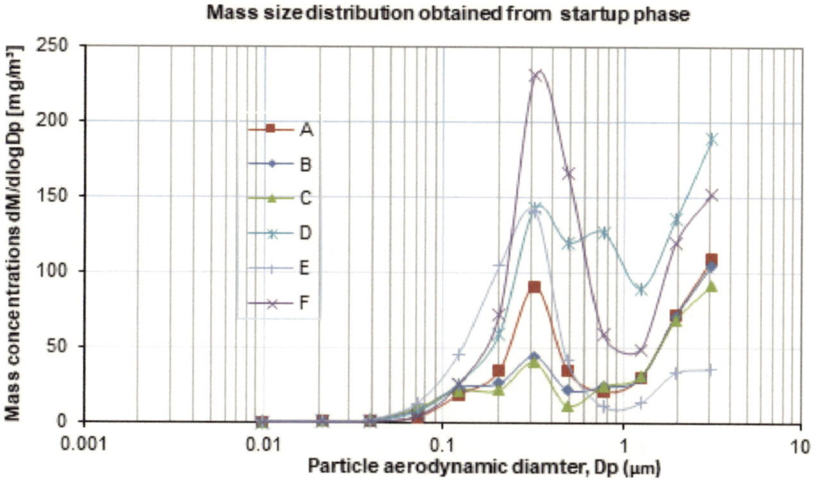

Figure 7.
Particle mass size distributions obtained from the startup phase of all the experiments [17].

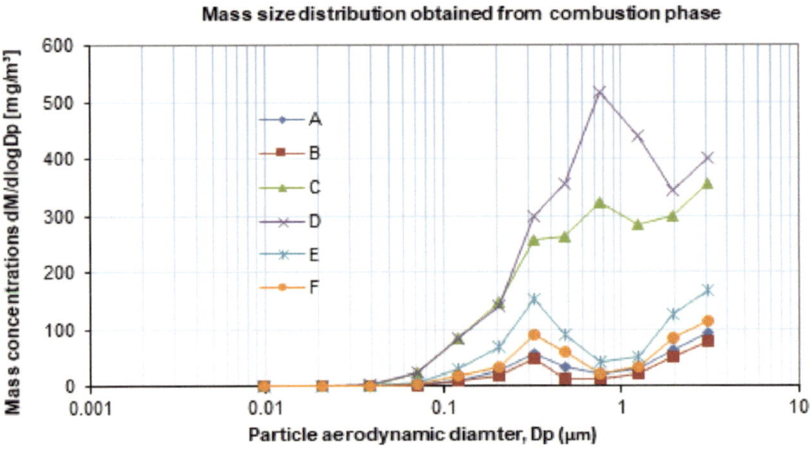

Figure 8.
Particle mass size distributions obtained from the combustion phase of all the experiments [17].

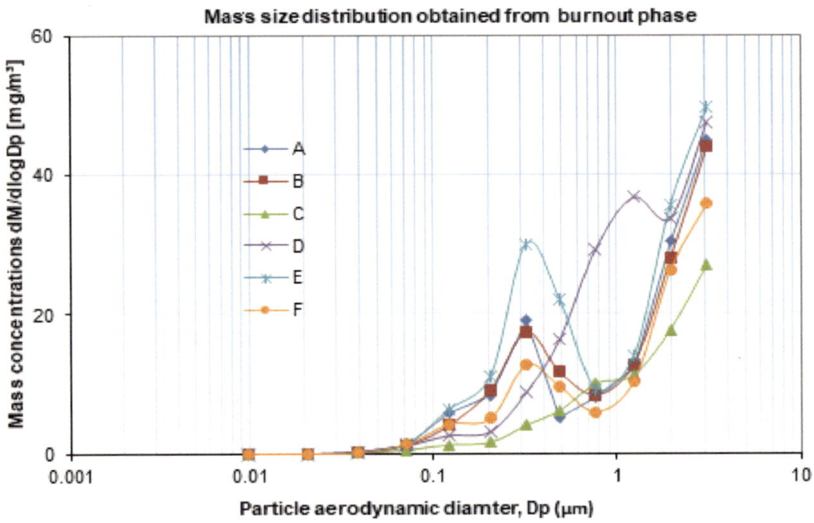

Figure 9.
Particle mass size distributions obtained from the burnout phase of all the experiments [17].

The duration of the main combustion phase varied from 3 h 50 min to 5 h 45 min for all the measurements. In the nominal load output for the combustion experiments C and D of the pellet stove, it can be seen from **Figure 10** that the combustion phase had the highest particle number concentrations, followed by the startup phase and the burnout phase, while for the combustion experiments E and F and part load experiments A and B had the highest particle number concentrations occurred in startup phase followed by the combustion phase and the burnout phase. This is because the configuration of the burner operated with the different fan speed used in the experiments.

Experiments operated with nominal load heat output, much lower particle number emissions obtained from the stove operated with high speed fan experiments E and F than the medium speed fan experiments C and D. This could be explained lower air excess factor λ = 2.8 for the medium speed fan experiments, while λ = 2.4 for the high speed fan experiments. Besides, the average flue gas temperature was 85°C for the medium speed experiments and 101°C for the high

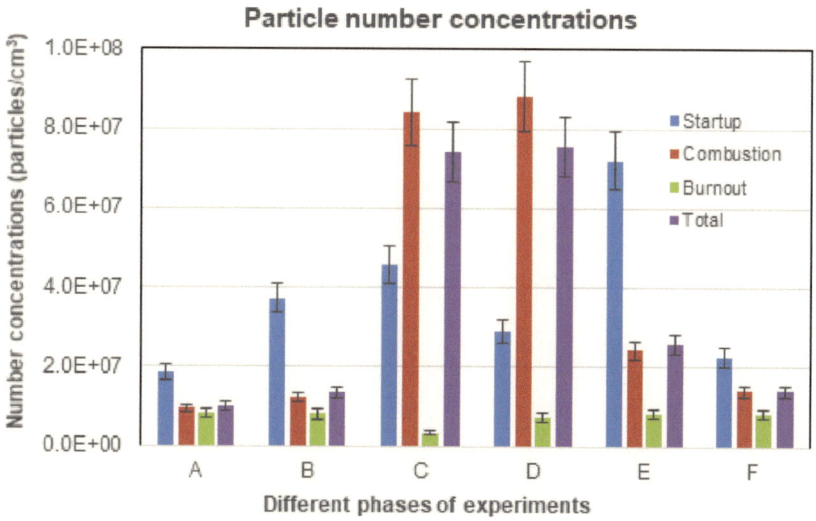

Figure 10.
Comparison of particle number concentrations [17].

speed fan experiments. The number concentrations from the main combustion of the part load experiments were much lower than nominal load experiments. This may be due to the difference of fuel consumption, fan speed of the screw, which regulates air flow into the combustion chamber and heat output. The average fuel consumption for part load experiments is about one-half lower than the nominal load experiments, which might impact on particle emissions.

The particle number concentrations obtained in this study can be compared with other studies. For example, Sippula et al. [27] investigated the effect on particle number concentrations from a top feed pellet stove with a capacity of 8 kW in nominal load output using an ELPI. Their results show that particle number emissions varied from 1.3×10^7 to 4.4×10^7 particles/cm^3, which is a little lower than the values obtained in our measurements in the combustion phase at nominal heat output. In another study, Bari et al. [29] studied particle number concentrations from a 5 kW pellet using a SMPS instrument. Their results show that particle number concentrations varied between 1.5×10^7 and 5.4×10^7 particles/cm^3, which is also little lower than the values obtained in our study. Since the fine particles are believed to be more harmful, more attention should be given to fine particle regulation. The EU standards describe the particle emissions in terms of mass concentration [46, 47], however, current research demonstrates that particle number emissions and particle size distributions are very important when considering particle impacts on air quality, climate, environment and human health [21, 48]. The emissions of particle concentration are strongly dependent on combustion conditions, fuel properties, combustion appliances, excess air, heat output of the combustion technology, etc. Small scale biomass combustion is generally considered as an important source of fine particles due to the lack of cleaning systems.

3.5 Number size distributions

Typical particle number size distributions obtained from the startup, combustion and burnout phases of all the combustion experiments are presented in **Figures 11–13**. **Figure 11** shows that the peak in particle number concentrations was observed at the particle size from 25 to 70 nm for the startup phases of all the combustion experiments. A uni-modal peak can be seen in the experiments C and D, while bimodal size distributions were appeared in the other combustion experiments.

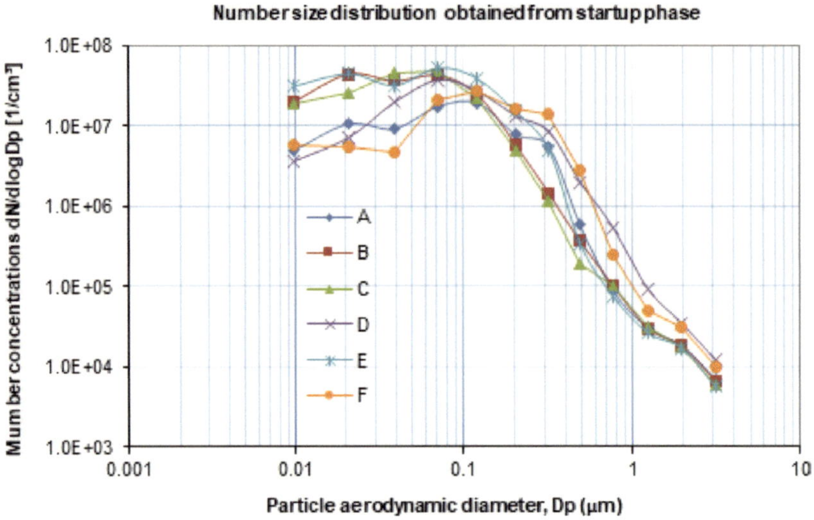

Figure 11.
Particle number size from distributions obtained from the startup phase of all the experiments [17].

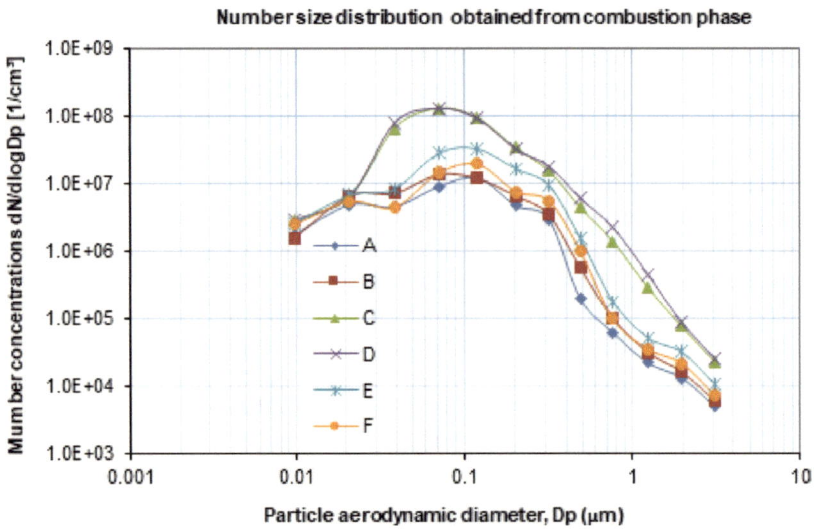

Figure 12.
Particle number size from distributions obtained from the combustion phase of all the experiments [17].

Similar particle number size distributions were observed in other studies. For example, Bari et al. [29] investigated particle number size distributions from a pellet stove of 5 kW nominal output using a SMPS. Their results show that the maximum number particle concentrations were found at the particle size within the diameter range from 55 to 90 nm. Boman et al. [49] investigated particle number size distributions from a pellet stove of capacity 5 kW using a SMPS and their results show that maximum particle at the particle size was about 70–80 nm.

Figure 12 presents the particle number size distributions graphs for the combustion phase and it can be seen that the emitted particles for all the experiments were very similar with the startup phase and peak particle number concentrations were at the particle size around 70–100 nm. In contrast to the startup phase, the maximum particle concentrations shifted to larger particle sizes. The measured particle number concentrations for combustion experiments C and D were significantly

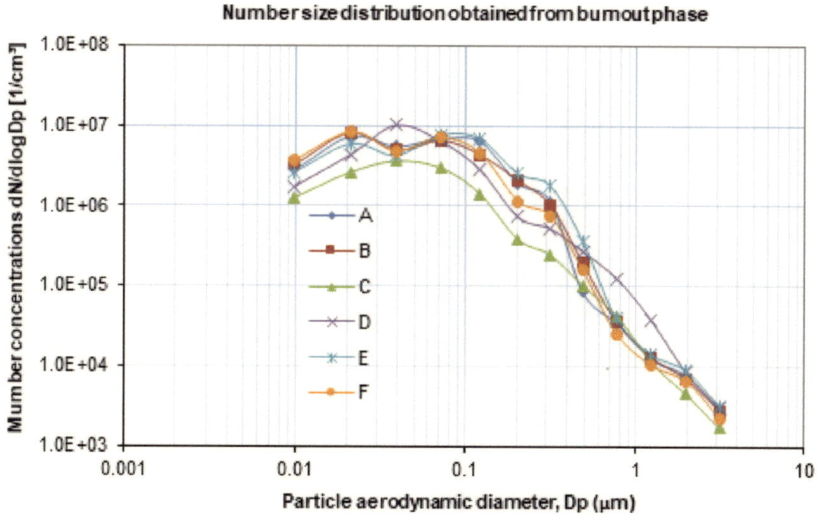

Figure 13.
Particle number size distributions obtained from burnout phase of all the experiments [17].

higher operated with medium fan speed than that high speed fan experiments followed by the low speed fan experiments.

Figure 13 presents the particle number size distributions for the burnout phase, where the maximum particle number concentrations were observed between 20 nm and 80 nm. The combustion experiments C and D had uni-modal size distributions, while the remaining experiments had bi-modal size distributions. The fine particles size less than 1 μm are formed from the easily volatile inorganic elements, released from the biomass fuels to the gas phase during combustion. Potassium, sulfur and chlorine are the most relevant element during the combustion of biomass fuels. These small size particles are considered very harmful for human health as they penetrate lower the alveolar region of the lung. Particles with diameter below 100 nm are the most important when considering the number distributions, but it contributes on only a very small fraction of the total mass. Fine particles originated from small scale biomass combustion mainly consist of ash, elemental carbon and organic material [9, 14]. Particle emissions are dominated by ash particles when the combustion quality is good, for example as in pellet combustion.

3.6 Performance analysis

The performance analysis in terms of combustion efficiency of the pellet stove was determined using an indirect method according to the standard EN 14785 that takes the thermal, chemical and radiation heat losses into consideration [35]. Efficiency was determined according to the difference between energy input and the sum of the losses. The thermal heat loss was evaluated on the basis of the difference between the temperature of the flue gas and the room temperature and the specific heat of the flue gas. The chemical heat loss is calculated from the CO and CO_2 concentrations of the flue gas. The radiation heat loss is taken as 0.2% according to the standard EN 14785. The formulas used for the calculation of the combustion efficiency and the different losses were discussed in the works [4, 7].

The combustion efficiency of the pellet stove as a function of different operational loads is compared with the required limit value of the standard NBN EN 14785 and presented in **Figure 14**. It can be seen from the figure that all the measurements both from nominal heat and part load output meet the standard. All

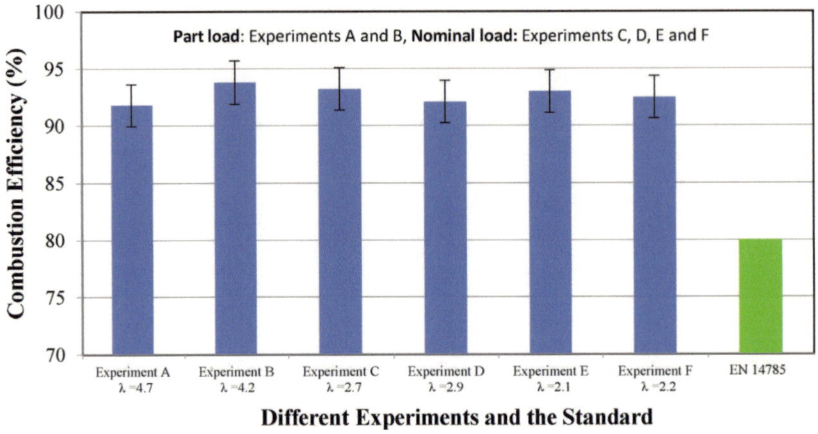

Figure 14.
Combustion efficiency compared with the standard [7].

the experiments have almost similar combustion efficiencies at both operational loads. The average combustion efficiency obtained from all the experiments for the low speed fan, medium speed fan and high speed fan was 92.8 ± 1.2, 92.4 ± 1.1 and 92.7 ± 1.2% respectively. The average thermal heat and chemical heat losses are estimated at 6.05 and 0.93% respectively for the low speed fan, 6.95 and 0.22% respectively for the medium speed fan and 6.95 and 0.14% respectively for the high speed fan, which influences the variations of the combustion efficiency as seen from **Figure 14**. Experiment B had the highest combustion efficiency from the part load operated with low speed fan, while experiments C and E had the higher efficiency from the nominal load output.

The combustion efficiency evaluated from the experiments in nominal load heat output can be compared with other works. For example, Sippula et al. [27] mentioned about 85% combustion efficiency from a 8 kW modern pellet stove in standard laboratory condition which is lower than this study.

4. Conclusions

A total of six combustion experiments on gaseous and particle emissions from a modern bottom feed pellet stove were conducted. Following conclusions can be drawn from this chapter.

- CO emissions in the burnout phase from all the experiments were significantly higher than that in the startup phase followed by the combustion phase. CO emissions in the burnout phase for experiments C to D were about 12 fold higher than in the startup phase and 75 fold higher than in the combustion phase. The air excess (λ) in the burnout phase for all the experiments was quite higher than that in the other two phases. The experimental results show that the impact of higher CO emissions in the startup and burnout phase has influence to increase the total CO emissions.

- The CO emissions obtained in the main combustion phase from the experiments conducted with high speed fan were lower than the medium speed fan followed by the low speed fan. Higher CO emission in the low speed fan was probably due to the higher air excess factor (about λ = 4.35), which gives lower combustion temperature, leading to high CO emissions.

- For the nominal load experiments, the particle mass fractions of PM_1 and $PM_{2.5}$ obtained from the combustion phase are significantly higher than those in the other two phases (startup and burnout phase). But, in the part load experiments, PM_1 emissions in the startup phase were relative higher than in the other phases.

- Particle mass size distributions analysis showed that all the experiments have maximum particle concentrations in the fine mode mainly at the particle size about 320 nm for the startup and combustion phase and at 300 nm for the burnout phase.

- The number concentrations from the main combustion of the part load experiments were much lower than nominal load experiments. This may be due to the difference of fuel consumption, fan speed of the screw, which regulate air flow into the combustion chamber and heat output.

- Analysis from the particle number size distributions showed that maximum particle emissions were found for all the experiments between 25 and 70 nm for startup phase, 70 and 100 nm for the combustion phase and 20 nm and 80 nm for the burnout phase.

Acknowledgements

The author(s) gratefully acknowledge the support of the Erasmus Mundus External Cooperation Window of the European Commission and the European Regional Development Fund. The author(s) also would like to express sincere thanks to the entire Department of Mechanical Engineering, VUB, especially Prof. Dr. Svend Bram and the combustion laboratory of a stove manufacturing plant, Belgium for their cooperation in conducting the experiments.

Conflict of interest

The author declares no conflict of interest.

Nomenclature

BLPI	Berner low pressure impactor
DLPI	Dekati low pressure impactor
ED	ejector diluter
EN	European norms
EU	European union
ELPI	electrical low pressure impactor
ELPI+	electrical low pressure impactor plus
Dp	particle diameter
DR	dilution ratio
Hr	hour
LHV	lower heating value
lpm	liter per minute
PAH	polycyclic aromatic hydrocarbons
PM	particulate matter
PRD	porous tube diluter

rpm	rotation per minute
SMPS	scanning mobility particle sizer
sec	second
VOC	volatile organic compounds
µm	micrometer
nm	nanometer
mg	milligram
min	minute
cm	centimeter
λ	air excess
°C	degree celsius

Author details

Md. Obaidullah[1,2*] and Jacques De Ruyck[2]

1 Centre for Energy Studies, Bangladesh University of Engineering and Technology (BUET), Dhaka, Bangladesh

2 Department of Mechanical Engineering, Vrije Universiteit Brussel (VUB), Brussels, Belgium

*Address all correspondence to: mdobaidullah@ces.buet.ac.bd

References

[1] Koppejan J, van Loo S. The Handbook of Biomass Combustion and co-Firing. London: Earthscan; 2010

[2] Kubica K, Paradiz B, Dilara P. Small combustion installation: Techniques, emissions and measure for emissions reduction. JRC Scientific and Technical Reports; 2007

[3] McDonald JD, Zielinska B, Fujita EM, Sagebiel JC, Chow JC, Watson JG. Fine particle and gaseous emission rates from residential wood combustion. Environmental Science & Technology. 2000;**34**:2080-2091

[4] Obaidullah M. Particle Emissions from Small Scale Biomass Combustion Appliences. Belgium: Vrije Universiteit Brussels; 2014

[5] Obaidullah M, Bram S, De Ruyck J. An overview of PM formation mechanisms from residential biomass combustion and instruments using in PM measurements. International Journal of Energy and Environment. 2018;**12**:41-50

[6] Nussbaumer T, Lauber A. Formation mechanisms and physical properties of particles from wood combustion for design and operation of electrostatic precipitators-18th European Biomass Conference and Exhibition. 18th European Biomass Conference and Exhibition, Lyon; 2010

[7] Obaidullah M, Bram S, De Ruyck J. Investigation on gaseous and particle mass emissions from automatically fired small scale heating system under laboratory conditions. The International Journal of Renewable Energy Development. 2018;7:111-121

[8] Yokoyama S, Matsumura Y. The Asian Biomass Handbook: A Guide for Biomass Production and Utilization. Japan: The Japan Institute of Energy; 2008. pp. 61-62

[9] Sippula O. Fine particle formation and emission in biomass combustion [PhD]. Finland: University of Eastern Finland; 2010

[10] Obernberger I, Brunner T, Barnthaler G. Fine particle emissions from Modern Austrian small scale biomass combustion plants. 15th European Biomass Conference and Exhibition. Germany 7-11 May, 2007. pp. 1546-1557

[11] Hinds WC. Aerosol Technology: Properties, Behavior, and Measurement of Airborne Particles. Los Angeles, California, USA: John Wiley & Sons; 1999

[12] Obaidullah M, Bram S, Verma V, De Ruyck J. A review on particle emissions from small scale biomass combustion. The International Journal of Renewable Energy Research (IJRER). 2012;**2**:147-159

[13] Guisson R, Marchal D. IEA Bioenergy Task 40-Sustainable International Bioenergy Trade Securing Supply and Demand Country Report Belgium. Belgium: VITO; 2009

[14] Tissari J, Sippula O, Kouki J, Vuorio K, Jokiniemi J. Fine particle and gas emissions from the combustion of agricultural fuels fired in a 20 kW burner. Energy & Fuels. 2008;**22**:2033-2042

[15] Lighty JS, Veranth JM, Sarofim AF. Combustion aerosols: Factors governing their size and composition and implications to human health. Journal of the Air & Waste Management Association: Taylor & Francis. 2000;**50**:1565-1618

[16] Nussbaumer T, Czasch C, Klippel N, Johansson L, Tullin C. Particulate Emissions from Biomass

Combustion in IEA Countries. Zurich, Switzerland: IEA Bioenergy Task 32; 2008

[17] Obaidullah M, Bram S, De Ruyck J. Measurements of particle emissions and size distributions from a modern residential pellet stove under laboratory conditions. International Journal of Systems Applications, Engineering & Development. 2019;**13**:1-9

[18] Johansson LS, Leckner B, Gustavsson L, Cooper D, Tullin C, Potter A. Emission characteristics of modern and old-type residential boilers fired with wood logs and wood pellets. Atmospheric Environment. 2004;**38**:4183-4195

[19] Tissari J. Fine particle emissions from residential wood combustion [PhD]. University of Kuopio: Finland; 2008

[20] Laumbach RJ, Kipen HM. Respiratory health effects of air pollution: Update on biomass smoke and traffic pollution. The Journal of Allergy and Clinical Immunology. 2012;**129**:3-11

[21] Bolling AK, Pagels J, Yttri KE, Barregard L, Sallsten G, Schwarze PE, et al. Health effects of residential wood smoke particles: The importance of combustion conditions and physicochemical particle properties. Particle and Fibre Toxicology. 2009;**6**:1-20

[22] Nussbaumer T, Klippel N, Johansson L. Survey on measurements and emission factors on particulate matter from biomass combustion in IEA countries. Proceedings of 16 th European Biomass Conference and Exhibition; 2008. pp. 2-6

[23] Obernberger I, Thek G. The Pellet Handbook: The Production and Thermal Utilisation of Pellets. London, UK; 2010

[24] Fiedler F, Persson T. Carbon monoxide emissions of combined pellet and solar heating systems. Applied Energy. 2009;**86**:135-143

[25] Obaidullah M, Bram S, De Ruyck J. Particle mass and gaseous emissions from small scale modern wood stoves. International Journal of Energy Applications and Technologies. 2019;**6**:57-64

[26] Boman C, Nordin A, Bostrom D, Ohman M. Characterization of inorganic particulate matter from residential combustion of pelletized biomass fuels. Energy & Fuels. 2004;**18**: 338-348

[27] Sippula O, Hytonen K, Tissari J, Raunemaa T, Jokiniemi J. Effect of wood fuel on the emissions from a top-feed pellet stove. Energy & Fuels. 2007;**21**: 1151-1160

[28] Gaegauf C, Wieser U, Macquat Y. Field investigation of nanoparticle emissions from various biomass combustion systems. Proceedings of International Seminar on Aerosol from Biomass Combustion Switzerland. Switzerland. 27 June 2001. pp. 81-85

[29] Bari MA, Baumbach G, Brodbeck J, Struschka M, Kuch B, Dreher W, et al. Characterisation of particulates and carcinogenic polycyclic aromatic hydrocarbons in wintertime wood-fired heating in residential areas. Atmospheric Environment. 2011;**45**: 7627-7634

[30] Bäfver LS, Leckner B, Tullin C, Berntsen M. Particle emissions from pellets stoves and modern and old-type wood stoves. Biomass and Bioenergy. 2011;**35**:3648-3655

[31] Qiu G. Testing of flue gas emissions of a biomass pellet boiler and abatement of particle emissions. Renewable Energy. 2013;**50**:94-102

[32] Johansson LS. Characterisation of particle emissions from small-scale biomass combustion [Licentiate thesis]. Sweden: Chalmers University of Technology; 2002

[33] Ohman M, Boman C, Hedman H, Nordin A, Bostrom D. Slagging tendencies of wood pellet ash during combustion in residential pellet burners. Biomass and Bioenergy. 2004;**27**:585-596

[34] Tissari J, Hytonen K, Sippula O, Jokiniemi J. The effects of operating conditions on emissions from masonry heaters and sauna stoves. Biomass and Bioenergy. 2009;**33**:513-520

[35] EN-14785. Residential space heating appliances fired by wood pellets: Requirements and test methods. Belgian Standards Institute (NBI); October 2006

[36] Marjamaki M, Keskinen J, Chen D-R, Pui DYH. Performance evaluation of the electrical low pressure impactor (ELPI). Journal of Aerosol Science. 2000;**31**:249-261

[37] Obaidullah M, Sarkar M, Bram S, De Ruyck J. Evaluation of Dilution Ratio from a Partial flow Dilution Tunnel. 7th IMEC &16th Annual Paper Meet. IEB: Dhaka; 2-3 Jan 2015

[38] Obaidullah M, Bram S, De Ruyck J. Investigation of optimal dilution ratio from a dilution tunnel using in particulate matter measurement. The International Journal on Engineering Technology and Sciences (IJETS). 2018;**5**:17-33

[39] Verma V, Bram S, Gauthier G, De Ruyck J. Evaluation of the performance of a multi-fuel domestic boiler with respect to the existing European standard and quality labels: Part-1. Biomass and Bioenergy. 2011;**35**:80-89

[40] Schmidl C, Luisser M, Padouvas E, Lasselsberger L, Rzaca M, Ramirez-Santa Cruz C, et al. Particulate and gaseous emissions from manually and automatically fired small scale combustion systems. Atmospheric Environment. 2011;**45**:7443-7454

[41] Gonzalez JF, Gonzalez-Garca CM, Ramiro A, Sabio E. Combustion optimisation of biomass residue pellets for domestic heating with a mural boiler. Biomass and Bioenergy. 2004;**27**:145-154

[42] Mediavilla I, Fernández M, Esteban L. Optimization of pelletisation and combustion in a boiler of 17.5 kW th for vine shoots and industrial cork residue. Fuel Processing Technology. 2009;**90**:621-628

[43] Roy MM, Corscadden KW. An experimental study of combustion and emissions of biomass briquettes in a domestic wood stove. Applied Energy. 2012;**99**:206-212

[44] Hukkanen A, Kaivosoja T, Sippula O, Nuutinen K, Jokiniemi J, Tissari J. Reduction of gaseous and particulate emissions from small-scale wood combustion with a catalytic combustor. Atmospheric Environment. Elsevier Ltd. 2012;**50**:16-23

[45] Obaidullah M, Bram S, De Ruyck J. Characteristics of Particle Mass Concentrations from Small Scale Biomass Combustion: A Review. VII International Conference on Energy and Environment for 21 Century. Las Villas, Cuba: Central University Marta Abreu (UCLV); 2012

[46] Villeneuve J, Palacios JH, Savoie P, Godbout S. A critical review of emission standards and regulations regarding biomass combustion in small scale units (<MW). Bioresource Technology. 2012;**111**:1-11

[47] Belgium RD. Royal Decree Regulating Minimum Performance Requirements and Pollutant Emission Levels for Solid Fuel-Fired Heating Appliances. Belgium: Belgian Institute for Standardization; 2006:1-79

[48] Vicente E, Alves C. An overview of particulate emissions from residential

biomass combustion. Atmospheric Research. 2018;**199**:159-185

[49] Boman C, Pettersson E, Westerholm R, Bostrom D, Nordin A. Stove performance and emission characteristics in residential wood log and pellet combustion, part 1: Pellet stoves. Energy & Fuels. 2011;**25**:307-314

Chapter 21

Analysis of Optimal Steady-State Operation of Power Systems with Embedded FACTS Devices: A Matlab-Based Flexible Approach

Jose Miguel García-Guzman, Néstor González-Cabrera, Luis Alberto Contreras-Aguilar, Jose Merced Lozano-García and Alejandro Pizano-Martinez

Abstract

This book chapter presents a flexible approach to incorporate mathematical models of FACTS devices into the Power Flow (PF) and the Optimal Power Flow (OPF) analysis tools, as well as into the standard OPF Market-Clearing (OPF-MC) procedure. The proposed approach uses the Matlab Optimization Toolbox because it allows to easily: (a) implement a given optimization model, (b) include different objective functions using distinct equality and inequality constraints and (c) modify and reuse an optimization model that has been previously implemented. The conventional OPF model is the main core of the proposed approach, which is easily implemented and adapted to include the mathematical models of FACTS devices. The resulting implementation of the OPF model featuring FACTS devices can be easily modified and adjusted to obtain the implementation of both the PF and the OPF-MC models which includes such devices. It should be mentioned that with the flexible approach proposed here, the complexity as well as the implementation time of optimized models featuring embedded FACTS devices is significantly reduced, since it is not necessary to define the expressions associated with the hessian matrix and the gradient vector. The flexibility and reliability of the proposed approach are demonstrated by means of several study cases using test as well as real power systems.

Keywords: FACTS devices, Matlab-based flexible approach, optimal power flow (OPF), steady-state operation, power systems

1. Introduction

The efficient operation and economics of electric power systems have always occupied an important place in the electric power industry [1]. Here, it is worth mentioning that highly desirable operative conditions cannot be successfully achieved by means of using the old electromechanically controlled devices, such as the mechanical phase shifter. These old devices execute control actions through mechanical actuators, and thus provide a poor high-speed control. Some of the consequences of this lack of speed and reliable control are associated with stability problems. Also, the power flow deviates through secondary transmission lines that are not the intended ones. In addition to the above mentioned issues, there are instability issues when the transmission resources are closely used to their thermal and economic limits. Not to mention, the high or low voltages resulting from such a poor high-speed control, in addition to many other issues [2]. Those poor operating conditions have given place for the need to develop faster controllers and a more efficient power system management. Fortunately, the fast development of power electronics based on new and powerful semiconductor devices has given rise to novel technologies such as FACTS controllers that greatly improve the operating conditions of power systems [3].

From a steady-state operation perspective, the control capabilities of FACTS devices allow to adjust the active and reactive power flows at their output terminals, as well as to achieve the local control of reactive power and voltages at the connected nodes [4]. In order to investigate the effectiveness of these control capabilities for alleviating a large variety of problems associated with the steady-state performance of power systems, the existing tools for the steady-state condition assessment have been upgraded to incorporate steady-state models of FACTS controllers. The Power Flow, the Optimal Power Flow and the OPF Market-Clearing procedures are the common tools used to assess the steady-state operation and market-clearing conditions of power systems. Accordingly, the upgrade of those three tools has received great attention. For instance, references [4–12, 4, 13–19, 20–24] are representative samples of the several publications where models of FACTS controllers are incorporated into the Power Flow, the Optimal Power Flow and the market-clearing procedures, respectively.

Certainly, the evolution of FACTS controllers is going to be progressive as function of time [2]. With the further development of technologies based on power electronics, the economic viability of these controllers will improve in such a way that more and more FACTS devices are expected to be designed and used in future applications [13]. In this sense, it is expected that modern power systems are going to be electronically controlled rather than mechanically controlled [2]. Consequently, to study the usefulness of new FACTS devices for improving the steady-state performance of power systems as well as to speed up research efforts, it would be needed the upgrade of the existing steady-state analysis tools in order to incorporate the corresponding mathematical models of such devices.

Bearing in mind the aforementioned prospective on FACTS controllers, in this book chapter is proposed a Matlab-based flexible approach that incorporates mathematical models of FACTS devices into: the power flow and the optimal power flow analysis tools; as well as into the standard OPF market-clearing procedure [25]. The flexibility of the proposed approach is achieved by taking advantage of both: (a) the tractability of the nonlinear continuous optimization theory to consider different objective functions that can be properly constrained and (b) the simplicity of the Matlab Optimization Toolbox to implement and solve an optimization model, as well as its ability to modify and reuse an optimization model that has already been implemented [26]. In order to show the flexibility of the proposed approach, first consider that the OPF model is the main core of this proposal, then the strategy to code the objective function, the equality and inequality constraints of the conventional OPF model are given in detail. Second, the implemented conventional OPF model is enlarged to include the steady-state models of FACTS controllers. Lastly, the resulting OPF implementation is modified and reused, to straightforwardly obtain the computational implementation of both the Power Flow and the standard OPF market-clearing models that use embedded FACTS controllers.

In this way, the proposed book chapter can reduce the computational implementation load of the FACTS model devices into one of the three previously mentioned procedures. Therefore, this proposal facilitates carrying out research in order to assess the effectiveness of FACTS devices for improving the steady-state performance of power systems. Arguably, the use of high-level programming languages and optimization toolboxes such as the ones provided by Matlab could reduce the computational efficiency of any approach that is implemented using such tools. However, for research purposes, the computational efficiency of the proposed approach is quite reasonable, since the solution of the distinct optimization models related to medium scale power systems is commonly obtained in a time lapse that varies from a few seconds to a maximum of 10 minutes. In addition, when the main objective is to investigate the effectiveness of FACTS devices for improving the steady-state performance of power systems, the computational efficiency requirement can be further relaxed.

This book chapter is organized as follows. Since the conventional OPF model is the core of this approach, its formulation is given in Section 2, as well as the steady-state models of the traditional power system components. These models are then considered in Section 3 to assemble the conventional OPF model. The Section 4 shows the incorporation of the FACTS models into the conventional OPF model. This last model can be straightforwardly extended in order to obtain the computational implementation of the PF and, the standard OPF market-clearing models in Sections 5 and 6, respectively. The applicability of the proposed approach for the steady-state operation of power systems using embedded FACTS devices is assessed by means of numerical examples in Section 7. Last but not least, the conclusions of this chapter book are stated in Section 8.

2. The optimal power flow general-formulation

The optimal power flow formulation is a non-linear optimization problem which consists in minimizing an objective function subject to a finite set of equality and inequality constraints. Mathematically, the OPF model is given by

$$\mathit{Minimize}\, f(y) \tag{1}$$

$$\mathit{Subject\ to}\ h(y) = 0 \tag{2}$$

$$g(y) \leq 0 \tag{3}$$

$$y^{\min} \leq y \leq y^{\max} \tag{4}$$

where $f(y)$ is the objective function to be minimized, $h(y)$ and $g(y)$ are a finite set of functions corresponding to the equality and inequality constraints, respectively. Whilst y, represents the full set of variables of the power system with a lower limit $y^{\min}$ and an upper limit $y^{\max}$. It must be mentioned that the set y contains of both the state variables as well as the control variables.

The OPF formulation is given in polar coordinates taking in consideration the model of each power system component. The most common components of a power system are: generators, loads, fixed shunt compensation elements, transmission lines and transformers; whose steady-state models are given in [4]. These models are adopted in the flexible approach proposed in the present book chapter.

3. Flexible approach to implement the OPF model

This section describes the flexible approach to implement the conventional OPF model as a computer program, which is developed based on structured programming along with the use of the **fmincon** optimization function of Matlab [26]. This proposal allows to obtain a computational implementation of the OPF model with the following remarkable features: (a) It considerably simplifies the computational complexity of the implementation of a conventional OPF, (b) it is very flexible when considering diverse physical laws and operating conditions, (c) it allows for the integration of FACTS models and (d) it has the tractability of easily being adapted to study electric power systems by means of different steady-state analyses. For example, it can be straightforwardly adapted to solve the conventional power flow problem and to carry out the analysis of electric power markets. In this way, steady-state analyses of distinct power systems featuring different electric components can be carried out by doing very few changes to the composition of the set of steady-state variables as well as the set of constraints and the objective function of the proposed implementation of the OPF model. The aforementioned features are further exploited in Sections 5, 6 and 7 in order to take into consideration: FACTS devices, the conventional power flow problem and the standard OPF market-clearing procedure, respectively.

3.1 The fmincon function of the Matlab optimization toolbox

The **fmincon** function of the Matlab Optimization Toolbox is used to solve the general OPF model which is a constrained nonlinear optimization or programming problem. That Matlab function attempts to find out the constrained minimum of a scalar objective function of several variables, starting with an initial estimation of the constrained minimal point. Here it is worth mentioning that the **fmincon** function is designed to solve problems where the objective and constraint functions are both continuous as well as their first derivatives. In order to obtain the optimal solution of the optimization problem at hand, the **fmincon** users can choose one of the following optimization options: the trust-region-reflective method, the active-set method, the interior-point method and the sequential quadratic programming approach [26]. **Figure 1** shows a schematic representation of the main input and output arguments of this function. It must be pointed out that, depending on to the requirements of the specific optimization problem, some empty input arguments must be provided.

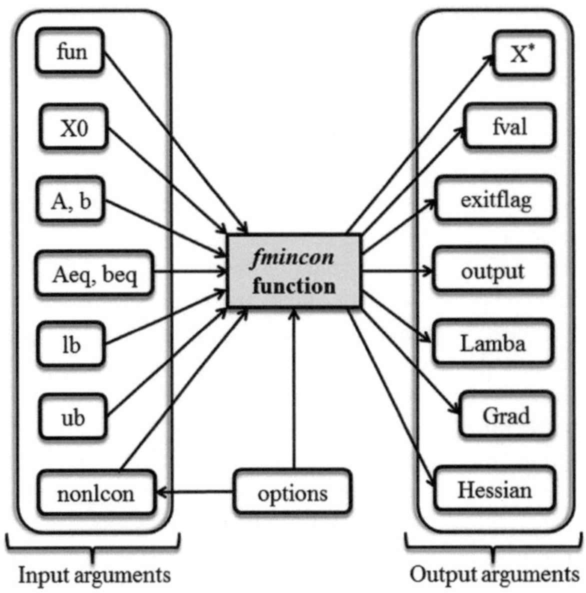

Figure 1.
Schematic representation of the main input and output arguments of ***fmincon*** *function.*

The next command line syntax of the **fmincon** function, including input and output arguments, is shown below [26]. Observe that the first line of code corresponds to the input arguments, whilst the second one corresponds to the output arguments.

```
x = fmincon(fun,x0,A,b,Aeq,beq,lb.,ub,nonlcon,options)
[x,fval,exitflag,output,lambda,grad,hessian] = fmincon(___)
```

In **Figure 1** and the syntax, the input argument **fun** is the handle of the M-function file containing the objective function to be optimized. $\mathbf{X_0}$ represents the initial conditions of the state vector X of the optimization model, which is given by (5). For the purpose of this work, this state vector ($\mathbf{X_0}$) stores the initial conditions of the steady-state variables of the power system. By the way, the lower and upper bounds of $\mathbf{X_0}$ are assigned to the arguments **lb** and **up**, respectively. The size of **lb** and **up** must be equal to the size of the vector $\mathbf{X_0}$. Next, the **nonlcon** input argument is the handle of the M-function file containing the nonlinear inequality as well as the equality constraints. In the same venue, the argument **options** is a structure where the user can define optional parameters for the optimization process, such as: the optimization algorithm, the convergence tolerance, the maximum amount of iterations, the tolerance on the constraint violation and the termination tolerance on the function value, among many other parameters. It is worth mentioning that the **options** structure has the parameter **LargeScale** that allows for a choice on which algorithm to use. If that parameter is set to **on**, then the gradient of the objective function as well as the constraints must be provided by the user. It must be highlighted, however, that in this work, the **LargeScale** parameter is set to **off** with the aim of simplifying and making more flexible the proposed computational implementation. Last but not least, the input arguments **A**, **b**, **Aeq** and **beq** are associated with the coefficients of the linear inequality and equality constraints. In this proposal, however, the OPF model only considers nonlinear equality and inequality constraints and, therefore, the above mentioned input arguments are considered as empty arrays.

Next, the output arguments are briefly described to provide more clarity on the approach proposed here. First, $\mathbf{X}^*$ stores the optimal value of the optimization model variables as they are assessed by the **fmincon** function. Second, the **fval** output returns the value of the objective function given by the optimal solution $\mathbf{X}^*$. Third, the output **argument** is a structure that contains information about the results of the optimization. Forth, **lambda** is an argument storing the value of the Lagrange multipliers at $\mathbf{X}^*$. Fifth, the arguments **hessian** and **grad** store the value of the Hessian matrix and the gradient vector of the objective function at $\mathbf{X}^*$, respectively. Last but not least, **exitflag** is an integer identifying the termination condition of the optimization algorithm; that is, if (a) exitflag > 0: the algorithm reached the optimal solution $\mathbf{X}^*$, if (b) exitflag $= 0$: either the maximum amount of function evaluations or iterations specified by the user was exceeded, thus the optimal solution has not been reached and if (c) exitflag < 0: the algorithm did not reach the optimal solution, therefore, there is an unfeasible solution.

A detailed way to define the set of steady-state variables, the objective function and the constraints sets of the OPF model is given below. Also, the explicit OPF formulation is given in polar coordinates in accordance to the model of each power system component, as it is described in the following subsections.

3.2 Flexible implementation of the OPF model

3.2.1 Steady-state variables

Within the proposed implementation, the set of the steady-state variables of the OPF model is treated as a vector containing such variables in the order shown in (5).

That vector is called the vector of steady-state variables, denoted by X. Note that T denotes the transpose of the vector X.

$$X = \left[\theta_1 \dots \theta_{nb} \;\; V_1 \dots V_{nb} \;\; P_{g1} \dots P_{gng} \right]^T \tag{5}$$

In (5), the number of nodes and generators of the power system are represented by nb and ng, respectively. θ_i and V_i represent the angle (grades) and magnitude (pu) of the phasor voltage in each node i of the power system, and P_{gi} is the power of the generator i. Note that the size of the steady-state vector is (2nb + ng, 1). It should be mentioned that in the implementation, the vector X also corresponds to the vector of the initial conditions of the state variables $\mathbf{X_0}$, which is also called by the **fmincon** function during the optimization process. That function returns the optimal value of the steady-state variables, which is stored in the vector X, that is, the result is rewritten in X. In this way, when the optimization process is finished, X is then the vector of the optimal solution represented by $\mathbf{X^*}$.

The vector $\mathbf{X_0}$, which represents the initial conditions of the state variables of the power system in the OPF model, is defined as follows: the magnitude and angle of the voltage in each node of the power system are initialized at 0 grades and 1.0 pu, respectively, whilst P_{g0} is the initial condition of active power at the generation nodes. It is worth mentioning that P_{g0} is set to the values of the active power obtained from the analysis of the Optimal Generation Dispatch (OGD), which in turn, neglects the losses in the transmission elements. Last but not least, observe that the initial conditions of the active power generation help to reduce the number of iterations associated with the OPF optimization process. The syntax of the vector X containing the state variables is the following:

```
X(1:nb,1)= V_angle;            %Phase angle of nodal voltage in radians
X(nb+1:2*nb,1)=V_mag;          %Magnitude of nodal voltage in pu
X(2*nb+1:2*nb+ng,1)=Pg;        %Active power generation
```

3.2.2 Objective function

The objective function considered in the OPF model corresponds to the total generation cost of active power, which is a nonlinear function given by

$$f(y) = \sum_{i=1}^{ng} a_i + b_i \left(P_{gi}\right) + c_i \left(P_{gi}\right)^2 \tag{6}$$

where the cost curve coefficients of each generator i are represented by ai, bi and ci. ng is the number of generation nodes including the slack node. P_{gi} is the individual generation power level. Observe that the objective function in (6) is coded into an M-function file with the name **objfun_OPF**. This last file is also called by the **fmincon** function through the input argument **fun**, using the syntax @(X) objfun_OPF(X). The general Matlab syntax for the objective function given by Eq. (6) is the following:

```
function f = objfun_OPF(___)
f=0;
for k=1:ng
    f=f+ai(k,1)+bi(k,1)*Pgi(k,1)+ci(k,1)*Pgi(k,1)^2;
end
```

It is worth mentioning that in the above M-function file is possible to: include, modify, define or write the objective function as it is desired. Therefore, the objective function (6) can be modified or even changed by a new function. This characteristic provides great flexibility to the proposal, including the possibility for handling several objective functions at the same time.

3.2.3 Equality constraints

The equality constraints of the OPF analysis are given by the balance energy equations for the power system. These restrictions must be unconditionally met because they establish the active and reactive power balance under steady-state conditions for all the nodes of the power system. If at least one constraint is not satisfied, then the solution of the OPF problem is not feasible. That set of equality constraints is given by

$$h(y) = \left\{ \begin{array}{l} P_{gi} - P_{li} - \sum\limits_{j \in i}^{nb} P_i^{inj,j} = 0 \\ \\ Q_{gk} - Q_{lk} - \sum\limits_{j \in k}^{nb} Q_k^{inj,j} = 0 \end{array} \right\} \quad \begin{array}{c} i = 1, 2, \dots, nb \\ k = 1, 2, \dots, nb \mid k \notin ng \end{array} \tag{7}$$

where i and k are the nodes in which the energy balance is satisfied. $P_i^{inj,j}$ and $Q_k^{inj,j}$ represent the active and reactive injected-power at node i and k, respectively. Note that the injection point is the *j-th* network element. Also note that the active-power and reactive-power (denoted by P_{gj} and Q_{gj}, respectively) are provided by the controllable complex power sources located at the *j-th* generation node ($j = i, k$). Last but not least, the complex power consumption is due to the active-power load P_{lj} and the reactive-power load Q_{lk}.

In the proposed flexible implementation, the set of equality constraints is written in an M-function file with the name **constraints_OPF**. This set of constrains is given by (8).

$$\begin{bmatrix} Ce_1 \\ \vdots \\ Ce_{nb} \\ Ce_{nb+1} \\ \vdots \\ Ce_{2nb} \\ Ce_{2nb+1} \end{bmatrix} = \begin{bmatrix} P_{g,1} \\ \vdots \\ P_{g,nb} \\ Q_{g,1} \\ \vdots \\ Q_{g,nb} \\ X_{sn} \end{bmatrix} - \begin{bmatrix} P_{l,1} \\ \vdots \\ P_{l,nb} \\ Q_{l,1} \\ \vdots \\ Q_{l,nb} \\ \theta_{sn} \end{bmatrix} - \begin{bmatrix} P_{inj,1} \\ \vdots \\ P_{inj,nb} \\ Q_{inj,1} \\ \vdots \\ Q_{inj,nb} \\ 0 \end{bmatrix} \tag{8}$$

In the above expression, sn is the subscript associated with the slack node, whilst X_{sn} stores the value of phase of the slack node in the corresponding component of the state vector X. Note that θ_{sn} is the value of the phase at which the angle of the slack node is fixed. It must be pointed out that in (7) and (8), the value of the reactive power Q_{gi} must always be zero for the *i-th* generation node. This is because the level of reactive power generation depends on the state variables of the power system. Henceforth, Q_{gi} lacks a scheduled value in the optimization model OPF. Furthermore, the constraint of reactive power balance is satisfied only at non-generation nodes. Then such constraint should be handled as a functional inequality constraint for generation nodes in order to establish the reactive power balance in such nodes, as it is explained below. Note that the way in which the vector of equality constraints (*Ce*) is written, provides the flexibility to add more constraints of this type. This in turn allows extending the OPF model to other similar models for the steady-state operation assessment of power systems. The general command line syntax for the vector of equality constraints, *Ce,* is the following:

```
function [ci,ce] = constraints_OPF(___)

% Nodal active power balance equality constraints (Pg-Pd_Pinj=0)
%i=1,2,...,nb
Ce(1:nb,1) = Pg_i(___) - Pl_i(___) - Pinj_i(___);

% Nodal reactive power balance equality constraints (Qg-Qd_Qinj=0)
%k=1,2,...,nb-ngen
Ce(nb+1:2*nb,1) = Qg_k(___) - Ql_k(___)- Qinj_k(___);

%Set fix the slack node angle
Ce(2*nb+1,1) = VP(sn) - V(sn,2);
```

It is important to note that in the above syntax, Pg_i, Pl_i and Pinj_i are implemented as M-function files. The same is valid for the code line that corresponds to the reactive power.

3.2.4 Inequality constraints

The physical and operational limits are mathematically modeled as inequality constraints. This is because these limits help to constrain the range of the practical steady-state operation of the power-system components. For a better handling of the inequality constraints, these are classified into two types: inequality constraints on variables and functional inequality constraints. The limits on the magnitude of the substation-voltage and the generation of active-power are represented by the inequality constraints given by (9).

$$Y = \left\{ \begin{array}{c} \theta_n{}^{\min} \le \theta_n \le \theta_n{}^{\max} \\ P_{gi}{}^{\min} \le P_{gi} \le P_{gi}{}^{\max} \\ V_n{}^{\min} \le V_n \le V_n{}^{\max} \end{array} \right\} \begin{array}{l} i = 1, 2, \ldots, ng \\ n = 1, 2, \ldots, nb \end{array} \tag{9}$$

In (9), the superscripts min and max represent the lower and upper bounds of the respective variable. As mentioned before, the lower bound is assigned to the input argument **lb**, and the upper bound is assigned to the input argument **up**; both arguments are called by **fmincon** function during the optimization process. The assignment of such limits to the respective previously mentioned input-arguments is shown in (10). The superscript T, on **lb** and **up** indicates the transposition of the respective vector.

$$\text{lb} = \left[\theta_1^{\min} \ldots \theta_{nb}^{\min} \quad V_1^{\min} \ldots V_{nb}^{\min} \quad P_{g1}^{\min} \ldots P_{gng}^{\min}\right]^T \tag{10}$$

$$\text{up} = \left[\theta_1^{\max} \ldots \theta_{nb}^{\max} \quad V_1^{\max} \ldots V_{nb}^{\max} \quad P_{g1}^{\max} \ldots P_{gng}^{\max}\right]^T \tag{11}$$

Clearly, the size of the vectors **lb** and **up** is equal to the size of the vector of the steady-state variables, *X*. As mentioned before, storing the inequality constraint-vectors into an M-file enables the convenient modification of such vectors by adding or removing constraints that are associated with other models of steady-state operation of power systems. This is shown in subsequent sections.

The level of reactive power generation is not a scheduled value because it is a function of the system variables. Therefore, the limits of the power (Q_{gi}) are modeled in the OPF problem as a functional inequality constraint as follows:

$$g(y) = \left\{ Q_{gi}^{\min} \le Q_{gi} \le Q_{gi}^{\max} \right\} \quad i = 1, 2, \dots, ng \tag{12}$$

The reactive power generation level Q_{gi} in (12) is given by the following function:

$$Q_{gi} = Q_{li} + \sum_{j \in i} - V_i^2 B_{ii} + V_i V_j \left[G_{ij} Sin\left(\theta_i - \theta_j\right) - B_{ij} Cos\left(\theta_i - \theta_j\right) \right] \tag{13}$$

In accordance with (13), the balance of reactive power at the *i-th* generation node is achieved only when the level of that power is inside its own bounds. If the level of the reactive power generation of a given generator is out of its upper or lower limit, then the corresponding inequality constraint (9) must be added to the active set in (12), becoming an equality constraint that helps in resetting the reactive power Q_{gi} at the violated limit $Q_{g,v}$.

$$Q_{gi} - Q_{g,v} = 0 \tag{14}$$

It should be noted that in the OPF model, the above expression avoids the violation of the limits of the reactive power generation. Also it allows to maintain the balance of reactive power at the generation nodes. Next observe that the functional inequality constraints are written in the same file of the equality constraints, as it is shown below.

$$\begin{bmatrix} Ci_1 \\ \vdots \\ Ci_{ng} \\ Ci_{ng+1} \\ \vdots \\ Ci_{2ng} \end{bmatrix} = \begin{bmatrix} -Q_{g,1} \\ \vdots \\ -Q_{g,ng} \\ Q_{g,1} \\ \vdots \\ Q_{g,ng} \end{bmatrix} + \begin{bmatrix} Q_{g,1}^{\min} \\ \vdots \\ Q_{g,ng}^{\min} \\ -Q_{g,1}^{\max} \\ \vdots \\ -Q_{g,ng}^{\max} \end{bmatrix} \tag{15}$$

The general Matlab syntax for the inequality constraints given by Eq. (15) is the following:

```
function [ci,ce] = constraints_OPF(___)

%Reactive power generation limits in generation nodes
ci(1:ngen,1) = -Qgi + Qgi_min;              %-Qg+Qlow<=0
ci(ngen+1:2*ngen,1) = Qgi - Qg_max;  % Qg-Qsup<=0
```

As mentioned above, the input argument **nonlcon** is the handle of the M-function file containing both the nonlinear equality and the functional inequality constraints. Therefore, the two sets of constraints are written in an M-function file whose name is **constraints_OPF**. The **fmincon** function calls this M-function file through the input argument **nonlcon**, using the syntax @(X)Constraints_OPF(X).

In addition to the M-functions: **objfun_OPF** and **constraints OPF**; the **fmincon** function also calls the vectors: **lb** and, **ub**; which contain the lower and upper limits of the state variables. The general syntax for the above is the following:

```
function [VP,fval,output,exitflag,lambda,grad]=OPF(___).
[X,fval,exitflag,output,lambda,grad,hessian] = fmincon(___);
[X] = fmincon(@(X) objfun_OPF(__), X, [], [], [], [], lb, ub,... @(X)
constraints_OPF(__),options);
```

The way in which the above M-files are written in Matlab facilitates the modification of the constraints, the objective function and the vector of state variables, providing flexibility to the proposed implementation. Furthermore, the development of this proposed methodology is based on structured programming, which increases its flexibility by allowing the inclusion of other electrical devices and/or different steady-state models for operational assessment of distinct power systems with minor code modifications.

3.3 Structured programming of the OPF model

The proposed flexible approach is developed using structured programming because in this way it is possible to develop computer programs that can be easily understood. Also this kind of programming is very useful when it is necessary to modify or extend an existing program. This is the case of the proposal presented here, where the OPF model is extended to consider several optimization applications of power systems with embedded FACTS controllers.

In that figure, it can be observed that the structure of the proposed implementation contains a main program. This one starts reading the M-file of the power system data, which includes a convergence tolerance, the MVA-base and the data of the most common components of a power system in accordance to Section 3 of this book chapter. After reading the M-file, in the main program, the power system data are converted to pu in order to work in the same units all entire system components, but also to avoid optimization scaling problems. Next, the program calls the respective functions for calculating the conductance and the susceptance of the transmission lines, the transformers and the shunt compensators. Observe that these quantities are used for calculating the injections of active and reactive power. This calculation involves the equality constraints associated with the power balance at each node of the power system. After that, the OGD M-function file is executed to obtain the solution of the Optimal Generation Dispatch. It must be pointed out that in the OGD M-function file, the **fmincon** function calls the objective function and, the OGD constraints through the argument **nonlcon** using a similar syntax to the one mentioned in the previous subsection. Then, the main program runs the OPF M-function file to solve the Optimal Power Flow model, where the **fmincon** function calls the M-function files containing the objective function and the constraints of the OPF model. Last but not least, the program computes the power flows in the network and prints results of the OPF steady-state solution. The output yielded by the proposed flexible approach include: (a) generators rating and the OPF dispatch, (b) the operating point computed by the OPF, (c) the power flow and power losses in transmission lines, transformers, and loads connected in shunt, (d) the total losses of the active and reactive power and (e) the CPU time of the optimization process.

4. Including FACTS models into the proposed flexible approach

Models of FACTS devices are easily integrated into the proposed flexible implementation by considering the following issues: the state variables and their limits; the injection of active and reactive power at the node where the controller is connected and the control equations or control modes of each FACTS device. Below, it is described in detail how the FACTS devices are integrated into the flexible approach.

4.1 State variables of the FACTS devices

The models of FACTS devices add one or more state variables to the OPF formulation; therefore, it is necessary to consider such variables in the proposed implementation. The state variables of FACTS devices are included in the set of steady-state variables (5), denoted by the state vector X. It is convenient to write these variables in the next position from P_{gi}, that is, in the position (2nb + ng + 1,1) of the vector X, in accordance to the structure of the proposed approach. When the model of a FACTS device is incorporated, it should be noted that the size of the vector X is increased from (2nb + ng, 1) to (2nb + ng + nvar, 1), where **nvar** is the number of state variables of each FACTS device. Similarly, the initial conditions of the state variables of FACTS devices are included in the vector $\mathbf{X_0}$ in a position following the initial condition of P_{gi}. The operating limits of FACTS devices are handled as inequality constraints on the respective sate variables. Therefore, the lower and upper bounds of the state variables for these devices are written in the position (2nb + ng + 1, 1) of the vector **lb** and **up**, respectively. Note that, when considering the model of any FACTS device, the size of these last vectors is extended to equal the size of the state vector X. The syntax of the vector X which contains both the power system state variables and the state variables added by FACTS devices is the following:

```
%Set of state variables added by i-th FACTS device
VP(2*nb+ngen+1:2*nb+ngen+nvar,1)=FACTS_i;
```

4.2 Power flow injections and the control equations of FACTS devices

The power balance in all nodes of a power system must be unconditionally satisfied. Therefore, the incorporation of the model of a FACTS device into the optimization model implies to consider the injection of active and reactive power provided by such device. Thus, the set of equality constraints (7) is extended to include the power flow injections provided by FACTS devices as follows:

$$h(y) = \left\{ \begin{aligned} & P_{gi} - P_{li} - \sum_{j \in i}^{nb} P_i^{inj,j} - \sum_{FACTS \in i}^{nb} P_i^{inj,\mathrm{FACTS}} = 0 \\ & Q_{gk} - Q_{lk} - \sum_{j \in k}^{nb} Q_k^{inj,j} - \sum_{FACTS \in i}^{nb} Q_i^{inj,\mathrm{FACTS}} = 0 \end{aligned} \right\} \tag{16}$$

where the terms $P_i^{inj,FACTS}$ and $Q_i^{inj,FACTS}$ represent the active and reactive power flow injections provided by FACTS devices. In the structure of the proposed flexible approach, shown in **Figure 2**, the power-flow injections are included in an M-function file whose name is P_injected. This last file is called by the M-function file **Constraints_OPF**, according to (8). Note that the size as well as the syntax of the vector, *Ce,* are not modified by the inclusion of the active and reactive power-flow injections corresponding to the FACTS devices.

In addition to the extended power-balance equality constraints set, given by (16), the inclusion of FACTS devices into the optimization model implies to include their respective control equations as elements of the following set of additional equality constraints given by

$$h_{ctrl}(y) = \{F_i - F_{spec}\} \tag{17}$$

where F is the parameter to be controlled by the i-th FACTS controller with the specified target value F_{spec}. Observe that such control parameter can either be the

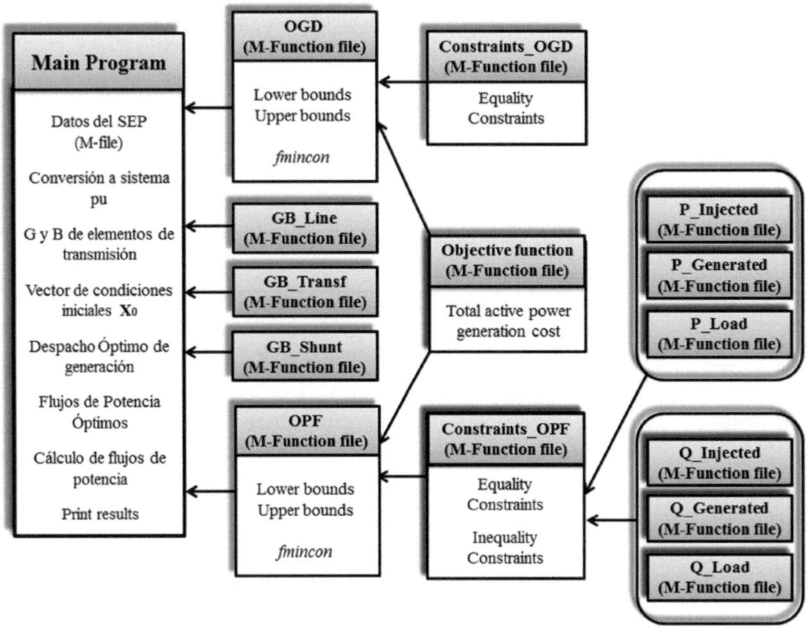

Figure 2.
Schematic representation of structured programming of the proposed flexible approach.

active/reactive power flow through the transmission line or the magnitude of the voltage at a specific node of the power system in accordance to the type of the FACTS device connected to the network at such node. It is noteworthy that when a device is connected to the network, its corresponding constraint is activated in the set given by (17) in order to enable the respective control action of such device during the process of solving the optimization model. Otherwise, that is, when no FACTS devices are involved, the equality constraints of the control modes remain inactive.

Thus, in the proposed flexible implementation, the vector *Ce* of equality constraints is extended to include the additional set of equality constraints. Mathematically, (8) is augmented to contain the equation given by (18)

$$[Ce_{2nb+1+nF}] = [F_i] - [F_{spec}] \tag{18}$$

Adding control equations increases the size of the vector *Ce*, given by (8), to (2nb + 1 + nF,1), where *nF* equals the number of control modes of FACTS devices in the optimization model. It is noteworthy that when the model of the FACTS device has operation equations such as in the case of the VSC-HVDC system; these equations must be considered as equality constraints associated with the corresponding device operation. Also note that such constraints are included in the optimization model given by (17), in the same way as the control constraints were included in that model. Last but not least, the vector *Ce* must be extended to include the above equality constraints. The control modes of the FACTS devices are implemented in M-function files in order to facilitate their handling and to provide flexibility to the proposed implementation. The general syntax for including the additional set of equality constraints associated with the control equations corresponding to the FACTS devices is the following:

```
ce(2*nb+2:2*nb+1+nF,1)=Control_FACTS(___);     %nF control modes
```

The previously mentioned procedure can be applied to integrate easily the model of any FACTS device into the proposed flexible approach. The proposed implementation has been structured to connect and disconnect any FACTS device belonging to the power system. This is achieved by writing a 0 or 1 in the data file of the corresponding FACTS controller, where 0 is for disconnecting and 1 is for connecting.

5. Solution of the conventional power flow problem

The conventional Power Flow analysis is aimed to determine the steady-state operation of an electric power system. This consists on computing the magnitude and angle of each voltage at all the nodes of the system. This is carried out in addition to determining the active and reactive power flow in the transmission elements of the power system. In this analysis, four quantities are associated with each node of the system which are the magnitude of the voltage, the angle of the phase and the active and reactive power-flow. Furthermore, the system nodes can be classified as: slack nodes, regulated or PV nodes and load or PQ nodes. The complex node voltage given in phase and magnitude can be specified at any slack node. Note that in the regulated nodes, the magnitude of the voltage and active power are known. Whilst the value of the active and reactive power is defined in any load node.

The power flow analysis is based on the power balance equations at each node of the power system which are given by the set of equality constraints in (7). The proposed approach in this chapter book allows obtaining the solution to conventional Power Flow problems by performing small changes in the OPF formulation. This can be achieved by converting the OPF model into a conventional PF model. In the OPF analysis is not important to identify the node types, as it is in the case of a PF analysis. However, both analyzes are based on the power balance equations for all nodes of the power system. Thus, it is possible to include the PV nodes in the OPF model by setting the magnitude of the voltage as well as the active power to a given value, only at the nodes where a generator is connected. Similarly, the slack node can be included by setting the magnitude of the voltage and the angle of the phase to a given value. In this way, the solution of the conventional PF analysis is obtained from the OPF model. The aforementioned procedure is accomplished by including an additional set of equality constraints in the optimization model. This allows for the setting of the magnitude of the voltage as well as the active power to certain values at the generation nodes. It also allows for the setting of the magnitude of the voltage and the angle of the phase to the stablished values at the slack nodes. This additional set of equality constraints that aims to obtain the solution of the PF problem is given by (19).

$$h_{OPF-PF}(y) = \left\{ \begin{array}{l} V_{sn} - V_{sn}^{spec} = 0 \\ \theta_{sn} - \theta_{sn}^{spec} = 0 \\ P_{gi} - P_{gi}^{spec} = 0 \\ V_{gk} - V_{gk}^{spec} = 0 \end{array} \right\} \begin{array}{c} i = 1, 2, \dots, ng \mid i \notin sn \\ k = 1, 2, \dots, ng \end{array} \tag{19}$$

In this way, the conventional Power Flow analysis is modeled as a problem of a constrained nonlinear optimization which is formulated using the OPF model. Therefore, the general model of the conventional power flow analysis considered in this work is given as follows:

$$Minimize\, f(y) \tag{20}$$

$$\text{Subject to } h(y) = 0 \tag{21}$$

$$g(y) \leq 0 \tag{22}$$

$$y^{\min} \leq y \leq y^{\max} \tag{23}$$

$$h_{OPF-PF}(y) = 0 \tag{24}$$

where the $h_{OPF\text{-}PF}$ is the set of equality constraints given by (19). This allows the inclusion of the node types in order to convert the OPF model into a conventional PF model.

Therefore in accordance to the conventional PF model given by (20)–(24), the proposed flexible approach solution to the Power Flow problem can be obtained by considering the full of the power flow model given by (6), (7), (9), (12), and (19). It is important to note that by including the above equality constraints, the resulting OPF implementation is modified and reused to straightforwardly obtain the computational implementation of the Power Flow analysis that considers embedded FACTS controllers. In this case, it must be considered the extended optimization model with FACTS devices given by (6), (9), (12), (16), (17) and (19).

In order to maintain the structure of the proposed implementation, the last set of equality constraints is written in the vector *Ce* of the M-function file named **Constraints_OPF** as follows:

$$\begin{bmatrix} Ce_{2nb+nF+2} \\ Ce_{2nb+nF+3} \\ Ce_{2nb+nF+4} \\ \vdots \\ Ce_{2nb+nF+(ng-1)} \\ Ce_{2nb+nF+ng} \\ \vdots \\ Ce_{2nb+nF+2(ng-1)} \end{bmatrix} = \begin{bmatrix} V_{sn} \\ \theta_{sn} \\ V_{g1} \\ \vdots \\ V_{gk} \\ P_{g1} \\ \vdots \\ P_{gi} \end{bmatrix} - \begin{bmatrix} V_{sn}^{spec} \\ \theta_{sn}^{spec} \\ V_{g1}^{spec} \\ \vdots \\ V_{gk}^{spec} \\ P_{g1}^{spec} \\ \vdots \\ P_{gi}^{spec} \end{bmatrix} \tag{25}$$

The syntax to include the constraints given by (19), which allows to obtain the solution of the Power Flow problem is the following:

```
Ce(2*nb+nF+2:2*nb+nF+2,1)=Vsn-Vsn_spec;                %Slack node.
Ce(2*nb+nF+3:2*nb+nF+3,1)=Angle_sn-Angle_sn_spec;      %Slack node.
Ce(2*nb+nF+4:2*nb+nF+ng-1,1)=Vgk-Vgk_spec;             %PV node.
Ce(2*nb+nF+ng:2*nb+nF+2*ng-2,1)=Pgi-Pgi_spec;          %PV node.
```

It should be noted that in the conventional PF model given by (20)–(24), the PQ nodes are considered in the same manner as in the OPF model; whereby it is not necessary to include equality constraints to set any variables in such nodes.

6. Solution of the OPF market clearing

The OPF market-clearing model to be solved for the Operator System Market is a non-linear constrained optimization problem, whose general formulation is given as follows [25]:

$$\text{Maximize } f(p) \tag{26}$$

$$\text{Subject to } g(y, p) = 0 \tag{27}$$

$$h_{\min} \leq h(y,p) \tag{28}$$

$$h(y,p) \leq h_{\max} \tag{29}$$

$$p_{\min} \leq p \tag{30}$$

$$p \leq p_{\max} \tag{31}$$

where the dependent variables, such as the nodal voltage, are represented by y. Whilst the control variables, for example, the supply bids P_S and power demand P_D, are denoted by p. The functions f, g and h are defined below.

6.1 Objective function

The objective function of the OPF market-clearing model represents the social welfare associated with the power system. That function is given by the difference between: the sum of the accepted demand bids P_{Di} times their corresponding bid prices C_{Di} (in \$/MWhr) and the sum of the accepted production bids P_{Sj} times their corresponding bids prices C_{Sj} (in \$/MWhr). Mathematically, this objective function is given by

$$f = \sum_{i=1}^{nl} C_{Di}(P_{Di}) - \sum_{j=1}^{ng} C_{Sj}(P_{Sj}) \tag{32}$$

In (32), the first and second term is also identified as f_D and f_S, respectively; *nl* represents the total demand bids, and *ng* is the total production bids.

6.2 Equality constraints

The set of equality constraints (33) represents the standard power flow equations

$$g(y,p) = g(\theta, V, k_G, P_S, P_D) = 0 \tag{33}$$

where the set y = $(\theta$, V, $k_G)$ and p = $(P_S$, $P_D)$. The variables θ and V are the complex-voltage phases and magnitudes at nodes of the power system, respectively. Whilst k_G is a scalar variable used to account for the system losses by means of either a unique or a distributed slack node. It is important to note that y represents the set of dependent variables to be optimized. The generator and load powers are defined as follows:

$$P_G = P_{G0} + P_S \tag{34}$$

$$P_L = P_{L0} + P_D \tag{35}$$

where P_{GO} and P_{LO} stand for the generator and load parameters that are not part of the market trading, that is, these parameters represent the negotiation of energy between particulars in a specific big industry and/or Generation Companies (GENCOS). Whilst PS and PD are the amounts of energy that can be traded in the electricity market by GENCOS and Energy Services Company (ESCOS). Finally, the reactive loads are assumed to have constant power factors which are computed by using the following expression:

$$Q_L = P_L \tan(\phi_L) \tag{36}$$

6.3 Inequality constraints

In this formulation, the set of inequality constraints is practically the same as in the OPF model, except by the addition of a new constraint on the state variables, which represents the limits on the demand bids (P_{Di}).

$$P_{Di}^{\min} \leq P_{Di} \leq P_{Di}^{\max} \tag{37}$$

where $P_{Di}{}^{min}$ and $P_{Di}{}^{max}$ are the lower and upper limits of the state variable that represents the demand bids.

The demand bids P_{Di} is a steady-state variable in the OPF market-clearing model; thereby, in order to solve this model by using the proposed flexible approach is necessary to include P_{Di} in the vector of steady-state variables X. It is convenient to write the variable P_{Di} in the next component of vector X, just after the vector-entries that represent the state variables of the FACTS devices. This corresponds to the position (2nb + ng + nvar+nl,1) of the vector X. This is done in this way in order to maintain the original structure of the proposed approach. The limits of the demand bids are handled as inequality constraints on the state variables. Therefore, the lower and upper bounds of the state variable P_{Di}, are written in the position (2nb + ng + nvar+nl, 1) of the vector **lb** and **up**, respectively.

The objective function given by (32) is coded into the M-function file: **objfun_OPF**. It is worth mentioning that (32) is used instead of (6). Also it is very important to point out that in order to maximize the objective function, it is necessary to write such function in a negative form. This is because the **fmincon** function achieves a minimization by default. Note that P_L and Q_L equations are coded in the M-function files: P_Load and Q_Load, respectively. These files are called by M-function file **Constraints_OPF**, as it is shown in **Figure 2**. In this way, they are considered within the set of equality constraints that establishes the active and reactive power balance under steady-state conditions for all the nodes of the power system.

According to the aforementioned discussion, a simple command line is presented below, which allows for modifying the conventional OPF model in order to obtain the implementation of the OPF-MC model.

```
%%%%%%%%% Vector of the state variables of the OPF-MC model %%%
%%%%%
X(2*nb+ng+1:2*nb+ng+nl,1)=Pload;          %Demand bids

%%%%%%%%% Definition of tan(phi), MVAb=100 MVA %%%%%%%%%
Pmax=Pmax/MVAb;                              % Pmax in pu
Qmax=Qmax/MVAb;      % Qmax in pu
tan_phi=Qmax./Pmax;   % Tan(phi)

%%%%%%%%% Obtaining QL = PL*tan(phi) (Eq. 36) %%%%%%%%%
function [QL]=Qload(__)
 PLoad=X(2*nb+ng+1:2*nb+ng+nl,1);
 QL=zeros(nb,1);
for k=1:nl                                   % nl: total demand bids
   send=node(k,1);      %Node where is connected the load
 QL(send,1)=QL(send,1)+PLoad(k,1)*tan_phi(k,1);
end

%%%%%%%%% Objetive function of the OPF Market-Clearing %%%%%
%%%%.
function f = objfun_OPF(___)
```

```
PLoad=X(2*nb+ng+1:2*nb+ng+nl,1);
f=0;
fd=0;
for k=1:PDi
 fd=fd+di(k,1)+ei(k,1)*PLoad(k,1)+fi(k,1)*PLoad(k,1)^2; %CDi(PDi)
 end
fs=0;
for k=1:ng
  fs=fs+ai(k,1)+bi(k,1)*Pgi(k,1)+ci(k,1)*Pgi(k,1)^2; %CSj(PSGj)
end
 f=fs-fd;
```

7. Numerical examples

Two numerical examples are tested (with the six-node power system and, an equivalent model of the Mexican Interconnected Power System - MIS.), to numerically illustrate the effectiveness and easy implementation of the proposed flexible approach to carry out the analysis of the OPF market-clearing, as well as the OPF and the conventional PF procedures. The numerical examples presented in this section were carried out in Matlab-R2018a, using a personal computer with a processor Intel Core i5-3210M CPU running at 2.50 GHz with a 6 GB of RAM.

In that way, the solution for the standard OPF market-clearing is obtained, in addition to solve both the OPF and the PF models. For both power systems, the numerical examples are designed as follows. First, the solution of standard OPF market-clearing model is obtained. Next from that solution, the value of the sum of the inelastic and the elastic loads is assigned to the corresponding load nodes in order to solve the OPF model. In other words, the Optimal Power Flow analysis is carried out by considering the value of the load obtained by OPF market-clearing study as an active load. Finally, the conventional PF analysis is executed by setting both the magnitude of the voltage and the active power generated at the PV nodes, to their corresponding values found by the OPF analysis. Similarly, within this last algorithm step, the voltage phases and magnitudes in the slack node are set to their corresponding values found by the OPF analysis. Here, it is worth mentioning that the value of the active load obtained in the market-clearing analysis is considered as a load. Also observe that the objective functions are the same in the three models mentioned above.

7.1 Testing the six nodes power system

The test case of the six node power system reported in [25] is reproduced in this section in order to take the corresponding results as a reference to determine the solution of the OPF and conventional PF models. The test power system consists of three GENCOS and ESCOS, respectively, which are interconnected through 11 transmission lines, as shown in **Figure 3**. The complete simulation data set for this system is given in [23, 25]. For the test cases belonging to this test power system, the convergence tolerance is set to 1×10^{-9}.

The steady-state solutions achieved by the proposal to: the standard OPF market-clearing, the OPF and the PF models are presented in **Table 1**. Here, it is worth mentioning that such solutions correspond to the case of not using FACTS devices.

As expected, **Table 1** shows that the solution of the three procedures is practically the same. This is because the numerical examples were designed to obtain

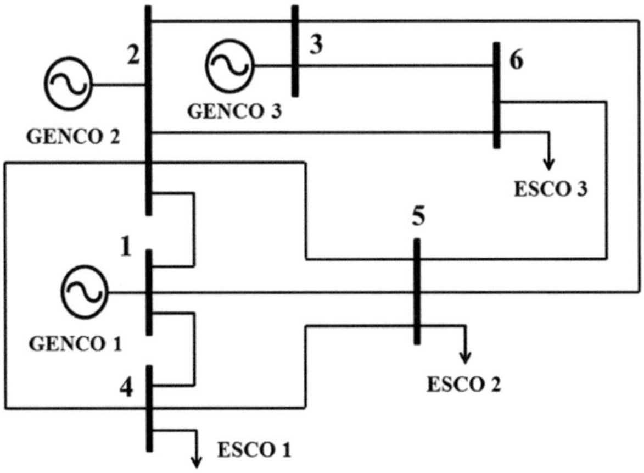

Figure 3.
Six nodes power system.

Parameter	Steady-state solution without FACTS			
	OPF market-clearing		Optimal power flow	Power flow
V_1 (pu-deg)	1.100∠0.0°		1.100∠0.0°	1.100∠0.0°
V_2 (pu-deg)	1.100∠-0.7°		1.100∠-0.7°	1.100∠-0.7°
V_3 (pu-deg)	1.100∠-2.0°		1.100∠-2.0°	1.100∠-2.0°
V_4 (pu-deg)	1.028∠-3.8°		1.028∠-3.8°	1.028∠-3.8°
V_5 (pu-deg)	1.017∠-5.0°		1.017∠-5.0°	1.017∠-5.0°
V_6 (pu-deg)	1.049∠-4.9°		1.049∠-4.9°	1.049∠-4.9°
P_{g1} (MW)	90.0^a	0.0^a	90.0	90.0
P_{g2} (MW)	140.0^a	25.0^a	165.0	165.0
P_{g3} (MW)	60.0^a	20.0^a	80.0	80.0
Q_{gtotal} (MVar)	188.58		188.58	188.58
Objective f ($/MWhr)	f_D	f_S	f_S	f_S
	3482.8	2885.0	2885.0	2885.0

^{a}*Data of left column correspond to* P_{GO} *and data of right column are* P_{BID}, *both in MW.*

Table 1.
Steady-state solutions of: the OPF market-clearing, the OPF and the PF procedures without FACTS devices.

those results. The active and reactive loads computed by the OPF market-clearing procedure for ESCO 1, 2 and 3 are: 115 MW-60 MVar, 110 MW-70 MVar and 99.767 MW-54.418 MVar, respectively. The CPU time for the case of the OPF market-clearing model is 2.3 s. Whilst for the case of the OPF and the PF models, such a time measure yields 2.1 and 1.5 s, respectively. Subsequently, the six nodes power system is modified to simultaneously include the TCSC and SVC controllers. The TCSC is commissioned to maintain the active power flow within the range of 5.62–7.0 MW across the transmission line connecting nodes 4 and 5. The first terminal of the TCSC controller is connected to node 4, whilst the second terminal is connected to a new substation called 6a. Note that the SVC is placed at node 5 to keep the magnitude of the voltage of this node at 1.02 pu. For the numerical examples, the state variables representing the thyristor's firing angle of the TCSC and the SVC devices. These angles are both initialized at 155° with a lower and an upper limit of 90 and 180°, respectively. The fixed values of the capacitive and inductive reactances of the controllers are X_C = 0.9375% and X_L = 0.1625%, respectively. These values are referred to a voltage base of 400 kV and a power base of 100 MVA. **Table 2** shows the corresponding steady-state solutions given by the three analyses for the case of a steady state operation using embedded FACTS devices.

Clearly, the value of the objective function is increased, due to the TCSC sets the active power flow through a transmission line at higher value than in the base case. This control action causes a redistribution of power flow in the system, increasing the active power losses and the accepted demand bids. Regarding the control performed by the SVC, there is a decrease in the generation of reactive power as it is shown in **Table 2**. This is due to the reactive power losses are keep to a minimum limit in the power system by the SVC controller. This controller supplies reactive power in order to control the magnitude of the voltage at the substations. Therefore, such controller serves as a source of reactive power into the power grid. Although the integration of these FACTS devices into the power system increases the power generation cost and decreases the value of the social welfare net, it is very important to install them in the power system. This is because such devices can improve the steady-state operation of the power system, which by the way can be evaluated using the proposal presented here.

When the FACTS devices are embedded in the power system, the CPU time required for obtaining the solution of the OPF market-clearing model is 2.3 s. Whilst for the OPF and the PF models, the steady-state solution is determined in 2.2 and 1.6 s, respectively.

Parameter	Steady-state solution with FACTS devices			
	OPF market-clearing		Optimal power flow	Power flow
V1 (pu-deg)	1.070∠0.0°		1.071∠0.0°	1.071∠0.0°
V2 (pu-deg)	1.065∠-0.6°		1.065∠-0.6°	1.065∠-0.6°
V3 (pu-deg)	1.065∠-1.9°		1.064∠-1.9°	1.064∠-1.9°
V4 (pu-deg)	0.997∠-4.1°		0.997∠-4.1°	0.997∠-4.1°
V5 (pu-deg)	1.020∠-5.7°		1.020∠-5.7°	1.020∠-5.7°
V6 (pu-deg)	1.019∠-5.0°		1.019∠-5.0°	1.019∠-5.0°
V6a (pu-deg)	0.996∠-4.0°		0.996∠-4.0°	0.996∠-4.0°
Pg1 (MW)	90.0[a]	0.0[a]	90	90
Pg2 (MW)	140.0[a]	25.0[a]	165	165
Pg3 (MW)	60.0[a]	20.0[a]	80	80
Qgtotal (MVar)	128.85		128.82	128.82
Objective f ($/MWhr)	f_D	f_S	f_S	f_S
	3,491.0	2,885.0	2,885.0	2,885.0

[a]*Data of left column correspond to* P_{G0} *and data of right column are* P_{BID}, *both in MW.*

Table 2.
Steady-state solutions of the OPF market-clearing, the OPF and the PF procedures with embedded FACTS devices.

7.2 The Mexican interconnected power system (MIS)

The system considered in this numerical example is a reduced model of the Mexican Interconnected System. The MIS is represented by an equivalent model composed of: 190 nodes, 46 generators, 90 loads, and 265 transmission lines operating at voltage levels ranging from 115 to 400 kV [27].

The limits for the magnitude of the voltage for all nodes are set to $0.94 \leq V_i \leq 1.07$ pu. Whilst the active and reactive power limits for all generators are set to: $200 \leq P_{gi} \leq 1000$ MW and $-250 \leq Q_{gi} \leq 350$ MVar, respectively. The cost functions of all generators are considered linear with a value of the linear coefficient between 0.0019 and 0.0020 $/MWhr. The MIS consists of two areas identified as Area A and Area B, interconnected through a relatively weak double-circuit tie line, as it is shown in **Figure 4**. For the numerical examples with the Mexican Interconnected System, the convergence tolerance is set to 1×10^{-6}.

The solution of the OPF market-clearing model for the base case, without FACTS controllers, gives: an accepted demand bid (f_D) of 197,788.409 $/MWhr; an accepted production bid (f_S) of 35,584.143 $/MWhr; the total active losses are 164.817 MW and a transferred active power from Area B to Area A of 195.34 MW. Whilst the solution of the Optimal Power Flow and the conventional Power Flow without FACTS devices gives: a generation cost of 35,632.303 $/MWhr; the total active losses are 194.26 MW and a transferred active power between the two referred areas of 208.66 MW. First, the MIS is modified to integrate a PtP VSC-HVDC system. The first voltage source of PtP VSC-HVDC is connected to node 182, whilst the second voltage source is connected to a new substation called 182a. The PtP VSC-HVDC system is commissioned to increase the active power flow through the inter-area link to 300 MW and to set the magnitude of the voltages at nodes 182 and 182a at 1 pu. The controller coupling transformer impedances are set at 0.001 + j0.01 pu. The lower and upper limits of the magnitude of both voltage sources are $0.95 \leq V_{C1} \leq 1.1$ pu. The amplitude modulation index bounds and their initial conditions are: $0.5 \leq M_{Ci} \leq 1.0$ and 0.9, respectively. The magnitude of DC-voltage bounds corresponding to both converters are: $2.0 \leq V_{DC} \leq 4.5$, where the

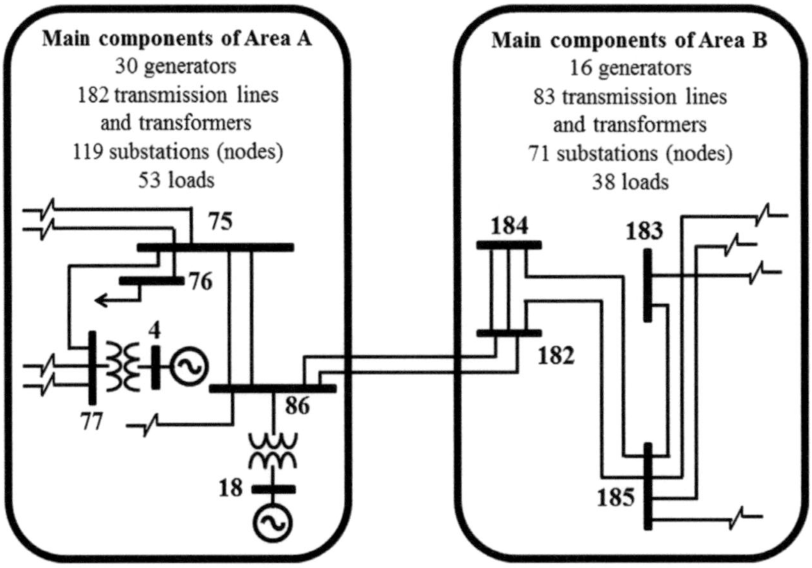

Figure 4.
Representative diagram of the Mexican interconnected system.

Study	P_{gtotal} (MW)	Q_{gtotal} (MVar)	P_{losses} (MW)	Objective f (\$/MWhr)
Market VSC-HVDC	9200.00[a]	1620.46	147.63	35,555.72[b]
	9305.64[a]			197,788.41[b]
OPF VSC-HVDC	18,580.60	2400.29	222.59	35,681.16
PF VSC-HVDC	18,580.59	2400.05	222.60	35,681.13
Market PST	9200.00[a]	1441.07	141.72	35,544.95[b]
	9299.72[a]			197,788.41[b]
OPF PST	18,599.21	2411.97	241.21	35,707.37
PF PST	18,599.19	2411.56	241.21	35,707.33

[a]*In both market studies, 9200 MW is* P_{GO} *and the value of MW in the lower row is* P_{BID}.
[b]*The value of the upper row corresponds to* f_S *and the lower row is* f_D.

Table 3.
Summary results of the steady-state studies with the VSC-HVDC and PST devices.

DC voltage is fixed at 3.0 pu. Whilst the DC link resistance is set to 0.0034 pu. The summary results of the OPF market-clearing, the Optimal Power Flow and, the conventional Power Flow, all of them including the PtP VSC-HVDC system embedded within the MIS system, are presented in **Table 3**.

Observe that instead of the VSC-HVDC, the Phase-Shifting Transformer (PST) is used to maintain the active power flow through the inter-area link at the same level in MW as in the above case. Also note that the primary and secondary winding impedances of the PST exhibit zero resistance. Then the primary and secondary inductive reactances are set to 0.0 and 0.05 pu, respectively. Next, the complex tap ratios are $T_v = Uv = 1.0\angle 0°$. Note that the control of the active power flow is carried out by means of a primary phase angle control. For implementing such control, the limits of the phase-shifter angle are set to $\pm 10°$. Last but not least, the summary results of the three steady-state cases previously considered in this book chapter are shown in **Table 3**, where a PST controller has been included.

The above table shows that when the FACTS controllers are integrated into the power system, the value of the objective function increases in the case of the OPF and the PF studies, but it decreases in the OPF market-clearing case. This is caused by the variations of the active power losses in the transmission elements, which in turn are due to the control of the active power flow performed by the VSC-HVDC and PST devices through the inter-area link. In the cases of the market-clearing studies using: (i) the PST and (ii) the VSC-HVDC; 20 and, 22 generators hit their limits of P_g, respectively. Whereas for the OPF and the PF studies using: (i) the PST and (ii) the VSC-HVDC; 13 and, 10 generators hit their limits of P_g, respectively.

Regarding the magnitudes of the nodal-voltages, 17 of them at power-substations hit their limits when the analysis of the market-clearing is performed using the PST and VSC-HVDC. Whereas in the OPF and the PF analysis featuring both FACTS devices, a total of four magnitudes of the substation-voltages hit their limits. It should be mentioned that the variables that violated their limits were either fixed at their upper or lower limits in a proper manner by the flexible approach proposed here. The CPU times required to obtain the solution of the market VSC-HVDC, the market PST, the OPF VSC-HVDC, the OPF PST, the PF VSC-HVDC and the PF PST are: 1503, 978, 66, 64, 34 and 33 s, respectively. From the results of the numerical examples presented in this section, it can be inferred that the proposed implementation reliably carries out the steady-state operation assessment of power systems using embedded FACTS devices.

8. Conclusions

A Matlab-based flexible approach to carry out the analysis of the steady-state operation of power systems using embedded FACTS controllers has been presented in this book chapter. The flexibility and reliability of this proposal has been demonstrated by means of several numerical examples. The referred numerical examples show that by using the proposed approach, it is possible to simulate the optimal operation of power systems in a simple way, because it avoids the implementation of complex programming algorithms typical of the majority of the optimization models that corresponds to power systems. For example, this is the case for the complexity related to the program code lines related to the hessian matrix as well as the gradient vector.

In this regard, it is also important to mention that the proposed flexible approach is easy to implement as a computer program in order to carry out the analysis of the optimal steady-state operation conditions corresponding to power systems featuring FACTS devices. This is because the proposed flexible approach greatly reduces the complexity as well as the implementation time of such optimization models.

Besides that, the flexibility of the proposed approach provides the possibility to modify and/or consider different objective functions as well as equality and inequality constraints. Therefore, in addition to solving the optimization models considered in this work, the steady-state model of any other FACTS device could also have been used. Not to mention that a large variety of steady-state optimization applications for power systems could also been readily implemented by using the flexible approach proposed here.

Last but not least, the computational efficiency of the proposal is quite reasonable at least for research and academic purposes, since most of the solutions to the optimization models related to medium scale power systems can be usually obtained within a few seconds to about 10–15 minutes. Furthermore, it is possible to infer that this proposed approach may lay a solid basis for researchers, teachers and students who seek to develop their own computational tool, whether it is for research, training or educational purposes.

Conflict of interest

The authors declare no conflict of interest.

Author details

Jose Miguel García-Guzman[1], Néstor González-Cabrera[2]*, Luis Alberto Contreras-Aguilar[3], Jose Merced Lozano-García[4] and Alejandro Pizano-Martinez[4]

1 Department of Electromechanical Engineering, Irapuato Superior Institute of Technology, Irapuato, México

2 Department of Electrical Energy, National Autonomous University of Mexico (UNAM), Mexico

3 Department of Electrical Engineering, University of Colima, Colima, México

4 Department of Electrical Engineering, University of Guanajuato, Salamanca, México

*Address all correspondence to: nestorgc@fi-b.unam.mx

References

[1] Wood AJ, Wollenberg BF, Sheble GB. Power Generation, Operation and Control. 3rd ed. New York, NY: John Wiley & Sons, Inc.; 2013. p. 656

[2] Hingorani NG. High power electronics and flexible AC transmission system. IEEE Power Engineering Review. 1988;**3**:4. DOI: 10.1109/MPER.1988.590799

[3] Hingorani NG, Gyugyi L. Understanding FACTS Concepts and Technology of Flexible AC Transmission Systems. New York, NY: John Wiley & Sons, Inc.; 1999. p. 452

[4] Acha E, Fuerte-Esquivel CR, Ambriz-Perez H, Angeles-Camacho C. FACTS: Modelling and Simulation on Power Networks. New York, NY: John Wiley & Sons, Inc.; 2004. p. 420

[5] Fuerte-Esquivel CR, Acha E. Newton-Raphson algorithm for the reliable solution of large power networks with embedded FACTS devices. IET Proceedings - Generation Transmission and Distribution. 1996; **447**:454. DOI: 10.1049/ip-gtd:19960531

[6] Gotham DJ, Heydt GT. Power flow control and power flow studies for systems with FACTS devices. IEEE Transactions on Power Systems. 1998; **60**:65. DOI: 10.1109/59.651614

[7] Fuerte-Esquivel CR, Acha E, Ambriz-Perez H. A thyristor-controlled series compensator model for the power flow solution of practical power networks. IEEE Transactions on Power Systems. 2000;**58**:64. DOI: 10.1109/59.852101

[8] Zhang XP. Advanced modeling of the multicontrol functional static synchronous series compensator (SSSC) in Newton power flow. IEEE Transactions on Power Systems. 2003; **1410**:1416. DOI: 10.1109/TPWRS.2003.818690

[9] Zhang XP. Multiterminal voltage-sourced converter-based HVDC models for power flow analysis. IEEE Transactions on Power Systems. 2004; **1877**:1884. DOI: 10.1109/TPWRS.2004.836250

[10] Bhowmick S, Das B, Kumar N. An indirect UPFC model to enhance reusability of Newton power-flow codes. IEEE Transactions on Power Delivery. 2008;**2079**:2088. DOI: 10.1109/TPWRD.2008.923105

[11] Bhowmick S, Das B, Kumar N. An advanced IPFC model to reuse Newton power flow codes. IEEE Transactions on Power Systems. 2009;**525**:532. DOI: 10.1109/TPWRS.2009.2016643

[12] Acha E, Kazemtabrizi B, Castro LM. A new VSC-HVDC model for power flows using the Newton-Raphson method. IEEE Transactions on Power Systems. 2013;**2602**:2612. DOI: 10.1109/TPWRS.2012.2236109

[13] Ying X, Song YH, Sun YZ. Power flow control approach to power systems with embedded FACTS devices. IEEE Transactions on Power Systems. 2002; **943**:950. DOI: 10.1109/TPWRS.2002.804919

[14] Orfanogianni T, Bacher R. Steady-state optimization in power systems with series FACTS devices. IEEE Transactions on Power Systems. 2003;**19**:26. DOI: 10.1109/TPWRS.2002.807110

[15] Berizzi A, Delfanti M, Marannino P, Pasquadibisceglie MS, Silvestri A. Enhanced security-constrained OPF with FACTS devices. IEEE Transactions on Power Systems. 2005;**1597**:1605. DOI: 10.1109/TPWRS.2005.852125

[16] Pizano-Martinez A, Fuerte-Esquivel CR, Ambriz-Perez H, Acha E. Modeling of VSC-based HVDC systems for a

Newton-Raphson OPF algorithm. IEEE Transactions on Power Systems. 2007; **1794**:1803. DOI: 10.1109/ TPWRS.2007.907535

[17] Zarate-Mihano R, Conejo AJ, Milano F. OPF-based security redispatching including FACTS devices. IET Generation Transmission and Distribution. 2008;**821**:833. DOI: 10.1049/iet-gtd:20080064

[18] Kazemtabrizi B, Acha E. An advanced STATCOM model for optimal power flows using Newton's method. IEEE Transactions on Power Systems. 2014;**514**:525. DOI: 10.1109/ TPWRS.2013.2287914

[19] Sahraei-Ardakani M, Hedman KW. Computationally efficient adjustment of FACTS set points in DC optimal power flow with shift factor structure. IEEE Transactions on Power Systems. 2016; **1733**:1740. DOI: 10.1109/ TPWRS.2016.2591503

[20] Lehmkoster C. Security constrained optimal power flow for an economical operation of FACTS-devices in liberalized energy markets. IEEE Transactions on Power Delivery. 2002; **603**:608. DOI: 10.1109/61.997946

[21] Palma-Behnke R, Vargas LS, Perez JR, Nunez JD, Torres RA. OPF with SVC and UPFC modeling for longitudinal systems. IEEE Transactions on Power Systems. 2004;**1742**:1753. DOI: 10.1109/TPWRS.2004.836260

[22] Gengyin L, Zhou M, Gao Y. Determination of total transfer capability incorporating FACTS devices in power markets. In: International Conference on Power Electronics and Drives Systems, IEEE; 28–31 November. Malaysia: IEEE; 2005. pp. 1-6

[23] Nasri A, Conejo AJ, Kazempour SJ, Ghandhari M. Minimizing wind power spillage using an OPF with FACTS devices. IEEE Transactions on Power Systems. 2014;**2150**:2159. DOI: 10.1109/ TPWRS.2014.2299533

[24] Sahraei-Ardakani M, Blumsack SA. Transfer capability improvement through market-based operation of series FACTS devices. IEEE Transactions on Power Systems. 2016; **3702**:3714. DOI: 10.1109/ TPWRS.2015.2508720

[25] Milano F, Cañizares CA, Conejo AJ. Sensitivity-based security-constrained OPF market clearing model. IEEE Transactions on Power Systems. 2005; **2051**:2060. DOI: 10.1109/ TPWRS.2005.856985

[26] The MathWorks, Inc. Matlab Optimization Toolbox: Users Guide Version 2. January 1999. Available from: http://www.mathworks.com

[27] Messina AR, Vittal V. Assessment of nonlinear interaction between nonlinearly coupled modes using higher order spectra. IEEE Transactions on Power Systems. 2005;**375**:383. DOI: 10.1109/TPWRS.2004.841240

Chapter 22

Distributed Sources Optimal Sites and Sizes Search in Large Power Systems

Mustafa Mosbah, Redha Djamel Mohammedi and Salem Arif

Abstract

The integration of renewable sources into the power system has now become an unavoidable necessity for these technical and economic advantages and for the protection of the environment. In this chapter, a study is given for the integration of the Distributed Source (DS) in an optimal way and this by looking for the best location (sites) and the best power to be injected (size). The optimization technique used is based on genetic algorithms under technical and safety constraints, with the aim of minimizing active network losses and maximizing voltage stability. These objective functions are handled as a single and multi-objective problem. This study is applied on the standard IEEE 30 bus network under the MATLAB code.

Keywords: distributed source, optimal site and size, active power losses, voltage stability improvement, genetic algorithm

1. Introduction

Non-conventional or clean production is increasing rapidly around the world, thanks to it's advantages such as reduced environmental impact, small size and it is a renewable energy [1]. The traditional power system is characterized by the unidirectional of the power flow because the energy comes from centralized sources [2]. These sources are generally based on fossil resources that are exhaustible and polluting for the environment.

The transmission power system spreads on long distances, which leads to losses in the lines by Joule effect on the one hand, and on the other hand, it requires huge investments and waste lands with long achievement times, which has led researchers to think about other solutions to face these problems [3].

Among these solutions is to have sources of electrical energy close to the consumers (local production). With the liberalization of the electricity market and the evolution of decentralized source (DS) technology in recent years, an increased trend towards their use has emerged. DS is defined as small producers based on renewable or conventional sources installed at different points in the power system, either at the transmission or distribution level [4]. The rate of DS integration tends to increase progressively in several countries [5].

In spite of the various advantages of DS, its sources present disadvantages when they are inserted into the power system, such as frequency instability caused by the intermittency of some renewable sources (e.g. photovoltaic and wind power) [6]. They also present some constraints, such as exceeding the thermal limits of power lines, increased Joule effect losses and the dysfunctioning of electrical protection devices and the exceeding of voltages at connection points. These constraints are due to the wrong choice of size (maximum power) and site of the DS, which requires the search for the best sites and adequate sizes [7].

A number of researchers have studied the problem of optimal site and size of DS to distribution networks [4, 8]. Example, in researchers [9] presented a technique a review of optimal DS placement in distribution network. DS site and size search techniques can be based on artificial intelligence techniques, metaheuristic techniques or deterministic techniques [10]. While most papers deal with this problem as a single-objective problem and some papers have focused on the problem of multi-objective optimizations [11]; these studies have considered a variety of objective functions including voltage profile improvement, losses minimization, reliability index minimization (SAIFI, SAIDI and END), fuel cost minimization, greenhouse gas emission minimization, maximization of DS penetration rate and maximization of voltage stability [12–18]. Some researchers have examined the problem of optimal sitting and size of DS to transmission networks [19–29].

Based on the literature search carried out, it was noted that the optimal integration of DS into distribution networks is largely discussed, however, there is less literature on the integration of DS into transmission networks.

The objective of this chapter is to study the problem of determining the Optimal Sites and Sizes of DS (OSSDS) in the transmission power system while taking into account the various constraints of the system (technical and security constraints). This is achieved by using a metaheuristic optimization technique such as the Genetic Algorithm (GA) technique. The objective functions considered in this work are the minimization of active power losses and the voltage stability improvement. These objective functions are treated as mono-objective and multi-objective (the objective functions are combined to a single objective function via weighting factors). This study has been applied on the IEEE 30 bus network under MATLAB code. For this reason, this chapter is organized according to the following plan: Section 2 will present the mathematical formulas of the OSSDS problem, i.e. the objective functions, the different constraints. Section 3 will present definitions on the method used (GA) and their application on OSSDS. Section 4 will give the description of the IEEE 30 node network and the limitations of the study framework. Section 5 will present the simulation, interpretation and analysis of the results obtained. Finally, the conclusion of the chapter and some perspectives.

2. Formulation of the OSSDS problem

The objective of this work is to research what is the optimal power to be injected by DS and the bus of their insertions that gives us the best performance of the power system considering the imposed constraints. To reach this objective we must define the fitness function and the constraints of equality and inequality which will be detailed in the following sections [30, 31].

2.1 Objective function (OF)

The OSSDS is formulated as a single and multi-objective problem using two objective functions.

2.1.1 First objective: minimization of active power losses

The first concern of the power system operator is to minimize active power losses in the line, this is expressed by the following equation:

$$\mathrm{TPLI_k} = \sum_{k=1}^{N_B} r_k I_k^2 \tag{1}$$

where r_k and I_k, are the resistance and the current of the transmission line k respectively. TPLI represent the Total Power Losses Index of the network.

2.1.2 Second objective: improvement of the line voltage stability index

To assure that the operating point of the power system is far from the voltage collapse point, the voltage stability index must be improved. Among the effective indices that have been proposed in the literature is the Line Voltage Stability Index (LVSI) [32]. The LVSI of the line between bus i and jis given by the following equation:

$$\mathrm{LVSI_k} = \frac{4X_{ij}Q_j}{\left[V_i \sin\left(\theta_{ij} - \delta_{ij}\right)\right]^2} \leq 1 \tag{2}$$

where X_{ij}, Q_j, V_i, θ_{ij}, and δ_{ij} are the line reactance, is the reactive power at receiving end, the voltage at receiving end, is the line impedance angle between bus i and j and voltage phase angle difference between bus i and busj, respectively.

The line which represents an LVSI value close to 1 will be considered, the most critical line and may lead to the network collapse.

For multiobjective optimization, the two objectives functions are combined in a single linear function using weighting factors. Mathematically, this is given in the following form:

$$F_{MO} = \alpha_1(\mathrm{TPLI_k}) + \alpha_2(\mathrm{LVSI_k}) \tag{3}$$

where α_1 and α_2 represent the weighting factors of active power losses index and the line voltage stability index, respectively.

2.2 Equality constraints

Equality constraints represent power balance equations between generation and demand. For a transmission power system in presence of DS, the active and reactive power equality constraints can be expressed as follows [33]:

$$\begin{cases} \sum_{i=1}^{N_G} P_{Gi} + \sum_{i=1}^{N_{DS}} P_{DSi} - \sum_{i=1}^{N_D} P_{Di} - \sum_{i=1}^{N_B} P_{li} = 0 \\ \sum_{i=1}^{N_G} Q_{Gi} + \sum_{i=1}^{N_{DS}} Q_{DSi} - \sum_{i=1}^{N_D} Q_{Di} - \sum_{i=1}^{N_B} Q_{li} = 0 \end{cases} \tag{4}$$

where (P_G, Q_G) is the active and reactive power of the conventional generator, (P_{DS}, Q_{DS}) is the active and reactive power produced by DS. This source is capable of delivering the active and reactive power. (P_D, Q_D) represent the active and reactive power demand at the load bus, (P_l, Q_l) represent the active and reactive power losses.

2.3 Inequality constraints

These constraints represent the physical limits of the lines, conventional generators and DS as also the security limits of the voltages of the network busses. They are expressed by the following equations:

$$S_{bi} \le S_{bimax} \text{ for } i = 1 \dots \dots \dots \dots \dots \dots N_B \tag{5}$$

$$P_{Gimin} \le P_{Gi} \le P_{Gimax} \text{ for } i = 1 \dots \dots \dots \dots \dots \dots N_G \tag{6}$$

$$Q_{Gimin} \le Q_{Gi} \le Q_{Gimax} \text{ for } i = 1 \dots \dots \dots \dots \dots \dots N_G \tag{7}$$

$$V_{imin} \le V_i \le V_{imax} \text{ for } i = 1 \dots \dots \dots \dots \dots \dots \dots N \tag{8}$$

$$S_{imin}^{DS} \le S_i^{DS} \le S_{imax}^{DS} \text{ for } i = 1 \dots \dots \dots \dots \dots \dots N_{DS} \tag{9}$$

$$\sum_{i=1}^{N_{DS}} P_{DSi} \le \tau * \sum_{i=1}^{N_D} P_{Di} \text{ for } i = 1 \dots \dots N_{DS} \text{ and } i = 1 \dots ..N_D \tag{10}$$

where S_{bi} is the apparent power that transits via the line between the nodes i and j. S_{bimax}, is the maximum limit of the line (thermal limit). P_{Gimin}, P_{Gimax}, P_{Gimin} and P_{Gimax} are the minimum and maximum active and reactive powers of the ith conventional generator. S_{min}^{DS} and S_{max}^{DS} are the minimum and maximum production powers of the DS. N, N_D, N_G, N_B and N_{DS} are the number of network busses, number of load busses, the number of generators, the number of branches and the number of DS, respectively. V_{imin} and V_{imax} are minimum and maximum voltage in bus i.

For reasons of power system security, the network operator has limited the penetration rate ($\tau_{\%}$) of DS. This limit is set depending on the robustness of the network (each network has its specific). This rate is calculated by the ratio of the total active power of the DS to the total active power demanded or the total power generated by conventional sources multiplied by one hundred; this is given by the following Equation [34]:

$$\tau_{\%} = \frac{\sum_{i=1}^{N_{DS}} P_{DSi}}{\sum_{i=1}^{N_D} P_{Di}} * 100 \tag{11}$$

2.4 Constraint processing

To optimize the site and size of the DS, it is important to mention that the control variables are generated within their allowable limits using a random strategy using a metaheuristic technique. During the optimization process, it is possible to come across solutions that are unfeasible due to exceeding the voltage limit, the thermal limit of the lines or the power limit of the reference generator, in this case the OF is penalized. This OF is reformulated as follows [33]:

$$F^P = \sum_{i=1}^{n_{OF}} f_i + k_P\left(P_{slack} - P_{slack}^{lim}\right)^2 + k_V \sum_{i=1}^{NL}\left(V_{L_i} - V_{L_i}^{lim}\right)^2 + k_s \sum_{i=1}^{NB}\left(S_{b_i} - S_{b_i}^{lim}\right)^2 \tag{12}$$

where k_p, k_V and k_s are the penalty factors of the active power produced by the reference generator (slack bus), the voltage at the load nodes and the penalty constrained of the thermal limit of the lines, respectively.

The limits of the various variables are determined by the following equation:

$$T_i^{lim} = \begin{cases} T_i^{max} & \text{if} \quad T_i > T_i^{max} \\ T_i^{min} & \text{if} \quad T_i < T_i^{min} \\ T_i & \text{if } T_i^{min} \leq T_i \leq T_i^{max} \end{cases} \tag{13}$$

T, represents the variables P_{slack}, V_{L_i} and S_{b_i}.

The values of the penalty factors are determined by empirical means. After several tests it is decided that, k_V=k_s =10,000 and k_p=1000.

From the above mathematical formulation, it can be noted that the use of such a model to solve real size problems is practically not possible with a classical approach. Consequently, in order to solve practical problems, it is necessary to use metaheuristic methods, that is, the genetic algorithms method.

3. Genetic algorithm method

GA is a metaheuristic optimization technique inspired by natural selection, and genetics developed by Holland-John, who conceived and realized an idea on how to transform the characteristics of natural evolution into a computer program [35]. The algorithm is based on a set of possible solutions randomly initialized in the search space. Individuals are represented by their design variables or by chromosome coding. Some solutions from the first population are used to form a new population based on genetic operators (crossover, mutation and selection). The goal is for the new population to be better than the previous one. The solutions that will be used to form new solutions are randomly selected according to their merit represented by an objective function specific to the problem posed, which should be minimized or maximized, so the better the individual, the greater his chances of surviving and reproducing, until the stop criterion is satisfied.

3.1 Application of the GA method

The goal of the GA method is to determine the optimal site and size of the DS to be integrated into the power system while minimizing the objective function under imposed constraints. Initially the vector of state variables and the vector of control variables are expressed as below:

The vector χ^T of state variables is composed of the active generation power at the reference bus P_{G_1} the load bus voltages V_{Li}, the reactive power produced by conventional generators Q_{G_i}, the apparent power transiting through transmission lines S_{li}. This vector χ^T can be expressed by:

$$\chi^T = \left[P_{G_1}, V_{L_1} \dots V_{L_{NL}}, Q_{G_1} \dots Q_{G_{NG}}, S_{l_1} \dots S_{l_{Nl}}\right] \tag{14}$$

The vector v^T of the control variables shows the active power outputs of the DS (size) S_{DS} and the site or location of the DS L_{DS}. Hence, this vector can be represented by:

$$v^T = [S_{DS_1}, S_{DS_2} \dots \dots .S_{DS_{NDS}}, L_{DS_1}, L_{DS_2} \dots ..L_{DS_{NDS}}] \tag{15}$$

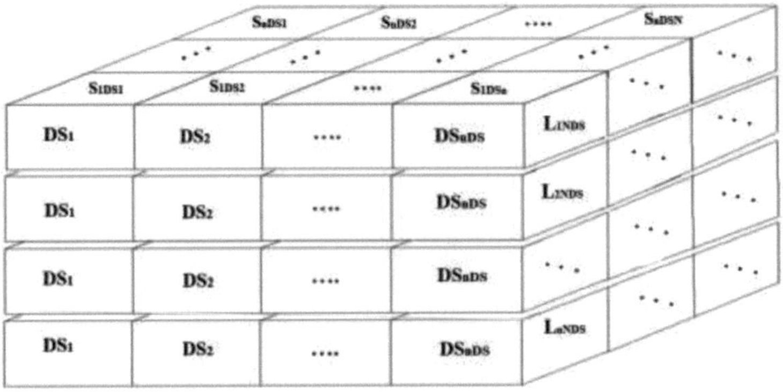

Figure 1.
Control variable vector structure.

Figure 1 indicates the chromosome structure employed in this study.

Step 1: In determining the site and size of DS, the genetic algorithm method was suggested. The main steps for researching the site and size of DS are as follows:

Step 2: Run a power flow and determine the various network parameters in the absence and presence of DS, using the Newton Raphson method.

Step 3: Select the GA parameters (number of generations, population size, crossover and mutation probability) and randomly generate the values of the sites and sizes between their limits using Eqs. E16, E17 and E18 (creation of the initial population P_0).

$$P_0 = [X_1, X_2 \dots \dots X_i \dots .X_{n\ population}] \tag{16}$$

$$L_{wi} = round(2 + rand(N_{bus} - 2)) \tag{17}$$

$$S_{wi} = P_{min}^{DSi} + rand(P_{max}^{DSi} - P_{min}^{DSi}) \tag{18}$$

where *round*(•) represents the value of a number that is rounded to the next integer, and *rand* represents a random number uniformly distributed between 0 and 1 [24].

Start the iteration meter t = 0.

Step 4: Run the power flow in the presence of the DS and evaluate the objective function for each individual.

Step 5: Examine all network constraints using Eqs. E4 to E8. If the latter are satisfied, go to the next step. If not, penalize the OF using Eq. E12 and go to the next step.

Step 6: Generate new populations following the laws of GA (crossover, mutation and selection), increase the generation meter (k = k + 1) and repeat steps 4 to step 6up to the stop criteria (maximum number of generations).

Step 7: Extract the best individual and show the results (site, size and various network parameters).

Figure 2 shows the flowchart of the search for the site and size of DS to be incorporated into the power system.

4. Description of the studied network

The IEEE 30 bus network is used by power system researchers to test the effectiveness of their programs simulation. This system is composed of 30 buss, 41 transmission lines, 04 transformers, 06 conventional generators with a total generating capacity of 435 MW with a reactive capacity of −95 MVAr to 520 MVAr and 21 loads with a total power demand of 283.4 MW and 126.2 MVAr [35]. Bus No. 1 is taken as the reference (slack bus). **Figure 3** shows the single line diagram of the network studied.

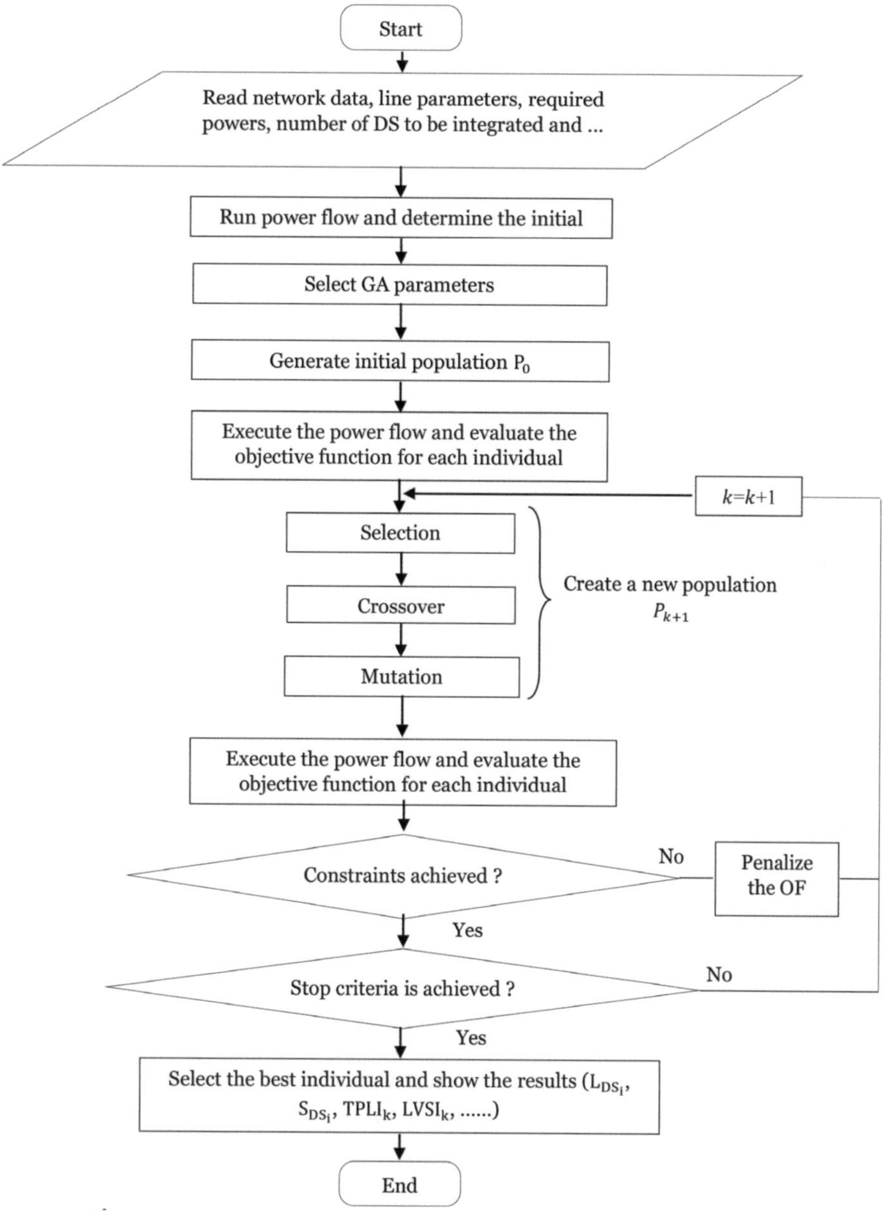

Figure 2.
DS site and size search flowchart.

4.1 Study framework

Before beginning the presentation of the simulation results, it is necessary to cite the limitations of the study framework under consideration. The Newton Raphson method is used via the MATPOWER software to calculate the power flow. For the GA method the existing Toolbox in the MATLAB library (GA function) will be used. The voltage limits of the network busses are limited between 0.95 pu- 1.1 pu. DS are considered as sources capable of delivering active and reactive power, using a power factor of 0.8. DS integration are modeled as PQ busses (negative loads). It is important to note that all load nodes are considered candidate busses for DS sites. It is important to note that the number of DS to be integrated into the network is

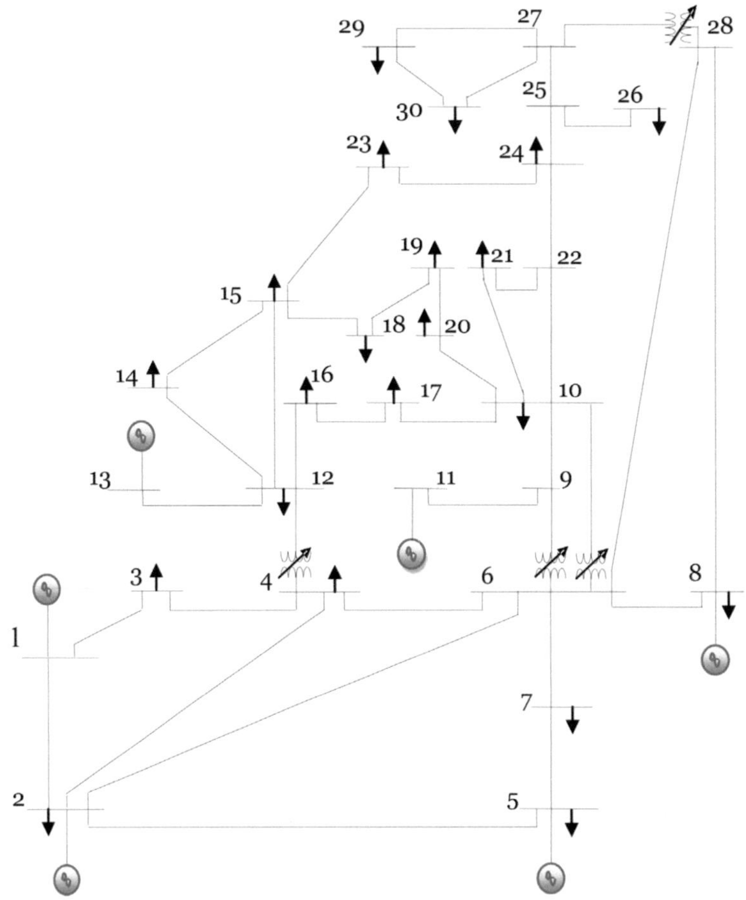

Figure 3.
Schematic diagram of the IEEE 30 bus network.

chosen in advance. After integration of a DS up to 10 DS, it is observed that, the best number from the viewpoint of improving the network parameters is 05 DS. For this reason, this study considers the insertion of five DS. The limits of the active power delivered by each DS are between 0 and 100 MW. And the penetration level limit of the total power delivered by the DS is $0.5 * \left(\sum_{i=1}^{N_D} P_{D_i}\right)$ MW that is $\sum_{i=1}^{N_{DS}} P_{DS_i} \leq 0.5 * \left(\sum_{i=1}^{N_D} P_{D_i}\right)$. After several tests on the studied network, it is noted that, the optimal parameters of the GA method are: the number of chromosomes (population size) is fixed at 100, the maximum number of generations is 50, the probability of mutation is 0.01 and the probability of crossover is 0.9. The multiobjective optimization considered in this study is based on the method of aggregate objectives through the weighting factors. It is important to note that, the choice of the values of these factors are dependent on the network operator and on the importance of the index to be improved. For this reason, the values of the weighting factors are set as follows: α_1=0.8 and α_2 =0.2. In this study four cases are considered: the first case shows the simulation results for the base case (Execute the optimal power flow without DS). Second case, presents the results of the minimization of active losses only. Third case, presents the results of the voltage stability index improvement. Fourth case, presents the simulation results of the multi-objective case, it is the minimization of the bi-objectives at the simultaneous (minimization of losses and voltage stability index improvement).

5. Simulation results and interpretations

After the execution of the elaborated program, the simulation results achieved are shown bellow. **Table 1**, represents all the simulation results obtained for the five cases studied, as also the physical power limits of conventional generators.

Parameters	Limits Max	Limits Min	Case 01	Case 02	Case 03	Case 04
P_{G1}(MW)	200	50	182.64	52.47	60.61	50.46
P_{G2}(MW)	80	20	50.58	50.58	50.58	50.58
P_{G3}(MW)	50	15	18.52	18.52	18.52	18.52
P_{G4}(MW)	35	10	18.09	18.09	18.09	18.09
P_{G5}(MW)	30	10	10.40	10.40	10.40	10.40
P_{G6}(MW)	40	10	13.26	13.26	13.26	13.26
Q_{G1}(MVAr)	200	−20	−8.56	16.94	6.16	17.97
Q_{G2}(MVAr)	100	−20	23.47	−8.88	−15.30	−8.98
Q_{G3}(MVAr)	50	−15	32.04	3.22	23.75	−5.10
Q_{G4}(MVAr)	60	−15	49.29	−5.34	−2.80	−4.82
Q_{G5}(MVAr)	50	−10	5.39	−10.00	−6.98	−7.92
Q_{G6}(MVAr)	60	−15	2.74	−14.26	−15.00	−12.29
Total Conventional Active Production	435	115	293.49	163.33	171.46	161.31
Total Conventional Reactive Production	520	−15	104.37	−18.32	−10.16	−21.14
Total Load (MW, MVAr)				(283.40, 126.20)		
DS Site(Best Location)	**	**	**	7	4	7
				9	12	17
				12	20	20
				18	23	23
				30	29	30
DS Size (MW, MVAr)			**	(46.96, 35.22)	(36.40, 27.30)	(67.52, 50.64)
				(38.58, 28.93)	(52.97, 39.73)	(27.48, 20.61)
				(5.63, 4.22)	(6.07, 4.55)	(10.36, 7.77)
				(18.09, 13.56)	(11.37, 8.52)	(4.60, 3.45)
				(13.60, 10.20)	(9.73, 7.30)	(15.05, 11.28)
TPLI (MW)			10.89	2.798	4.620	2.945
LVSI			0.181	0.158	0.067	0.149
Loading Parameter (pu)			2.1	4.4	5.7	5.2

*** without DS integration.*

Table 1.
Summary of network simulation results for the various cases.

Figure 4, shows the situation of apparent power transmitted via transmission lines in all the cases studied. **Figure 5** shows the voltage profile of several different cases.

From the results obtained by running the optimal power flow without DS (case 01), it is found that the total active losses are 10.89 MW, the LVSI is 0.181 and loading parameter (LP) is 2.1 pu.

In the case of the minimization the TPLI only (case 2), it can be seen that the total active losses are reduced to 2.798 MW, which represents a reduction rate of 74.3%, the LVSI has been improved with 12.7% and loading parameter represents a rate of 52.27%, that is after the integration of 05 DS with a total active power of 122.86 MW and reactive power of 92.13 MVAr distributed to the different sites (busses 7, 9, 12, 18 and 30) respecting all the constraints of the network.

When improving the LVSI only (case 3), the simulation results show that busses 4, 12, 20, 23 and 29 are selected as the best sites for DS power installations producing a total power of 116.54 MW and 87.4 MVAr. This implementation has enhanced the LVSI parameter from the value 0.181 to 0.067, which represents a rate of 62.9%, the losses are reduced to 4.62 MW, which is 57.5% and a loading parameter improvement rate of 63.15%.

The multiobjective optimization (case 04) proved that the results give a compromise between the different values of the minimized objective functions. In this case, the TPLI is improved by 72.9%, LVSI has been minimized by 17.67% and loading parameter has been increased by 59.61%. The advantage of Multi-objective

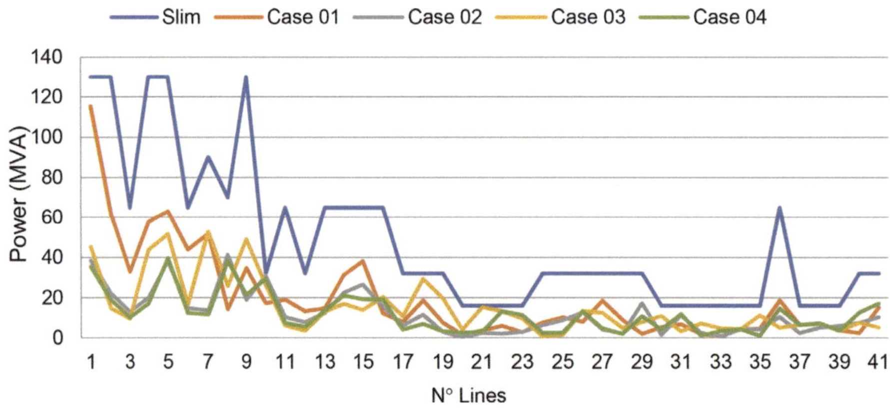

Figure 4.
Apparent power of the transmission lines for several cases.

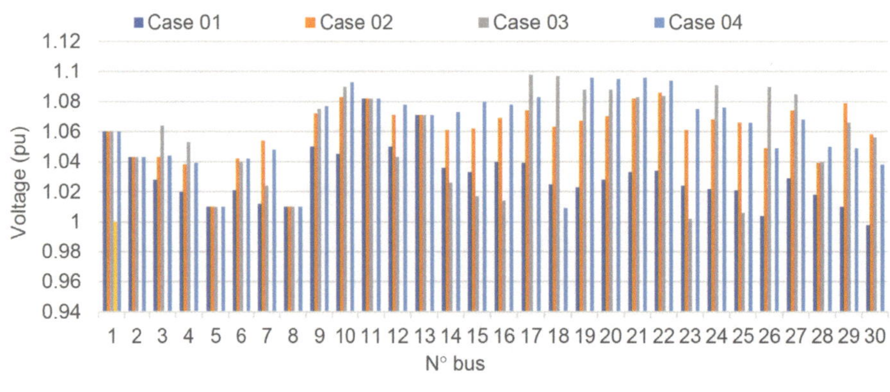

Figure 5.
Voltage values of the network bus for several cases.

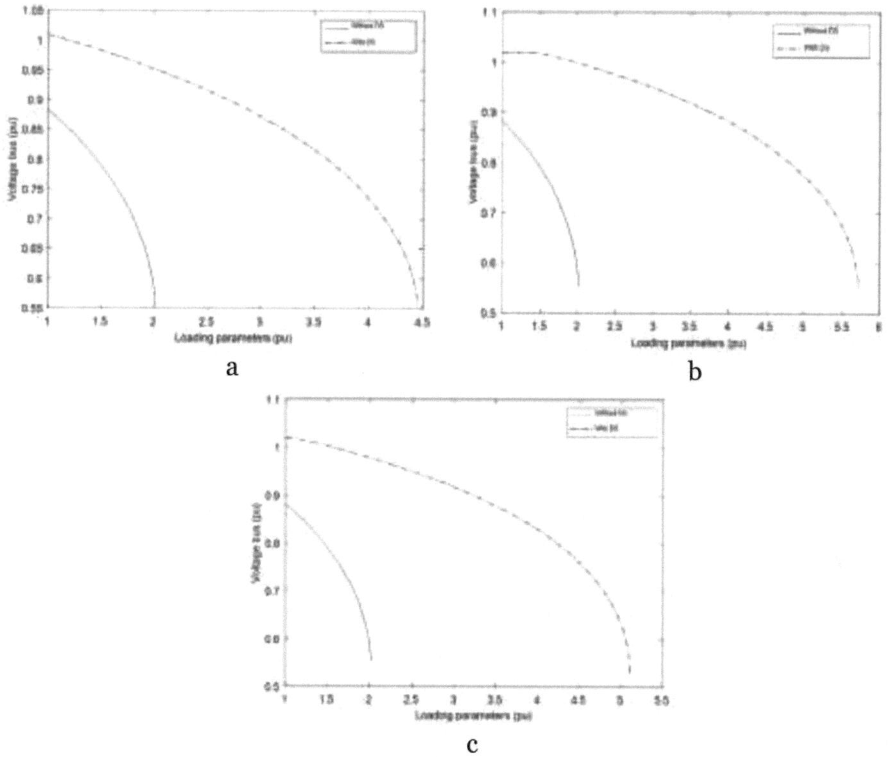

Figure 6.
(a) LP in case 2, (b) LP in case 3 (c) LP in case 4.

optimization consists in improving several parameters of the electrical network and having the best compromise.

Figure 4 shows that all line powers are below the thermal limit power, which demonstrates the consideration of line stresses.

Figure 5 shows that all the bus voltages are included between the limits 0.95 pu and 1.1 pu. **Figure 6** illustrates the PV curves before and after the integration of the DS for the three cases.

Figure 6 shows the positive influence of DS integration on the voltage stability margin of the IEEE 30 bus network. This figure also shows that case 3 represents the best value of LM, because the objective function only maximizes the voltage stability.

6. Conclusion

This chapter solves the problem of searching for the optimal location and size of DS in the transmission power system considering the required constraints, using the GA technique. In this work, various objective functions are targeted, minimizing active losses and voltage stability improvement. The last ones are treated as a mono and multi objective problem. The simulations carried out have shown that the results are dependent on the minimized objective function. The results achieved show the efficiency of the GA method. The perspective of this study is to include the constraint related to the transient stability of the rotor angles of conventional generators, in order to maintain the stability of the system during DS disconnection.

Acknowledgements

I thank very sincerely the members of the laboratory of LACoSERE, Electrical Department of the Amar Telidji University, Laghouat, Algeria and laboratory of LAADI, Electrical Department of Djelfa University, Djelfa, Alegria for their participations in the realization of this work.

Appendix

The data (lines, consumptions, generators) of the IEEE 30 bus network used in this study are shown below [36]. **Table 2** shows the bus data of the IEEE network studied. **Table 3** represents the line data. **Table 4** shows the data of the conventional generators.

Bus i	Type	P_d(MW)	Q_d(MVAr)	G_s	B_s	Area	V_m	V_a	BaseKV	Zone	V_{max}	V_{min}
1*	3	0	0	0	0	1	1.06	0	132	1	1.1	0.95
2	2	21.7	12.7	0	0	1	1.043	−5.48	132	1	1.1	0.95
3	1	2.4	1.2	0	0	1	1.021	−7.96	132	1	1.1	0.95
4	1	7.6	1.6	0	0	1	1.012	−9.62	132	1	1.1	0.95
5	2	94.2	19	0	0	1	1.01	−14.37	132	1	1.1	0.95
6	1	0	0	0	0	1	1.01	−11.34	132	1	1.1	0.95
7	1	22.8	10.9	0	0	1	1.002	−13.12	132	1	1.1	0.95
8	2	30	30	0	0	1	1.01	−12.1	132	1	1.1	0.95
9	1	0	0	0	0	1	1.051	−14.38	1	1	1.1	0.95
10	1	5.8	2	0	19	1	1.045	−15.97	33	1	1.1	0.95
11	2	0	0	0	0	1	1.082	−14.39	11	1	1.1	0.95
12	1	11.2	7.5	0	0	1	1.057	−15.24	33	1	1.1	0.95
13	2	0	0	0	0	1	1.071	−15.24	11	1	1.1	0.95
14	1	6.2	1.6	0	0	1	1.042	−16.13	33	1	1.1	0.95
15	1	8.2	2.5	0	0	1	1.038	−16.22	33	1	1.1	0.95
16	1	3.5	1.8	0	0	1	1.045	−15.83	33	1	1.1	0.95
17	1	9	5.8	0	0	1	1.04	−16.14	33	1	1.1	0.95
18	1	3.2	0.9	0	0	1	1.028	−16.82	33	1	1.1	0.95
19	1	9.5	3.4	0	0	1	1.026	−17	33	1	1.1	0.95
20	1	2.2	0.7	0	0	1	1.03	−16.8	33	1	1.1	0.95
21	1	17.5	11.2	0	1	1	1.033	−16.42	33	1	1.1	0.95
22	1	0	0	0	0	1	1.033	−16.41	33	1	1.1	0.95
23	1	3.2	1.6	0	0	1	1.027	−16.61	33	1	1.1	0.95
24	1	8.7	6.7	0	4.3	1	1.021	−16.78	33	1	1.1	0.95
25	1	0	0	0	0	1	1.017	−16.35	33	1	1.1	0.95
26	1	3.5	2.3	0	0	1	1	−16.77	33	1	1.1	0.95
27	1	0	0	0	0	1	1.023	−15.82	33	1	1.1	0.95
28	1	0	0	0	0	1	1.007	−11.97	132	1	1.1	0.95

Bus i	Type	P_d(MW)	Q_d(MVAr)	G_s	B_s	Area	V_m	V_a	BaseKV	Zone	V_{max}	V_{min}
29	1	2.4	0.9	0	0	1	1.003	−17.06	33	1	1.1	0.95
30	1	10.6	1.9	0	0	1	0.992	−17.94	33	1	1.1	0.95
Base MVA = 100												

Table 2.
Bus data of IEEE 30 bus power system.

Bus_i	Bus_j	R(pu)	x(pu)	b(pu)	RateA (S_{lim} MVA)	RateB	RateC	Ratio	Angle	Status
1	2	0.0192	0.0575	0.0528	130	0	0	0	0	1
1	3	0.0452	0.1652	0.0408	130	0	0	0	0	1
2	4	0.057	0.1737	0.0368	65	0	0	0	0	1
3	4	0.0132	0.0379	0.0084	130	0	0	0	0	1
2	5	0.0472	0.1983	0.0418	130	0	0	0	0	1
2	6	0.0581	0.1763	0.0374	65	0	0	0	0	1
4	6	0.0119	0.0414	0.009	90	0	0	0	0	1
5	7	0.046	0.116	0.0204	70	0	0	0	0	1
6	7	0.0267	0.082	0.017	130	0	0	0	0	1
6	8	0.012	0.042	0.009	32	0	0	0	0	1
6	9	0	0.208	0	65	0	0	0.978	0	1
6	10	0	0.556	0	32	0	0	0.969	0	1
9	11	0	0.208	0	65	0	0	0	0	1
9	10	0	0.11	0	65	0	0	0	0	1
4	12	0	0.256	0	65	0	0	0.932	0	1
12	13	0	0.14	0	65	0	0	0	0	1
12	14	0.1231	0.2559	0	32	0	0	0	0	1
12	15	0.0662	0.1304	0	32	0	0	0	0	1
12	16	0.0945	0.1987	0	32	0	0	0	0	1
14	15	0.221	0.1997	0	16	0	0	0	0	1
16	17	0.0524	0.1923	0	16	0	0	0	0	1
15	18	0.1073	0.2185	0	16	0	0	0	0	1
18	19	0.0639	0.1292	0	16	0	0	0	0	1
19	20	0.034	0.068	0	32	0	0	0	0	1
10	20	0.0936	0.209	0	32	0	0	0	0	1
10	17	0.0324	0.0845	0	32	0	0	0	0	1
10	21	0.0348	0.0749	0	32	0	0	0	0	1
10	22	0.0727	0.1499	0	32	0	0	0	0	1
21	22	0.0116	0.0236	0	32	0	0	0	0	1
15	23	0.1	0.202	0	16	0	0	0	0	1
22	24	0.115	0.179	0	16	0	0	0	0	1
23	24	0.132	0.27	0	16	0	0	0	0	1

Bus_i	Bus_j	R(pu)	x(pu)	b(pu)	RateA (S_{lim} MVA)	RateB	RateC	Ratio	Angle	Status
24	25	0.1885	0.3292	0	16	0	0	0	0	1
25	26	0.2544	0.38	0	16	0	0	0	0	1
25	27	0.1093	0.2087	0	16	0	0	0	0	1
28	27	0	0.396	0	65	0	0	0.968	0	1
27	29	0.2198	0.4153	0	16	0	0	0	0	1
27	30	0.3202	0.6027	0	16	0	0	0	0	1
29	30	0.2399	0.4533	0	16	0	0	0	0	1
8	28	0.0636	0.2	0.0428	32	0	0	0	0	1
6	28	0.0169	0.0599	0.013	32	0	0	0	0	1

Table 3.
Line data of IEEE 30 bus power system.

Bus	P_g (MW)	Q_g (MVAr)	Q_{max} (MVAr)	Q_{min} (MVAr)	V_g	S_{Base}	Status	P_{max} (MW)	P_{min} (MW)
1	0	0	200	−20	1.06	100	1	200	50
2	50.5846	0	100	−20	1.045	100	1	80	20
5	18.5227	0	50	−15	1.02	100	1	50	15
8	18.0865	0	60	−15	1.029	100	1	35	10
11	10.4038	0	50	−10	1.06	100	1	30	10
13	13.2553	0	60	−15	1.06	100	1	40	12

Table 4.
Generator data of IEEE 30 bus power system.

Author details

Mustafa Mosbah[1*], Redha Djamel Mohammedi[2] and Salem Arif[1]

1 Laboratoire d'Analyse et de Commande des Systèmes d'Energie et Réseaux Electriques (LACoSERE), Amar Telidji University, Laghouat, Alegria

2 Laboratoire d'Automatique Appliquée et Diagnostic Industriel (LAADI), Djelfa University, Djelfa, Alegria

*Address all correspondence to: m.mosbah@lagh-univ.dz

References

[1] Y. Bakelli, M. Mosbah, A. Kaabeche, and M. Acimi, 'Voltage stability improvement by optimal location of wind sources', 2019 1st International Conference on Sustainable Renewable Energy Systems and Applications (ICSRESA), 2019.

[2] N. Hadjsaid, J. F. Canard, and F. Dumas, "Dispersed generation impact on distribution networks," IEEE Comput. Appl. Power, vol. 12, pp. 22–28, Apr. 1999.

[3] N. Hadjsaïd, J. C. Sabonnadière, Electrical distribution networks. Wiley ISTE, 2011.

[4] R. O. Bawazir, N. S.Cetin, "Comprehensive overview of optimizing PV-DG allocation in power system and solar energy resource potential assessments, Energy Reports, Vol. 6, pp. 173–208, 2020.

[5] V. Mytilinou, A. J. Kolios, "A multi-objective optimisation approach applied to offshore wind farm location selection" Journal of Ocean Engineering and Marine Energy vol. 3, pp. 265–284, 2017.

[6] P. Barker, T. Leskan, H. Zaininger and D. Smith, "Integration of distributed resources in electric utility systems: current interconnection practice and unified approach" EPRI TR-111489, Nov 1998.

[7] W. El-khattam, et al, "Optimal investment planning for distributed generation in a competitive electricity markets" IEEE Tran Power Sys, vol. 19, pp. 1674–84, 2004.

[8] P. Kayal, C.K. Chanda, "Placement of wind and solar based DGs in distribution system for power loss minimization and voltage stability improvement" Int J Electr Power Energy Syst, vol. 53, pp. 795–809 2013.

[9] U.Sultana, al, "A review of optimum DG placement based on minimization of power losses and voltage stability enhancement of distribution system" Renewable and Sustainable Energy Reviews, vol. 63, pp. 363–378, 2016.

[10] A. R. Jordehi, "Allocation of distributed generation units in electric power systems: A review" Renewable and Sustainable Energy Reviews, Vol.56, pp.893–905, 2016.

[11] R. S. Al Abri, E. F. El-Saadany, Y. M. Atwa, "Optimal placement and sizing method to improve the voltage stability margin in a distribution system using distributed generation" IEEE Transactions on Power Systems" vol. 28, pp. 326–334, 2012.

[12] M. Dehghani, Z. Montazeriand O. P. Malik, "Optimal sizing and placement of capacitor banks and distributed generation in distribution systems using spring search algorithm" International Journal of Emerging Electric Power Systems, vol. 21, 2020.

[13] K. H.Truong et al, "A Quasi-Oppositional-Chaotic Symbiotic Organisms Search algorithm for optimal allocation of DG in radial distribution networks" Applied Soft Computing, vol. 88, 2020.

[14] Adel A. Abou El-Ela, R. A. El-Sehiemy, A. S. Abbas, "Optimal placement and sizing of distributed generation and capacitor banks in distribution systems using water cycle algorithm" IEEE Systems Journal, vol. 12, pp. 3629–3636, 2018.

[15] S. H. Lee, J. W. Park, "Optimal placement and sizing of multiple DGs in a practical distribution system by considering power loss"IEEE Transactions on Industry Applications, vol. 49, pp. 2262–2270, 2013.

[16] A. Picciariello, et al, "Distributed generation and distribution pricing: why do we need new tariff design methodologies?" Electr Power Sys Rev, vol. 6, pp. 119–370, 2015.

[17] M. S. Syed, S. V. Chintalapudi, S. Sirigiri, "Optimal power flow solution in the presence of renewable energy sources" Iranian Journal of Science and Technology, Transactions of Electrical Engineering, vol. xx, pp. xxx–xxx, 2020.

[18] Phung DangHuy, al, "Optimal placement, sizing and power factor of distributed generation: A comprehensive study spanning from the planning stage to the operation stage" Energy,vol. 195, pp. 110–117, 2020.

[19] A. Tamimi, A. Pahwa, S. Starrett, "Maximizing wind penetration using voltage stability based methods for sizing and locating new wind farms in power system" In: Power and Energy Society General Meeting, vol. xx, pp. 1–7, 2010.

[20] S. Ghosh, et al, "Optimal sizing and placement of distributed generation in a network system" Electrical Power and Energy Systems,vol.32, pp. 849–856, 2010.

[21] R.K. Singh, S.K. Goswami, "Optimum allocation of distributed generations based on nodal pricing for profit,loss reduction, and voltage improvement including voltage rise issue" Electrical Power and Energy Systems, vol. 32,pp. 637–644, 2010.

[22] M. Mosbah., A. Khattara, M. Becherif, S. Arif, "Optimal PV location choice considering static and dynamic constraints" Int. J. Emerg Electr Power Syst, vol. 18, 2016.

[23] A. Shuaibu Hassan, Y. Sun, Z. Wang, "Multi-objective for optimal placement and sizing DG units in reducing loss of power and enhancing voltage profile using BPSO-SLFA" Energy Reports, vol. 6, pp. 1581–1589, 2020.

[24] A. S. O. Ogunjuyigbe, T. R. Ayodele, O. O. Akinola, "Impact of distributed generators on the power loss and voltage profile of sub-transmission network" J. Electr. Syst. Inf. Technol, vol. 3, pp. 94–107, 2016.

[25] H. C. Nejad et al, "Reliability based optimal allocation of distributed generations in transmission systems under demand response program" Electric Power Systems Research, vol. 176, 2019.

[26] C. Wang, M. H. Nehrir, Analytical approaches for optimal placement of distributed generation sources in power systems, IEEE Trans. Power Syst, vol. 19, pp. 2068–2076, 2004.

[27] S. Kabir, O. Krause, S. Bartlett, "Impact of large-scale photovoltaic system on short and long term voltage stability in sub transmission networl, Australas. Univ. Power Eng. Conf. AUPEC (October) (2013) 1–6.

[28] A. A. Sadiq, S. S. Adamu, M. Buhari, "Optimal distributed generation planning in distribution networks: A comparison of transmission network models with FACTS" Engineering Science and Technology, an International Journal, vol. 22, pp. 33–46, 2019.

[29] A. Ogunjuyigbe, T. Ayodele, O. Akinola, "Impact of distributed generators on the power loss and voltage profile of sub-transmission network" J. Electr. Syst. Inf. Technol, vol. 3, pp. 94–107, 2016.

[30] M. Mosbah, et al, "Optimal location and size of wind source in large power system for losses minimization" International Conference in Artificial Intelligence in Renewable Energetic Systems ICAIRES 2019: Smart Energy Empowerment in Smart and Resilient Cities, vol. 102, pp 566–574, 2019.

[31] R. Bawazir, N. Çetin M. Mosbah, S. Arif, "Genetic algorithm for improving voltage stability by optimal integration of wind source" International Conference on Electrical Engineering (ICEE) September 25–27, Istanbul, Turkey, 2020.

[32] M. Moghavvemi, F. Omar, "Transmission, and Distribution, "Technique for contingency monitoring and voltage collapse prediction" vol. 145, pp. 634–640, 1998.

[33] M. Mosbah, R. D. Mohammedi, S. Arif, "Differential evolution method for optimal size and localization of photovoltaic in the algerian transmission power system" In: Proceedings of 2019 Algerian Large Electrical Network Conference (CAGRE-2019), pp. 1–6, 2019.

[34] M. Reza et al, "Impacts of distributed generation penetration levels on power systems transient stability" In Proceedings of the IEEE Power Engineering Society General Meeting, Denver, USA, June 2004.

[35] J. H. Holland "Adaptation in Nature and Artificial Systems" The University of Michigan Press, 1975.

[36] The University of Washington Electrical Engineering, "Power system test case archive, the IEEE 30-bus test system data," October 2017.

Chapter 23

New Generation Aero Combustor

Jushan Chin and Jin Dang

Abstract

The purpose of this study is to identify the technology for next generation aero combustors, and to propose totally new combustor design approaches. Next generation aero combustors need very high combustion air fraction, that brings idle lean blow out (LBO) problem. The present study suggests several measures to solve this problem, including: pilot and main two concentric combustion zones with separation, aerodynamic design to have main air slipping by pilot combustion zones, etc. For high fuel air ratio (FAR) combustor, the present authors propose using angled main fuel co-axial air plain jet injection. Make use of different penetration to meet the need for low power and high power conditions. For low emissions combustor, the present authors use small scale close contact fuel-air mixing with fuel staging to have low emissions at the same time to have good idle, good high altitude ignition, etc. Brand new cooling designs are proposed for outliner and inner liner. This chapter is mainly a survey of present author's own research. The results of this study will provide guideline for the development of next generation aero combustors.

Keywords: low emissions combustor, high FAR combustor, fuel air module design, liner cooling design, lean direct mixing combustion, idle LBO

1. Introduction

What are new generation aero combustors?

Aero gas turbine engine combustors have been developed over 80 years. It does not matter if it is a civil engine combustor, or military engine combustor. They are all developed under one line, that is, towards higher performance, higher reliability, and lower fuel consumption.

On 1977, International Civil Aviation Organization (ICAO) published a document named "Control of Aircraft Engine Emissions". Since then aero combustors have entered a new era, it is that of low emissions combustors. The requirements for low emissions are different for civil aero combustor versus military aero combustor. For civil aero combustors, their emissions are regulated by ICAO Committee on Aviation Environmental Protection (CEAP) [1]. The standard has been developed from CEAP 1, CEAP 2, CEAP 4, CEAP 6, now it is CEAP 8. It is getting more and more restrictive. Now-a-days, any civil aero engine thrust higher than 26.7 KN must be in accordance with the CEAP 8 standard. Their emissions of nitrous oxide (NOx), carbon monoxide (CO), unburnt hydrocarbon (UHC), and smoke shall all be controlled. Actually, an aero engine company is required to report the percentage their engine combustor will produce of each type of emission lower than the corresponding CEAP specified.

Because of the requirement of continuous improvement for reduction of fuel consumption, civil aero engines have development in two aspects. An aero engine as a propulsion unit has propulsion efficiency. This is to increase bypass ratio. On the other aspect, an aero engine is also a thermal engine, it has thermal efficiency. The way to improve its thermal efficiency is to increase the pressure ratio of the engine (at the same time, increase turbine inlet temperature appropriately). Thus, for several decades engine pressure ratio has been going up all the time, from nearly 10, then 20, 30, 40, 50, and pressure ratio 60 aero engines has been certified, will be in service soon. New generation civil aero engines will achieve a high pressure ratio of 70.

Conclusion: a new generation civil aero combustor is a high pressure low emissions combustor.

For a military aero engine, the most important development target is to have a higher thrust-to-weight ratio. In order to improve thrust-to-weight ratio, the engine shall have higher turbine inlet temperature (or higher combustor fuel air ratio, FAR) and increase the engine pressure ratio appropriately. Thus, the military aero combustor FAR has been increased from lower than 0.02 to 0.03, to 0.038, to 0.046. The new generation military aero combustor will have FAR, 0.051. Notice that a high FAR combustor is also called a high temperature rise combustor.

Conclusion: new generation military aero combustor is high FAR combustor.

2. Design of high FAR combustor

For a military aero combustor, there is no 30% power condition, no 85% power condition, and no maximum cruise condition. There is a ground idle condition, maximum power condition (or 100% power condition) and cruise conditions at different altitudes and different Mach numbers. There is a high altitude idle condition. Particularly there is low altitude not-so-low Mach number penetration dash condition. In this condition, the combustor inlet pressure may be even higher than at the take-off condition. The inlet air temperature is very close to the take-off condition and FAR is only a little lower than take-off condition. At this condition, liner wall temperature higher than that at take-off condition is possible.

Don Bahr [2] reported that there are two major problems for high temperature rise combustor design; they are idle lean blow out (LBO) and liner cooling. According to present author's experience, it is true that these two issues are critical for high FAR combustor design and development. But there are other issues too. The design reported in this chapter is mainly from reference [3].

2.1 Idle LBO

For military aero combustor idle LBO issues, the present author proposed several design approaches. One approach is that, there is a concentric "twin" combustion zone concept. That is, the pilot fuel air combustion zone is at the center, main fuel air combustion zone is surrounding the pilot fuel air combustion zone, with some separation. Significantly, a reduction of the main air quenching effect on pilot fuel air combustion is performed. This is the way to improve idle LBO. The annular combustor has several fuel air modules, the fuel air module configuration is shown in **Figure 1**. This multiple swirler dome design is very different from reference [4]. The pilot fuel air module consists of a pressure swirl fuel nozzle and an axial air swirler. The main air module has a distance radially away from pilot module. This distance is from pilot module exit diameter to the inner diameter of main module annular exit. In this design it is 0.95 in.

The second design approach for solving idle LBO issue is the combustion zone aerodynamics. As shown in **Figure 1**, the main air module consists of two portions, one third is non-swirling air on inner side and two thirds is swirling air. The pilot module air is having weak swirling. Together with the separation distance, the combustion zone aerodynamics is shown in **Figure 2**. The combustion aerodynamics is low swirling. But in the present design, there is combination of swirling air and non-swirling air, which is rather different from reference [5]. The key feature is that the main air flow is just slipping and passing through pilot air recirculation without mixing with pilot air, that is the most important factor for reducing main air quenching effect on idle condition pilot fuel air combustion.

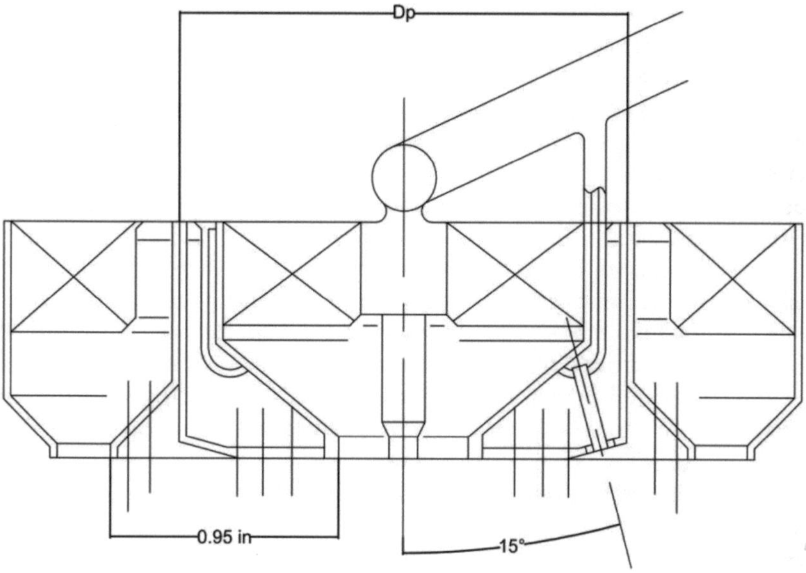

Figure 1.
Fuel-air module [3].

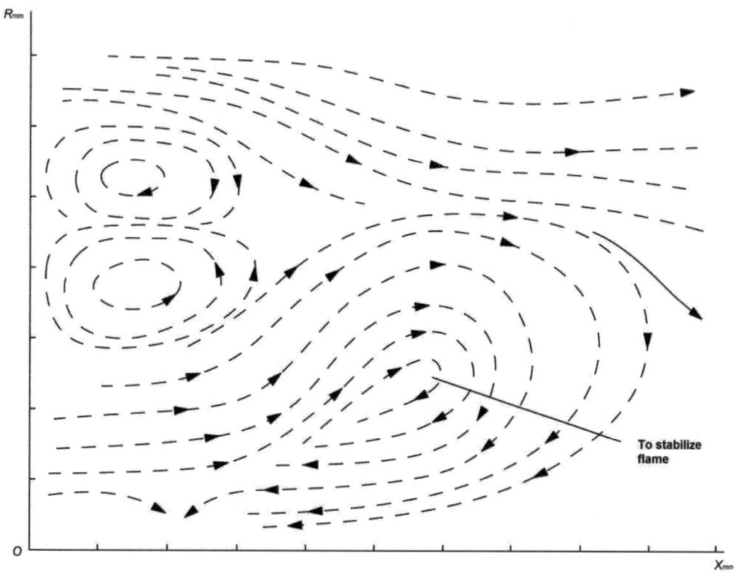

Figure 2.
Air flow pattern [3].

The third design approach for solving idle LBO issue is design of pilot fuel air combustion. Pilot fuel air combustion, pilot air alone module and pilot fuel nozzle are designed at idle condition, not at maximum condition. For high FAR combustor, its idle condition FAR is higher than idle FAR for civil combustor. At the idle condition, if only the pilot fuel is working, when approaching flame out, the pilot fuel nozzle pressure drop will be very low, which is harmful for LBO (it is not practical that at idle condition pilot fuel nozzle pressure drop is extremely high, such as higher than 200 psig). Thus, it needs main fuel to be open to work together with pilot fuel combustion. At the idle condition, its fuel flow is split at 70% pilot fuel and 30% main fuel. 70% idle fuel flow together with pilot module air to form an idle pilot fuel combustion at equivalence ratio 1.2 and based on this design criteria to determine pilot air fraction within the combustion air. At idle condition, 30% main fuel with co-flowing air to provide idle main fuel combustion equivalence ratio 1.2 to determine co-flowing air amount. At idle condition, the main fuel injection pressure drop is very low, main fuel jet spray with co-flowing air is collapsed with pilot fuel combustion. Pilot fuel nozzle operation is designed at idle. At idle condition pilot fuel nozzle pressure drop is 120 psig. Based on this design approach to determine pilot fuel nozzle flow number (FN). There is a flow divider valve between the pilot fuel nozzle and the main fuel injector. The crack pressure for this flow divider valve is a critical design parameter. It needs to have an initial choice, then after both pilot fuel and main fuel design finished for the whole power condition range, it may be modified several times. Pilot fuel nozzle spray angle is 90 degree. The pilot air module inlet swirler is a thin curved blade low swirling axial swirler. The inlet effective flow area (ACd) is much higher than exit ACd to let the exit be the flow metering device.

High FAR combustor idle LBO design is related to many aspects. There are several design choices which must be balanced, making many times modification to have an all-round good solution. These design choices are:

- Idle condition, main fuel-pilot fuel division
- Idle condition, pilot fuel combustion equivalence ratio (determine pilot combustion air fraction).
- Idle condition main fuel and coaxial flowing air FAR ratio (determine co-flowing air amount)
- Idle condition, pilot fuel nozzle pressure drops (determine pilot nozzle flow number)
- Idle condition, main fuel injection pressure drops
- Flow divider valve crack pressure (affect at maximum condition, difference of injection pressure drop between pilot nozzle and main fuel injector, affect required maximum pump pressure capability)
- Maximum condition pilot fuel and main fuel division (determine maximum condition pilot fuel combustion and main fuel combustion equivalence ratio)

The final design shall have good idle LBO, appropriate maximum condition pilot fuel and main fuel combustion equivalence ratio (none of them may exceed 1.2), and a maximum fuel nozzle pressure drop which is not higher than 800 psig.

Conclusion: with these design measures, a high FAR combustor idle LBO problem is solved [3].

2.2 Design of maximum condition

Design of maximum condition is to reach both pilot fuel and main fuel combustion near stoichiometric combustion to determine the total combustion air fraction [3]. For example, if the combustor FAR is 0.051, combustion air is 75% to have combustion FAR 0.051/0.75 = 0.068, which is an equivalence ratio of one. Then the main fuel combustion air fraction is total combustion air minus the pilot combustion air fraction.

Main fuel air combustion design is concentrated on main fuel injection. As shown in **Figure 1**, main fuel is co-axial air flowing plain jet fuel injection. There is an angle, between main injection center line and module center line, which is 15 degree in this design. This injection angle is a critical design choice. At the low power condition, main fuel injection pressure drop is very low, main fuel will not penetrate far away, thus main fuel will burn with pilot fuel combustion. At maximum power condition, main fuel injection pressure drop is very high, the spray will penetrate radially out to meet main air to form direct mixing combustion. Here the **design idea is to make use of the change of penetration to suit low power condition and high-power condition** [3]. That is the reason why the main fuel injection must be with an angle relative to module center line. Main fuel injection pressure drop at maximum condition is very high, hopefully higher than 600 psig. Somewhere between 40% to 60% power condition, there will be a situation the main fuel spray will detach from the pilot fuel combustion to form a separate main fuel air combustion zone. Main fuel penetration and mixing with main air is critical to form non-visible smoke, to have non-luminous flame, to have high efficiency at maximum power condition. Because of the need for high penetration, the main fuel injector is not a hole, but a **section of straight tubing of diameter 0.03 in**. Notice that the present author does not call main fuel injection as co-axial air blast atomization, but rather air co-flowing plain jet injection. In studies on such injection, it was found that under very low liquid injection pressure drop, it is air blast atomization, at medium injection pressure drop, it is air assist atomization, with very high liquid injection pressure drop, it is air retard atomization. Air retard atomization is a new term. It is a case where air does not help atomization of a finer drop size, but hurts atomization by becoming a coarser drop size. Actually, this is good. This is because with very fine droplets, the spray cannot penetrate far out, while for main fuel combustion, at high power condition, penetration is more important than drop size. But it is not the more penetration the better, for fuel injection the present author shall take atomization, penetration, dispersion and fuel air mixing all four aspects into consideration.

Notice that the circumferential distribution of main fuel injectors may be uniform or may be non-uniform. For example, main fuel injector positioning may start other than 12 o'clock position. The different main fuel injector circumferential arrangement is for minor adjustment of exit radial temperature and FAR profile because there is no dilution air.

3. Other issues

3.1 Combustion efficiency

High FAR combustor has a combustion efficiency issue. Some designer mentioned that for high FAR combustor, at maximum power condition, its efficiency can only be 98%, it is because of chemical dissociation. The present author has studied chemical dissociation. It will have significant effect on efficiency at higher temperature. For aviation kerosene and air combustion, at stoichiometric fuel air ratio, the effect of chemical dissociation will not be so much. Thus, for well-organized high FAR combustion, its combustion efficiency shall be higher than 98%.

3.2 Exit distribution

Notice here the title is only exit distribution, not exit temperature distribution. In reference [6], it was reported that for high FAR combustor exit temperature radial profile is different from exit FAR radial profile, as shown in **Figure 3**.

If **Figure 3** is a true situation in the engine, that is very harmful. This is because the over rich combustion gas entering turbine, meeting with turbine cooling air, will result in additional burning which can totally destroy local turbine cooling. But this was some work done more than 20 years ago. At that time the combustion organization for high FAR combustor was poor. Combustion was rather non-uniform. From present design, the high FAR combustion is well organized, there will not be such severe difference. The temperature defined radial profile and FAR defined radial profile will be of same shape. But the point reported in reference [6] is very important. It shows that to delete some extremely rich pockets in the combustor exit is of very high importance. It also shows in addition to temperature defined exit radial profile, there shall be FAR defined exit radial profile.

3.3 Visible smoke

High FAR combustor must avoid visible smoke. This requirement is not only for the maximum condition, it is for all operational conditions. For smoke reduction, fuel additives method cannot be used [7]. To avoid maximum condition visible smoke, design at maximum condition combustion fuel air ratio is stoichiometric, particularly pilot fuel combustion shall not be over rich, avoid any possible local over rich pocket, and the whole combustion FAR shall be uniform. In reference [2], Don Bahr reported the number one issue for high temperature rise combustors is the contradiction between high power condition visible smoke and idle LBO. As mentioned in this chapter, if the idle LBO problem is to be solved, then the design may significantly increase combustion air fraction, such as for combustor FAR 0.051, combustion air is 75%. Then **the combustor cannot have primary air**

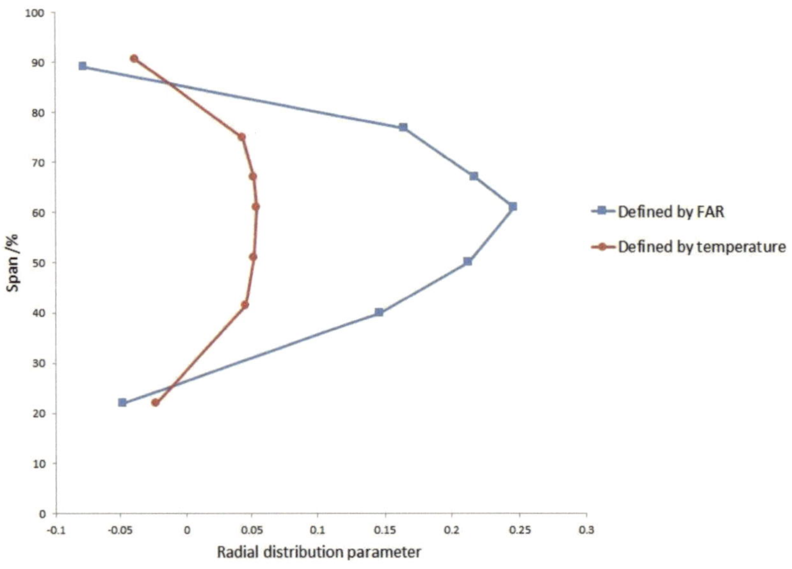

Figure 3.
Difference between temperature defined and FAR defined radial distribution parameter [6].

DOI: http://dx.doi.org/10.5772/intechopen.93916

holes. Also, there is no dilution air holes. That brings liner configuration greatly changed. Also needs to use other design measures to have exit distribution adjusted.

3.4 High altitude ignition

As combustion air fraction is significantly increased, there is a high altitude ignition issue. For a high FAR combustor, it is required to have 35000 ft reliable ignition. It is more severe than civil aero combustor requirement, which usually has a 30000 ft ignition. Also, it is not only required to have ignition, it must provide engine with quick pull-up. That is after high altitude ignition, the combustion efficiency must be appropriately good for pull-up.

Two major design measures for high altitude ignition are:

a. Enlarge the liner cross sectional area. Liner cross sectional area is, **at least** equal to 12 times combustion air ACd

b. Using small FN pilot fuel nozzle. That is the reason why at idle condition pilot fuel nozzle injection pressure drop at least 120 psig

3.5 NO_2 issue

A high FAR combustor has a special issue, that is exhaust nitrogen dioxide (NO_2). This is an environmental issue, NO_2 is toxic. When NO_2 exhausted to atmosphere it will combine with water vapor to form nitrous acid (HNO_2) and nitric acid (HNO3), they are volatile micro matter. But for military combustor, it is related to visible exhaust.

NO_2 is a brown color gas, at 50 ppm volume concentration, it is visible. It has been seen in previous aero engine operation with afterburner working. Thus, the design requirement is that combustor exhaust raw NO_2 concentration lower than 50 ppm (not converted to 15% oxygen concentration). In combustor, chemical reaction mainly generates NO, but under some conditions NO will be converted to NO_2. Chemical reaction NO plus HO2 will become NO_2 plus OH. That will be the case when high temperature combustion gas meets cold air temperature 1100 degree F. Particularly if in combustor there is some UHC, UHC will accelerate the NO_2 formation reaction. Notice that if soot particle combined with NO_2 their visible concentration limit will be lower than when each of them counting separately.

To control NO_2 from a high FAR combustor, the combustor designer needs to do three things:

- Manage to reduce the total NOx level [8]. This is very difficult. At stoichiometric combustion, NOx is at 1000 ppm level. Even with rather low NO_2 over NOx ratio, which is about 8%, NO_2 is at 80 ppm level

- Avoid direct contact between high temperature combustion gas with cold cooling air

- Try to reduce UHC

3.6 Cooling

Reference [9] is important for liner cooling. The author reported it is not right try to make use of air flow passing through liner wall material absorbing heat to solve liner cooling issues, which was the way Lamilloy developer used.

For high FAR combustor, liner cooling is another very big issue. The present author cooling design is effusion cooling with brand new cooling hole configuration. Experiments have proven they are much more effective than conventional cooling configuration.

3.6.1 Outer liner cooling

The outer liner cooling configuration is shown in **Figure 4** [3]. The same configuration may also be used for tubular combustor. It is a tangential hole, but **not totally tangential,** it is **a compound angle tangential hole**. As shown in **Figure 5**, the axial direction angle is to prevent the upstream air jets impinging the downstream cooling air jets. 15 degree angle is only an example. The designed axial direction angle shall be based on axial spacing and circumferential spacing. It is Arctan (H/3*S), as shown in **Figure 5**. From cooling hole center line to the wall inner surface there is a short distance. The minimum distance is half hole diameter. Of course, if the hole is perfectly tangential to the inner wall that is the best. Because it will be the most compact air flow. Depending on liner wall manufacturing, this distance may be more than half hole diameter. For a machined tubular liner, it is half the hole diameter plus 0.005 in. For a large diameter liner formed by sheet metal rolled and welded, this distance may be half hole diameter plus 0.02 in, depending on the liner roundness to avoid laser drilling blind hole.

Such cooling configuration design is based on the present author's long-time cooling study. The most important concept for liner cooling is not how to have cooling air passing liner wall internal passage absorbing more heat, such as Lamilloy (or Transply). That is no good. It is **how to form a compact cooling air layer sticking on wall surface**. The present author's cooling design is based on such concept. The compound angle tangential cooling hole will form such a thin, compact air layer

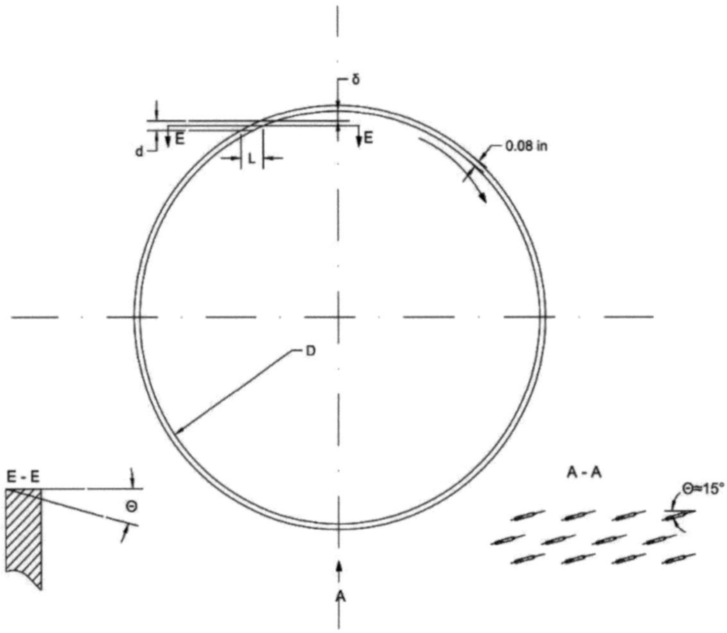

Figure 4.
Cooling hole configuration for annular or tubular liner [3].

sticking on wall surface, to **force the convective heat transfer from hot wall to the lower temperature cooling air layer, instead of from gas to wall.**

Of course, the longer cooling hole length is helpful for liner cooling. As shown in **Figure 6**, by author's own calculation, if liner diameter 25 in, cooling hole diameter 0.02 in, hole center line to inner wall surface is 0.03 in, cooling hole length, as defined by the hole center line length, is 0.8 in. This is good for liner cooing. It shall be stressed that the major advantage of this cooling design is not the long hole length. Instead, it is the cooling effectiveness. For a properly designed and manufactured cooling hole arrangement, very often the cooling effectiveness is 100%. From video an **air layer is spirally flowing around the liner surface.** That is why the cooling effectiveness is very good. The major weakness of Lamilloy is not its air passage inside of liner material is too short, it is the hole exit flow vertical to the wall surface, there will be more mixing of cooling air with the combustion gas, thus cooling effectiveness is low.

For outer liner or tubular liner compound angle tangential cooling hole the discharge coefficient can be 0.86. The discharge coefficient depends on laser drilling. If the laser drilled hole likes something "dog bite", then the discharge coefficient will be lower. If laser drilling technology is available for even smaller holes, it is desirable to have smaller hole but more number of holes.

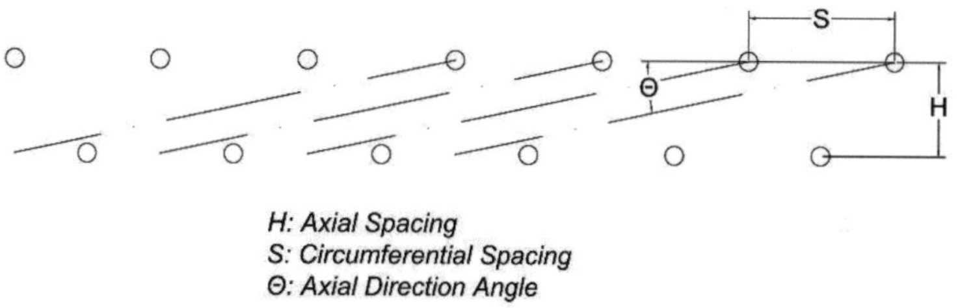

Figure 5.
Design of the axial direction angle in compound angle tangential inlet cooling hole configuration [3].

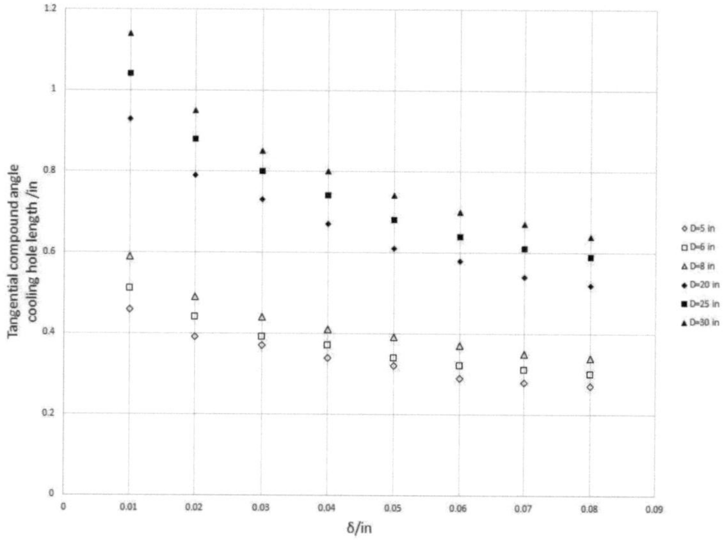

Figure 6.
Tangential compound angle cooling hole length changed with the distance from hole centerline to liner inner surface for tubular and outer annular liner, hole diameter 0.02".

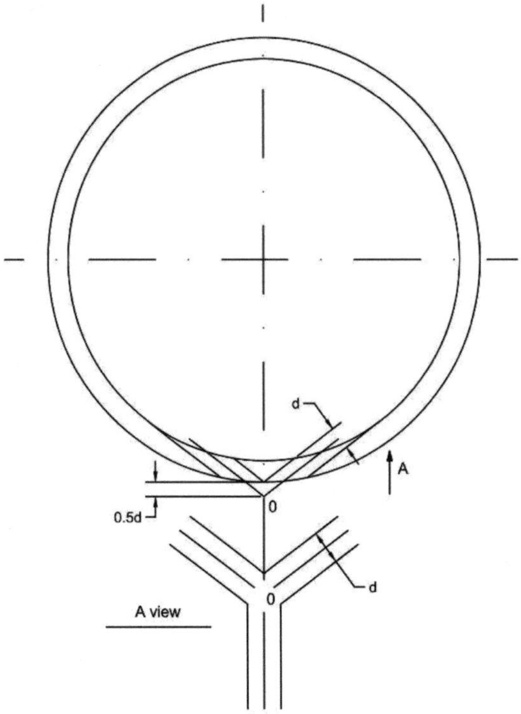

Figure 7.
Cooling hole configuration for inner liner [3].

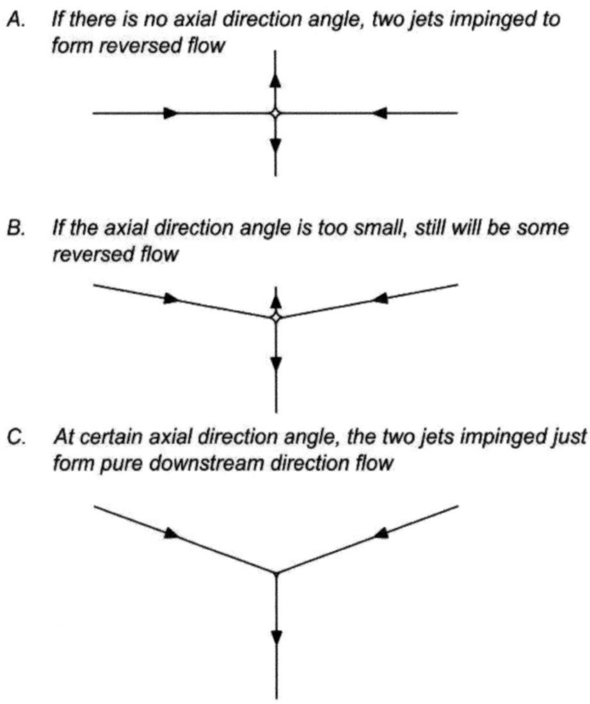

Figure 8.
For inner liner cooling hole, there must be axial direction angle [3].

3.6.2 Inner liner cooling

Inner liner cooling is very different from outer liner cooling. Because if using the same configuration, the cooling air will flow away from the wall surface to mix with combustion gas, which is very bad for cooling result.

Inner liner cooling configuration design is based on one concept that **two compound angle tangential flowing air jets impinged at wall surface to form a pure axial direction cooling air layer**. The design is shown in **Figure 7** [3]. The impinged two jets must have axial direction angle. As shown in **Figure 8** this axial direction angle must be larger than a certain value to avoid the impinged jets forming reversed flow. The minimum axial direction angle depends on liner diameter. Of course, no need to have axial direction angle greater than the necessary one.

In a conventional combustor there are machined cooling air rings, which has one advantage, which is that the cooling air is flowing in the axial direction. However, the machined ring has a lip, which is life limiting factor. Now the newly designed inner liner cooling configuration has truly axial direction flow cooling air, but without a life limiting factor.

Such inner liner cooling configuration is very good. But it brings some challenges to laser drilling. Particularly drilling very small holes.

Such cooling designs are suitable for high FAR combustor, suitable for low emissions combustor, actually suitable for all kind gas turbine combustors.

4. High pressure low emissions combustor design

For a very high pressure civil aero combustor, it cannot use lean pre-vaporized-premixed (LPP) combustion. Because for an engine of pressure ratio 70, its combustor inlet conditions (inlet temperature and inlet pressure) are so high that the autoignition delay time is extremely short, that will not provide any useful reduction of NOx, but suffer very high risk of auto-ignition. From reference [10], auto-ignition delay time for aviation fuel at high pressure and high inlet temperature were obtained (pressure at 40 atm, temperature at 900 K, not so high as up to pressure ratio 70). The present author derived a calculation model, the pre-ignition chemical reaction and heat release were correlated by fitting the prediction to the experimental data. Then using the model to predict auto-ignition delay time for pressure 70 engine combustor inlet condition. The delay time is 0.31 msec. With a safety factor of 2, the usable premixing time is 0.155 msec. That is really no meaning to design an LPP combustion. Thus, it can only be non-premixing combustion. But for low NOx, at high power condition, fuel and air shall still be well mixed. Then it must be **direct mixing combustion**. Actually, for a high FAR combustor, it is also direct mixing combustion. The difference between these two direct mixing combustions is that, for a low emissions combustor, it is **lean direct mixing (LDM) combustion**, for high FAR combustor, it is **stoichiometric direct mixing combustion**.

4.1 LDM vs. LDI

More than 20 years ago, someone suggested using lean direct injection (LDI) combustion. The suggestion was very simple, only a sketch. Actually, direct fuel injection was not a new concept. All conventional aero combustors (except vaporizer combustor) have fuel direct injection. There was no explanation from LDI suggestion, how to design a combustor for low emissions. Later on, several fuel air module configurations were proposed and tested. Unfortunately, the emissions were no good. Even more trouble is the fact that these fuel air module configurations cannot be integrated on to engine combustor.

Present author proposed lean direct mixing concept. The concept has put stress on one thing that the major design approach shall be concentrated on how to improve high power condition fuel air direct mixing. And the fuel air module proposed by the present author, is based on basic mixing concept, with realistically mechanical design, it can truly be integrated on to engine combustor.

Conclusion: for high pressure low emissions combustor, it is LDM, not LDI.

4.2 Mixing concept

For good fuel air mixing, fuel and air shall have close contact. There must be small scale mixing. Thus, the fuel air module shall be of small size. From many years combustion research, combustor design and development, combustor test, the present author has defined **one very good and simple small size fuel air module**. It is one **single axial flow air swirler, and in the center position**, there is a simple pressure swirl fuel **nozzle**. Such combination will offer good efficiency and flame stabilization at low conditions (such as idle condition) with near stoichiometric combustion, while at high power condition, it is very close to premixing situation, if it is lean burn, it offers low emission index (EI) of NOx. There is a fundamental reason. At a high power condition, fuel injection pressure is high, atomization spray drop size is fine, with high air temperature, fuel evaporation is rather quick. Then with appropriate air swirling, fuel air mixing is good. **The flame zone cannot distinguish if it as a pre-vaporized, premixed fuel air mixture in a premixed module, or if it is mixed out of module during the flowing process before reaching the flame zone**. For combustion these two cases do not have a significant difference. That is the reason why at a high power condition, with high pressure, such small size one single axial air swirler plus one simple pressure swirl fuel nozzle will provide very good NOx output. Because of its small size, that means the whole combustor must have large number of modules. In this case, fuel air modules arrangement on dome and their installation on combustor must have good design.

4.3 Direct mixing fuel air module design

The direct mixing fuel air module for high pressure low emissions combustor is shown in **Figure 9** [11]. The design is very simple. At the inlet there is an axial swirler of low swirling strength, geometrical swirling angle is 35 degree, with thin swirler curved blade of thickness 0.045 in. It is suggested using 0.88 for swirler discharge efficient. In the center there is a simple pressure swirl fuel nozzle. The module has a convergent section of half angle 45 degree at exit. The fuel nozzle exit surface is at the throat section (0.02 in) downstream side. There is a very short divergent section of half angle 75 degree. It is not for aerodynamics, such as a "swirl-venturi". It is for a structural purpose, in the module floating design. The module exit has a cross sectional area ACd smaller than the inlet ACd. That makes the exit a metering device.

4.4 Arrangement of fuel air modules on dome

The arrangement of large number fuel air modules on dome is shown in **Figure 10**. All fuel nozzles are of same size, all air-alone modules are aerodynamically the same and all the same size. Air-alone modules are staying on dome, only

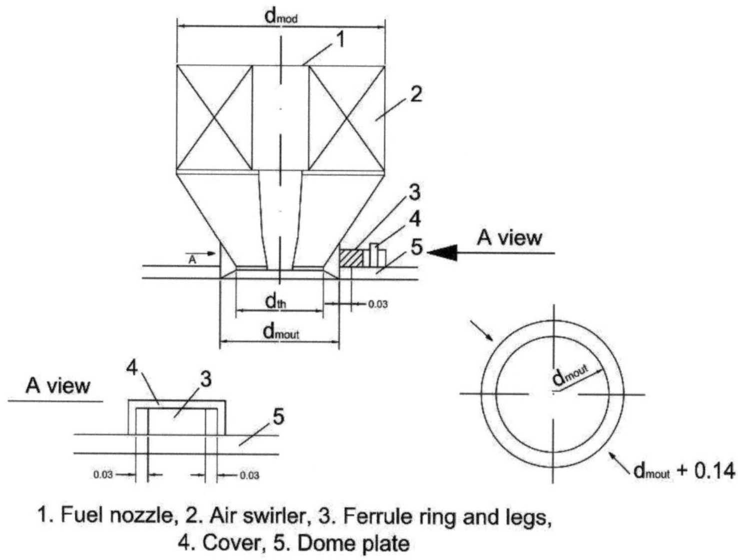

Figure 9.
Fuel-air module [11].

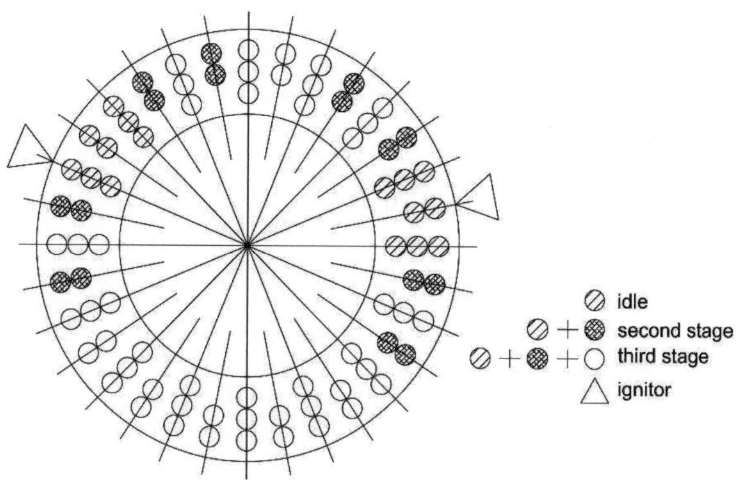

Figure 10.
On dome fuel-air module arrangement [11].

fuel nozzles are installed from air casing. This design idea is from the GE LM 6000 industrial low emissions combustor [12]. In the LM 6000, there are large number of premixed fuel (natural gas) and air modules. As they are premixed modules, the size is big, thus the opening on air casing is rather large. For the present design, the module is not premixed, only fuel nozzle is passing through air casing, the opening can be much smaller.

In **Figure 10**, altogether there are 80 fuel air modules. Distributed in three rows. From outer liner towards inner liner, they are 32, 32, 16 modules. The whole combustor is having three stages, as shown in **Figure 10**, first stage is idle condition, 16 modules are working. Second stage is from 20% power condition to 50% power condition, 32 modules working. Third stage is from 50% power condition to 100% power condition, all modules are working.

4.5 Installation of fuel nozzles on dome

Fuel nozzles are not installed individually. They are installed in cluster nozzles. There are two types of cluster nozzle. One is three fuel nozzles in one cluster, the other is two fuel nozzles in one cluster. Because the fuel nozzles in one cluster are all in the same stage, so inside the cluster stem there is only one fuel line. That will simplify the cluster nozzle. The cluster of three fuel nozzles is shown in **Figure 11**. From liner axial direction, the cluster is close to one fuel nozzle size. The opening on air casing is not big, as shown in **Figure 12**, it is only an ellipse of one inch times two inches, much smaller than the opening on LM 6000 air casing. It is because of two reasons: first, in this design only the fuel nozzle passes through the air casing. The air-alone module does not pass through the air casing. In the LM 6000, fuel injector and air module together pass through the air casing. Second, in this design in one cluster, nozzles are all in the same stage, that makes cluster simplified and reduces the size. During installation, in one three nozzle cluster, three nozzles shall be installed onto dome to match three air module center holes, that requires very accurate manufacturing and accurate assembly. The same situation for two nozzle cluster installation. This is the reason as shown in **Figure 9**, there is a floating design. It allows the air-alone module moving in either direction 0.03 in. In this design, only the middle row air-alone modules are welded with dome without floating design, other air modules are all with floating design to make the assembly easier.

4.6 Air distribution and liner cross sectional area

For a high pressure low emissions combustor, its air flow distribution is combustion air 75% and cooling air 25%. The liner layout has no primary air holes, no dilution air holes. Liner cross sectional area is designed by 12 times combustion air ACd, or it can also be designed as liner average Mach number 0.02. The Mach number is defined by that, air flow rate is combustion air, sonic velocity is defined by inlet air temperature, air density is defined by inlet air pressure, inlet air temperature.

Notice the design choices will always need several times modification, make good balance between the following items:

- Liner average diameter (affect total number of modules and the circumferential spacing between modules)
- Liner annular height (affect module radial spacing)
- Module detailed design (affect module inlet diameter and module exit diameter)
- For module installation, the required minimum floating distance (may affect the locations of modules with floating design)

The key item is the total number of fuel air modules. That will affect the ACd for each air module and module size, also affect fuel nozzle flow number and whole fuel air module arrangement on dome.

4.7 Fuel nozzle design

Fuel nozzle is designed for the maximum condition. At the 100% condition, the nozzle injection pressure drop is between 400 psig and 800 psig. At the idle

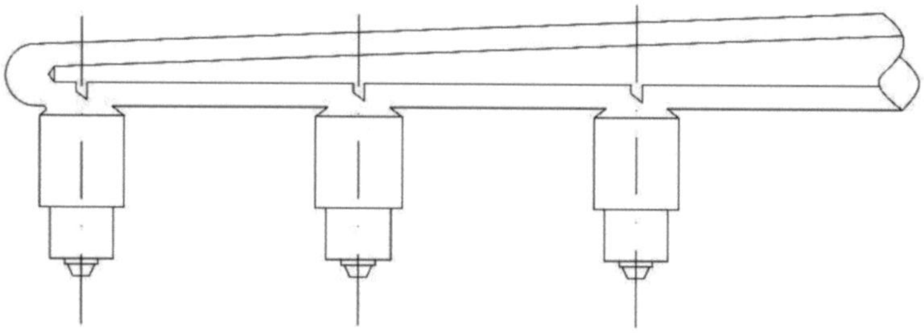

Figure 11.
Cluster of 3 nozzles [11].

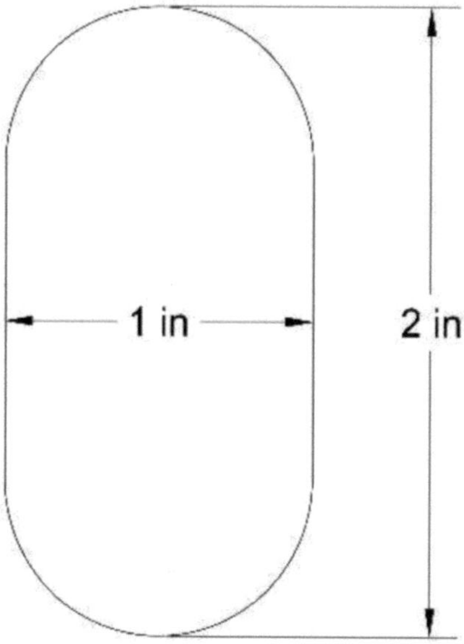

Figure 12.
Opening on air casing for cluster fuel nozzles [11].

condition, fuel nozzle pressure drop is 120 psig. At all operational conditions, such as before one new stage is opened, nozzle pressure drop shall not be higher than 800 psig and just after a new stage is opened, nozzle pressure drop is not less than 100 psig.

Since it is for a high pressure low emissions combustor, there needs to be a fuel pump with very high pressure capacity.

The control of fuel staging is simple on-off valve.

4.8 90-degree sector combustor design

90 degree sector combustor design is shown in **Figure 13**, which is a rectangular shape, not a fan shape. Previously a 90 degree sector was usually cut from a full annular combustor, the cost was a whole full annular combustor. And the traversing gear to measure exit temperature distribution is complicated. From present author's experience, using a rectangular 90 degree sector will not affect the combustor development, its aerodynamics will not be affected. Two side walls are water cooled with no effect on air distribution. The combustion data will be taken from the middle 60 degree sector. The rectangular 90 degree sector design idea is taken from reference [13]. In reference [13], the recommended 90 degree rectangular sector is shown in **Figure 14**, which was designed by the present author.

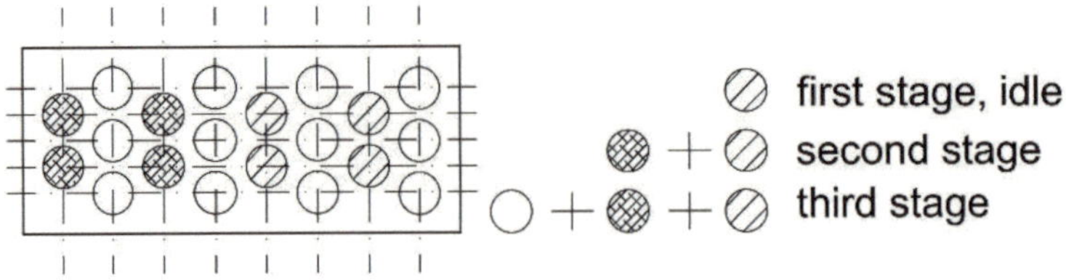

Figure 13.
Sector combustor of 20 fuel-air modules.

Figure 14.
Sector test combustor (shown with 3 fuel injectors) [13].

5. Combustor performances

This high pressure low emissions combustor design has good combustion performances. First, the high power condition EI NOx is good, which is very close to a well-developed LPP system. The high pressure EI NOx can be easily obtained by single module tubular combustor test.

High altitude ignition is good, because as shown in **Figure 10**, the fuel nozzle spray for ignition is rather close to ignitor and there is no other air in-between spray and ignitor.

Flame stabilization is good. Idle LBO is good for two reasons:

- The combustion equivalence ratio for working fuel air module is 1.2
- As shown in **Figure 10**, at idle condition there are 8 modules working as a group in one combustion zone. There is not much non-working module air quenching effect. Particularly there are two working module flames which are protected by the surrounding flames. They support each other, make the idle LBO a very good one.

For the 30% power condition, as the designed working modules combustion equivalence ratio is close to one, which is good flame stabilization for the storm weather heavy rain test.

As it is non-premixed, lean direct mixing combustion with low swirling, there will not be severe combustion instability.

For exit temperature distribution, the pattern factor is low. Because there is small scale mixing, it is close to uniform heat release combustion.

For the exit radial profile, this design has its natural feature for good radial profile. If there needs some minor adjustment, the designer may easily move the two fuel nozzle cluster and related air modules radially just a little bit.

One additional advantage for this combustor design is that, for high pressure low emissions combustor development, very often there is a lack of such a high pressure and high air flow rate test facility to run up to the 100% power condition. When there are such facilities, the running cost is very high. For this combustor design, only 1.25% of 100% power condition air flow is required for single module tubular combustor test. That is, the designer can run a large number of tests to verify the effect of pressure, air temperature, FAR on EI NOx to correlate an equation of

$$\text{EI NOx} = f\ (P,\ T,\ \text{FAR}) \tag{1}$$

That gives the designer opportunity to study the combustion in depth.

6. Summary of this chapter

- The new generation civil aero combustor is a high pressure (such as 70 atm) low emissions combustor. The new generation military aero combustor is a high FAR (such as 0.051) combustor. For combustion organization, they are both direct mixing combustion. For the civil combustor, it is a lean direct mixing combustion concept. For the military combustor, it is a stoichiometric direct mixing combustion concept. They both require a high combustion air fraction, such as 75%. The liner will have no primary holes, no dilution holes. Cooling air is reduced to 25%. There must be advanced cooling technology. The liner design for the new generation aero combustor has been reported in this chapter. New generation aero combustor also needs some more technology development.

- As 75% air coming into liner through dome, there needs to be a new inlet diffuser design. The new diffuser design for the new generation aero combustor will be reported somewhere else by the present author

- The new generation aero combustor needs a new liner material, which is ceramic matrix composites (CMC). Such technology is available. For further development, two problems need to be solved. One is how to drill tiny small cooling holes on CMC. The other is how to connect CMC with metal parts. For the second problem, there needs a transition region, from metal gradually change to CMC, then CMC part can easily weld or connect to metal parts

- The new generation aero combustor will require to be fabricated with additive manufacturing (or 3-D printing manufacturing). Such technology is available. It needs further development to be used for liner dome parts, in addition to fuel nozzles, currently made by such technology

- New generation aero combustor shall have laser ignition technology. This technology is not available now. It will be particularly useful for high altitude ignition. Such technology shall be developed in the future.

Author details

Jushan Chin[1*] and Jin Dang[2]

1 Retired, AIAA Associate Fellow, California, USA

2 Fossil Energy Research Corp., California, USA

*Address all correspondence to: jschin2016@gmail.com

References

[1] Secretariat IC. Annex 16—Environmental Protection Volume II—Aircraft Engine Emissions. ISBN 978-92-9231-123-02008; 2008.

[2] Bahr DW. Technology for the Design of High Temperature Rise Combustors. Journal of Propulsion and Power. 1987 Mar;3(2):179-86.

[3] Chin J. Suggestions on High Temperature Rise Combustor. In AIAA Propulsion and Energy 2019 Forum 2019 (p. 4327).

[4] KRESS E, Taylor J, Dodds W. Multiple Swirler Dome Combustor for High Temperature Rise Applications. In 26th Joint Propulsion Conference 1990 (p. 2159).

[5] Johnson MR, Littlejohn D, Nazeer WA, Smith KO, Cheng RK. A Comparison of the Flowfields and Emissions of High-swirl Injectors and Low-swirl Injectors for Lean Premixed Gas Turbines. Proceedings of the Combustion Institute. 2005 Jan 1;30(2):2867-74.

[6] Van Erp CA, Richman MH. Technical Challenges Associated with the Development of Advanced Combustion Systems. In RTO Meeting Proceedings 1999.

[7] Liscinsky D, Colket M, Hautman D, True B. Effect of Fuel Additives on Particle Formation in Gas Turbine Combustors. In 37th Joint Propulsion Conference and Exhibit 2001 (p. 3745).

[8] Sturgess, G, Zelina, J, Shouse, D, Roquemore, W. Emissions Reduction Technologies for Military Gas Turbine Engines. Journal of Propulsion and Power 2005; 21(2):193-217.

[9] Andrews GE, Asere AA, Hussain CI, Mkpadi MC, Nazari A. Impingement/Effusion cooling: Overall Wall Heat Transfer. In Turbo Expo: Power for Land, Sea, and Air 1988 Jun 6 (Vol. 79214, p. V004T09A036). American Society of Mechanical Engineers.

[10] Guin C. Characterization of Autoignition and Flashback in Premixed Injection Systems. In RTO Meeting Proceedings 1999.

[11] Chin J. Design of Aero Engine Lean Direct Mixing Combustor. In AIAA Propulsion and Energy 2018 Forum 2018 (p. 4921).

[12] Leonard, G, Stegmaier, J. Development of an Aeroderivative Gas Turbine Dry Low Emissions Combustion System. Journal of Engineering for Gas Turbines and Power 1994; 116(3):542-546.

[13] NASA Glenn Research Center. Technology Readiness Levels. CAEP/6-IP/4 Appendix A.